Study Guide and Student Solutions Manual
for

PRINCIPLES OF MODERN CHEMISTRY

SECOND EDITION

BY OXTOBY NACHTRIEB

Wade A. Freeman
University of Illinois at Chicago

Saunders College Publishing

Philadelphia Fort Worth Chicago San Francisco
Montreal Toronto London Sydney Tokyo

Printed in the United States of America.

Oxtoby/Nachtrieb: Study Guide and Student Solutions Manual for PRINCIPLES OF MODERN CHEMISTRY, 2/E

ISBN # 0-03-053503-4

123 987654321

FOREWORD

The second edition of the textbook *Principles of Modern Chemistry,* by David W. Oxtoby and Norman H. Nachtrieb, presents a thorough introduction to University chemistry in three major parts:

• The first eleven chapters cover the macroscopic principles;

• The next six chapters cover modern theories of atomic structure and chemical bonding;

• The final six chapters apply the principles of the first seventeen chapters to descriptive chemistry with emphasis on important chemical processes, problems, and trends.

Each chapter ends with a list of concepts and skills and an extensive problem set. The first portion of each problem set is organized according to the topical sections of the chapter. The problems are paired–they treat the same or closely related material. If the student can solve one of them he or she should be able to solve the other. The final portion of the problem sets consists of unpaired problems ("Additional Problems"). These draw on the concepts and skills covered in the chapter both singly and in combinations. **Students cannot expect to succeed in a chemistry course without working problems.**

A good recipe for learning chemistry is to study the text until it seems to be comprehended, try the problems, and then restudy based on success (or lack of it) with the problems. This Guide has been written to help the student complete this recipe faster and more effectively. It summarizes definitions, concepts, and equations, gives additional insights into the material presented in the text, and provides hints on problem-solving.

Each numbered chapter in this Guide corresponds to the same numbered chapter in the text. The first part of each chapter is a review of concepts and skills with particular emphasis on the type and point of the problems that arise from them. Concepts are referenced to the problems that use them.

In the second portion of each chapter, this Guide gives detailed solutions to all the odd-numbered problems. The solutions mention alternative avenues of attack on the problems and point out common pitfalls.

The serious student will try to solve odd-numbered paired problems on his or her own, reviewing concepts and checking results against the answers in this Guide. The even-numbered paired problems can then be used as a test. Finally, the unpaired problems will allow a final honing of skills. The detailed solutions should be used as practical illustrations of problem-solving techniques as well as a guide to study.

The second edition of *Principles of Modern Chemistry* contains a total of 1630 problems, which is about 80 percent more than number in the first edition. This is so many problems that it is unlikely that students will be asked to solve (or should solve) them all in a University chemistry course. Students however should not ignore unassigned or apparently duplicative problems, but instead study the solutions given here to confirm their understanding.

Wade A. Freeman

Contents

1 Stoichiometry and the Atomic Theory of Matter 1

2 Chemical Periodicity and Inorganic Reactions 29

3 The Gaseous State 49

4 Condensed Phases and Solutions 76

5 Chemical Equilibrium 104

6 Acid-Base Equilibria 132

7 Dissolution and Precipitation Equilibria 176

8 Thermodynamic Processes and Thermochemistry 203

9 Spontaneous Change and Equilibrium 236

10 Electrochemistry 268

11 Chemical Kinetics 303

12 Fundamental Particles and Nuclear Chemistry 327

13 Quantum Mechanics and the Hydrogen Atom 349

14 Many-Electron Atoms and Chemical Bonding 373

15 Molecular Orbitals and Spectroscopy 392

16 Coordination Complexes 411

17 Solids and Liquids 427

18 Chemical Processes 458

19 The Lithosphere 463

20 Ceramics and Semiconductors 478

21 Sulfur, Nitrogen, and Phosphorus 485

22 The Halogen Family and the Noble Gases 495

23 Organic Chemistry and Biochemistry 507

Appendices 526

Chapter 1

Stoichiometry and the Atomic Theory of Matter

Chapter 1 presents the fundamental concepts of stoichiometry (chemical arithmetic) as they developed historically.

Chemistry and the Natural Sciences

Chemistry occupies a central role among the natural sciences. Its many areas of overlap with other fields provide fertile sources of new discoveries.

Progress in chemistry and other sciences involves a blend of theory and experiment. The formulation of theories is the construction of models to explain experimental results. Such models are always subject to revision in the light of newly discovered facts. Theories are always held provisionally. Experimentation amounts to asking nature a carefully-crafted question. A good experiment gets an unambiguous reply. Experimental facts, once verified by repetition, endure.

The Composition of Matter

Substances and Mixtures

Mixtures consist of two or more pure substances that can in principle be separated from each other by ordinary physical means. Mixtures are **homogeneous** (having properties that are uniform from region to region throughout the sample) or **heterogeneous** (having identifiable regions with different properties). If all efforts by physical means to separate a material into portions having different properties fail, then the material is a **substance.**

Analysis and **synthesis** are the poles of a major dichotomy in chemistry. Analysis means taking things apart; synthesis means putting them together. Chemists do both. An analytical chemist separates the component substances from mixtures and then subjects these substances to elemental analysis. A synthetic chemist constructs new substances that have theoretical or practical interest and subjects them to analysis to verify their composition.

Elements

Elements are substances that cannot be decomposed into two or more simpler substances by ordinary physical or by ordinary chemical means. **Compounds** are substances that contain two or more different elements.

Every element has a name and a symbol. The symbol consists either of a capital letter or a capital letter followed by a lower case letter. **It is essential to learn the correctly spelled names and symbols of the elements in proper correspondence.** This task cannot usually be completed overnight. It is less tedious if it is done in conjunction with a study of the structure of the periodic table of the elements (Chapter 2). It is wise to learn the names and symbols of the *common* elements first. See Figure 2-7 (text page 52) for indication of the relative abundance of the elements in nature.

The Atomic Theory of Matter

The Law of Conservation of Mass

Matter is neither created nor destroyed in chemical reactions. Reactions in which the products appear to weigh more or less than the reactants typically have picked up a reactant from the surroundings (like oxygen from the air) or evolved a gas into the surroundings. When proper experimental design eliminates these confounding factors the law of conservation of mass always holds for chemical processes. Furthermore, one chemical element is never transmuted into another element in chemical reactions. **The mass of the elements is conserved in chemical change element by element.** The law of conservation of matter can be of real help in practical problem-solving. See **problems 1-53, 1-61, 1-81,** and **2-83.**

The Law of Definite Proportions

The law of definite proportions states that in a pure chemical compound the proportions by mass of the constituent elements are fixed and definite, independent of the history of the sample. For example, truly pure vitamin C is the same whether isolated from a natural source or in a synthetic laboratory and has the same definite proportions of its constituent elements (see **problem 1-5**). The law has direct use in solving problems. See **problem 1-63.**

The composition of some solids, **nonstoichiometric** compounds, violates the law of definite proportions. See Chapter 15. Thus the solid compound of nominal formula $K_2Pt(CN)_4$ can, depending on its method of preparation, be isolated as $K_{1.3}Pt(CN)_4$. The law of definite proportions, however, applies rigorously to all gaseous and liquid compounds.

A **chemical formula** specifies unequivocally the definite composition of a compound. It tells the number and kind of atoms in a compound. Common problems are:

- To calculate the mass percent of an element in a compound from the chemical formula. The answer always comes from multiplying the ratio of the atomic mass of the element in question to the molecular mass of the compound by the number of atoms of interest and then converting to percent. **Example:** Calculate the mass percent of oxygen in WO_3. **Solution:** The molecular mass of WO_3 is 183.85 + 3(15.9994) or 231.85. The mass *fraction* of oxygen is 3(15.9994) / 231.85 or 0.20702. The mass percent of oxygen is 100 times larger: 20.702 percent. It is not necessary to know what the symbols stand for in the formula to do this problem. All that is required is finding the symbols on a copy of the periodic table and reading off the atomic masses. Sometimes these problems can involve a lot of arithmetic (see **problem 1-75**).

- To calculate the mass percent of some *group* of elements in a compound such as the mass percent of water in $CuSO_4{\cdot}5H_2O$. The "·5" in this formula means that it contains 5 molecules of loosely-bound water. The percentage of water in the compound is then $(5 \times 18.015/249.69) \times \% = 36.07\%$ where the 249.69 is the molecular mass of the copper-containing compound and 18.015 is the molecular mass of water. This is an elaboration of the first type of problem.

- To tell the elemental composition of a compound that has been designated with a complex formula (such as $(NO_2)_2C_6H_3CH_2CH_2COOH$). The answer comes by adding up the subscripts for each element with due respect for the parentheses. In the example the composition is $C_9H_8N_2O_6$. Formulas are written in this way to convey structural information (connections among atoms) in addition to compositional information see **problem 1-53**.

The Atomic Theory of Dalton

Atomic theory postulates a lower limit to the subdivisibility of matter. Dalton called this lower bound of the graininess of matter the **atom.** The postulates of Dalton's 19th-century theory are stated and reviewed in the light of modern understanding in the answer to **problem 1-71** on page 24 of this Guide. Acceptance of the law of definite proportions obliges acceptance of such a view. See **problem 1-5** for an application of the law of definite proportions.

The Law of Multiple Proportions

The composition of a compound is given by its chemical formula. Subscripts added to the symbols of the elements give the relative *number* of atoms of each kind in the compound. The order in which the elements are written in a chemical formula has no fundamental significance, but there are many conventions: thus H_2O is far more common than OH_2. If the atoms listed in a chemical formula are bound together by forces strong enough to maintain them as a group for a reasonable period of time, then that group is called a **molecule.** There is no way to tell just by looking at a chemical formula whether independent molecules of the specified composition actually exist. If it is not known by other means that the group represented in a formula is a molecule, then the subscripts must be taken to give simply the relative proportions of the different elements. The noncommittal term **formula unit** is used when it is not known that there really are independent molecules of the given composition. The simplest possible molecule is a **diatomic** molecule, which contains exactly two atoms.

Often two elements can combine in different proportions to give more than one compound. When this occurs, the masses of one element that combine with a fixed mass of the other stand to each other in the ratio of small whole numbers. This is the **law of multiple proportions.** See **problem 1-7**. The solution to this problem shows quite clearly the use of unit-factors (see Appendix B, text page A-10) in chemical calculations. The use of unit-factors to guide a complex calculation involving both chemical and non-chemical units is shown in **problem 1-23.**

The mass ratios in the four oxides of vanadium (VO, V_2O_3, VO_2, V_2O_5) are an excellent example of the law of multiple proportions. See the solution to **problem 1-11** for full details.

The Law of Combining Volumes and Avogadro's Hypothesis

Joseph Gay-Lussac established experimentally that when two gases react the volumes that combine stand in the ratio of simple integers (if the two samples of gas are at the same temperature and pressure). The theme of small whole-number ratios introduced with the law of definite proportions recurs in Gay-Lussac's work.

Avogadro offered a theoretical explanation of Gay-Lussac's results. He hypothesized that equal volumes of different gases at the same conditions contain equal numbers of particles. Thus, 1 L of oxygen contains the same number of molecules as 1 L of argon provided the volumes are measured under the same conditions.

The particles in Avogadro's hypothesis are not the same as the atoms of Dalton's atomic theory. The elements hydrogen, oxygen, nitrogen, fluorine, chlorine, bromine and iodine, which are all gases at or near room conditions, exist as diatomic molecules: H_2, O_2 N_2, F_2, Cl_2, Br_2, I_2. It is best to memorize these elements. They are the set of the elements having names ending in "-gen" or "-ine." Avogadro's hypothesis affords a method (Cannizzaro's method) for the determination of the relative masses of the molecules of gases.

Atomic Masses

The absolute masses of atoms cannot be measured directly. Fortunately it is sufficient in chemistry to know the relative masses of atoms on a scale defined by assigning some reference element an arbitrary relative mass. The internationally accepted scale today assigns ^{12}C atoms an arbitrary relative mass of exactly 12. ^{12}C refers to an **isotope** (see below) of carbon. The periodic table of the elements tabulates relative atomic masses on this scale. Consult the inside front cover of the text and of this Guide for atomic masses organized according to the periodic table; see the inside back covers for an alphabetic listing. Relative atomic masses are very frequently used in problems. As ratios, relative atomic masses do not have units. Once a reference standard is assigned, several methods are available for determining relative atomic masses. One of these is Cannizzaro's method.

Cannizzaro's Method

The method is based on the Avogadro's hypothesis and is therefore limited to gases. An arbitrary molecular mass is assigned to one gas. Cannizzaro originally chose hydrogen as the standard gas. His arbitrary selection of 2 as its molecular mass was nearly equivalent to the currently accepted relative molecular mass of H_2. Then, the densities of that gas and of others at set conditions of temperature and pressure are tabulated. The ratio of the density of any gas on the list to the density of the reference gas equals the ratio of the molecular masses of the two. The densities of gases depend on the temperature and pressure. The **standard temperature and pressure, STP** for the comparison of gas densities are 0° C and 1 atmosphere See **problem 1-45** and **1-82.**

Careful elemental analysis of pure compounds would by itself give the relative atomic masses of the constituent elements, *if* the number of atoms of each kind of element in the compound were known. This is a big if when there is no easy way to know subscripts in the formulas of compounds. Fortunately Cannizarro's method gives estimates of relative molecular masses. These estimates allow computation of approximate formulas. Approximate formulas are as good as exact because atoms combine in small whole-number ratios. Errors vanish in the rounding-to-whole-number process. Study the solutions to **problems 1-45 and 1-47** as examples of this logic.

Mass Spectrometry and Isotopes

In a mass spectrometer positively charged ions are accelerated by an electric field and then passed through a magnetic field. More massive ions curve less in the magnetic field and the ions in a mixture are separated according to mass. Careful control of the strength of the magnetic and electric fields allows accurate determination of the ratio of charge to mass of the various ions. Because the charge of the ions is known, the instrument gives accurate values of the masses of the ions.

The advent of the mass spectrometer completely outmoded analytical chemical methods in the determination of atomic masses. No one has weighed anything to determine an atomic mass for 60 years or more.

Mass spectrometry revealed the presence of **isotopes.** Atoms that have essentially the same chemical properties but different masses are isotopes. Isotopes are designated by the proper chemical symbol prefixed with a superscript that is the **mass number** of the isotope. The mass number is the integer nearest to the relative atomic mass of the isotope. Because $^{17}O^+$ ions, for example, are heavier than $^{16}O^+$ ions, the mass spectrometer separates them. In addition the instrument measures the **fractional abundance** of each isotope. The fractional abundance of an isotope is the fraction of the atoms of the element that are of that isotopic kind. The masses of the isotopes of an element, when combined in a weighted average with fractional abundances as weighting factors, give the effective atomic mass of the element. Fractional abundances are often quoted as percent abundances. Divide by 100 to convert a percent to a fraction. Always use fractional abundances in numerical problem-solving. **Problem 1-15** gives details about weighted averages; **problem 1-17** is related problem.

Counting and Weighing Molecules

The Mole

The essence of the concept of the mole is to substitute the convenience of weighing for the tedium of counting. When factories inventory small parts (like screws and washers), they often determine the mass of a set number (for example, 1000) and then weigh the rest of the supply. The number of thousands of washers is the total mass of washers divided by the mass per thousand. The process is identical in chemistry. Items as small as atoms are impossible to count individually so the count-by-weighing approach is particularly important. The set number in chemistry is **Avogadro's number:** 6.022×10^{23}. Avogadro's number of "elemental entities" is a mole of those entities.

The mole is a unit for **amount of substance,** or **chemical amount.** Formally it is the number of elementary entities in exactly 12 g of ^{12}C. The entity in the case of ^{12}C is the carbon atom. Moles of molecules, ions and atoms, which are all different types of elemental entity, are both possible and common. Other units for amount of substance are the dozen (12), the gross (144), the pair (2). A mole contains 6.022×10^{23} elementary particles, which is a lot more than 12, 144 or 2 but not any different in principle. The mole is a fundamental unit in the International System (SI) of units (see Appendix B, text page A-10). This is so because chemical interactions proceed in small whole-number ratios of moles of the various substances, a simplicity that is obscured by all other measures of amount.

It is perfectly acceptable to modify the term "mole" by adding prefixes to specify various powers of 10; the millimole (1 mmol $= 10^{-3}$ mol) and the kilomole (1 kmol $= 10^3$ mol) are most common.

The **molar mass** of an element or compound is the mass of exactly one mole of its atoms or molecules. The following is exceedingly useful in solving problems:

The molar mass of an element is its relative atomic mass expressed in grams; the molar mass of a compound is its molecular mass (or formula mass) expressed in grams.

These statements allow conversion at will between the mass of a sample of a substance and the chemical amount of the substance. To convert mass in grams to chemical amount in moles, *divide* by the molar mass; to convert chemical amount in moles to mass in grams, *multiply* by the molar mass. The unit conversion method (text page A-12) is an error-proof guide to deciding when to divide or multiply by the molar mass. To use the unit conversion method, simply remember that the units of molar mass are grams (of substance) per mole g mol^{-1}. A close study of **problem 1-73** helps in understanding this aspect of the mole concept.

Density and Molecular Size

The density of a sample is its mass divided by its volume. This holds whether the sample be a solid, a liquid or a gas. Units of gram per cubic centimeter are prevalent in measurements of densities in chemistry, Many problems require manipulation of units of density (see **problem 1-27**). The density of a sample can be seen as providing a conversion factor that makes it possible to measure choose to measure either volume or mass, whichever is more convenient.

The density of all samples depends on the temperature and pressure. The dependence is very strong for gases and less strong but still real for solids and liquids (see **problem 2-57** and Figure 17-39 on text page 665).

When the molar mass of a substance is divided by its density, the result has the units of volume per mole. It is V_m, the **molar volume**. The molar volumes of gases are much larger than those of liquids and solids because gases are much less dense. The low density of gases compared to liquids and solids is explained by reasoning that in liquids and solids the molecules (or other constituent particles) are essentially in contact, but in gases the molecules are separated by large distances.

At given set of conditions, all gases have approximately the same molar volume. The molar volume, like the density depends strongly on conditions of temperature and pressure:

At STP the volume of one mole of any gas is close to 22.4 L.

Standard conditions of temperature and pressure are 0°C and 1 atm. This figure, the **STP molar volume of a gas**, can exert an unhealthy fascination for some who use it promiscuously in trying to solve problems. It applies to *gases* and has no meaning for liquids or solids. It is correct for gases *only* under specific conditions. It is only approximately correct for any particular gas as show by the following more exact STP molar volumes:

Gas	STP Molar Volume	Gas	STP Molar Volume
Ar	22.401 L mol^{-1}	H_2	22.410 L mol^{-1}
He	22.398	N_2	22.413
Ne	22.430	O_2	22.414

In view of these variations in the molar volume of gases at STP, it is a mistake to quote or use more than three significant digits in the quantity 22.4 L mol^{-1}.

Chemical Formulas and Chemical Equations

A chemical equation represents in brief form what happens in a chemical transformation. People write chemical equations, and the symbols on the paper obviously are incapable of influencing what goes on in real life in a beaker or flask. Equations are *models* of the reality of chemical reactions.

A chemical equation gives the formulas of the **reactants** on the left and the formulas of the **products** on the right, linking them by an arrow to indicate the change. In a **balanced chemical equation** the number of atoms of each kind of element represented on the left-hand side equals the number of atoms of the same element shown on the right-hand side. This is **material balance** and is an immediate consequence of the law of conservation of matter. Also, the algebraic sum of all of the electrical charges represented on the left-hand side must equal the sum of the charges on the right-hand side.

In a chemical reaction, mass is conserved, element by element; charge is conserved.

Balancing equations is a common and essential exercise in chemistry.

- Balance is achieved using **coefficients** in front of the formulas representing the molecules.

- Arbitrarily changing the chemical formulas of reactants and products is not allowed in balancing equations but is a common error. Misreading chemical formulas, carelessly altering or omitting subscripts and superscripts during transcription, omitting one or more compounds entirely during transcription, and incorporation into formulas of numbers from coefficients are other common reasons for failure properly to balance chemical equations. For example the formula $[Co(NH_3)_6]^{3+}$ includes $3 \times 6 = 18$ H atoms and has a +3 electrical charge. Taking the formula to represent fewer H atoms or to have an zero electrical charge leads to disaster in balancing equations.

- Many chemical equations can be balanced by inspection. The steps for this process are set forth on text page 31: Consider the compound with the most atoms or type of atoms first; assign it a coefficient of 1. Assign coefficients to the other species to achieve balance. See **problems 1-49 and 4-31.**

- The coefficients in chemical equations may refer to moles or molecules of substances, depending on context. Specifying a fraction of a molecule or atom of a substance is absurd, many chemists eliminate any fractional coefficients from their balanced equations, writing

$$H_2O + N_2O_5 \rightarrow 2\,HNO_3 \quad \text{instead of} \quad 1/2\,H_2O + 1/2\,N_2O_5 \rightarrow HNO_3$$

However both versions are acceptable and represent the necessary balance.

Mass Relationships in Chemical Reactions

Stoichiometry is chemical arithmetic. Balanced chemical equations provide quantitative mass relationships among the reactants and products in a reaction. Chemical formulas provide similar mass relationships among the elements comprising a compound. Stoichiometry concerns the use of these relationships. Typical practical problems in stoichiometry:

Computation of a Molar Mass From a Formula. To compute the molar mass of any substance from its formula add up the atomic masses of all of the atoms represented in the formula. The units of the answer are always grams per mole (g mol^{-1}).

Yield Problems. Given the mass of a reactant and a balanced chemical equation, the task is to determine the maximum possible amount of one or more products if all of the reactant is consumed by the reaction. First calculate the chemical amount (moles) of the given reactant; from this calculate the chemical amounts of all the other substances by setting up **chemical conversion factors**. and multiplying with them in such a way as to change units from moles of the known reactant to moles of the reactant or product of interest. Chemical conversion factors come from the coefficients in the balanced chemical equation representing the reaction. For example, the balanced chemical equation $2\,N_2 + O_2 \rightarrow 2\,H_2O$ gives six factors:

$$\frac{2 \text{ mol } N_2}{1\ O_2} \quad \frac{1 \text{ mol } O_2}{2\ N_2} \quad \frac{2 \text{ mol } N_2}{2\ N_2O} \quad \frac{2 \text{ mol } N_2O}{2\ N_2} \quad \frac{1 \text{ mol } O_2}{2\ N_2O} \quad \frac{2 \text{ mol } N_2O}{1\ O_2}$$

Observe that the second factor is the reciprocal of the first, the fourth the reciprocal of the third, and the sixth the reciprocal of the fifth. The different balanced equation $2\,N_2 + 3\,O_2 \rightarrow 2\,N_2O_3$ gives six different factors:

$$\frac{2 \text{ mol } N_2}{3\ O_2} \quad \frac{3 \text{ mol } O_2}{2\ N_2} \quad \frac{2 \text{ mol } N_2}{2\ N_2O_3} \quad \frac{2 \text{ mol } N_2O_3}{2\ N_2} \quad \frac{3 \text{ mol } O_2}{2\ N_2O_3} \quad \frac{2 \text{ mol } N_2O_3}{3\ O_2}$$

Once the chemical amount (moles) of a substance is determined then multiplication by the molar mass (in g mol^{-1}) quickly gives its mass (grams). See **problems 1-21** and **1-49.**

Variations of this problem recognize the fact that in real procedures some product may be lost to side-reactions or during purification. The yield of a product computed in this way is a **theoretical yield.** The **actual yield** refers to the weighed quantity of product finally isolated from a reaction. Actual yield is strictly an experimental quantity. It is possible to perform a chemical reaction with great success but have an actual yield of zero (by accidentally throwing away the product). **Percent yield** is the actual yield of product divided by its theoretical yield then multiplied by 100 percent.

Determination of Empirical Formulas. Elemental analysis can give the mass percent or mass fraction of all of the various elements in a sample of a compound. A common problem is to compute the empirical formula from such data. The crucial step is to imagine that some convenient amount of the compound (usually 100 g) is present. Next, get the mass of each element and then the chemical amount of each element. Finally, figure out the ratio of these chemical amounts. It is worth verifying that the amount of compound chosen has no effect on the answer. To do, rework for example **problem 1-41** assuming 200 g of the compounds. One converts the mass of an element to its chemical amount by *dividing* by the molar mass of that element. The empirical formula of the compound has as its subscripts the smallest whole numbers that have the same ratios as the results of these divisions.

Limiting Reagent Problems. In general a chemical reaction consumes one of the reactants before the others. The first reactant to run out is the **limiting reactant.** A error-proof method of determining which reactant is limiting is to compute the yield of a product (it does not matter which) assuming that all reactants but one are present in *unlimited supply.* Repeat the computation making the same assumption for each of the different reactants. The reactant that gives the smallest yield of product when the answers are compared is the limiting reactant. All others are **in excess.** See **problem 1-65.** A quicker way to determine the limiting reactant is to divide the mass of each reactant by the coefficient that the reactant has in the balanced chemical equation. The *smallest* result corresponds to the limiting reactant.

Detailed Solutions to Odd-Numbered Problems

1-1 Table salt is primarily sodium chloride with a small number of additives; wood is a heterogeneous mixture; mercury is a substance (elemental); air is a homogeneous mixture of several gases; water is a substance (a compound, H_2O); seawater is a homogeneous mixture of many compounds. Sodium chloride is a substance (a compound, NaCl); mayonnaise is a heterogeneous mixture.

1-3 The chemist is writing about *substances.* Mixtures of substances can be separated (resolved) into the individual compounds by physical means.

1-5 As provided by the law of definite proportions, the compound ascorbic acid has exactly the same chemical composition regardless of source. Therefore, the ratio of carbon to oxygen in the natural sample (from lemons) is identical to the ratio in the laboratory sample. The laboratory sample contains 40.00 g of O for every 30.00 g of C, which is a 1.333 to 1 ratio of masses. In the sample isolated from lemons there is accordingly 16.9 g of oxygen, computed as $(1.333/1) \times 12.7$ g.

1-7 a) In MoS_3 there are 3 sulfur atoms per molybdenum atom. Since each sulfur atom has 1/3.0 times the mass of the molybdenum atoms, it follows that 1.0 g of sulfur combines with 1.0 g of Mo in MoS_3.

In MoS_4 there are 4 sulfur atoms per molybdenum atom. Suppose that the mass of a sulfur atom is x. Then the mass of molybdenum atoms is $3.0x$. These facts can be used in a series of unit conversions:

$$\left(\frac{4 \text{ atom S}}{1 \text{ atom Mo}}\right) \times \left(\frac{x \text{ g S}}{1 \text{ atom S}}\right) \times \left(\frac{1 \text{ atom Mo}}{(3.0x) \text{ atom Mo}}\right) = \frac{1.3 \text{ g S}}{1.0 \text{ g Mo}}$$

Note that the x's cancelled out.

In Mo_2S_3 the sulfur-to-molybdenum ratio by atoms is 3/2. The mass ratio is therefore $3/2 \times 1/3 = 1/2$ or 0.50 g S per 1.0 gram Mo.

b) The mass of molybdenum per one gram of sulfur in each of the above compounds is simply the reciprocal of the mass ratio calculated in the previous part.

MoS_3	MoS_4	Mo_2S_3
1.0 g Mo/1.0 g S	0.75 g Mo/1.0 g S	2.0 g Mo/1.0 g S

1-9 a) The law of combining volumes states that the volumes of gases combining to form a substance are in the ratio of small whole numbers. The problem describes the reverse of combination, namely the breakdown of a compound into two gases, hydrogen and oxygen. Still, the ratio of the number of particles of hydrogen to the number of particles of oxygen in the liquid compound must equal the ratio of the volume of gaseous hydrogen to the volume of gaseous oxygen. Since the ratio 14.4 mL/14.4 mL is 1 to 1, the simplest chemical formula is H_1O_1, or HO. Note the assumption in writing this formula that the particles of gaseous hydrogen (H_2 molecules) contain the same number of atoms as the particles of gaseous oxygen (O_2 molecules).

b) The formula HO is just one of many possible answers. Any formula H_nO_n in which the number of atoms of H equals the number of atoms of O is also correct.

1-11 The problem asks for the *relative* number of atoms of oxygen for a given mass of vanadium in four compounds, so the answer must be a ratio. A good first step is to determine what mass of oxygen is present in each compound in combination with some convenient mass of vanadium, for example 1.000 g. This idea is quite general: to "reduce to one" for comparison. From the percentages given in the table in the problem, the first compound contains 23.90 g of O for every 76.10 g of V. To see this, simply imagine 100.0 g of the compound and ask how many grams of the constituent elements must be present. Putting the result on the basis of 1.000 g of vanadium means simply dividing 23.90 g of O by 76.10 g of V. The quotient is 0.3140 g O/ g V. For the second, third and fourth compounds, similarly constructed ratios come out 0.4710 g O/g V, 0.6281 g O/g V, and 0.7851 g O/g V, respectively. The ratios increase going down the table, showing the increasing oxygen-richness of the compounds.

The next step is to compare the ratios. To compare the first two ratios, divide the second by the first: 0.4710/0.3140 = 1.500. The relative amount of oxygen in the second compound is 1.500 times larger than in the first. The second compound therefore must have 1.500 times more atoms of oxygen than the first. Compare the third and fourth compounds to the first in exactly the same way, forming the ratios 0.6281/0.3140 and 0.7851/0.3140, which equal 2.000 and 2.500 respectively. The relative numbers of atoms of oxygen for a given mass of vanadium in these four compounds are therefore 1 to $1\frac{1}{2}$ to 2 to $2\frac{1}{2}$. This is the same as 2 to 3 to 4 to 5. Suppose that an arbitrary mass, say 50.942 g, of vanadium is chosen as the mass to consider. (This choice is actually a bit sly because 50.942 is the relative atomic mass of vanadium.) The masses of oxygen that would chemically combine with 50.942 g of V in the four compounds are, going down the table: 16.00 g, 24.00 g, 32.00 g and 39.99 g. Division of these numbers each by the smallest, 16.00 g, gives the ratios 1 to $1\frac{1}{2}$ to 2 to $2\frac{1}{2}$, just as before. Because 16.00 g of oxygen is 1.000 mol of oxygen atoms, the numbers of moles of oxygen atoms combined with 1.00 mol of V (50.942 g) in the four compounds are 1, $1\frac{1}{2}$, 2, and $2\frac{1}{2}$.

1-13 The formula GaAs states that there is only atom of Ga for every atom of As. According to the mass percentages, there is 51.80 g of As for every 48.20 g of Ga. Hence an atom of As weighs $51.80/48.20 = 1.0747$ times more than an atom of Ga. An equivalent statement is that an atom of Ga weighs $48.20/51.80 = 0.93050$ times the mass of an atom of Ga. If As has a relative atomic mass of 100, the relative atomic mass of Ga equals 93.05.

1-15 The atomic mass of naturally-occurring Si is the *weighted* mean (weighted average) of the atomic masses of the three isotopes listed. What does weighting an average imply? The *un*-weighted mean (symbolized $\bar{n}$) of the mass of the three isotopes would be:

$$\bar{n} = \frac{1}{3}(27.97693) + \frac{1}{3}(28.97649) + \frac{1}{3}(29.97376)$$

Weighting corresponds to replacing the 1/3's in this expression with values telling each isotope's *true* contribution to the total. These values are the abundances. Fractional abundances (which add up to exactly 1.00) rather than percent abundances (which add up to 100.0) must be used. The weighted mean is

$$0.9221(27.97693) + 0.0470(28.97649) + 0.0309(29.97376) = 28.086$$

1-17 The relative atomic mass of natural boron is the weighted mean of the relative masses of the two isotopes

$$A_{\text{boron}} = A_{10}p_{10} + A_{11}p_{11}$$

where A_{10} and A_{11} represent the atomic masses of the ^{10}B and ^{11}B isotopes, respectively, and p_{10} and p_{11} represent the fractional abundances. With one exception, all of the quantities in this equation are known:

$$10.811 = (10.013)(0.1961) + A_{11}(0.8039) = 1.9635 + 0.8039A_{11}$$

Solving gives $A_{11} = 11.01$.

1-19 The mass of a single iodine atom is

$$\left(\frac{126.90447 \text{ g I}}{1 \text{ mol I}}\right) \times \left(\frac{1 \text{ mol I}}{6.022137 \times 10^{23} \text{ atom I}}\right) = 2.107300 \times 10^{-22} \frac{\text{g}}{\text{atom}}$$

1-21 a) P_4O_{10}: $4(30.974) + 10(15.999) = 283.89$.
b) BrCl: $79.904 + 35.453 = 115.36$.
c) $Ca(NO_3)_2$: $40.08 + 2\big(14.01 + 3(16.00)\big) = 164.09$.
d) K_2MnO_4: $2(39.098) + 54.938 + 4(15.999) = 197.13$.
e) $(NH_4)_2SO_4$: $2\big(14.007 + 4(1.0079)\big) + 32.06 + 4(15.999) = 132.13$.

1-23 Find the number of seconds in 80 years:

$$80 \text{ yr} \times \left(\frac{365.25 \text{ day}}{1 \text{ yr}}\right) \times \left(\frac{24 \text{ hr}}{1 \text{ day}}\right) \times \left(\frac{3600 \text{ s}}{1 \text{ hr}}\right) = 2.53 \times 10^9 \text{ s}$$

The extra 0.25 in 365.25 days takes leap years into consideration. Next figure the chemical amount of gold atoms counted out:

$$2.53 \times 10^9 \text{ s} \times 1\frac{\text{atom Au}}{1 \text{ s}} \times \left(\frac{1 \text{ mol Au}}{6.022 \times 10^{23} \text{ atoms Au}}\right) = 4.20 \times 10^{-15} \text{ mol Au}$$

Finally, calculate the mass of this amount of gold:

$$4.20 \times 10^{-15} \text{ mol Au} \times \left(\frac{197 \text{ g Au}}{1.00 \text{ mol Au}}\right) = 8.3 \times 10^{-13} \text{ g Au}$$

The most sensitive analytical balances detect down to about 10 μg (10^{-5} g). Highly sophisticated quartz crystal microbalances detect as little as about 100 pg (10^{-10} g). After a lifetime of counting, the mass of the counted atoms is still too small to detect.

1-25 According to the formula, there are 51 atoms of all kinds in a single molecule of vitamin A. Use this fact to find out how many atoms there are in 1.000 mol of vitamin A:

$$1.000 \text{ mol vit A} \times \left(\frac{N_0 \text{ molecule}}{1 \text{ mol}}\right)\left(\frac{51 \text{ atom}}{1 \text{ molecule}}\right) = 51.00 N_0 \text{ atom}$$

Now compute the chemical amount of vitamin A_2 that contains this number of atoms:

$$51.00 N_0 \text{ atom} \times \left(\frac{1 \text{ molecule } A_2}{49 \text{ atom}}\right)\left(\frac{1 \text{ mol } A_2}{N_0 \text{ molecule}}\right) = 1.041 \text{ mol } A_2$$

1-27 The volume of a "flask" of mercury is the volume per unit mass of mercury multiplied by the mass of mercury. The volume per unit mass is the reciprocal of the density:

$$\left(\frac{1 \text{ cm}^3 \text{ Hg}}{13.6 \text{ g Hg}}\right) \times (34.5 \times 10^3 \text{ g Hg}) = 2540 \text{ cm}^3 \text{ Hg}$$

1-29 The correct answer must be on the order of 10^{23} atoms because a volume of 15.0 cm^3 of corundum is on the ordinary human scale. The given volume of Al_2O_3 is multiplied by a series of conversion factors:

$$15.0 \text{ cm}^3 \text{ Al}_2\text{O}_3 \times \left(\frac{3.97 \text{ g Al}_2\text{O}_3}{1 \text{ cm}^3 \text{ Al}_2\text{O}_3}\right) \times \left(\frac{1 \text{ mol Al}_2\text{O}_3}{101.96 \text{ g Al}_2\text{O}_3}\right)$$

$$\times \left(\frac{6.022 \times 10^{23} \text{ Al}_2\text{O}_3 \text{ units}}{1 \text{ mol Al}_2\text{O}_3}\right) \times \left(\frac{2 \text{ atom Al}}{1 \text{ Al}_2\text{O}_3 \text{ unit}}\right) = 7.03 \times 10^{23} \text{ atom Al}$$

1-31 Assume there is one mole of $ClF_2O_2PtF_6$. Rewrite the formula as $PtClO_2F_8$ The mass of the one mole is 414.52 g, a result obtained by multiplying the molar masses (in g mol^{-1}) of the various elements by their subscripts in the formula and adding the several results (See problem 1-21). The mass percentage of each element is the mass that it brings to this quantity of matter divided by the mass of the whole and multiplied by 100%:

$$\text{for Cl}: \frac{1 \text{ mol}(35.453 \text{ g mol}^{-1})}{414.52 \text{ g}} \times 100\% = 8.553\% \text{ Cl}$$

$$\text{for F}: \frac{8 \text{ mol}(18.998 \text{ g mol}^{-1})}{414.52 \text{ g}} \times 100\% = 36.67\% \text{ F}$$

$$\text{for O}: \frac{2 \text{ mol}(15.999 \text{ g mol}^{-1})}{414.52 \text{ g}} \times 100\% = 7.720\% \text{ O}$$

$$\text{for Pt}: \frac{1 \text{ mol}(195.08 \text{ g mol}^{-1})}{414.52 \text{ g}} \times 100\% = 47.06\% \text{ Pt}$$

Although platinum ties with chlorine as the least prevalent element in the compound on the basis of number of atoms, it is by far the most prevalent on the basis of mass.

1-33 The mass percentage of hydrogen in each of the compounds can certainly be calculated as in problem 1-31, and the resulting numbers used to get the required order. The answers are 11.19% for H_2O, 15.35% for $C_{12}H_{26}$, 9.742% for N_4H_6 and 12.68% for LiH. This method is a lot of work. A faster way is to estimate the relative hydrogen content of each compound. Do this by adding up the masses of the non-hydrogen atoms and dividing by the number of hydrogens. Exact arithmetic is not necessary. Thus, in H_2O, there are $16/2 = 8$ units of non-hydrogen mass per hydrogen atom. In $C_{12}H_{26}$ there are $144/26 \approx 6$ such units; in N_4H_6 there are $56/6 \approx 9$ such units; in LiH there are 7.9 such units. The compound that is richest in hydrogen has the smallest amount of non-hydrogen mass per hydrogen atom. The desired order is therefore: $N_4H_6 < H_2O < LiH < C_{12}H_{26}$.

1-35 Calculate the *fraction* (not percentage) by mass of hydrogen (H) in the compound C_4H_{10} (butane) by the method of problem 1-31 and multiply the result by 0.0130, the fraction of butane in "Q-gas". This fraction-of-a-fraction method works because helium contains no hydrogen:

$$\left(\frac{10 \times (1.008) \text{ g H}}{(4 \times 12.011) + (10 \times 1.008) \text{ g butane}}\right) \times \left(\frac{0.0130 \text{ g butane}}{1 \text{ g Q gas}}\right) = \frac{0.00225 \text{ g H}}{1 \text{ g Q gas}}$$

Multiply by 100% to give the desired percentage: 0.225% H.

1-37 The empirical formula of a compound is the smallest whole-number ratio of atoms of different kinds (or moles of atoms of different kinds) in the compound. First, calculate the chemical amount of each element from the given masses:

$$16.58 \times 10^{-3} \text{ g O} \times \left(\frac{1 \text{ mol O}}{15.999 \text{ g O}}\right) = 1.036 \times 10^{-3} \text{ mol O}$$

$$8.02 \times 10^{-3} \text{ g P} \times \left(\frac{1 \text{ mol P}}{30.97 \text{ g O}}\right) = 2.59 \times 10^{-4} \text{ mol P}$$

$$25.40 \times 10^{-3} \text{ g Zn} \times \left(\frac{1 \text{ mol Zn}}{65.38 \text{ g Zn}}\right) = 3.885 \times 10^{-4} \text{ mol Zn}$$

Next, divide through by the smallest number of moles to put the quantities on a basis of 1 for comparison.

$$\frac{\text{mol O}}{1 \text{ mol P}} = \frac{1.036 \times 10^{-3}}{2.59 \times 10^{-4}} = 4.00 \quad \text{and} \quad \frac{\text{mol Zn}}{1 \text{ mol P}} = \frac{3.885 \times 10^{-4}}{2.59 \times 10^{-4}} = 1.50$$

This gives the formula $Zn_{1.5}PO_4$. But the empirical formula is defined as the smallest whole-number ratio of moles of elements in a compound. Simply multiply all subscripts by 2 to give the empirical formula $Zn_3P_2O_8$, which is often written $Zn_3(PO_4)_2$ to shown how the atoms are organized in the compound.

1-39 The percentages of Fe and Si in the crystalline grain in the fulgurite as determined by the analysis apply to any arbitrary amount of compound. In a 100.0 g sample there would be 46.01 g Fe and 53.99 g Si. Compute the chemical amounts of the two elements in such a sample:

$$46.01 \text{ g Fe} \times \frac{1 \text{ mol Fe}}{55.847 \text{ g Fe}} = 0.8239 \text{ mol Fe; } 53.99 \text{ g Si} \times \frac{1 \text{ mol Si}}{28.086 \text{ g Si}} = 1.922 \text{ mol Si}$$

The two chemical amounts are in the ratio of 2.333 mol of Si to 1.000 mol of Fe. This is expressed by the formula $FeSi_{2.333}$. Multiplying through by 3 gives the correct empirical formula Fe_3Si_7.

1-41 Consider the two cases separately. In 100.0 g of the first compound, there is 90.745 g of Ba and, by subtraction, 9.255 g of N. Compute the chemical amounts of the two elements:

$$90.745 \text{ g Ba} \times \left(\frac{1 \text{ mol Ba}}{137.33 \text{ g Ba}}\right) = 0.66078 \text{ mol Ba}$$

$$9.255 \text{ g N} \times \left(\frac{1 \text{ mol N}}{14.007 \text{ g N}}\right) = 0.6607 \text{ mol N}$$

The two elements are present in the same chemical amount: they are present in a 1-to-1 molar ratio. Thus, the empirical formula is BaN. In 100.0 g of the second compound, there is 93.634 g of Ba and, by subtraction, 6.366 g of N. The chemical amounts are:

$$93.634 \text{ g Ba} \times \left(\frac{1 \text{ mol Ba}}{137.33 \text{ g Ba}}\right) = 0.68182 \text{ mol Ba}$$

$$6.366 \text{ g N} \times \left(\frac{1 \text{ mol N}}{14.007 \text{ g N}}\right) = 0.4545 \text{ mol N}$$

Dividing both these chemical amounts by the smaller establishes that the two elements are in a 1.500-to-1 molar ratio. Thus, the empirical formula is Ba_3N_2.

1-43 a) The compound gives 0.692 g of H_2O and 3.381 g of CO_2. Determine the masses of elemental H and elemental C in these amounts of H_2O and CO_2:

$$0.692 \text{ g H}_2\text{O} \times \left(\frac{2.016 \text{ g H}}{18.015 \text{ g H}_2\text{O}}\right) = 0.0774 \text{ g of H}$$

$$3.381 \text{ g CO}_2 \times \left(\frac{12.01 \text{ g C}}{44.01 \text{ g CO}_2}\right) = 0.9226 \text{ g of C}$$

b) The masses of C and H in the CO_2 and H_2O add up to 1.000 g. The compound therefore contains no other elements.
c) The compound is 7.74 percent H and 92.26 percent C by mass.
d) The 1.000 g of compound contains 0.0774 g of H and 0.9226 g of C. To determine the empirical formula of the compound, convert these quantities to chemical amounts (in moles) and determine their ratio:

$$0.0774 \text{ g H} \times \left(\frac{1 \text{ mol H}}{1.008 \text{ g H}}\right) = 0.0767 \text{ mol of H}$$

$$0.9226 \text{ g C} \times \left(\frac{1 \text{ mol}}{12.01115 \text{ g C}}\right) = 0.0768 \text{ mol of C}$$

The C and H are present in a 1-to-1 molar ratio: $C_1H_1 \equiv CH$.

1-45 At STP, 1.00 mol of a gas occupies 22.4 L. This fact allows calculation of the chemical amount of gas present in this sample. It is:

$$0.174 \text{ L gas} \times \left(\frac{1 \text{ mol gas}}{22.4 \text{ L gas}}\right) = 7.77 \times 10^{-3} \text{ mol gas}$$

The mass of the gas (which is 1.55 g) divided by this chemical amount gives the molar mass of the gas. It is 200 g mol^{-1}. If the molecular formula of this gas were the same as its empirical formula, the molar mass would be 50 g mol^{-1}. As it is, the molar mass is four times large. Hence. the molecular formula is four times the empirical formula or C_4F_8.

1-47 a) The unknown binary compound is gaseous. According to Avogadro's hypothesis, at any particular set of conditions the vapor densities of gaseous compounds are directly proportional to their molar masses. The density of hydrogen (H_2) at STP is 0.0900 g L^{-1} (see text, page 17). This is the molar mass of H_2 (2.016 g mol^{-1}) divided by its molar volume at STP (22.4 L). The density of the unknown is 2.77 g L^{-1} at STP. The unknown is denser than H_2 and yet has the same number of molecules per liter under the same conditions, Hence the molecules of the unknown must weigh more than H_2 molecules. Indeed, the relative molecular mass of the unknown is greater than that of H_2 by a factor of (2.77/0.0900). (This amounts to a summary statement of Cannizzaro's method for determining relative atomic masses.) The relative molecular mass of the unknown is:

$$\left(\frac{2.77 \text{ g L}^{-1}}{0.0900 \text{ g L}^{-1}}\right) \times 2.016 = 62.0$$

b) A mass of 1.21 g of water is formed in the burning of 0.500 L of the gaseous compound. This water captures all of the hydrogen present. Like other molecular compounds, H_2O contains its constituent elements in definite proportion The definite proportion of hydrogen in water is 2.016 parts in 18.0153 by mass. The first number is two times the relative atomic mass of H; the second the relative atomic mass of O added to two times the relative atomic mass of H. Therefore the 0.500 L sample of gas contains:

$$1.21 \text{ g H}_2\text{O} \times \left(\frac{2.016 \text{ g H}}{18.0153 \text{ g H}_2\text{O}}\right) = 0.135 \text{ g H}$$

Exactly 0.500 L of the gaseous compound contains 0.1354 g of H at STP. The molar volume of the gaseous compound is 22.4 L at STP. It follows that one mole of the gaseous compound contains:

$$\frac{22.4 \text{ L}}{0.500 \text{ L}} \times 0.1354 \text{ g H} = 6.07 \text{ g H}$$

The molar mass of H is 1.008 g mol^{-1}. It follows that 6.07 g of H amounts to 6.02 mol of H. There are 6.02 mol of H for every 1.00 mol of the unknown gas, which is the same as saying that there are six atoms of H per molecule of unknown.
c) The one molar volume (22.4 L at STP) of gaseous compound that contains 6.02 g of H has this mass:

$$22.4\ \text{L} \times \left(\frac{2.77\ \text{g}}{1\ \text{L}}\right) = 62.0\ \text{g}$$

The gas is binary, (has only two kinds of atoms in it)—H and the other, call it Z. The mass of Z in 22.4 L of the unknown gas (at STP) is 62.0 − 6.02 = 55.98 g. If there is only one mole of Z in the 22.4 L, then the relative atomic mass of Z is 55.98. This is the maximum value of the atomic mass of Z because there cannot be less than one mole of Z in one mole of unknown (just as there cannot be less than one atom of an element in one molecule of a compound).
d) The 55.98 g of Z could contain more than one mole of Z. If the 55.98 g contained two moles of Z, then the relative atomic mass of Z would be 28.0; if the 55.98 g contained three moles of Z, the atomic mass of Z would be 18.7, and so forth. There would seem to be a large range of possible atomic masses for Z. In the following table the subscripts of Z get larger and larger, corresponding to ever smaller atomic masses of Z.

Formula	Atomic Mass of Z	Formula	Atomic Mass of Z
ZH_6	56.0	Z_2H_6	28.0
Z_3H_6	18.7	Z_4H_6	14.0
Z_5H_6	11.2	Z_6H_6	9.33
Z_9H_6	6.22	$Z_{14}H_6$	4.00
		$Z_{56}H_6$	1.00

Compare the atomic masses in this table with authentic values from the periodic table. the identity of the element Z. If the subscript of Z is 2 or 4, numbers quite close to the atomic masses of Si and N come out. A subscript of 56 (last in the table) gives Z an atomic mass of 1.00, and so make Z equal H. But then the substance would no longer be a binary compound, having instead the formula H_{62}. A subscript of 14 gives an atomic mass of 4.00, but He, which has 4.00 as its relative atomic mass, forms no known compounds. None of the other possible atomic masses of Z occurs on the periodic table.
e) The compound is Si_2H_6 (silane) or N_4H_6 (tetrazane). Both substances exist, but Si_2H_6 is much more stable.

1-49 The equations are balanced by inspection.
a) $3\,H_2 + N_2 \rightarrow 2\,NH_3$
b) $2\,K + O_2 \rightarrow K_2O_2$
c) $PbO_2 + Pb + 2\,H_2SO_4 \rightarrow 2\,PbSO_4 + 2\,H_2O$
d) $2\,BF_3 + 3\,H_2O \rightarrow B_2O_3 + 6\,HF$

e) $2\,KClO_3 \rightarrow 2\,KCl + 3\,O_2$
f) $CH_3COOH + 2\,O_2 \rightarrow 2\,CO_2 + 2\,H_2O$
g) $2\,K_2O_2 + 2\,H_2O \rightarrow 4\,KOH + O_2$
h) $3\,PCl_5 + 5AsF_3 \rightarrow 3\,PF_5 + 5\,AsCl_3$

1-51 a) According to the equation $Mg + 2\,HCl \rightarrow H_2 + MgCl_2$, the reaction produces 1 mol of H_2 for every 1 mol of Mg consumed. Diatomic hydrogen has a relative molecular mass of $2 \times 1.00797 = 2.01594$, and Mg has a relative atomic mass of 24.305. Therefore, the 1 mol Mg $\rightarrow$ 1 mol H_2 relationship implies that 24.305 g Mg yields 2.01594 g H_2. The problem is to compute the mass of Mg that yields 1.000 g of H_2. It is good problem-solving technique to estimate the answers to problems before starting detailed calculations. Thus, 1.000 g is about half of 2.01594 g, so the answer should be about half of 24.305 g Mg. More exactly:

$$1.000\text{ g H}_2 \times \left(\frac{24.305\text{ g Mg}}{2.01594\text{ g H}_2}\right) = 12.06\text{ g Mg}$$

b) The equation states that 1 mol of I_2 arises from every 2 mol of $CuSO_4$:

$$2\,CuSO_4 + 4\,KI \rightarrow 2\,CuI + I_2 + 2\,K_2SO_4$$

Write down a train of conversions, starting with the 1.000 g of I_2:

$$1.000\text{ g I}_2 \times \left(\frac{1\text{ mol I}_2}{253.809\text{ g I}_2}\right) \times \left(\frac{2\text{ mol CuSO}_4}{1\text{ mol I}_2}\right) \times \left(\frac{159.602\text{ g CuSO}_4}{1\text{ mol CuSO}_4}\right)$$

The answer is 1.258 g $CuSO_4$. The reaction wastes 2 mol of I (in CuI) for every mole of I_2 it makes, which would be bad if the aim of doing the reaction were to make I_2.
c) According to the balanced equation, 1 mole of $NaBH_4$ yields 4 moles of H_2. Some might dispute that such a reaction is possible, arguing that no reaction can transform the 4 H atoms of $NaBH_4$ into the 8 H atoms of 4 H_2. As the equation however shows, the extra H comes from the other reactant, water. Write a series of conversion factors:

$$1.000\text{ g H}_2 \times \left(\frac{1\text{ mol H}_2}{4\text{ mol H}_2}\right) \times \left(\frac{37.833\text{ g NaBH}_4}{1\text{ mol NaBH}_4}\right) = 4.692\text{ g NaBH}_4$$

1-53 Comparison of the subscripts on the C's in the two compounds indicates that 12 moles of K_2CO_3 must form for every mole of $K_2Zn_3[Fe(CN)_6]_2$ that is treated. Convert the mass of K_2CO_3 to moles and use the above relationship to convert to moles of $K_2Zn_3[Fe(CN)_6]_2$:

$$18.6\text{ g K}_2\text{CO}_3 \times \left(\frac{1\text{ mol K}_2\text{CO}_3}{138.2\text{ g K}_2\text{CO}_3}\right) \times \left(\frac{1\text{ mol K}_2\text{Zn}_3[\text{Fe(CN)}_6]_2}{12\text{ mol K}_2\text{CO}_3}\right)$$

$$\times \left(\frac{698.2 \text{ g } K_2Zn_3[Fe(CN)_6]_2}{1 \text{ mol } K_2Zn_3[Fe(CN)_6]_2}\right) = 7.83 \text{ g } K_2Zn_3[Fe(CN)_6]_2$$

1-55 Write a balanced chemical equation to learn the relationship between the chemical amount of Si_2H_6 consumed and the chemical amount of SiO_2 formed. Then use the fact that at STP one mole of any ideal gas will occupy 22.4 L:

$$6\, Si_2H_6 + 21\, O_2 \rightarrow 12\, SiO_2 + 18\, H_2O$$

$$25.0 \text{ cm}^3 \times \frac{1 \text{ L}}{1000 \text{ cm}^3} \times \frac{1 \text{ mol } Si_2H_6}{22.4 \text{ L}} \times \frac{12 \text{ mol } SiO_2}{6 \text{ mol } Si_2H_6} \times \frac{60.08 \text{ g } SiO_2}{1 \text{ mol } SiO_2} = 0.134 \text{ g } SiO_2$$

1-57 a) Use a series of unit factors. Note that the first factor is the reciprocal of the molar mass of sulfur and uses kilograms, not the more usual grams.

$$2.00 \text{ kg S} \times \left(\frac{1 \text{ mol S}}{0.03207 \text{ kg S}}\right) \times \left(\frac{2 \text{ mol } H_2S}{3 \text{ mol S}}\right) \times \left(\frac{22.4 \text{ L } H_2S}{1 \text{ mol } H_2S}\right) = 931 \text{ L } H_2S$$

b) Use similar trains of unit factors:

$$2.00 \text{ kg S} \times \frac{1 \text{ mol S}}{0.03207 \text{ kg S}} \times \left(\frac{1 \text{ mol } SO_2}{3 \text{ mol S}}\right) \times \frac{0.06407 \text{ kg } SO_2}{1 \text{ mol } SO_2} = 1.33 \text{ kg } SO_2$$

$$2.00 \text{ kg S} \times \left(\frac{1 \text{ mol S}}{0.03207 \text{ kg S}}\right) \times \left(\frac{1 \text{ mol } SO_2}{3 \text{ mol S}}\right) \times \left(\frac{22.4 \text{ L } SO_2}{1 \text{ mol } SO_2}\right) = 466 \text{ L } SO_2$$

1-59 The key factor is the recognition from the balanced equation that the chemical amount of NO(g) consumed is equal to the chemical amount of $NaNO_2$ produced.

$$50.0 \text{ kg of } NaNO_2 \times \left(\frac{1 \text{ mol } NaNO_2}{0.068985 \text{ kg } NaNO_2}\right) \times \left(\frac{1 \text{ mol NO}(g)}{1 \text{ mol } NaNO_2}\right)$$

$$\times \left(\frac{22.4 \text{ L NO}(g) \text{ (at STP)}}{1 \text{ mol NO}(g)}\right) = 1.62 \times 10^4 \text{ L NO}(g)$$

1-61 a) The small whole-number ratios built into formulas of balanced equations always refer to chemical amount, never to mass. The balanced equation in this problem states that the chemical amounts of XCl_2 and XBr_2 are equal. To use this fact one must convert from the mass of XBr_2 to the chemical amount of XBr_2. The conversion requires the molar mass of XBr_2. Similarly, one needs to go from the mass of XCl_2 to the chemical amount of XCl_2, a conversion that requires the molar mass of XCl_2. Computing the molar masses of the two compounds requires the molar mass of X.

This quantity is unknown so let it be represented by x. Then the molecular mass of XBr_2 is $(2 \times 79.909 + x)$, and the molecular mass of XCl_2 is $(2 \times 35.453 + x)$. The chemical amounts of the two compounds are:

$$1.500 \text{ g } XBr_2 \times \left(\frac{1 \text{ mol } XBr_2}{(159.818 + x) \text{ g } XBr_2}\right) = \left(\frac{1.500}{159.818 + x}\right) \text{ mol } XBr_2$$

$$0.890 \text{ g } XCl_2 \times \left(\frac{1 \text{ mol } XCl_2}{(70.906 + x) \text{ g } XCl_2}\right) = \left(\frac{0.890}{70.906 + x}\right) \text{ mol } XCl_2$$

But the chemical amounts of the XBr_2 and XCl_2 are the same. Hence:

$$\left(\frac{1.500}{159.818 + x}\right) = \left(\frac{0.890}{70.906 + x}\right)$$

This equation is easily solved for x, the atomic mass of the unknown element. Its equals 58.8.

b) Checking the relative atomic masses in the periodic table shows that the unknown element is mostly likely nickel.

Another approach to this problem is to recognize that the mass of X in the sample cannot be changed by the chemical reaction. Let y equal this mass. Then $(1.500 - y)$ g is the mass of Br originally present and $(0.890 - y)$ g is the mass of Cl combined with X after the reaction. According to the chemical equation, the chemical amount of Br in the system before the change equals the chemical amount of Cl after the change: $n_{\text{Cl}} = n_{\text{Br}}$. The chemical amount of a substance in moles is its mass in grams divided by its molar mass in g mol^{-1}. In this problem:

$$n_{\text{Cl}} = \frac{(0.890 - y) \text{ g}}{35.453 \text{ g mol}^{-1}} \quad \text{and} \quad n_{\text{Br}} = \frac{(1.500 - y) \text{ g}}{79.909 \text{ g mol}^{-1}}$$

Setting these two expressions equal to each other and solving for y reveals that 0.403 g of X was present both before and after the reaction. The chemical formula XCl_2 guarantees that this 0.403 g of X must amount to exactly one-half the chemical amount of Cl present in (0.890 – 0.4035) g of Cl. The latter is:

$$\frac{(0.890 - 0.4035) \text{ g}}{35.453 \text{ g mol}^{-1}} = 0.0137 \text{ mol Cl}$$

The 0.403 g of X is accordingly (0.0137/2) or 0.00686 mol of X. The sample has 0.4035 g of X per 0.00686 mol of X or 58.8 grams of X per 1 mol of X; that is, the atomic mass of X is 58.8.

1-63 Problems of this sort appear to give insufficient information for solution. Problem 2-83 is similar. What is overlooked is that both NaCl and KCl furnish only a set, definite proportion of their mass to the total mass of chlorine. These fractions are readily available from the atomic masses and formulas of KCl and NaCl. Thus, in addition to the immediately obvious relationship: $x + y = 1.0000$ g where x and y are the masses in grams of the NaCl and KCl respectively, there is the relationship:

$$\text{mass of Cl from NaCl} + \text{mass of Cl from KCl} = \text{mass of Cl in AgCl}$$

Computing the molecular masses of NaCl, KCl and AgCl and using the law of definite proportions to get the fraction of each compound that is Cl:

$$\left(\frac{35.453}{58.4428}\right) x + \left(\frac{35.453}{74.555}\right) y = \left(\frac{35.453}{143.323}\right) 2.1476$$

If both sides of this equation are divided by 35.453:

$$\frac{x}{58.4428} + \frac{y}{74.555} = \left(\frac{1}{143.323}\right) 2.1476$$

Combining this equation with $x + y = 1.0000$ and solving gives:

$$x = 0.4250 \text{ g} \quad \text{and} \quad y = 0.5750 \text{ g}$$

The mass percentages of NaCl and KCl in the original mixture of NaCl and KCl are then 42.50% and 57.50% respectively.

1-65 The first step is to write the balanced chemical equation representing the reaction:

$$\text{HCl}(g) + \text{NH}_3(g) \rightarrow \text{NH}_4\text{Cl}(s)$$

Next, to note that one mole of HCl gas weighs 36.46 g and one mole of NH_3 weighs only 17.03 g. The equation shows the two gases react in a 1-to-1 molar ratio. It takes fewer heavy molecules than light molecules to make up a specific mass. Therefore, given equal masses of two reactants, the one with the heavier molecules will be used up first. Because the molecules of HCl are heavier, it is the limiting reactant: all of it will be used up to produce NH_4Cl and then the reaction will stop, leaving excess NH_3. The mass of NH_4Cl produced will be:

$$2.00 \text{ g HCl} \times \left(\frac{1 \text{ mol HCl}}{36.46 \text{ g HCl}}\right) \times \left(\frac{1 \text{ mol NH}_4\text{Cl}}{1 \text{ mol HCl}}\right) \times \left(\frac{53.49 \text{ g NH}_4\text{Cl}}{1 \text{ mol NH}_4\text{Cl}}\right) = 14.7 \text{ g NH}_4\text{Cl}$$

Since 20.0 g of matter was present originally, the mass of left-over NH_3 is $(20.0 - 14.7) = 5.3$ g. Such reasoning works easily in this relatively simple case. It becomes harder to carry through in cases in which reactants are consumed in 2 to 3 or 5 to 4 or even more complex ratios and in which the initial amounts of the reactants are unequal. The general approach is to compute the mass of product that would form first based on the given amount of one reactant and assuming an unlimited supply of the second. Then repeat the computation of yield, this time based on the given amount of the second reactant and assuming an unlimited supply of the first. Comparison of the results quickly identifies the limiting reactant: the one giving *less* product. In this problem, 10.0 g of NH_3 and unlimited HCl would give 31.4 g of NH_4Cl. On the other hand, 10.0 g of HCl and unlimited NH_3 would give only 14.7 g of NH_4Cl. The HCl is the limiting reactant.

1-67 a) If 8.0% of the cookies are broken (and eaten) then 92.0% can be sold. Let x be the number of cookies. Then $0.92x = 200$ dozen cookies, and $x = 217.4$ dozen cookies. This is 2609 cookies.
b) 217.4 dozen − 200 dozen = 17.4 dozen cookies eaten. This is 209 cookies.

1-69 a) Soft-wood chips: wood is a complex mixture of many substances. Water: H_2O is a pure compound. Sodium hydroxide: NaOH is a pure compound.
b) Since the iron vessel was sealed, no matter was able to enter or escape, including gases. Therefore, all of the original mass is still contained in the vessel—no more, no less. The total mass is $17.2 + 150.1 + 22.43 = 189.7$ kg.

1-71 Dalton's postulates were:
1. Matter consists of indivisible atoms. We now know that atoms are not completely indivisible. Some elements (such as uranium and radium) are radioactive, and the atoms spontaneously decompose to different atoms and subatomic particles (see Chapter 12).
2. All atoms of a given chemical element are identical in mass and in all other properties. Dalton had no way of knowing about isotopes. Atoms of a given chemical element can have different masses. For example, the element hydrogen has three isotopes.
3. Different chemical elements have different kinds of atoms, and in particular, such atoms have different masses. This statement (so far) needs no modification or extension.
4. Atoms are indestructible and retain their identity in chemical reactions. Atoms are not indestructible. This can be split to give new kinds in atoms in particle accelerators and in nuclear bombs. Atoms are still not known to change their identity in chemical reactions.
5. The formation of a compound from its elements occurs through combining atoms of unlike elements in small whole-number ratios. Certain compounds have compositions that vary within a range. They are nonstoichiometric compounds (see Chapter 17). The law of definite proportions is strictly true for gaseous and liquid compounds but not for all solid compounds.

1-73 a) The search for a solution follows one guiding principle: no matter what units are chosen to express the masses, the masses of atoms of ^{32}S and of P are the same in the distant galaxy as here. The problem then becomes one of proper conversion of units. By earthly definitions there are N_0 (6.0220×10^{23}) atoms of per mole of ^{32}S. Therefore:

$$\frac{N_0 \text{ atom } ^{32}\text{S}}{31.972 \text{ g } ^{32}\text{S}} \times \left(\frac{4.8648 \text{ g } ^{32}\text{S}}{1 \text{ marg } ^{32}\text{S}}\right) \times \left(\frac{32.000 \text{ marg } ^{32}\text{S}}{1 \text{ elom of } ^{32}\text{S}}\right) = \frac{2.9321 \times 10^{24} \text{ atom } ^{32}\text{S}}{1 \text{ elom of } ^{32}\text{S}}$$

"Ordagova's number" is 2.932×10^{24}. The number has units of elom^{-1} (reciprocal eloms).

b) On earth, the molar mass of phosphorus is 30.9738 g mol^{-1}. A series of unit-conversions works to convert this to marg elom^{-1}:

$$\frac{30.9738 \text{ g P}}{1 \text{ mol P}} \times \frac{1 \text{ marg P}}{4.8648 \text{ g P}} \times \frac{1 \text{ mol P}}{N_0 \text{ atom P}} \times \frac{N_{or} \text{ atom P}}{1 \text{ elom P}} = \frac{30.9738 N_{or} \text{ marg P}}{4.8648 N_0 \text{ elom P}}$$

But note from part a) that:

$$N_{or} = \frac{N_0(4.8648)(32.000) \text{ atom}}{31.972 \text{ elom}}$$

Substituting this value for N_{or} into the expression just preceding gives:

$$\frac{30.9738}{4.8648 N_0} \times \left(\frac{N_0(4.8648)(32.000)}{31.972}\right) = \frac{30.9738(32.000) \text{ marg P}}{31.972 \text{ elom P}} = 31.001 \frac{\text{marg P}}{1 \text{ (elom P)}}$$

The N_0 and "4.8648" cancel away showing that the answer is independent of the definition of a "marg" in terms of a gram and of an "elom" in terms of a mole. This makes sense because the people of the planet in the distant galaxy would hardly need our definitions to do their chemistry.

1-75 a) The problem requires more significant figures than the computations in problems 1-21 and 1-31 but follows the same principles. The relative molecular mass of the human parathormone is 13,932.24. The mass percentages are: C, 59.571%; H, 6.4967%; N, 12.5668%; O, 18.833%; S, 2.532%.

1-77 (a) The word "binary" means that the only elements in the three compounds are oxygen and the metal "M." The first compound contains 13.38 g of O for every 86.62 g of M, which is 0.15457 g of O per gram of M. The second and third compounds have 0.1029 g of O and 0.07721 g of O per gram of M, respectively.

b) If the first compound of MO_2, then the second is "$MO_{4/3}$" because the second compound has almost exactly 2/3 the oxygen per quantity of M as the first. This formula is improved by multiplying both subscripts by 3 to clear the fraction. The answer is M_3O_4. The third compound is MO if the first is MO_2 because it has almost exactly 1/2 as much oxygen per quantity of M as the first.

c) Continue the assumption that the first compound is MO_2. Compute the amount of metal that combines with 2 mol of O (2×15.9994 g of O) in each of the first compounds. It is 207.2 g M. Because there is 1 mole of M per 2 mol of O. the relative atomic mass of M is 207.2. Consulting the periodic table establishes that M is lead (Pb).

1-79 Imagine there is 1.000 g of each of the first oxides. There is then 0.6960 g of Mn and 0.3040 g of O. Dividing by respective relative atomic masses gives the relative numbers of moles:

$$\mathrm{Mn}_{\frac{0.6960}{54.94}}\mathrm{O}_{\frac{0.3040}{16.00}} \Longrightarrow \mathrm{Mn}_{0.01267}\mathrm{O}_{0.01900} \Longrightarrow \mathrm{Mn}_{1.000}\mathrm{O}_{1.500} \Longrightarrow \mathrm{Mn_2O_3}$$

Repeat the procedure for the second compound:

$$\mathrm{Mn}_{\frac{0.6319}{54.94}}\mathrm{O}_{\frac{0.3681}{16.00}} \Longrightarrow \mathrm{Mn}_{0.01150}\mathrm{O}_{0.02301} \Longrightarrow \mathrm{Mn}_{1.000}\mathrm{O}_{2.000} \Longrightarrow \mathrm{MnO_2}$$

1-81 The only product of the reaction that contains nitrogen is *m*-toluidine, and the only reactant that contains nitrogen is 3′-methylphthalanilic acid. The mass of nitrogen coming from the reactant must equal the mass of nitrogen ending up in the product, since nitrogen (like all other elements) is conserved in chemical reactions. The *m*-toluidine (empirical formula C_7H_9N) is 13.1% nitrogen by mass (calculated as in problem 1-31). The 5.23 g of *m*-toluidine therefore contains 0.685 g of nitrogen. The 3′-methylphthalanilic acid consists of 5.49% nitrogen by mass (as given in the problem). The issue thus becomes that of finding the mass of 3′-methylphthalanilic acid that contains 0.685 g of nitrogen. Let this mass be x. Then $0.0549x = 0.685$ g. Solving gives x equal to 12.5 g 3′-methylphthalanilic acid.

1-83 a) Write an unbalanced equation to represent what the problem tells about the process.

$$\mathrm{C_{12}H_{22}O_{11} + O_2 \rightarrow C_6H_8O_7 + H_2O}$$

Balance this equation as to carbon by inserting the coefficient 2 in front of the citric acid. Then balance the hydrogens by putting a 3 in front of the water (of the 22 H's on the left, 16 appear in the citric acid and the rest appear in the water). Consider now the oxygen. The right side has $(2 \times 7) + (3 \times 1) = 17$ O's. On the left side the sucrose furnishes 11 O's so the remaining 6 must come from 3 molecules of oxygen.

$$\mathrm{C_{12}H_{22}O_{11} + 3\,O_2 \rightarrow 2\,C_6H_8O_7 + 3\,H_2O}$$

b) The balanced equation provides the information to write the third term in the following:

$$15.0 \text{ kg sucrose} \times \left(\frac{1 \text{ kmol sucrose}}{342.3 \text{ kg sucrose}}\right) \times \left(\frac{2 \text{ kmol citric acid}}{1 \text{ kmol sucrose}}\right)$$

$$\times \left(\frac{192.13 \text{ kg citric acid}}{1 \text{ kmol citric acid}} \right) = 16.8 \text{ kg citric acid}$$

It is perfectly acceptable to create and use conversion factors like "1 kilomole sucrose / 342.3 kg sucrose". Also, it was *not* necessary to have the *entire* balanced equation to work part b). Knowing that the reaction produces two moles of citric acid for every one mole of sucrose that it consumes is sufficient. The O_2 and H_2O could have been left unbalanced.

1-85 The equations show that in the Solvay process each mole of NH_3 takes up a mole of CO_2 and in effect holds it for NaCl to attack. Each mole of $NaHCO_3$ formed by this attack gives 1/2 mole of Na_2CO_3. Thus, for each mole of NH_3 1/2 mole of Na_2CO_3 can form. In the following units "M" stands for mega (see Table B-3, text page A-11):

$$\frac{1 \text{ Mmol } NH_3}{22.4 \text{ ML } NH_3} \times \frac{1 \text{ Mmol } Na_2CO_3}{2 \text{ Mmol } NH_3} \times \frac{105.98 \text{ Mg } Na_2CO_3}{1 \text{ Mmol } Na_2CO_3} = \frac{2.37 \text{ Mg } Na_2CO_3}{\text{megaliter } NH_3}$$

A metric ton equals a megagram so the answer is 2.37 metric tons.

1-87 a) Begin by computing the theoretical yield of the process. The key conversion factor in this is the equivalence of 1 mol of C_2H_2 to 1 mol of $CaCO_3$. This fact is read from the balanced equations. Then:

$$10 \times 10^6 \text{ g } CaCO_3 \times \left(\frac{1 \text{ mol } CaCO_3}{100.1 \text{ mol } CaCO_3} \right) \times \left(\frac{1 \text{ mol } C_2H_2}{1 \text{ mol } CaCO_3} \right)$$

$$\times \left(\frac{22.4 \text{ L } C_2H_2 \text{ (STP)}}{1 \text{ mol } C_2H_2} \right) = 2.24 \times 10^6 \text{ L } C_2H_2 \text{ at STP}$$

The percent yield is the actual yield divided by the theoretical yield and multiplied by 100%:

$$\text{percent yield} = \frac{2 \times 10^6 \text{ L}}{2.24 \times 10^6 \text{ L}} \times 100\% = 89.4\%$$

b) Start with the given volume of acetylene and set up another series of conversions. All volumes are at STP:

$$2 \times 10^6 \text{ L } C_2H_2(g) \times \left(\frac{1 \text{ mol } C_2H_2}{22.4 \text{ L } C_2H_2(g)} \right) \times \left(\frac{26.04 \text{ g } C_2H_2}{1 \text{ mol } C_2H_2} \right) = 2.32 \times 10^6 \text{ g } C_2H_2$$

1-89 a) The balanced equations are:

$$C_3N_3(OH)_3 \rightarrow 3\,HNCO \quad \text{and} \quad 8\,HNCO + 6\,NO_2 \rightarrow 7\,N_2 + 8\,CO_2 + 4\,H_2O$$

b) The problem can be solved by a series of unit conversions that use the molar masses of the compounds involved and coefficients from the two balanced equations:

$$1.70 \times 10^{10} \text{ kg } NO_2 \times \left(\frac{1 \text{ mol } NO_2}{0.046 \text{ kg } NO_2}\right) \times \left(\frac{8 \text{ mol HNCO}}{6 \text{ mol } NO_2}\right)$$

$$\times \left(\frac{1 \text{ mol } C_3N_3(OH)_3}{3 \text{ mol HNCO}}\right) \times \left(\frac{0.129 \text{ kg } C_3N_3(OH)_3}{1 \text{ mol } C_3N_3(OH)_3}\right) = 2.1 \times 10^{10} \text{ kg } C_3N_3(OH)_3$$

Chapter 2

Chemical Periodicity and Inorganic Reactions

The Periodic Table

The periodic table provides an organizing framework for the understanding of a large number of chemical facts and relationships. The periodic table was originally derived from the study of recurring patterns in the chemical and physical properties of the elements when they were set down in order of their relative atomic masses. The patterns are now ascribed to recurring similarities in the arrangement of the electrons within the atoms of the elements.

The chemical reactivity of the elements involves the transfer or sharing of valence electrons (see below) between and among atoms. The periodic table shows the number of valence electrons that each element can furnish. In that way it provides the basis for writing Lewis electron-dot structures, which are very helpful in understanding chemical bonding.

Periodic trends in dozens of physical properties have been established. Numerical values of several physical properties are given in Appendix F (text page A-38). The graphing of periodic trends in physical properties and the prediction of both chemical and physical properties using the periodic table are standard problems (see **problems 2-3, 2-5 and 2-53**). Periodic trends are found up and down the columns of the table, which define **groups** of the elements, and across the rows of the table, which are **periods**. See **problem 2-55.**

The **representative elements** comprise 8 groups, each designated by a Roman numeral. Group I is the **alkali metals**; Group II is the **alkali earths**; Group VI is the **chalcogens**, Group VII is the **halogens**; Group VIII is the **noble gases.** Groups can also be named after the topmost element in the column (example: the Nitrogen Group). A periodic table of the representative elements alone has some symmetry that the full table lacks:

I	II	III	IV	V	VI	VII	VIII
H							He
Li	Be	B	C	N	O	F	Ne
Na	Mg	Al	Si	P	S	Cl	Ar
K	Ca	Ga	Ge	As	Se	Br	Kr
Rb	Sr	In	Sn	Sb	Te	I	Xe
Cs	Ba	Tl	Pb	Bi	Po	At	Rn
Fr	Ra						

This version of the periodic table omits more than half of the elements (compare it to the full table in Figure 2-7, text page 52. The heavy step line marks the boundary between **metals** and **nonmetals.** Metals lie to the left, except H, which is a nonmetal. The abbreviated table shows plainly that almost exactly half of the representative elements are metals and half are nonmetals. It is important to identify an element as a metal or nonmetal because the distinction is used in naming compounds. The distinction is based on a constellation of properties including metallic luster, conducting ability for both electricity and heat, and malleability. Elements among the above that have boxes adjoining the line have properties intermediate between metallic and nonmetallic. They are **metalloids.**

Among the representative elements, the group number equals the number of valence electrons (see below). The representative elements are also called the **main-group** elements.

The Other Elements

The elements that are not representative elements are grouped as the transition-metal elements, the lanthanide elements, and the actinide elements. All of the elements in these categories are metals. The **transition-metal** elements include 10 groups of three or four metals each. The groups appear in the table between Group II and III in the 4th, 5th, 6th, and 7th rows. The **lanthanide** elements are lanthanum, element 57 in the table, and the 14 elements immediately subsequent. The lanthanide elements (also called the rare earths) are generally quite similar in their chemistries. The **actinide** elements are actinium, element 89 in the table, and the 14 elements that follow. They are all radioactive. Their nuclear instability means they are mostly produced artificially.

Atomic Structure

Every atom consists of a small positively charged **nucleus** surrounded by a swarm of negatively charged electrons. The net charge on every atom is zero because the total negative charge of the electrons exactly equals the total positive charge of the nucleus. Atoms of different elements have different positive charges on their nuclei and accordingly hold different numbers of electrons. The number of electrons in an atom is called its **atomic number.**

The **electron** is the fundamental unit of electricity. It has a very small mass, and is exceedingly mobile. Chemical interactions between atoms derive from the transfer of electrons or the sharing of electrons between and among atoms. The chemistry of an atom depends entirely on how many electrons it has.

Ions and Ionic Compounds

Not all of an atom's electrons participate in chemical bonding. Electrons are organized in **shells** surrounding the nucleus. Those in inner shells **(core electrons)** are not directly involved in interactions with other atoms. The outermost, partially filled shell **(the valence shell)** of an atom holds the electrons that are active in chemical bonding. These are the **valence electrons.**

The **Lewis electron-dot model** of an atom represents the valence electrons with dots. The dots go around the chemical symbol at its four sides. If there are more than four valence electrons, dots for them are paired with those already present. It does not matter which location gets the first dot or the first pair of dots. The maximum number of dots surrounding an atom is usually eight. The number of valence electrons (dots) to use in a Lewis dot symbol of a representative element equals the group number of the element.

Atoms often gain or lose one or more valence electrons. A positively charged **ion** results when electrons are lost. A negatively charge ion results when electrons are gained. These are **cations** and **anions** respectively. Lewis dot symbols for ions are arrived at by removing or adding the proper number of dots from the symbol for the atom. Also, the net charge is indicated by a right superscript. For a +1 cation and a −1 anion, the superscripts are a simple plus and minus sign. For other cations and anions the superscript is, for example, Ca^{2+}. Putting the sign first in the superscript (as in "Ca^{+2}") should be avoided.

Atoms tend to lose, or gain electrons to form ions that have the same number of electrons as a noble-gas atom The noble-gas atoms, with the exception of helium, have eight electrons in their valence shells. The above statement is therefore abbreviated by saying atoms tend to lose or gain electrons to attain a valence **octet** of electrons. This rule explains the formulas of binary **ionic compounds** that form between the metallic elements on the left side of the periodic table and the nonmetallic elements on the right side of the periodic table. Sodium chloride (salt) is the archetypal ionic compound. For this reason ionic compounds are often called **salts**.

Ionic compounds are named by giving the name of the cation followed by a space and then the name of the anion. Monatomic cations are named after the element from which they are derived. In the case of some metals that have more than one ion, Roman numerals in parentheses after the name of the metal indicate the charge on the ion. Cations are mostly derived from metals. Simple anions are mostly derived from nonmetals. A monatomic anion is named by adding the suffix *-ide* to the stem of the name of the element from which it derives.

There are many **polyatomic** anions, particularly oxoanions. It is best to memorize the names of the common inorganic polyatomic anions. See Table 2-2 (text page 56).

The formulas of ionic compounds can be determined from the principle of charge neutrality: the total positive charge on the cations must equal the total negative charge on the anions. See **problem 2-79.**

Covalent Compounds and Their Lewis Structures

Atoms of many elements do not have a big tendency to form anions or cations in their compounds, but still tend to attain valence octets in interactions with other atoms. Such atoms get valence octets by *sharing* electrons. **Covalent bonds** result from the sharing of electrons. Compounds in which the bonding is predominately by means of the sharing of electrons are **covalent compounds**. The **Lewis structure** of a molecule shows the sharing of electrons by positioning shared pairs of electrons between the symbols of the atoms. Electrons that are not shared are **lone pairs.** They make no contribution to the bonding, but stay on their original atoms. Lewis structures are written so as to follow the **octet rule,** which states that the electrons surrounding every atom of a main-group element should be surrounded by eight (should "see" eight) electrons. The only exception is hydrogen,which should see two electrons in covalent bonding, according to the octet rule.

Lewis structures are drawn not only for molecules, but also for **molecule-ions**, which are groups of atoms that as a whole have lost or gained electrons.

Whenever more than two atoms are bonded in a molecule, the question of **connectivity** (order in which atoms are linked) arises. A Lewis structure *always* states the connectivity of atoms. The Lewis structure of water shows that the order of the three atoms is H—O—H and not H—H—O. Writing the three atoms in a straight line is equivalent in a Lewis structure to writing them with an angle at the oxygen atom. One does *not* read bond angles from a Lewis structure. Lewis structures are formulated in the following steps:

1. Sum up the total number of valence electrons in the molecule or molecule-ion under consideration. **Example**: ClO_2^+. Each O atom has 6 valence electrons (oxygen is in Group VI). and the Cl atom has 7 valence electrons (chlorine is in Group VII). The +1 charge on the group as a whole removes one electron: $2(6) + 7 - 1 = 18$ valence electrons.

2. Count the number of connections among the atoms. Note that joining p atoms together requires at least $(p-1)$ connections. An actual structure may use more than this many links but it can never use fewer.

3. Determine the "skeleton" of the molecule or molecule-ion. This is the pattern by which the atoms are joined together. A molecular or empirical formula does not reveal which atoms are located next to which others. Indeed some sets of atoms may arrange themselves in more than one skeleton. However, there are ways to get a skeleton in most cases:

   (a) The chemical formula of the molecule as written may specify the order of the connections among the atoms. **Example:** The linkage of the atoms in the compound dichlorine heptoxide, Cl_2O_7, is obscure. Writing the formula as $O_3ClOClO_3$ suggests:

   ```
         O       O
     O   Cl  O   Cl  O
         O       O
   ```

   (b) Hydrogen atoms are always on the outside of Lewis structures. In Lewis structures H atoms can form only one covalent bond. Therefore they are incapable of lying between two other atoms in a structure. **Example:** H_2O_2. The skeleton is [H O O H], not [H O H O] or anything else with interior hydrogen atoms.

   (c) Molecules and molecular ions tend to be clusters with any unique atom at the center of the cluster. **Example:** SF_4. The sulfur atom is the unique atom. Therefore, place it at the center of a cluster with all four fluorine atoms bonded to it. Naturally, various chain arrangements are possible.

4. Put the valence electrons around the symbols for the elements in the skeleton in such a way that all the elements attain a noble-gas electron configuration, if possible. For most elements this means that the elements should see an octet

of electrons–four pairs. For hydrogen, each atom should see a single pair of electrons.

There are numerous structures for which it is *not* possible to complete the last step in a simple manner. To see why, look closely at what it means to satisfy the octet rule on every atom in a Lewis structure. If there were no sharing of electrons, the octet rule would require eight times the number of non-hydrogen atoms plus two times the number of hydrogen atoms. However *every single bond reduces the number of electrons required because shared electrons are counted twice.* Since the minimum number of connections among the atoms is one less than the number of atoms, the number of valence electrons required for single-bonding with octets (duets for the hydrogen atoms) is:

$$N_{e^-} = [8 \times N_{\text{non-H}}] + [2 \times N_{\text{H}}] - [2 \times (p - 1)]$$

where p is the number of atoms. For example, for HNO_3, substitution in the formula gives:

$$N_{e^-} = [8 \times 4] + [2 \times 1] - [2 \times (4 - 1)] = 26$$

When the test number computed in this way equals the actual number of electrons available, it is possible to insert a shared pair of electrons at every connection between atoms, to fill in octets all around (except on H) and get a satisfactory Lewis structure. Unfortunately, two other cases arise:

- the number of valence electrons required by the theory may *exceed* the actual number of valence electrons available.

- the number of valence electrons required by the theory may *be less than* the number of valence electrons that must be used.

In the first case, the answer is to invoke multiple bonding, which is the sharing of four or six electrons between some atoms. The sharing of four electrons between two atoms is a **double bond;** the sharing of six electrons between two atoms is a **triple bond.** The more that atoms share electrons, then the fewer electrons are required to give every atom an octet. The octet rule can still be satisfied. The formation of *rings* of atoms also allows the use of fewer electrons. See **problem 2-19.**

In the second case, the octet rule fails. There are too many electrons to be accommodated according to the rule, yet the electrons must be accommodated. It is then concluded that there must be **valence-shell expansion** on at least one atom. An atom with an expanded valence shell sees more than eight valence electrons. When there is valence-shell expansion the octet rule should be abandoned for the central atom but preserved for the outer atoms. Thus, in SF_6 the central sulfur atom has

12 electrons around it, and in ICl_7, the central iodine has no fewer than 14 electrons around it. For other cases see **problems 2-65 and 2-75.**

Valence-shell expansion is quite common for elements from the third and subsequent period. The octet rule is well obeyed only for second-row elements.

The ability to write a Lewis structure that satisfies the octet rule is useful in predicting whether molecules will be stable. The impossibility of writing such a structure suggests that the molecule will be unstable. The rule is not however iron-clad. In addition to molecules having atoms with expanded valence shells, there are:

Octet-Deficient Molecules. A few molecules (BF_3 and BeF_2 are the main examples) do *not* have octets on the central atom despite the fact that by invoking multiple bonding the octet rule could be satisfied.

Odd-Electron Molecules. Many "odd molecules" (those with an odd number of valence electrons) exist. Yet, satisfactory Lewis structures cannot be drawn unless the number of valence electrons is even. The Lewis approach is therefore defective, or at best incomplete. See **problem 2-63.**

Formal Charges

The calculation of the **formal charge** on the atoms in a Lewis structure enables the selection of the best structure in cases in which more than one Lewis structure is possible. The best Lewis structure keeps the formal charges on all atoms in the structure near zero or at zero if possible. See **problem 2-59.**

The formal charge is *not* equal to the actual charge on an atom in a molecule or molecule-ion, but is instead determined by the following formal procedure. First, draw the Lewis structure. Then:

$$\text{Formal charge} = \text{Group Number} - N_{e^-\text{'s in lone pairs}} - \frac{1}{2}N_{e^-\text{'s in bonds}}$$

This formulas means: **the formal charge on an atom in a Lewis structure equals the number of valence electrons the atom started with minus the number it retains for itself minus the number the atom gets if all of its bonds are broken with a 50:50 distribution of the shared electrons.** In the Lewis structure:

$$:\ddot{O}{=}\ddot{S}{-}\ddot{\underset{\cdot\cdot}{O}}:$$

the oxygen atom on the left has a formal charge of zero:

$$\text{Formal charge} = \text{Group No.} - N_{e^-\text{'s in lone pairs}} - \frac{1}{2}N_{e^-\text{'s in bonds}} = 6 - 4 - 2 = 0$$

The oxygen atom has 4 electrons in lone pairs. If its double bond to the S atom is broken with a 50:50 split of the electrons, it comes out with 2 more electrons. This equals its original 6. The oxygen at the right has 6 electrons in three lone pairs. An even break of its bond to the S atom gives it another electron. It originally had 6 electrons, so its formal charge is $6 - 6 - 1 = -1$. If the 3 covalent bonds to the central S atom are broken, the S atom gets back 3 electrons to go with the 2 in its lone pair. Because it is 1 electron short of the 6 it started with, it has the formal charge $6 - 2 - 3 = +1$. The Lewis structure of SO_2 represented along with the various formal charges is:

$$:{}^{0}\ddot{O}{=}\ddot{S}^{+}{-}\ddot{\underset{..}{O}}^{-}:$$

The algebraic sum of the formal charges on all of the atoms in a molecule or ion must equal the actual charge on that species. This provides a useful check against errors in writing Lewis structures and in computing formal charge.

Resonance

Often it is possible to draw distinct Lewis structures based on the same skeleton but having a different distribution of electrons. Lewis structures related in this way are **resonance structures.** The rules for constructing Lewis structures oblige a decision as to the exact location of every electron: localized in a lone pair on the first atom, localized as part of a double bond between two other atoms, and so forth Resonance is invoked when there is no reason for preferring one such location over another. The structures:

$$:\ddot{O}{=}\underset{..}{O}{-}\ddot{\underset{..}{O}}: \quad \longleftrightarrow \quad :\ddot{\underset{..}{O}}{-}\underset{..}{O}{=}\ddot{O}:$$

are resonance structures for the O_3 molecule. Both localizations of the double bond (to the right and to the left) are inadequate representations of the bonding. Neither is preferable to the other. The actual bonding is a mixture of the resonance contributors. There is a 3/2 bond on each side. The word "resonance" is sometimes misunderstood to mean that the electrons flip rapidly back and forth between the two sites. This is not the case. The electron distribution is a stationary intermediate hybrid of what the extremes represented by the resonance structures. See **problem 2-61.**

Not all resonance structures necessarily contribute equally to the final picture of the bonding in a molecule or molecular ion. Structures that violate the octet rule or require the build-up of large formal charges on the atoms are then not so much wrong but simply very low-percentage contributors. Every Lewis structure that displays the correct number of valence electrons is, in principle, defensible as a low-percentage resonance contributor.

Oxidation Numbers and Inorganic Nomenclature

Oxidation Numbers

The **oxidation number** of an atom in a compound is the charge that the atom would have if all of the electrons in its bonds are suddenly deemed either to belong entirely to it or else entirely to the atoms with which the electrons are shared. Oxidation numbers are an artificial device for keeping track of electrons. The oxidation number of an atom in a compound has little or nothing to do with its true charge. Nonmetals tend to gain electrons in the formation of ionic compounds; nonmetals are usually assigned negative oxidation numbers in compounds. Metals tend to lose electrons in the formation of ionic compounds; metals usually are assigned positive oxidation numbers in compounds: The full rules for assigning oxidation numbers:

1. The oxidation numbers of the atoms in a neutral molecule must add up to zero; those in an molecule-ion must add up to the charge on the ion.

2. Alkali metal atoms have oxidation number +1 and alkali earth atoms +2 in their compounds.

3. Fluorine always has an oxidation number of −1 in its compounds. Other halogens have an oxidation number of −1 in their compounds except in compounds with oxygen or with other halogens, in which they may have positive oxidation numbers.

4. Hydrogen has an oxidation number of +1 in its compounds except those with metals from Group I and Group II, in which it has an oxidation number of −1.

5. Oxygen has an oxidation number of −2 in compounds except those with fluorine and those containing O—O bonds.

These rules allow the assignment of oxidation numbers in a very large number (but not all) compounds and molecule-ions. See **problem 2-25.**

Oxidation numbers are used in chemical nomenclature and in discussing oxidation-reduction reactions. They are not at all the same as formal charges, which find use only in conjunction with Lewis structures. The term **oxidation state** is synonymous with oxidation number.

Naming Binary Covalent Compounds

The naming of binary covalent compounds resembles the naming of ionic compounds. If a pair of elements forms only one covalent compound, the name of the compound consists of the name of the element appearing first in the formula followed by the name of the second element with *-ide* added to its root. (The first element should be located either to the left or down from the second in the periodic table.) Often, a given pair of elements forms more than one binary covalent compound. In such cases either:

1. Specify the number of atoms of each element in the molecular formula with Greek prefixes. *Di-* stands for two. *Tri, tetra, penta, hexa, hepta* stand for three through seven, in order.

2. Treat the compound as if it were ionic and place the oxidation number of the first-named element as a Roman numeral in parentheses after the name of that element. See **problem 2-29.**

Inorganic Reaction Chemistry

Reactions are classified into three broad categories that give some organization to the very broad realm of chemical reactivity. The types are **dissolution-precipitation** reactions, **acid-base** reactions, and **oxidation-reduction** reactions. The text introduces the three types briefly at this stage and considers further aspects of each in chapters Chapter 7, 6, and 10, respectively.

Dissolution-Precipitation. Many inorganic solids are **strong electrolytes.** They dissolve in water to give solutions that are good conductors of electricity. The mechanism of these dissolution reactions is the formation of aquated ions, which are readily able to conduct an electric current. An aquated ion is symbolized by adding "(aq)" to the formula for the ion; an undissolved solid is symbolized by adding an "(s)" to the formula.

Solubilities vary widely. $LiClO_3$ is tremendously soluble in cold water, but AgCl is essentially insoluble. Potassium perchlorate is an intermediate case. There is always an upper limit to the capacity of water (or any other solvent) to take on more solute. Whenever this limit of solubility is exceeded the solute tends to **precipitate** from the solution. Precipitation reactions are the reverse of dissolution reactions.

The formation of a precipitate is very striking when two solutions containing ions which form an insoluble combination are mixed. The insoluble product forms, often quite quickly, and settles out of the solution. A **net ionic equation** represents such an event by showing only the ions that actually react. The other ions (which must be present but play no part in the precipitation) are **spectator ions.** See **problem 2-33** for writing net ionic equations.

Acid-Base. Acids are substances that increase the concentration of the aquated hydrogen ion ($H^+(aq)$) when they are placed in water. Bases are substances that increase the concentration of the aquated hydroxide ion ($OH^-(aq)$) when they are placed in water. Even the purest water contains small concentrations of $H^+(aq)$ and $OH^-(aq)$ because the reaction

$$H_2O \rightarrow H^+(aq) + OH^-(aq)$$

forms them. When an acidic solution and a basic solution are mixed, **neutralization,** the reverse of the above reaction, occurs.

Acid and base are more broadly defined in the Lewis definition. A **Lewis acid** is an electron-pair acceptor and a **Lewis base** is an electron-pair donor. See **problem 2-41.** The generalization moves beyond strictly aqueous systems and allows convenient use of Lewis structures. If a Lewis base donates lone-pair electrons to a Lewis acid in such a way that the two finish up sharing the pair of electrons, then a new bond has formed. See **problem 2-39.** Such a bond is a **coordinate covalent bond.**

Oxidation-Reduction. Oxidation-reduction equations represent chemical reactions in which at least one element changes its oxidation number. See **problems 2-43, 2-45. Oxidation** is an increase in oxidation number. Atoms (or groups of atoms) are oxidized when they lose electrons. **Reduction** is a decrease in oxidation number. Atoms (or groups of atoms) MOre than one element may be oxidized or reduced in the same redox reaction. See **problem 2-73.**

Detailed Solutions to Odd-Numbered Problems

2-1 a) The element M has a valence of 2, since it forms compounds MCl_2 and MO. It belongs to Group II (alkaline-earth metals).
b) The relative molecular mass of MCl_2 is the sum of the relative masses of the three constituent atoms: $x + 2(35.453) = x + 70.906$ where x stands for the relative atomic mass of the element M. The fraction of mass that is chlorine is 0.447. So,

$$0.447 = \frac{70.906}{(x + 70.906)}$$

Solving gives $x = 87.7$. Reference to a table of relative atomic masses identifies the element M as strontium.

2-3 According to the periodic law, the properties of scandium can be expected to be intermediate between those of calcium and titanium. Using the averages of the numerical data, we have:

	Predicted	Observed
Melting point	1250°C	1541°C
Boiling point	2386°C	2831°C
Density	3.03 g cm^{-3}	2.99 g cm^{-3}

The observed facts are from Appendix F.

2-5 Antimony is in Group V: SbH_3; bromine is in Group VII: HBr; tin is in Group IV: SnH_4; selenium is in Group VI: H_2Se.

2-7 It is best just to memorize the patterns of nomenclature of simple inorganic compounds. **a)** Al_2O_3: aluminum oxide **b)** Rb_2Se: rubidium selenide **c)** $(NH_4)_2S$: ammonium sulfide **d)** $Ca(NO_3)_2$: calcium nitrate **e)** Cs_2SO_4: cesium sulfate **f)** $KHCO_3$: potassium hydrogen carbonate (potassium bicarbonate).

2-9 a) Silver cyanide: AgCN **b)** Calcium hypochlorite: $Ca(OCl)_2$ **c)** Potassium chromate: K_2CrO_4 **d)** Gallium oxide: Ga_2O_3 **e)** Potassium superoxide: KO_2 **f)** Barium hydrogen carbonate: $Ba(HCO_3)_2$.

2-11 The phosphate ion has the formula PO_4^{3-}. It has a charge of −3. Trisodium phosphate is therefore Na_3PO_4. The systematic name for this ionic compound would be sodium phosphate.

2-13 a) SO_4^{2-}: In the Lewis structure given in the problem the four oxygens have −1 formal charges and the sulfur has a +2 formal charge. A singly bonded O always has a formal charge of −1.
b) $S_2O_3^{2-}$: In this Lewis structure for the thiosulfate ion the central sulfur has a +2 formal charge. The three oxygen atoms and the peripheral sulfur have −1 formal charges.
c) SbF_3: All atoms have formal charges of zero in the given structure.
d) SCN^-: The sulfur atom in this structure of the thiocyanate ion has a formal charge of −1; the C and N have formal charges of zero.

2-15 a) In this structure, the oxygen atoms both possess a formal charge of 0. Since there is no net charge, Z must also have a formal charge of 0. Therefore, it has 4 valence electrons and belongs to group IV. CO_2 is an example.
b) Each of the six peripheral oxygens has a formal charge of −1, while the bridging oxygen has a formal charge of 0. Since the entire molecule has no net charge, the formal charge on each Z is +3. Element Z has 7 valence electrons and is therefore a halogen. An example is Cl_2O_7 (dichlorine heptoxide).

c) One of the oxygen atoms is neutral but the other has a −1 formal charge. Since this ion has a −1 net charge, Z must have a formal charge of 0. Therefore, it belongs in group V (5 valence electrons). An example is the nitrite ion NO_2^-.

d) Three of the oxygen atoms carry −1 formal charges, while the other oxygen atom and the hydrogen atom have zero formal charges. Since the whole ion has a −1 net charge, Z must have a formal charge of +2. Element Z therefore has 6 valence electrons and comes from group VI. An example is the hydrogen sulfate ion, HSO_4^-.

2-17 **a)** H—S—H **b)** H—As—H (with H bonded below As)

c) H—O—Cl: **d)** :Cl—O—Cl:

2-19 A Lewis structure for S_8:

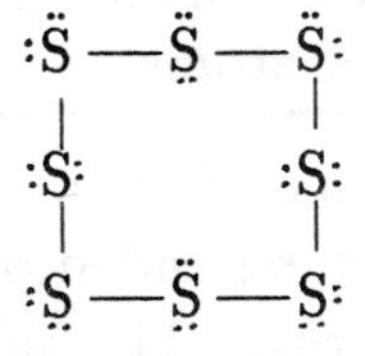

```
:S —S —S:
 |      |
:S:    :S:
 |      |
:S —S —S:
```

Each sulfur obeys the octet rule and each has a formal charge of 0.

2-21 a) The nitrogen and boron atoms would both get 4 valence electrons if all the bonds were broken with uniform distribution of the electrons. The nitrogen atom is supposed to have five valence electrons, so its formal charge is +1. The formal charge is −1 on the boron atom, and zero on the rest of the atoms.

b) The formal charge on the single-bonded O atom is −1. All other formal charges are zero.

c) The azide ion is linear. This fact is given to rule out cyclic Lewis dot structures in which each N is bonded to both others. The Lewis structures are:

$$\left(: \ddot{N} :: N :: \ddot{N} : \leftrightarrow : \ddot{N} : N ::: N : \leftrightarrow : N ::: N : \ddot{N} : \right)^-$$

In the structure on the left, the formal charges (from left to right) are −1, +1, −1; in the center structure, they are −2, +1 and 0; in the structure on the right, they are 0, +1, −2. Computation of formal charges for the structure on the left is discussed in Example 2-2 (text page 59).

d) The bicarbonate ion has 24 valence electrons. The C atom contributes 4, the H atom contributes 1, and the 3 O atoms contribute 6 valence electrons each. The final electron comes from outside to make the net charge −1. The resonance Lewis structures are:

```
H:Ö                      H:Ö
    \                        \
     C—Ö:        ⟷           C=Ö:
    //                       /
 :Ö                       :Ö
```

The double-headed arrow links the two resonance structures. The overall charge on the ion is −1. Formal charges on all atoms are zero with the exception of the O atoms that are singly bonded to C atoms and not bonded to H. These oxygen atoms have a formal charge of −1. The sum of the formal charges in both of the resonance structures is −1, equal to the net charge on the HCO_3^- ion. Resonance structures differ only in the positions of the electrons. One common wrong answer to this problem includes a third structure in which the oxygen atom on the upper left shares two pairs of electrons with the central carbon atom and the H atom is moved to avoid putting a +1 formal charge onto that oxygen atom. This structure is *not* a true resonance structure, because an atom as well as electrons has moved. Resonance structures are always built around the some skeleton of atoms.

2-23 a) PF_5: each atom has a formal charge of 0. The fluorine atoms all obey the octet rule. The phosphorus atom has an expanded octet of 10 electrons.
b) SF_4: each fluorine obeys the octet rule. The sulfur has an expanded octet. All of the formal charges are 0.
c) XeO_2F_2: the fluorine atoms and the oxygen atoms obey the octet rule. The xenon has an expanded octet. Each atom has a formal charge of 0.

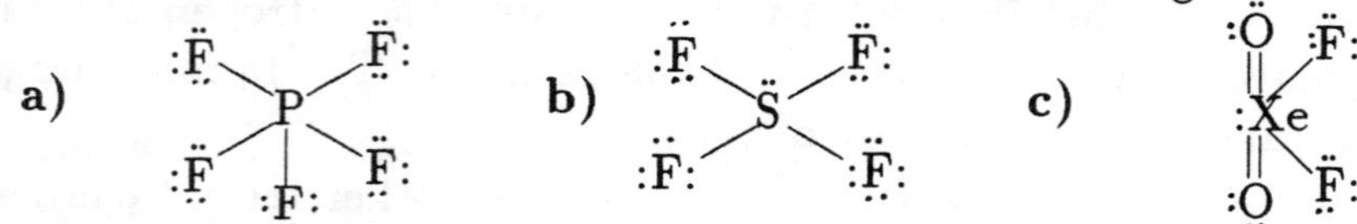

2-25 The oxidation numbers are determined by the standard rules:

$SrBr_2$	Sr +2	Br −1		
$Zn(OH)_4^{2-}$	Zn +2	O −2	H +1	
SiH_4	Si −4	H +1		
$CaSiO_3$	Ca +2	Si +4	O −2	
$Cr_2O_7^{2-}$	Cr +6	O −2		
$Ca_5(PO_4)_3F$	Ca +2	P +5	O −2	F −1
KO_2	K +1	O −1/2		
CsH	Cs +1	H −1		

2-27 a) SiO_2 **b)** $(NH_4)_2CO_3$ **c)** PbO_2 **d)** P_2O_5 **e)** CaI_2 **f)** $Fe(NO_3)_3$.

2-29
a) Copper(I) sulfide and copper(II) sulfide **b)** Sodium sulfate
c) Tetraarsenic hexoxide or arsenic(III) oxide **d)** Zirconium(IV) chloride
e) Dichlorine heptoxide or chlorine(VII) oxide **f)** Gallium(I) oxide

---

2-31 Tetraphosphorus decoxide is P_4O_{10} (tetra means 4; deca means 10), and phosphorus pentachloride (penta means 5) is PCl_5. These two substances appear on the left side of the equation. With $POCl_3$ as the sole product, the balanced equation is $P_4O_{10} + 6\,PCl_5 \rightarrow 10\,POCl_3$.

2-33 Break up each soluble reactant and product into its component ions and cancel out the spectator ions:
a) $Ag^+(aq) + Cl^-(aq) \rightarrow AgCl(s)$
b) $K_2CO_3(s) + 2\,H^+(aq) \rightarrow 2\,K^+(aq) + CO_2(g) + H_2O(l)$
c) $2\,Cs(s) + 2\,H_2O(l) \rightarrow 2\,Cs^+(aq) + 2\,OH^-(aq) + H_2(g)$ is unchanged.
d) $2\,MnO_4^-(aq) + 16\,H^+(aq) + 10\,Cl^-(aq) \rightarrow 5\,Cl_2(g) + 2\,Mn^{2+}(aq) + 8\,H_2O(l)$

2-35 When a salt forms in an acid-base reaction, the cation derives from the base and the anion from the acid:
a) $Ca(OH)_2(aq) + 2\,HF(aq) \rightarrow CaF_2(s) + 2\,H_2O(l)$
calcium hydroxide + hydrofluoric acid → calcium fluoride + water
b) $2\,RbOH(aq) + H_2SO_4(aq) \rightarrow Rb_2SO_4(aq) + 2\,H_2O(l)$
rubidium hydroxide + sulfuric acid → rubidium sulfate + water
c) $Zn(OH)_2(aq) + 2\,HNO_3(aq) \rightarrow Zn(NO_3)_2(aq) + 2\,H_2O(l)$
zinc hydroxide + nitric acid zinc nitrate + water
d) $KOH(aq) + HCH_3COO(aq) \rightarrow KCH_3COO(aq) + H_2O(l)$
potassium hydroxide + acetic acid → potassium acetate + water

---

2-37 The reaction is $H_2S + 2\,NaOH \rightarrow Na_2S + 2\,H_2O$. Sodium sulfide is the salt produced by this neutralization reaction. Without the hint, NaHS (sodium hydrogen sulfide) might be answered.

2-39 a) The slaking of lime: $CaO(s) + H_2O(l) \rightarrow Ca(OH)_2(s)$.
b) The reaction can be seen as a Lewis acid-base reaction. The CaO is the Lewis base. It donates a pair of electrons (located on the oxide ion) to a hydrogen atom in the H_2O molecule; the result is a new O—H bond.

2-41 a) The fluoride ion (F^-) has a negative charge. In the Brønsted-Lowry system, an acid is a donor of a positive entity (the H^+ ion). By plus-minus symmetry then, an acid in this scheme is a fluoride-ion acceptor.
b) In $ClF_3O_2 + BF_3 \rightarrow ClF_2O_2{\cdot}BF_4$, BF_3 accepts a F^- ion from ClF_3O_2, so BF_3 is the acid, and ClF_3O_2 is the base.
In $TiF_4 + 2\,KF \rightarrow K_2[TiF_6]$, TiF_4 accepts an F^- ion from KF, so TiF_4 is the acid, and KF is the base.

2-43 a) $2\ \overset{+3}{\mathrm{P}}\mathrm{F_2I}(l) + 2\ \overset{0}{\mathrm{Hg}}\ (l) \rightarrow \overset{+2}{\mathrm{P_2}}\mathrm{F_4}(g) + \overset{+1}{\mathrm{Hg_2}}\mathrm{I_2}(s)$

b) $2\,\mathrm{K}\overset{+5}{\mathrm{Cl}}\overset{-2}{\mathrm{O_3}}(s) \rightarrow 2\,\mathrm{K}\overset{-1}{\mathrm{Cl}}(s) + 3\ \overset{0}{\mathrm{O_2}}(g)$

c) $4\ \overset{-3}{\mathrm{N}}\mathrm{H_3}(g) + 5\ \overset{0}{\mathrm{O_2}}(g) \rightarrow 4\ \overset{+2}{\mathrm{N}}\overset{-2}{\mathrm{O}}\ (g) + 6\,\mathrm{H_2}\overset{-2}{\mathrm{O}}\ (g)$

d) $2\ \overset{0}{\mathrm{As}}(s) + 6\,\mathrm{NaO}\overset{+1}{\mathrm{H}}\ (l) \rightarrow 2\,\mathrm{Na_3}\overset{+3}{\mathrm{As}}\mathrm{O_3}(s) + 3\ \overset{0}{\mathrm{H_2}}(g)$

2-45 Refer to the rules on oxidation states. Neither hydrogen (oxidation state +1) nor oxygen (oxidation state −2) changes oxidation state in this reaction. The gold loses $3\,e^-$ per atom, passing from the zero to the +3 oxidation state, It is oxidized. The Se atom in H_2SeO_4 passes from the +6 to the +4 oxidation state; it gains $2\,e^-$ and so H_2SeO_4 is reduced. Note that only half of the H_2SeO_4 reacted is actually reduced.

2-47 Formation of barium hydride from the elements: $\mathrm{Ba}(s) + \mathrm{H_2}(g) \rightarrow \mathrm{BaH_2}(s)$.
Reaction with water: $\mathrm{BaH_2}(s) + 2\,\mathrm{H_2O}(l) \rightarrow 2\,\mathrm{H_2}(g) + \mathrm{Ba^{2+}}(aq) + 2\,\mathrm{OH^-}(aq)$.
Reaction with hydrochloric acid:
$\mathrm{BaH_2}(s) + 2\,\mathrm{HCl}(aq) \rightarrow \mathrm{Ba^{2+}}(aq) + 2\,\mathrm{H_2}(g) + 2\,\mathrm{Cl^-}(aq)$.
Barium hydride reacts with water to produce Ba^{2+} and OH^- ions. The sulfate ion (formed previously when the $ZnSO_4$ dissolved in the water) reacts with the Ba^{2+} as soon as it forms to produce insoluble $BaSO_4$. Also, the zinc(II) ions react with the OH^- as it forms. Thus, three reactions go on nearly simultaneously:
$\mathrm{BaH_2}(s) + 2\,\mathrm{H_2O}(l) \rightarrow \mathrm{Ba^{2+}}(aq) + 2\,\mathrm{OH^-}(aq) + 2\,\mathrm{H_2}(g)$
$\mathrm{Ba^{2+}}(aq) + \mathrm{SO_4^{2-}}(aq) \rightarrow \mathrm{BaSO_4}(s)$,and $\mathrm{Zn^{2+}} + 2\,\mathrm{OH^-}(aq) \rightarrow \mathrm{Zn(OH)_2}(s)$.

2-49 a) MgO is the base anhydride of magnesium hydroxide $Mg(OH)_2$.
b) Cl_2O is the acid anhydride of hypochlorous acid HOCl.
c) SO_3 is the acid anhydride of sulfuric acid H_2SO_4.
d) Cs_2O is the base anhydride of cesium hydroxide CsOH.

2-51 a) $\mathrm{H_2Te}(aq) + \mathrm{H_2O}(l) \rightarrow \mathrm{H_3O^+}(aq) + \mathrm{HTe^-}(aq)$ (acid-base reaction)
b) $\mathrm{SrO}(s) + \mathrm{CO_2}(g) \rightarrow \mathrm{SrCO_3}(s)$ (Lewis acid-base reaction)
c) $2\,\mathrm{HI}(aq) + \mathrm{CaCO_3}(s) \rightarrow \mathrm{CaI_2}(aq) + \mathrm{H_2O}(l) + \mathrm{CO_2}(g)$ (acid-base reaction)
d) $\mathrm{Na_2O}(s) + 2\,\mathrm{NH_4Br}(s) \rightarrow 2\,\mathrm{NaBr}(aq) + \mathrm{H_2O}(l) + 2\,\mathrm{NH_3}(g)$ (acid-base reaction)

2-53 (a) Take the boiling point of zinc to be approximately midway between the boiling point of its neighbors in the periodic table copper and gallium:

$$T_b(\mathrm{Zn}) = \frac{2467°\mathrm{C} + 2403°\mathrm{C}}{2} = 2485°\mathrm{C}$$

This does not compare at all well with the experimental value of 907°C.
b) Gold and thallium are both solids whereas mercury is a liquid at room temperature. It is reasonable to suggest that the boiling point of mercury is likewise much lower than that of gold or thallium. The elements in the zinc group probably all have unexpectedly low boiling points.

c) Based on the idea just stated, the boiling point of Cd should be a lot lower than those of Ag and In.

2-55 a) Element 111 would come directly beneath gold (Au) in the column headed by copper in the periodic table. It would be expected to be a metal resembling gold in its properties.
b) Element 117 would be the next-to-last element in the seventh period. It would come beneath At in Group VII in the periodic table. Element 117 would be metallic, but still probably would form a compound with potassium of formula K(Uus) where Uus is the symbol of element 117.

2-57 a) Ammonium phosphate is $(NH_4)_3PO_4$; potassium nitrate is KNO_3; ammonium sulfate is $(NH_4)_2SO_4$.
b) $(NH_4)_3PO_4$. First calculate the molecular masses by summing up the individual atomic masses of each element in the compound multiplied by the number of times it appears in the compound. Next, compute ratios of the individual masses of each element in the compound to the total molecular mass. and convert to the basis of percent:

Compound	Molecular Mass	Percent N	Percent P	Percent K
$(NH_4)_3PO_4$	149.09	28.18	20.78	0
KNO_3	101.10	13.85	0	38.67
$(NH_4)_2SO_4$	132.14	21.20	0	0

2-59 a) A Lewis structure for O—P—Cl in which the octet rule is obeyed for all atoms and all atoms have a formal charge of zero is $\ddot{O}=\ddot{P}—\ddot{Cl}$:
b) A Lewis structure for O_2PCl in which all of the formal charges are 0 is:

:Ö=P—C̈l: (with :Ö= also double bonded to P)

Although all of the formal charges are 0, the octet rule is violated for the phosphorus, which has an expanded octet. Two structures for O_2PCl in which the octet rule is not violated for any atom are:

$^{-1}$:Ö—P^{+1}—C̈l: (with :Ö= double bonded to P) ⟷ :Ö=P^{+1}—C̈l: (with $^{-1}$:Ö— single bonded to P)

The trouble now is that there exists a separation of formal charge between the phosphorus and one of the oxygens.

2-61 The resonance structures of nitryl chloride are:

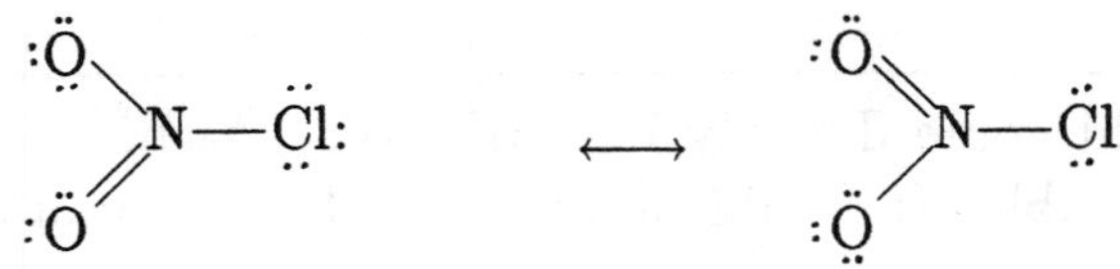

2-63 The molecule of nitrogen dioxide has 17 valence electrons, so at least one atom must see an odd number of electrons. If the central nitrogen atom has the odd number of electrons, the only way to have zero formal charges on all atoms is for the central nitrogen atom to have more than an octet. If one of the oxygen atoms has the odd number of electrons, zero formal charges are attained with that oxygen having less than an octet. The better structure will therefore put the odd number of electrons on an oxygen atom.

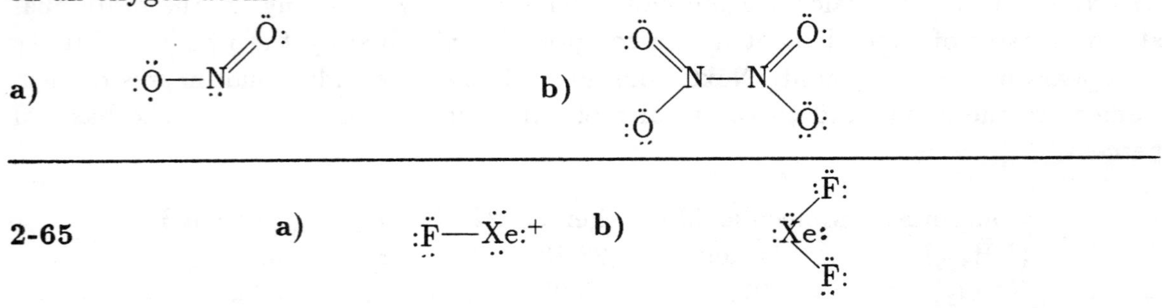

2-67 a) The preparation of acetylene from calcium carbide and water is:

$$\overset{+2}{Ca}\overset{-1}{C_2}(s) + 2\,\overset{+1}{H_2}\overset{-2}{O}(l) \rightarrow Ca^{2+}(aq) + 2\,\overset{-2}{O}\overset{+1}{H}{}^-(aq) + \overset{-1}{C_2}\overset{+1}{H_2}(g)$$

where the oxidation numbers of the elements are all given. None changes, so the reaction is not a redox reaction.

b) The reaction can be considered to be an acid-base reaction. The water is the Lewis acid and CaC_2 is the Lewis base.

2-69 From the formula Bi_5^{3+} we conclude that the oxidation number of the bismuth is $+3/5$. Because the oxidation number of F is always -1 by convention, the oxidation number of the As is $+5$. The elevated dot in the formula means that SO_2 is loosely associated with this salt in its solid state. The oxidation number of S in SO_2 is $+4$, and the oxidation number of O is -2.

2-71 The reaction is $2\,FeO(s) + CaO(s) + Si(s) \rightarrow 2\,Fe(l) + CaSiO_3(l)$. In this reaction Si is oxidized from zero to $+4$ and Fe(II) is reduced from $+2$ to zero.

2-73 The equation with oxidation numbers shown for each element is:

$$2\,\overset{+1}{H}\overset{+6}{Xe}\overset{-2}{O_4}{}^-(aq) + 2\,\overset{-2}{O}\overset{+1}{H}{}^-(aq) \rightarrow \overset{+8}{Xe}\overset{-2}{O_6}{}^{4-}\ (aq) + \overset{0}{Xe}(g) + \overset{0}{O_2}(g) + +2\overset{+1}{H_2}\overset{-2}{O}(l)$$

The species that is reduced is $HXeO_4^-$, because the Xe goes from the +6 oxidation state in $HXeO_4^-$ to the 0 oxidation state in Xe. Clearly however, $HXeO_4^-$ is also oxidized in this reaction, as the Xe in it passes from the +6 oxidation state to the +8 oxidation state in XeO_6^{4-}. Also, oxygen in the −2 oxidation state (in the hydroxide ion) is oxidized to elemental oxygen.

2-75 Ordinary single electron-pair bonds link six of the H atoms to the four B atoms, accounting for 12 out of 22 valence electrons. Another such link connects the two B atoms, accounting for another 2 electrons. The remaining 8 valence electrons form four bonds that are spread over three centers (from H to B to H). The lines in the diagram connecting B's with non-terminal H's are *not* therefore to be taken to stand for a pair of electrons each.

2-77 The acid anhydride of HNO_3 is formed by dehydrating the compound to N_2O_5 with tetraphosphorus pentoxide: $P_4O_{10} + 4\,HNO_3 \rightarrow 4\,HPO_3 + 2\,N_2O_5$.

2-79 a) From the formula Ba_2XeO_6 we determine that the anion is XeO_6^{4-}. Charge balance between this −4 anion and the +3 americium cation in a salt requires the formula $Am_4(XeO_6)_3$.

b) The formula of the corresponding acid would be H_4XeO_6.

c) Acid anhydrides are formed by removing H_2O from the parent acid. Removal of two H_2O's from the parent acid (H_4XeO_6) gives the acid anhydride, XeO_4.

2-81 The reactions would both be Lewis acid-base reactions with CaO as the base and the oxide of sulfur as the acid:

$CaO + SO_2 \rightarrow CaSO_3$ (calcium sulfite)

$CaO + SO_3 \rightarrow CaSO_4$ (calcium sulfate)

2-83 Apply the principle of conservation of mass. There is 5.00 g of sodium present both before and after the reaction. The heating causes the sodium to combine chemically with oxygen. Let x equal the mass of oxygen taken up to form Na_2O_2 (sodium peroxide) and y equal the mass of oxygen taken up to form NaO_2 (sodium superoxide), the two solid products. The total mass of the two solids is 9.25 g, of which 4.25 g is obviously oxygen, so:

$$x + y = 4.25$$

Sodium peroxide has a molar mass of 77.99 g mol^{-1}. Based on its chemical formula, it is 58.96% Na and 41.04% O by mass. Similarly, sodium superoxide (with a molar mass of 54.99 g mol^{-1}) is 41.81% Na and 58.19% O by mass. If there is x g of O in the Na_2O_2, then there is $(58.96/41.04)x$ g of Na in the Na_2O_2, according to the fixed composition of the compound. Similarly, if there is y of O in the NaO_2, then there is $(41.81/58.19)y$ g of Na in the NaO_2. The total mass of Na is 5.00 g. Hence:

$$\left(\frac{58.96}{41.04}\right)x + \left(\frac{41.81}{58.19}\right)y = 5.00$$

The two equations in x and y are easily solved giving x equal to 2.71 g and y equal to 1.54 g. The required percentages are:

$$\frac{2.71}{4.25} \times 100\% = 63.8\% \text{ peroxide O:} \quad \frac{1.54}{4.25} \times 100\% = 36.2\% \text{ superoxide O}$$

Chapter 3

The Gaseous State

The Chemistry of Gases

Unlike solids and liquids, gases expand to fill the container they occupy. The molecules of gases interact only weakly and are on the average far apart from each other in comparison to their own diameters. Gases form chemically:

- By thermal decomposition of many solids (such as mercury(II) oxide, calcium carbonate, and ammonium hydrogen carbonate). Look for the formula or partial formula of a known stable gas embedded in the formula of a solid. If such an embedded formula can be found, the solid may well decompose when heated to give that gas (see **problem 3-3**).

- By the direct reaction of many elements with oxygen to give oxides.

- By the action of acids on many ionic solids. When ionic solids are dissolved in water and then mixed, the escape of a water-insoluble gas tends to drive the reaction toward the products.

Pressure and Temperature of Gases

Pressure, temperature, and volume are the three physical properties or variables that are used the most in describing the physical state of gases:

Pressure is force divided by area. Intuitively, a force is a push. Even a weak push driving a sharp pin against the skin causes pain because the area of the point is small, and enormous pressure exists directly under it. The SI unit of force is the **newton**, which equals a kg m s^{-2}. From the definition of pressure, the SI unit of pressure is the newton per square meter or newton·meter^{-2}, also called the **pascal** (see Table B-2, text page A-11). The newton is a reasonably sized unit of force for chemical applications, but the square meter is quite a large area. Consequently, the pascal is too small for common applications. Many other pressure units are defined and in use. Some important ones are related to each other and the pascal as follows:

$$101\,325 \text{ pascal} = 1 \text{ atmosphere} = 760 \text{ torr} = 1.01325 \text{ bar} = 14.6960 \text{ psi}$$

A torr is the pressure exerted by a column of mercury one millimeter high when the temperature of the mercury is 0°C. The diameter of such a column does not matter. A large-diameter column of mercury exerts a large force but spreads it over a large area. A small column of mercury exerts less force, but because the area at its base is proportionately less, the pressure at its base is the same. This fact is the basis for the operation of a **barometer** and is summarized:

$$P = \rho g h$$

where ρ is the density of the liquid in the column, g is the acceleration of gravity, and h is the height of the column. The acceleration of gravity at the surface of the earth is 9.80665 m s^{-2}. See **problems 3-5, 3-7,** and **3-57.**

The everyday pressure of the earth's atmosphere fluctuates in the neighborhood of one standard atmosphere (atm). A psi is a pound of force per square inch.

The pressure *of* a gas is equal to the pressure *on* that gas. When the internal pressure of a sample of gas differs from the external pressure, then the gas either expands or is compressed.

Volume is the three-dimensional region available to the sample of gas under consideration. The obvious SI unit for volume is the cubic meter. Because this unit is inconveniently large, additional units of volume are in common use. The most important in chemistry are the cubic decimeter, which is given the special name of the **liter,** and the cubic centimeter. The following relationships should be memorized:

$$\frac{1}{1000} \text{ m}^3 = 1 \text{ dm}^3 = 1 \text{ liter (L)} \quad \text{and} \quad \frac{1}{1000} \text{ L} = 1 \text{ mL} = 1 \text{ cm}^3$$

The volume of a sample is very easily measured and consequently is often studied as a function of other variables.

Boyle's law states that the pressure of a quantity of gas is inversely proportional to its volume when the gas is held at a constant temperature.

$$P = C\left(\frac{1}{V}\right)$$

A useful operational version of Boyle's law is:

$$P_1V_1 = P_2V_2 \quad \text{for a fixed amount of gas at constant temperature}$$

where the subscripts refer to the variables before and after some change.

Temperature is intimately connected with **Charles's law.** Charles's law states that the volume of a quantity of gas is linearly dependent on its temperature when the gas is held at a constant pressure. All gases (at low enough pressures) expand by 1/273.15 of their volume at 0°C for every degree Celsius that they are heated:

$$V = V_0\left(1 + \frac{t}{273.15°\text{C}}\right)$$

This finding by Charles suggested the definition of the absolute scale of temperature upon which the volume of the gas is directly proportional to the temperature:

$$V \propto T \quad \text{(fixed pressure and fixed amount of gas)}$$

The **Kelvin temperature scale** is an absolute scale having degrees (named kelvins) of the same size as Celsius degrees but having its zero at −273.15°C. Hence:

$$T\,(\text{Kelvin}) = t\,(\text{Celsius}) + 273.15$$

The Ideal Gas Law

At sufficiently low densities, all gases follow the **ideal-gas law,** an equation combining Boyle's law, Charles's law, and the fact that the volume of a gas depends directly on its amount at a given temperature and pressure. The ideal-gas law is:

$$\boldsymbol{PV = nRT}$$

In this equation, V represents the volume of the gas, P the pressure, T the absolute temperature, and n the chemical amount. The R is a constant. **The absolute temperature must be used whenever any form of the ideal gas law is used.** The constant R is the **universal gas constant**. It is a fundamental constant of nature, like the speed of light. The units of the gas constant are:

$$\text{energy}\cdot\text{temperature}^{-1}(\text{chemical amount})^{-1}$$

In the SI, $R = 8.315$ J $\text{K}^{-1}\text{mol}^{-1}$. Another value is 0.0820578 L atm $\text{mol}^{-1}\text{K}^{-1}$. The first value is mostly used in calculations involving the energy of molecular motion and in chemical thermodynamics. The second is more useful in calculations involving different states of gases.

• It is legitimate and often advantageous to recast the units of R. Alternates are 8.315×10^{-3} kJ $K^{-1}mol^{-1}$ and 82.057 cm^3 atm $K^{-1}mol^{-1}$. See **problems 3-35** and **3-49**.

A comparison of the different values of R shows that the L atm must be a unit of energy. It is (force/area) × volume which equals force × distance. Think of a piston being forced along a cylinder by the expansion of a gas.

Problems Based on the Ideal-Gas Law

A limitless number of possible problems use the ideal-gas equation. Many of these have their roots in practical laboratory work, and others are more fanciful. Whatever the problem, it will fall into one of two broad categories:

Problems Involving a Change of Conditions. The temperature, pressure and volume of a gas are stated. One or two of the variables are then changed and the resulting value of the third variable is required. Such problems are usually easy. A reading of the problem allows construction of a table of P_1, V_1, T_1 and P_2, V_2, T_2. Then, given that:

$$R = \frac{PV}{nT} \qquad \text{it follows that} \qquad \frac{P_1V_1}{n_1T_1} = \frac{P_2V_2}{n_2T_2}$$

and the rest is substitution. Cancellation of a variable that does not change between the initial and final states is a very common motif (see **problems 3-11, 3-19, 3-21,** and **3-63**). Problems in which the chemical amount of the gas changes along with or instead of P, V and T also occur. See **problems 3-65** and **3-67**.

Problems Involving an Unknown in the Gas Law. The problem will in some guise give the values of three out of four of the variables in the ideal-gas equation and ask the computation of the fourth. All that is required is access to a suitable value for R, the universal gas constant, substitution, and some care with the units. Important variations on this theme do not give explicit values of variables but instead give a quantity relating two of them. For example, knowing the ratio n/V, the molar density of a gas, is just as good as knowing both n and V if P or T must be computed. See **problems 3-67** and **4-59**.

Mixtures of Gases

According to **Dalton's law** of partial pressures, each gas in a mixture of gases exerts a **partial pressure** equal to the pressure that it would exert if it were present by itself under the same conditions. The sum of the partial pressures of the components of a mixture of gases equals the total pressure:

$$P_{\text{tot}} = P_{\text{A}} + P_{\text{B}} + P_{\text{B}} + \cdots$$

where P_{A} is the partial pressure of the first component, P_{B} is the partial pressure of the second, and so forth. Dalton's law implies that a molecule in a mixture of gases interacts with *unlike* molecules to just the same extent it interacts with *like* molecules. In other words, the molecules of the several gases in a gas mixture that follows Dalton's law effectively interpenetrate each other. The partial pressure of a component gas in a mixture of ideal gases is the total pressure multiplied by the mole fraction of that component. The **mole fraction** is the number of moles of the gas in question divided by the total number of moles of all gases present. See **problem 3-65** for a use of Dalton's law.

A common application of Dalton's law occurs when gases are collected over volatile liquids, *e.g.*, the collection of oxygen gas over water. The O_2 gas is admixed with water vapors. The total pressure of the sample is then:

$$P_{\text{tot}} = P_{\textbf{oxygen}} + P_{\textbf{water vapor}}$$

To get the pressure that the oxygen would exert if it were pure, the contribution of the water vapor must be subtracted. An example is **problem 4-3.** The vapor pressure of water depends only on the temperature (see Chapter 4) and is widely tabulated.

The Kinetic Theory of Gases

The postulates of the **kinetic theory of gases** are brief:

1. A pure gas has many identical molecules of negligible size. That is, the molecules are approximately point masses.

2. The molecules of a gas are in chaotic motion with a distribution of speeds.

3. There are no interactions among the molecules except during elastic collisions. (An elastic collision is a collision in which no translational energy is lost.)

4. The collisions of molecules with the walls of the container are elastic .

From these postulates and the ideal-gas law, it is possible to derive a relationship that links the speeds of molecules and the temperature. (see text page 104-106). The derivation uses Newton's laws of motion; the result is:

$$\frac{1}{3}N_0 m\bar{u^2} = RT$$

where $\bar{u^2}$ is the **mean-square speed** of the molecules of the gas and m is the mass of the individual molecule. The mean-square speed is the average of the squares of the speeds of the molecules. According to the equation, the mean-square speed of the molecules in an ideal gas depends entirely on the temperature and not at all on the identity or mass of the molecules. The mass of the individual molecules multiplied by Avogadro's number equals the molar mass $\mathcal{M}$ of the molecules. Therefore:

$$\bar{u^2} = \frac{3RT}{\mathcal{M}}$$

An important aspect of the derivation is the difference between **speed** and **velocity.** The velocity v of an object is a vector. It has both magnitude and direction. The speed u of an object is how fast it moves without regard to direction. Speed has magnitude alone. A car proceeding too fast gets a speeding ticket. A car proceeding (even slowly) the wrong way on a one-way street gets a velocity ticket.

Like any vector, velocity can be expressed in terms of a triple of components, v_x, v_y, and v_z, relative to a set of Cartesian coordinate axes. The square of the speed is:

$$u^2 = v_x^2 + v_y^2 + v_y^2$$

• The addition of velocities must take their directions into account. The **momentum** of a particle equals its velocity times its mass. Therefore, the addition of momenta must take their directions in account.

Distribution of Molecular Speeds

The speeds of the molecules in a gaseous substance **at thermal equilibrium** are distributed across a wide but statistically predictable range. The prediction is made by the **Maxwell-Boltzmann speed distribution:**

$$\boldsymbol{f(u) = 4\pi \left(\frac{m}{2\pi k_B T}\right)^{3/2} u^2 \exp(-mu^2/2k_B T)}$$

where k_B is a new constant **(Boltzmann's constant)** equal to R/N_0, m is the mass of the molecules, and T is the absolute temperature.

To learn what $f(u)$ means take the function apart and study it. Note first the "exp" means to raise the base e to a power equal to what follows in parentheses. Now, suppose that m and T are fixed. Two factors involve the molecular speed u: $\exp(-mu^2/2k_BT)$ and u^2. The quantity $-mu^2/2k_BT$ gets more and more negative if u increases. (and with k_B a constant, see below). Therefore $\exp(-mu^2/2k_BT)$ gets smaller and smaller as u increases If this exponential part alone controlled the distribution of molecular speeds, the molecules in a gas would scarcely be moving. But the exponential part is multiplied by u^2, which of course increases as u increases. As a result, $f(u)$ grows as u rises from zero, reaches a maximum, and finally diminishes as the effect of the exponential term becomes dominant. The following table shows this for nitrogen gas (N_2) at T equal to 273.15 K:

u (m s^{-1})	u^2 (m^{2}s^{-2})	$\exp(-mu^2/2k_BT)$	$f(u)$ (s m^{-1})
0	0	1.000	0.0
50	2.5×10^3	0.985	8.5×10^{-5}
100	1.0×10^4	0.940	3.3×10^{-4}
200	4.0×10^4	0.781	1.1×10^{-3}
300	9.0×10^4	0.574	1.8×10^{-3}
400	1.5×10^5	0.373	2.1×10^{-3}
500	2.5×10^5	0.214	1.9×10^{-3}
1000	1.0×10^6	2.10×10^{-3}	7.3×10^{-5}
3000	9.0×10^6	7.81×10^{-25}	2.4×10^{-25}

At low speeds u^2 dominates. At high speeds the exponential term crushes the total function $f(u)$ to zero. The two leading factors in the Maxwell-Boltzmann distribution are constants for a given gas at a given temperature. In this example their combined value is 3.457×10^{-8} s^3 m^{-3}.

The units of $f(u)$ are s m^{-1}, the reciprocal of the units of speed. When $f(u)$ is multiplied by Δu, a range of speed, all of the units cancel out. This reveals $f(u)\Delta u$ as a **probability distribution,** a pure number telling the chance that a molecule will have a speed between u and $(u + \Delta u)$.

Example: What is the chance that a molecule in a sample of gaseous N_2 held at 273.15 K has a speed between 390 and 410 m s^{-1}?

Solution: Δu is 20 m s^{-1}, $f(u)$ at 400 m s^{-1} is 2.1×10^{-3} s m^{-1} (from the table). Therefore, $f(u)\Delta u = 0.042$. This is the chance that an individual N_2 molecule is in the desired range of speeds. It is as well the fraction of all the molecules with speeds in the 390 to 410 m s^{-1} range. The answer makes the approximation that $f(u)$ is constant across the 390 to 410 m s^{-1} range. It can be improved by computing $f(u)$ at various speeds between 390 and 400 m s^{-1}, using shorter Δu's that bracket these speeds, and adding up the subtotals. See Appendix C (text page A-22) for details on this operation, which corresponds to estimating an area under the Maxwell-Boltzmann speed distribution curve. Also, see **problem 3-71.**

The Maxwell-Boltzmann distribution is formed in such a way that the sum of all possible probabilities is 1.

The **root-mean-square speed** u_{rms} is the square root of the mean of the squares of the speeds of a collection of molecules. It is defined by the equation:

$$u_{rms} = \sqrt{\frac{3k_BT}{m}}$$

The **Boltzmann constant** k_B has the units of (energy temperature^{-1}). It equals 1.38×10^{-23} J K^{-1}. As always, T and m stand for the absolute temperature of the gas and the mass of the molecules. If m is in kg then u_{rms} comes out in m s^{-1}. Common gases at room temperature have root-mean-square molecular speeds on the order of 10^3 m s^{-1}. At high temperatures, average speeds are higher (see **problem 3-69a).**

The Boltzmann constant is equal to the universal gas constant divided by Avogadro's number: $k_B = R/N_0$. This means that k_B/m can be replaced by $R/\mathcal{M}$ where $\mathcal{M}$ is the mass of a mole (Avogadro's number) of molecules and vice versa (see **problem 3-69a).** Therefore the root-mean-square speed is also:

$$u_{rms} = \sqrt{\frac{3RT}{\mathcal{M}}}$$

If R in this equation is 8.315 J K^{-1}mol^{-1} then $\mathcal{M}$ must have the units kg mol^{-1} (and not the more usual g mol^{-1}) in order for the units of the speed to come out as m s^{-1}. To prove this, recall that a joule is a kg m^{2}s^{-2} and see **problem 3-37.**

The Maxwell-Boltzmann distribution curve is not symmetrical about its maximum. Consequently two other representative speeds of molecules, differing from u_{rms}, arise:

$$\bar{u} = \sqrt{\frac{8k_BT}{\pi m}} = \sqrt{\frac{8RT}{\pi\mathcal{M}}} \quad \text{and} \quad u_{mp} = \sqrt{\frac{2k_BT}{m}} = \sqrt{\frac{2RT}{\mathcal{M}}}$$

The **average speed** $\bar{u}$ is the arithmetic mean of the speeds of the molecules in a sample of gas. It is always less than u_{rms} because squaring the speeds, the procedure used in computing a root-mean-square speed, gives extra emphasis to larger speeds. The **most probable speed** u_{mp} is always less than both u_{rms} and $\bar{u}$. It is the speed at which the Maxwell-Boltzmann distribution curve reaches its maximum.

Applications of the Kinetic Theory

Wall Collisions, Effusion and Diffusion

The rate of collisions of the molecules of a gas with a section of a wall of its container is:

$$Z_w = \frac{1}{4}\frac{N}{V}\bar{u}A = \frac{1}{4}\frac{N}{V}\sqrt{\frac{8RT}{\pi\mathcal{M}}}A$$

where A is the wall area under consideration and N/V is the **number density** of the gas. The number density of a sample of gas equals its molar density (in, for example, mol L^{-1}) multiplied by Avogadro's number (N_0). The units of Z_w are time^{-1}, that is, Z_w gives the number of collisions per unit time. See **problem 3-43.**

Graham's law of effusion follows from this expression. Effusion is the escape of a gas from a container through a small hole into a vacuum. Graham's law states that the rate of effusion of a gas at constant temperature and pressure is inversely proportional to the square root of its molecular mass:

$$\text{rate of effusion} \propto \sqrt{\frac{1}{\mathcal{M}}} \quad (\text{constant } T \text{ and } P)$$

It is often applied to mixtures of two gases effusing through a small orifice. The ratio of the rate of effusions is then:

$$\frac{\text{rate of effusion of A}}{\text{rate of effusion of B}} = \frac{N_{\text{A}}}{N_{\text{B}}}\sqrt{\frac{\mathcal{M}_{\text{B}}}{\mathcal{M}_{\text{A}}}}$$

The gas that emerges is enriched by the lighter component by the **enrichment factor** $\sqrt{\mathcal{M}_B/\mathcal{M}_A}$ if B is heavier than A. The rate of diffusion, which is physically a different process, is also inversely proportional to the square root of the molar mass of the diffusing gas (see **problem 3-47**).

The typical problem involving Graham's law requires the computation of the molecular mass of an unknown gas by comparison of its rate of effusion to the rate of effusion of a known gas under the same conditions. Trouble arises in problems based on Graham's law because the "rate of effusion" of a gas may be stated in a variety of ways: the number of molecules that escape per unit time, the volume of gas that escapes per unit time, and the mass of gas that escapes per unit time. See **problem 3-73.** Effusion data may also be given in terms of the inverses of the preceding: the time it takes a fixed number of molecules to escape, the time it takes a fixed volume of gas to escape, and the time it takes a fixed mass of gas to escape. Rates of diffusion can be stated in all of these units and also in terms of the distance a gas diffuses down a tube per unit time and its inverse, the time it takes a gas to diffuse a fixed distance.

Frequency of Molecular Collisions

If the molecules of a gas were mathematical points, would always miss each other and could collide only with the walls of the container. In real gases the molecules have a small effective diameter d. There is then a calculable quantity Z_1, the rate at which a typical molecule experiences collisions:

$$Z_1 = \sqrt{2}\frac{N}{V}\pi d^2 \bar{u} = 4\frac{N}{V}d^2\sqrt{\frac{\pi RT}{\mathcal{M}}}$$

The units of Z_1 in the SI are s^{-1}. The derivation of this expression assigns a diameter d to the molecules of the gas and imagines an individual molecule sweeping out a cylinder over some fixed time. The frequency of collisions of this molecule with others varies with the square of the diameter of its cylinder. A bigger molecule is harder to avoid. Because a faster moving molecule sweeps out a longer cylinder in the fixed time, the frequency of collisions is also directly proportional to the root-mean-square speed of the molecules. Finally, the greater the density of the gas, then the more encounters the test molecule experiences per unit time, making the frequency of molecular collisions directly proportional to the number density of the gas.

Mean Free Path and Diffusion

The **mean free path** λ of a gas is the average distance that its molecules travel between collisions. It is the product of the average speed of the molecules ($\bar{u}$) and the average time between their collisions. The average time between collisions is the reciprocal of the rate of collision. Therefore:

$$\lambda = \bar{u}\left(\frac{1}{Z_1}\right)$$

It is wise to understand the reasoning behind this simple form before going on with the following manipulation. After substituting for Z_1, $\bar{u}$ cancels out:

$$\lambda = \frac{1}{\sqrt{2}\pi d^2 N/V}$$

Thus, the mean free path of a gas is independent of its temperature and its molar mass, a result that many find surprising. See **problems 3-49** and **3-75** for calculations using this formula. At higher densities, the mean free path of a gas becomes comparable to its molecular diameter. At this point, deviations from ideality (next section) become important.

Diffusion in a gas can be described in terms of the mean-square displacement $\overline{\Delta r^2}$ and the root-mean-square displacement $\sqrt{\overline{\Delta r^2}}$ of the molecules over the course of time. Any molecule follows a the path of a drunken pedestrian (a random walk) as the chaotic buffeting of collisions with the other molecules constantly redirects it. The mean-square displacement is the average of the squares of such displacements over many molecules. In the absence of gas currents, the mean-square displacement equals a constant multiplied by the elapsed time:

$$\overline{\Delta r^2} = 6Dt$$

where D is the **diffusion constant.** The diffusion constant depends on the temperature, density, molar mass, and molecular diameter of the gas:

$$D = \frac{3}{8}\sqrt{\frac{RT}{\pi \mathcal{M}}}\,\frac{1}{d^2\,N/V}$$

As in the formulas for the root-mean-square speed and other molecular speeds in gases, the molar mass $\mathcal{M}$ must be in kg mol^{-1} if R in J K^{-1}mol^{-1} is used. Also, the number density should have units of m^{-3}, and the molecular diameter should be in meters. See **problem 3-49**. Then, the units of D, which have the form (length)^2·(time)$^{-1}$, come out to m^{2}s^{-1}. When this unit is multiplied by a time, the result is a distance squared.

Real Gases

The ideal-gas equation is an equation of state. It relates some of the state variables that describe the physical behavior of a sample of gas. It is not the only possible equation of state, although it is certainly the simplest.

More elaborate equations of state take into consideration the definite, if small, size of the molecules comprising the system and the existence of intermolecular attractions. Such equations of state fit observed P-V-T data better.

The Van der Waals Equation of State

One comparatively simple equation of state is the **van der Waals equation**:

$$\left(\boldsymbol{P} + \boldsymbol{a}\frac{\boldsymbol{n}^2}{\boldsymbol{V}^2}\right)(\boldsymbol{V} - \boldsymbol{nb}) = \boldsymbol{nRT}$$

The van der Waals equation is the ideal gas equation modified by two correction factors. An additive correction (an^2/V^2) operates on the observed pressure of the gas. Intermolecular attractions tend to reduce pressures from what they would be in the ideal case (no attractions) and the addition of a-factor makes up for the reduction.

A subtractive correction (nb) is applied to the actual volume available to the motion of the molecules of the gas to account for the volume excluded to free motion by the existence of other molecules.

The strength of the intermolecular attractions in a gas and the size of the excluded volume both depend on the identity of the gas. Therefore the values of the van der Waals constants a and b differ from gas to gas. They are experimentally determined. See Table 3-3, text page 116.

In working the van der Waals equation, it is easy to compute P given everything else and also easy to compute T given everything else (see **problems 3-51 and 3-53**). It is however hard to compute V given all the other variables. The iterative approach (Appendix C, text page A-17) helps.

Detailed Solutions to Odd-Numbered Problems

3-1 The gases produced in the decomposition of ammonium hydrosulfide are ammonia,

NH_3, and hydrogen sulfide, H_2S. The chemical equation for the decomposition is $NH_4HS(s) \rightarrow NH_3(g) + H_2S(g)$

3-3 The easiest way to generate ammonia from ammonium bromide is to mix the ammonium bromide with a base. Ammonium bromide, NH_4Br, dissolves in water to form ammonium ion, $NH_4^+(aq)$ and bromide ion $Br^-(aq)$ The $NH_4^+(aq)$ ion will react with bases to form ammonia, $NH_3(g)$: $NH_4Br(aq) + NaOH(aq) \rightarrow NH_3(g) + NaBr(aq) + H_2O(l)$. By heating the solution to the boiling point, one can drive the gaseous ammonia from solution (gases are typically less soluble in hot solvents than in cold solvents).

3-5 Because water is considerably less dense than mercury it takes a much longer column of it to balance the pressure of the atmosphere. A pressure of 1.00 atm exerted by air is balanced in a barometer by a column of mercury 76.0 cm high. The density of mercury is 13.6 g cm^{-3}, and the density of water is only 1.00 g cm^{-3}. Therefore the column of water must be longer in proportion to its lesser density:

$$76.0 \text{ cm} \times \left(\frac{13.6 \text{ g cm}^{-3}}{1.00 \text{ g cm}^{-3}}\right) = 1.03 \times 10^3 \text{ cm} = 10.3 \text{ m}$$

This is nearly 34 feet. Such water barometers have been built. The problem can also be solved by direct substitution in the formula $P = \rho g h$:

$$h = \frac{P}{\rho g} = \frac{101\,325 \text{ Pa}}{(1.00 \times (10^3 \text{kg m}^3)(9.807 \text{ m s}^{-2})} = 10.3 \text{ m}$$

Note the careful conversion of the pressure to pascals, the SI unit, and the density of water to a combination of SI base units so that correct cancellation of units takes place.

3-7 Convert the pressure from atmospheres to pascals (Pa) and then use the barometric formula $P = \rho g h$ to compute the depth (h) of water that exerts the same pressure. In this computation, g is the acceleration of the earth's gravity (9.807 m s^{-2}) and ρ is the density of water. It is safe to take the density as a constant 1.0×10^3 kg m^{-3} throughout the oceans—the density of even the deep sea is not too much affected by dissolved salts or by the compression of overlying layers. The pressure is:

$$P = \frac{101\,325 \text{ Pa}}{1 \text{ atm}} \times 414 \text{ atm} = 4.19 \times 10^7 \text{ Pa}$$

The depth of water that exerts this pressure at its base is

$$h = \frac{P}{\rho g} = \frac{4.195 \times 10^7 \text{ kg m}^{-1}\text{s}^{-2}}{(1.0 \times 10^3 \text{ kg m}^{-3})\ 9.807 \text{ m s}^{-2}} = 4.3 \times 10^3 \text{ m}$$

A pascal equals a kg m^{-1}s^{-2}, according to Table B-2 text page A-11). One meter is 3.28 feet, so the depth of 4 300 m is equivalent to 14 000 feet.

3-9 The pascal (Pa) is a newton per square meter (N m^{-2}). One standard atmosphere is defined as 101 325 newton per square meter or 101 325 Pa. The enormous pressure of 172.00 MPa (megapascals) is 1.720×10^8 Pa, which is converted to atmospheres as follows:

$$1.720 \times 10^8 \text{ Pa} \times \left(\frac{1 \text{ atm}}{101\,325 \text{ Pa}}\right) = 1.6975 \times 10^3 \text{ atm}$$

We can convert the pressure in pascals to bars simply by multiplying it by the factor (1 bar/10^5 Pa). The answer is 1.7200×10^3 bar.

3-11 The temperature does not change. Therefore, assume N_2 to behave as an ideal gas, and use Boyle's law in the form $P_1V_1 = P_2V_2$. The initial pressure P_1 is 3.00 atm, the initial volume V_1 is 2.00 L, P_2 is the final pressure (this is the desired answer) and V_2 is the final volume. If the volumes of the valve and associated plumbing are negligibly small, $V_2 = 2.00 + 5.00 = 7.00$ L. Hence (3.00 atm)(2.00 L) = P_2(7.00 L), and $P_2 = 0.857$ atm.

3-13 This is an application of Charles's law $V_1T_2 = V_2T_1$. The V_1 is given (4.00 L), and it is stated that the temperature is doubled, that is, $T_2 = 2T_1$. Accordingly, $V_2 = 8.00$ L. If the absolute temperature is doubled at constant pressure, the volume of an ideal gas also doubles.

3-15 Again, use Charles's Law. In this case $V_1 = 17.3$ gill and V_2 is required. The temperatures are given in degrees Fahrenheit and must be converted to kelvins before being used. Always convert temperature to the Kelvin scale when working gas-law problems:

$$T_1 = 37.8 + 273.15 = 310.95 \text{ K} \quad \text{and} \quad T_2 = -17.8 + 273.15 = 255.35 \text{ K}$$

Now use Charles's Law:

$$V_2 = \left(\frac{T_2}{T_1}\right) V_1 = \left(\frac{255.35 \text{ K}}{310.95 \text{ K}}\right) 17.4 \text{ gills} = 14.3 \text{ gills}$$

3-17 The same mass of CaC_2 will produce the same chemical amount of $C_2H_2(g)$ regardless of the temperature. Since the pressure is the same (1 atm) in both cases, this is a Charles's law problem where $V_1 = 64.5$ L, $T_1 = 50°C$, $T_2 = 400°C$, and V_2 is unknown. Convert the temperatures to kelvins by adding 273.15 and substitute in Charles's law:

$$V_2 = \left(\frac{T_1}{T_2}\right) V_1 = \left(\frac{673 \text{ K}}{323 \text{ K}}\right) 64.5 \text{ L} = 134 \text{ L}$$

3-19 The actual pressure inside the tire is 30 + 14.7 psi = 44.7 psi. Assuming that the air in the tire behaves ideally:

$$\frac{P_1V_1}{n_1T_1} = \frac{P_2V_2}{n_2T_2}$$

where the subscripts refer to the values of the variables before and after the change from 0°C to 32°C. Because the tire does not expand $V_1 = V_2$. Heating does not change the quantity of air inside the tire, so $n_1 = n_2$. Cancelling these quantities gives:

$$\frac{P_1}{P_2} = \frac{T_1}{T_2}$$

The temperatures are $T_1 = 273$ K and $T_2 = 305$ K. Also, $P_1 = 44.7$ psi. Substituting and rearranging:

$$P_2 = P_1\left(\frac{305\text{ K}}{273\text{ K}}\right) = 44.7\text{ psi}\left(\frac{305}{273}\right) = 49.9\text{ psi}$$

It was unnecessary to convert the pressure to a metric unit. The units of P_2 are automatically the same as the units of P_1. The conversion to an absolute temperature *was* necessary. If the temperature had not been converted from the Celsius scale in this case, it would have meant dividing by zero. Gauge pressure is 14.7 psi *less* than the actual pressure. The gauge pressure of the tire at 32°C is thus 49.9 – 14.7 = 35.2 psi. The answer is larger than the original gauge pressure, which fulfills common-sense expectation.

3-21 a) Let state 1 (represented by subscript 1) be the original state of the air before it is put in the bottle. Assuming that the ideal-gas law applies, this means $P_1V_1 = n_1RT_1$. Let state 2 be the state of the air after it is packaged in the bottle: $P_2V_2 = n_2RT_2$. The chemical amount of air does not change between state 1 and state 2 nor does the temperature. This means $n_1RT_1 = n_2RT_2$ and

$$P_1V_1 = P_2V_2$$

The above amounts to a derivation of Boyle's law from the ideal-gas equation. Substituting 20.6 L and 1.05 L for V_1 and V_2 respectively and 1.01 atm for P_1 gives $P_2 = 19.8$ atm.

b) Let the state of the bottled air in the European laboratory be state 3. The pressure in the bottle P_3 is higher in Europe than the pressure in Greenland because the temperature T_3 (294 K) is higher than T_2 (253 K). Remember to convert temperature to kelvin before using any gas law. Neither the volume of the bottle nor the chemical amount of the air in the bottle changes during the trip to Europe. This means $n_2 = n_3$ and $V_2 = V_3$. Rearranging the ideal-gas equation as applied to states 2 and 3 gives:

$$\frac{P_2}{RT_2} = \frac{n_2}{V_2} \quad \text{and} \quad \frac{P_3}{RT_3} = \frac{n_3}{V_3} \quad \text{so that} \quad \frac{P_2}{T_2} = \frac{P_3}{T_3}$$

Substitution of the known temperatures and $P_2 = 19.8$ atm (from part a) gives $P_3 = 23.0$ atm as the answer.

3-23 The information about the density of the gas is nearly valueless because gas densities are strongly dependent on the pressure and temperature, but neither is given. We might choose to assume room temperature and pressure, but STP is also a reasonable assumption. The density of the H_2Te is given. From it we can compute the molar volume of the gas:

$$\frac{1 \text{ L}}{6.234 \text{ g}} \times 129.615 \text{ g mol}^{-1} = 20.791 \text{ L mol}^{-1}$$

If H_2Te behaves ideally, then:

$$\frac{V}{n} = \frac{RT}{P}$$

Solving for T gives:

$$T = \frac{P}{R}\left(\frac{n}{V}\right) = \frac{1 \text{ atm}}{0.08206 \text{ L atm mol}^{-1}\text{K}^{-1}}\left(20.791 \text{ L mol}^{-1}\right) = 253.4 \text{ K}$$

. This is $-19.8°$C.

3-25 a) The other product of the reaction is sodium chloride:

$$2\,\text{Na}(s) + 2\,\text{HCl}(g) \rightarrow \text{H}_2(g) + 2\,\text{NaCl}(s)$$

b) First, calculate the chemical amount of $H_2(g)$ produced from the complete reaction of 6.24 g of Na(s).

$$6.24 \text{ g Na} \times \left(\frac{1 \text{ mol Na}}{22.99 \text{ g Na}}\right) \times \left(\frac{1 \text{ mol H}_2}{2 \text{ mol Na}}\right) = 0.1357 \text{ mol H}_2$$

Next, convert the temperature into kelvins (323 K), rearrange the ideal-gas equation to give V explicitly and substitute the known values: $n = 0.1357$ mol and $P = 0.850$ atm:

$$V = \frac{nRT}{P} = \frac{(0.01357 \text{ mol})(0.08206 \text{ L atm mol}^{-1}\text{K}^{-1})(323 \text{ K})}{0.850 \text{ atm}} = 4.23 \text{ L}$$

3-27 First calculate the chemical amount of NaCl being reacted:

$$2.5 \times 10^6 \text{ g NaCl} \times \left(\frac{1 \text{ mol NaCl}}{58.44 \text{ g NaCl}}\right) = 4.28 \times 10^4 \text{ mol NaCl}$$

According to the balanced equation for the reaction of NaCl with H_2SO_4, 1 mol of HCl forms per 1 mol of NaCl consumed. Therefore, 4.28×10^4 mol NaCl will produce

4.28×10^4 mol HCl. Next, use the ideal-gas equation to compute the volume of the gaseous HCl. Note the conversion of the temperature from the Celsius to the absolute scale:

$$V = \frac{nRT}{P} = \frac{(4.28 \times 10^4 \text{ mol})(0.08206 \text{ L atm mol}^{-1}\text{K}^{-1})(823 \text{ K})}{0.970 \text{ atm}} = 3.0 \times 10^6 \text{ L}$$

3-29 According to the balanced chemical equation, there is a 3-to-2 molar ratio between the O_2 formed and $KClO_3$ consumed. This fact forms a crucial conversion in the following train of unit conversions, which solves the problem:

$$87.6 \text{ g KClO}_3 \times \left(\frac{1 \text{ mol KClO}_3}{122.54 \text{ g KClO}_3}\right) \times \left(\frac{3 \text{ mol O}_2}{2 \text{ mol KClO}_3}\right) = 1.072 \text{ mol O}_2$$

Now, use the ideal-gas equation to compute the volume. Note the conversion of the temperature from the Celsius to the absolute scale:

$$V = \frac{nRT}{P} = \frac{(1.072 \text{ mol})(0.08206 \text{ L atm mol}^{-1}\text{K}^{-1})(286.4 \text{ K})}{1.04 \text{ atm}} = 24.2 \text{ L}$$

3-31 The mole fraction of SO_3 can be calculated directly from the data given in the problem,

$$X_{SO_3} = \frac{\text{mol SO}_3}{\text{total moles}} = \frac{17.0 \text{ mol}}{26.0 \text{ mol} + 83.0 \text{ mol} + 17.0 \text{ mol}} = \frac{17.0 \text{ mol}}{126.0 \text{ mol}} = 0.135$$

The partial pressure of the SO_3 is simply its mole fraction times the total pressure,

$$P_{SO_3} = (0.135)(0.950 \text{ atm}) = 0.128 \text{ atm}$$

3-33 a) The mole fraction of CO in the mixture is,

$$X_{CO} = \frac{\text{mol CO}}{\text{total moles}} = \frac{10.0 \text{ mol}}{10.0 \text{ mol} + 12.5 \text{ mol}} = \frac{10.0 \text{ mol}}{22.5 \text{ mol}} = 0.444$$

b) Since there is a 1-to-1 molar ratio between CO and CO_2 in the reaction, the formation of 3.0 mol of CO_2 must be accompanied by the consumption of 3.0 moles of CO. Therefore, the chemical amount of CO left in the mixture at this point in the reaction is 7.0 mol. The chemical amount of O_2 left at the same point is 12.5 mol – (3.0 mol/2) = 11.0 mol. The factor 2 comes from the 2-to-1 molar relationship between the CO_2 and O_2). The mixture of gases at this point in the reaction consists of 7.0 mol of CO, 11.0 mol of O_2, and 3.0 mol of CO_2. The mole fraction of CO is therefore:

$$X_{CO} = \frac{\text{mol CO}}{\text{total moles}} = \frac{7.0 \text{ mol}}{7.0 \text{ mol} + 11.0 \text{ mol} + 3.0 \text{ mol}} = \frac{7.0 \text{ mol}}{21.0 \text{ mol}} = 0.33$$

3-35 a) Assume the water vapor mixed with air can be treated as a mixture of ideal gases. Then for the water vapor the equation:

$$P_{\text{water}} = n_{\text{water}} \left(\frac{RT}{V}\right)$$

holds, according to Dalton's law of partial pressures. The volume is 1.0 cm^3, the temperature is (20 + 273.15) K, and P_{water} is 0.0230 atm. Substituting these values together with $R = 82.057$ cm^3 atm K^{-1}mol^{-1} into the previous equation gives: gives $n_{\text{water}} = 9.56 \times 10^{-7}$ mol.

Each mole of H_2O contains Avogadro's number, N_0, of molecules so the 1.0 cm^3 of saturated air contains:

$$9.56 \times 10^{-7} \text{ mol} \times \left(\frac{6.022 \times 10^{23} \text{ molecules}}{1 \text{ mol}}\right) = 5.76 \times 10^{17} \text{ molecules of water}$$

Rounding off to two significant figures as required gives 5.8×10^{17} molecules.

b) Because 1.0 cm^3 of air holds only about 10^{-6} mole of water, it will require much more than 1 cm^3 to hold 0.50 mol of water:

$$0.50 \text{ mol } H_2O \times \left(\frac{1.0 \text{ cm}^3 \text{ sat. air}}{9.56 \times 10^{-7} \text{ mol } H_2O}\right) = 5.23 \times 10^5 \text{ cm}^3 \text{ sat. air}$$

After converting to liters and rounding off to two significant figures, the answer is 5.2×10^3 L of saturated air.

3-37 a) The root-mean-square speed of a molecule is given by:

$$u_{\text{rms}} = \sqrt{\frac{3k_BT}{m}} = \sqrt{\frac{3RT}{\mathcal{M}}}$$

where T is the absolute temperature and $\mathcal{M}$ is the molar mass. For H_2 at 300 K:

$$u_{\text{rms}} = \sqrt{\frac{3(8.315 \text{ J K}^{-1}\text{mol}^{-1})(300 \text{ K})}{0.002016 \text{ kg mol}^{-1}}} = 1.93 \times 10^3 \text{ m s}^{-1}$$

The analysis of the units is worth separate study:

$$\sqrt{\frac{\text{J K}^{-1}\text{mol}^{-1} \text{ K}}{\text{kg mol}^{-1}}} = \sqrt{\frac{\text{kg m}^2 \text{ s}^{-2} \text{ K}^{-1} \text{ mol}^{-1} \text{ K}}{\text{kg mol}^{-1}}} = \sqrt{\text{m}^2 \text{ s}^{-2}} = \text{m s}^{-1}$$

b) For sulfur hexafluoride, SF_6, at 300 K:

$$u_{\text{rms}} = \sqrt{\frac{3(8.315 \text{ J K}^{-1}\text{mol}^{-1})(300 \text{ K})}{0.14607 \text{ kg mol}^{-1}}} = 226 \text{ m s}^{-1}$$

The heavier SF_6 molecule has a root-mean-square speed about 8.5 times slower than the lighter H_2 molecule at the same temperature. The ratio of the speeds should equal to the reciprocal of the square root of the ratio of the molar masses.

3-39 The root-mean-square speed of a helium atom at the surface of the sun (6000 K) is:

$$u_{\text{rms}} = \sqrt{\frac{3(8.315 \text{ J K}^{-1}\text{mol}^{-1})(6000 \text{ K})}{0.004003 \text{ kg mol}^{-1}}} = 6100 \text{ m s}^{-1}$$

In an interstellar cloud at 100 K:

$$u_{\text{rms}} = \sqrt{\frac{3(8.315 \text{ J K}^{-1}\text{mol}^{-1})(100 \text{ K})}{0.004003 \text{ kg mol}^{-1}}} = 790 \text{ m s}^{-1}$$

"Comparison" probably means to take the ratio of the two rms speeds rather than to calculate the actual speeds. Getting the ratio is simpler that the previous calculation because $3R$ and $\mathcal{M}$ cancel out:

$$\frac{u_{\text{rms}} \text{ (near sun)}}{u_{\text{rms}} \text{ (interstellar)}} = \sqrt{\frac{6000 \text{ K}}{100 \text{ K}}} = 7.7$$

The root-mean-square speed of a helium atom is only about 8 times faster at 6000 K than at 100 K.

3-41 The root-mean-square speed of an atom in gaseous sodium at 0.00024 K is:

$$u_{\text{rms}} = \sqrt{\frac{3(8.315 \text{ J K}^{-1}\text{mol}^{-1})(0.00024 \text{ K})}{0.02299 \text{ kg mol}^{-1}}} = 0.51 \text{ m s}^{-1}$$

3-43 Compute the chemical amount of air that leaks into the 500 cm^3 bulb during the one-hour period immediately after it is sealed by using the ideal-gas equation. Use the pressure observed at the one-hour time and the known volume and temperature of the container:

$$n = \frac{PV}{RT} = \frac{(1.00 \times 10^{-7} \text{ atm})(0.500 \text{ L})}{(0.082057 \text{ L atm mol}^{-1}\text{K}^{-1})(300 \text{ K})} = 2.0311 \times 10^{-9} \text{ mol}$$

Because a mole contains 6.022×10^{23} particles, this amounts to 1.22×10^{15} molecules. The rate of the leak was 1.22×10^{15} molecules per 3600 s, which equals 3.4×10^{11} molecules s^{-1}.

This outside air enters into the vessel by means of collisions of its molecules with the area of the tiny hole. Therefore, use the formula for the rate of wall collisions by a gas (equation 3-12, text page 110). The rate of the leak is Z_w in this formula. The density of the outside air can be computed using the ideal-gas equation and the known temperature and pressure of the outside air:

$$n/V = \left(\frac{1.00 \text{ atm}}{(0.082057 \text{ L atm mol}^{-1}\text{K}^{-1})(300 \text{ K})}\right) = 4.062 \times 10^{-2} \text{ mol L}^{-1}$$

Multiplying this answer by N_0 makes it a number density: 2.46×10^{22} molecule L^{-1}, which is equal to 2.46×10^{25} molecule m^{-3}. Solve text equation 3-12 for A the area of the wall:

$$A = 4\frac{1}{N/V}\sqrt{\frac{\pi\mathcal{M}}{8RT}}Z_w$$

Insert numbers for the several quantities on the right, taking care with units:

$$A = 4\left(\frac{1}{2.46 \times 10^{25}\ \text{m}^{-3}}\right)\sqrt{\frac{3.1416 \times 0.0288\ \text{kg mol}^{-1}}{8(8.315\ \text{J K}^{-1}\text{mol}^{-1})(300\ \text{K})}}(3.4 \times 10^{11}\ \text{s}^{-1})$$

The area of the hole is 1.18×10^{-16} m^2. Because the hole is circular, its radius is:

$$r = \sqrt{\frac{A}{\pi}} = \sqrt{\frac{1.18 \times 10^{-16}}{3.1416}} = 6.1 \times 10^{-9}\ \text{m} = 6.1\ \text{nm}$$

3-45 The ratio of the rates of effusion of two gases is given by

$$\frac{\text{rate of effusion of A}}{\text{rate of effusion of B}} = \frac{N_A/V}{N_B/V}\sqrt{\frac{\mathcal{M}_B}{\mathcal{M}_A}}$$

Let the methane be A and the unknown gas be B. The ratio of N_A/V to N_B/V equals 1 because the two gases are held under identical conditions in the same container. Taking the molar mass of methane as 16.04 g mol^{-1}, we write:

$$\frac{1.30 \times 10^{-8}}{5.41 \times 10^{-9}} = \sqrt{\frac{\mathcal{M}_B}{16.04\ \text{g mol}^{-1}}}$$

Solving for $\mathcal{M}_B$ gives 92.6 g mol^{-1}.

3-47 The rate of diffusion of a gas is inversely proportional to the square root of its molecular mass. One stage of diffusion of a mixture of $^{235}UF_6$ and $^{238}UF_6$ enriches the product mixture by a factor of $\sqrt{352/349}$ or 1.0043 in the lighter gas, $^{235}UF_6$. The 352 in this expression is the molecular mass of $^{238}UF_6$ (the atomic mass of ^{238}U plus six times the atomic mass of fluorine), and the 349 is the molecular mass of $^{235}UF_6$. Fortunately, fluorine has only one naturally occurring isotope so only these two molecular masses have to be considered.

The problem calls for enrichment from 0.72 percent $^{235}UF_6$ to 95 percent $^{235}UF_6$. The ratio (n_{235}/n_{238}) equals the ratio of the number of molecules of the lighter gas $^{235}UF_6$ to the number of molecules of the heavier gas $^{238}UF_6$. Each stage of diffusion achieves a slight enrichment in the lighter gas: each stage multiplies the original light-heavy ratio by the factor 1.0043. At the start, the ratio is 0.72/99.27 = 0.007253. After x stages, the ratio must rise to the required final value of 95/5 = 19. Hence:

$$(0.007253)(1.0043)^x = 19$$

Dividing through by 0.007253 and then taking the logarithm of both sides:

$$x \log(1.0043) = \log\left(\frac{19}{0.007253}\right) \quad \text{from which} \quad x = 1830$$

3-49 The key is to recognize that the pressure of the krypton is directly proportional to its number density. This fact follows from the ideal gas law:

$$P = \frac{n}{V}RT = \frac{1}{N_0}\left(\frac{N}{V}\right)RT$$

where the quantity on the right reflects the fact that the number of molecules of gas (N) divided by Avogadro's number (N_0) equals the number of moles of gas. The mean free path (λ) of the molecules is:

$$\lambda = \frac{1}{\sqrt{2}\pi d^2(N/V)} \quad \text{from which} \quad N/V = \frac{1}{\sqrt{2}\pi d^2 \lambda}$$

where d is the molecular diameter. Substituting for N/V in the preceding equation for P gives:

$$P = \frac{1}{N_0}\left(\frac{1}{\sqrt{2}\pi d^2 \lambda}\right)RT$$

Meanwhile, the volume of the spherical vessel is 1.00 L (1.00×10^{-3} m^3) and $V = (4/3)\pi r^3$ where r is the vessel's radius. The radius of the vessel is therefore 0.0620 m and its diameter is 0.124 m.
Now set λ equal to 0.124 m, fulfilling the condition that the mean free path be comparable to the diameter of the vessel. From the statement of the problem, T is 300 K and d is 3.16×10^{-10} m. The gas constant equals 8.206×10^{-5} m^3 atm mol^{-1} K^{-1} (note the carefully chosen units of R). Substitution gives:

$$P = \frac{1}{6.022 \times 10^{23}\ \text{mol}^{-1}}\left(\frac{1}{\sqrt{2}\pi(3.16 \times 10^{-10}\ \text{m})^2(0.124\ \text{m})}\right)$$

$$\times(8.206 \times 10^{-5}\ \text{m}^3\ \text{atm mol}^{-1}\ \text{K}^{-1})(300\ \text{K}) = 7.4 \times 10^{-7}\ \text{atm}$$

The number density (N/V) of the krypton is needed to calculate the diffusion constant. At this P and T, the density of the krypton is:

$$\frac{n}{V} = \frac{P}{RT} = \frac{7.4 \times 10^{-7}\ \text{atm}}{0.08206\ \text{L atm mol}^{-1}\text{K}^{-1}(300\ \text{K})} = 3.00 \times 10^{-8}\ \text{mol L}^{-1}$$

There are Avogadro's number of molecules of the gas per mole so the number density of the gas is Avogadro's number times this: $N/V = 1.81 \times 10^{16}$ L^{-1} This is equivalent to 1.81×10^{19} m^{-3}, because there are 1000 liters in a cubic meter. The diffusion constant of a gas is given by:

$$D = \frac{3}{8}\sqrt{\frac{RT}{\pi \mathcal{M}}}\left(\frac{1}{d^2 N/V}\right)$$

Substitute into this formula using values strictly in SI units:

$$D = \frac{3}{8}\sqrt{\frac{8.315 \text{ J K}^{-1}\text{mol}^{-1}(300 \text{ K})}{\pi(0.08380 \text{ kg mol}^{-1})}}\left(\frac{1}{(3.16 \times 10^{-10} \text{ m})^2(1.81 \times 10^{19} \text{ m}^{-3})}\right)$$

The answer is 20 m^2 s^{-1}.

3-51 The van der Waals equation of state is:

$$\left(P + a\frac{n^2}{V^2}\right)(V - nb) = nRT$$

and the chemical amount of O_2 in 6.80 kg is:

$$n_{O_2} = (6.80 \times 10^3 \text{ g}) \times \left(\frac{1 \text{ mol } O_2}{32.0 \text{ g } O_2}\right) = 212.5 \text{ mol } O_2$$

The a and b constants for O_2 are 1.360 atm L^2 mol^{-2}, and 0.031834 L mol^{-1}, respectively. Insert the volume in liters and these two values into the van der Waals equation. Use R in units that will cancel out the units in use (L atm mol^{-1}K^{-1} in this case). The result (omitting the units) is:

$$\left(P + \frac{1.360(212.5)^2}{(28.0)^2}\right)\left(28.0 - 212.5(0.031834)\right) = 212.5(0.08206)293$$

$$(P + 78.33)(21.24) = 5109 \quad \text{hence} \quad P = 162 \text{ atm}$$

This pressure is equivalent to 2.38×10^3 psi, since 14.696 psi equals 1 atm.

3-53 a) Rearranging the ideal gas law gives $P = nRT/V$. In this case:
$n = 50.0/44.0 = 1.136$ mol $\quad T = 298.15$ K
$V = 1.00$ L $\quad R = 0.08206$ L atm mol^{-1}K^{-1}
Substitution gives $P = 27.8$ atm.
b) The van der Waals equation includes terms (a and b) that depend on the identity of the gas. For $CO_2(g)$, a is 3.592 atm L^{2}mol^{-2} and b is 0.04267 L mol^{-1}. The equation is:

$$\left(P + a\frac{n^2}{V^2}\right)(V - nb) = RT \quad \text{so:} \quad P = \frac{RT}{V - nb} - a\frac{n^2}{V^2}$$

Substitution gives P equal to 24.6 atm.

3-55 The earth is covered by an ocean of air that exerts a pressure of 730 mm Hg all over its surface. Imagine the air replaced by an ocean of mercury. To exert the same pressure the mercury ocean would need to be only 730 mm deep. The volume of such a mercury ocean would be, to a close approximation, the surface area of the earth times its depth d: $V = 4\pi r^2 d$. With $r = 6.370 \times 10^6$ m (converting from km) and $d = 730 \times 10^{-3}$ m, the mercury would have a volume of 3.72×10^{14} m^3. The mass of a sample is its volume times its density. The density of mercury at ordinary temperatures is about 13.6 g cm^{-3} which, remembering that 10^6 cm^3 equals 1 m^3 and that 1000 g equals 1 kg, converts to 13.6×10^3 kg m^{-3}. The mass of the imagined ocean of mercury is therefore 5.06×10^{18} kg. This equals the mass of the atmosphere as well because the atmosphere exerts the same pressure at the surface of the earth as the ocean of mercury would.

3-57 The mercury in the barometer is less dense at 35°C than it is at 0.0°C, so the column of mercury that the atmosphere supports is longer than it would be if the barometer were put into a freezer. The actual pressure is accordingly less than 760.0 torr = 1 atm. In fact, from the equation $P = \rho g h$ the height h of the column of liquid in a barometer is inversely proportional at a given pressure P to the density of the liquid that is in it. Hence, the height that this column would have at 0.0°C is:

$$h_{\text{at } 0^\circ\text{C}} = h_{\text{at } 35^\circ\text{C}} \times \left(\frac{13.5094}{13.5955}\right) = 760.0 \text{ mm} \times 0.993667 = 755.19 \text{ mm}$$

This height in a mercury barometer at 0.0°C means a pressure of 755.19 torr, which is 0.9937 atm. The problem can also be solved by direct substitution into the barometric formula:

$$P = \rho g h = (13.5094 \times 10^3 \text{kg m}^3)(9.80665 \text{ m s}^{-2})(0.7600 \text{ m} = 1.00686 \times 10^5 \text{ Pa}$$

Note the changes in units: the height is in meters and the density of mercury is in a combination of SI base units so that correct cancellation of units takes place. Dividing by (101 325 Pa atm^{-1}) gives 0.9937 atm.

3-59 Reasoning by analogy to the textbook statements of Charles's law and Boyle's law, we conclude that Amontons's law must state that at constant volume, the pressure of a sample of a gas is directly proportional to its absolute temperature.

3-61 a) 1005 moles of helium displaces 1005 moles of air since the pressure and temperature of the gases inside and outside of the balloon are the same. The masses of these gases are:

$$1005 \text{ mol He} \times \left(\frac{4.003 \text{ g He}}{1 \text{ mol He}}\right) = 4023 \text{ g He}$$

$$1005 \text{ mol air} \times \left(\frac{29.0 \text{ g air}}{1 \text{ mol air}}\right) = 29100 \text{ g air}$$

The answer is the difference between these two masses, which is 2.51×10^4 g.
b) No gas enters or leaves the balloon during the ascent, so the answer is the same as in part a). The balloon is much larger at high altitude than at ground level and displaces the same mass of air despite the lower density of the air at high altitude.

3-63 A quick way to solve this problem is to think in terms of proportions. From the ideal-gas law, a given volume contains moles of gas in *inverse* proportion to its absolute temperature, as long as the pressure is constant. This means the higher the temperature, the lower the amount of gas that is in the volume. Also, the amount of products of a chemical reaction is in direct proportion to the amount of reactants. Raising the temperature at which the HCl(g) is collected from 323.15 to 773.15 K (from 50 to 500°C) is an increase by a factor of 2.392. The number of moles of HCl(g) produced in the high-temperature experiment is less by the same factor. But the number of moles of HCl(g) produced is in direct proportion to the amount of NaCl reacted. The amount of NaCl used in the high-temperature experiment is therefore less by a factor of 2.392. The answer is simply 10.0 kg divided by 2.392. It equals 4.18 kg NaCl.
Another approach is actually to compute the "certain volume" and then find the chemical amount (number of moles) of gas it contains at both 323.15 K and 773.15 K. First, find the number of moles of HCl(g) that can form from 10.0 kg of NaCl:

$$10.0 \times 10^3 \text{ g NaCl} \times \left(\frac{1 \text{ mol NaCl}}{58.44 \text{ g NaCl}}\right) \times \left(\frac{1 \text{ mol HCl}}{1 \text{ mol NaCl}}\right) = 171.12 \text{ mol HCl}$$

The "certain volume" occupied by this HCl at 50°C (323.15 K) is:

$$V = \frac{nRT}{P} = \frac{171.12 \text{ mol}(0.08206 \text{ L atm mol}^{-1}\text{K}^{-1})(323.15 \text{ K})}{1 \text{ atm}} = 4538 \text{ L}$$

At 500°C (773.15 K) this volume contains fewer moles:

$$n_{500°\text{C}} = \frac{PV}{RT} = \frac{(1 \text{ atm})(4538 \text{ L})}{(0.08206 \text{ L atm mol}^{-1}\text{K}^{-1})(773.15 \text{ K})} = 71.52 \text{ mol HCl}$$

Finally, compute the mass of NaCl required to produce this amount of HCl:

$$71.52 \text{ mol HCl} \times \left(\frac{1 \text{ mol NaCl}}{1 \text{ mol HCl}}\right) \times \left(\frac{58.44 \text{ g NaCl}}{1 \text{ mol NaCl}}\right) = 4179.6 \text{ g NaCl}$$

The rounds to 4.18×10^3 g. The second calculation is certainly slower, but probably surer, particularly for beginners. Now, suppose that the problem simply stated that the pressure was the same in the two experiments (not telling what it was). None of the intermediate numbers in the second method could be computed but the first method would work the same!

3-65 The total volume of the system is 12.00 L, the sum of the volumes of the three containers (assuming that the volumes of the connecting tubes are negligible). Assuming that the three gases behave ideally in their containers, their chemical amounts are:

$$n_{O_2} = \frac{2.51 \times 5.00}{RT} \qquad n_{N_2} = \frac{0.792 \times 4.00}{RT} \qquad n_{Ar} = \frac{1.23 \times 3.00}{RT}$$

where the units are understood to be liters, atmospheres and moles. The total pressure after the gases mix is also given by the ideal-gas equation, assuming Dalton's law holds. In writing the expression, the total chemical amount of the mixed gas is the sum of the chemical amounts of the three components:

$$P_{tot} = \left(\frac{2.51 \times 5.00}{RT} + \frac{0.792 \times 4.00}{RT} + \frac{1.23 \times 3.00}{RT}\right)\left(\frac{RT}{12.00}\right)$$

The RT's cancel out, and the pressure is easily calculated. It is 1.63 atm.

3-67 Assume that the gases behave ideally before the catalyst is introduced and that they behave ideally after the reaction goes to completion, too. To *react* is, on the other hand, quite non-ideal behavior. A simplifying step is to imagine that the temperature and volume of the system are such that the total chemical amount of gases starts at 0.100 mol. Under this assumption, the reaction: $C_2H_2(g) + 2\,H_2(g) \rightarrow C_2H_6(g)$ decreases the total chemical amount of gas to 0.042 mol. This is true because chemical amount is directly proportional to the pressure if T and V do not change. Let x represent the original chemical amount of $C_2H_2(g)$ and y the original chemical amount of $H_2(g)$. Before reaction there is no $C_2H_6(g)$ so:

$$x + y = 0.100 \text{ mol}$$

The reaction produces exactly x mol of $C_2H_6(g)$ as it consumes $2x$ mol of $H_2(g)$ and x mol of $C_2H_2(g)$. Since $C_2H_2(g)$ is the limiting reagent, the reaction stops dead when the x mol of C_2H_2 gas is used up. At this point, the vessel contains x mol of $C_2H_6(g)$, the product, and $(y - 2x)$ mol of left-over $H_2(g)$. Hence:

$$x + (y - 2x) = 0.042 \text{ mol}$$

Solving the two simultaneous equations for x, we find it to equal 0.029. There is 0.029 mol of $C_2H_2(g)$. The original mole fraction of $C_2H_2(g)$ is therefore $0.029/0.100 = 0.29$. Had we assumed the system to be big enough to hold for example 100 moles of gases all of the numbers in this computation, except the answer, would have been 1000 times bigger.

3-69 a) The average kinetic energy of an atom of deuterium, which is given in the problem, is depends only on the absolute temperature (assuming ideal-gas behavior):

$$KE_{avg} = 8 \times 10^{-16} \text{ J} = \frac{3}{2}k_B T = \frac{3}{2}(1.38 \times 10^{-23} \text{ J K}^{-1})T$$

Hence $T = 3.86 \times 10^7$ K. The atomic mass of ^{2}H is not needed in this part of the problem.

b) The average kinetic energy of the particles in a gas equals $1/2m\bar{u^2}$. Write this relationship for the ^{1}H and divide it by a similar relationship for the ^{2}D:

$$\frac{1/2m_H\bar{u_H^2}}{1/2m_D\bar{u_D^2}} = \frac{32 \times 10^{-16}\ \text{J}}{8 \times 10^{-16}\ \text{J}}$$

It follows that:

$$\sqrt{\frac{\bar{u_H^2}}{\bar{u_D^2}}} = \sqrt{\frac{32}{8}}\sqrt{\frac{2.015}{1.0078}} = 2.83$$

3-71 The Maxwell-Boltzmann distribution is:

$$f(u) = 4\pi\left(\frac{m}{2\pi k_B T}\right)^{3/2} u^2 \exp(-mu^2/2k_B T)$$

The quantity $f(u)\Delta u$ equals the probability that a molecule in a gas at thermal equilibrium has a speed between u and $u + \Delta u$ (see text page 108):

$$\frac{\Delta N}{N} = f(u)\Delta u$$

The gas in this case is O_2, for which $m = 0.0320$ kg mol^{-1}. The temperature is 300 K, and the specified range of speed is 500 to 510 m s^{-1}. Evaluate $f(u)$ at $u = 500$ m s^{-1}:

$$f(u) = 4\pi\left(\frac{0.0320\ \text{kg mol}^{-1}}{2\pi(8.315\ \text{J K}^{-1}\text{mol}^{-1})(300\ \text{K})}\right)^{3/2} (500\ \text{m s}^{-1})^2$$

$$\exp\left(\frac{-0.032\ \text{kg mol}^{-1}(500)^2\ \text{m}^2\text{s}^{-2}}{2(8.315\ \text{J K}^{-1}\text{mol}^{-1})(300\ \text{K})}\right) = 1.843 \times 10^{-3}\ \text{s m}^{-1}$$

The value of $f(u)$ changes over the range of u. The hint proposes a method for dealing with this change, which is a 2.5 percent decrease as shown in the following table:

u (m s^{-1})	$f(u)$ (s m^{-1})	u (m s^{-1})	$f(u)$ (s m^{-1})
500	1.843×10^{-3}	506	1.816×10^{-3}
501	1.839×10^{-3}	507	1.812×10^{-3}
502	1.834×10^{-3}	508	1.807×10^{-3}
503	1.830×10^{-3}	509	1.802×10^{-3}
504	1.825×10^{-3}	510	1.797×10^{-3}
505	1.821×10^{-3}	—	—

The desired probability $f(u)\Delta u$ can be visualized as the area under the distribution curve between 500 and 510 m s^{-1}. See Appendix C, text page A-23. This area has a width of 10 m s^{-1} and a smoothly changing height (see the table). Approximate it by a series of 10 narrow columns of width 1 m s^{-1} and heights given by the first ten values of $f(u)$ in the table. The sum of these ten areas is 1.823×10^{-2}. Note that all units cancel out. The desired probability is thus 1.823 percent.

3-73 Let the unknown gas be called Z and have molar mass $\mathcal{M}_Z$. Convert the rates of effusion of the oxygen and Z from g min^{-1} to mol min^{-1} so that Graham's law can be applied. using 32.0 g mol^{-1} as the molar mass of $O_2(g)$. The two rates of effusion come out:

$$\frac{3.25 \text{ g min}^{-1}}{32.0 \text{ g mol}^{-1}} = 0.1016 \text{ mol min}^{-1}; \quad \frac{1.96 \text{ g min}^{-1}}{\mathcal{M}_Z \text{ g mol}^{-1}} = (1.96/\mathcal{M}_Z) \text{ mol min}^{-1}$$

Write Graham's law as a comparison of the two gases (equation 3-13, text page 111):

$$\frac{\text{rate of effusion of O}_2}{\text{rate of effusion of Z}} = \frac{N_{O_2}}{N_Z}\sqrt{\frac{\mathcal{M}_Z}{\mathcal{M}_{O_2}}}$$

The numbers of molecules in the vessel are the same in the two experiments because the two are carried out as the same temperature and pressure. Hence N_{O_2} and N_Z cancel out of the preceding equation. Inserting the two rates gives:

$$\frac{0.1016 \text{ mol min}^{-1}}{(1.96/\mathcal{M}_Z) \text{ mol min}^{-1}} = \frac{0.1016\mathcal{M}_Z}{1.96} = \sqrt{\frac{\mathcal{M}_Z}{32.0 \text{ g mol}^{-1}}}$$

Solution of the last equation gives $\mathcal{M}_Z$ equal to 11.6 g mol^{-1}. It is not necessary to know T and P as long as they were the same in the two experiments. A smaller mass of the unknown gas than of oxygen effuses in the given time. Despite this, the molar mass of the unknown is truly less than the molar mass of oxygen. The result is easier to accept when it is noted that 0.169 mol of the unknown effuses, but only 0.102 mol of oxygen effuses.

3-75 The mean free path is given by:

$$\lambda = \frac{1}{\sqrt{2}\pi d^2(N/V)}$$

Substituting the ideal-gas law in the form $N/V = N_0P/RT$ gives:

$$\lambda = \frac{(R/N_0)T}{\sqrt{2}\pi d^2 P} = \frac{k_B T}{\sqrt{2}\pi d^2 P}$$

Solving for P gives:

$$P = \frac{k_B T}{\sqrt{2}\pi d^2 \lambda}$$

All the quantities on the right side of this equation are known. Substitute them:

$$P = \frac{(1.38 \times 10^{-23} \text{ J K}^{-1})(300 \text{ K})}{\sqrt{2}\pi(3.1 \times 10^{-10} \text{ m})^2(0.1 \text{ m})} = 0.097 \text{ Pa} = 9.6 \times 10^{-7} \text{ atm}$$

3-77 Although molecules of UF_6 are much heavier than those of H_2, the root-mean-square speed of a UF_6 molecule is much slower than that of an H_2 molecule. Pressure of a gas is due to the force exerted by its molecules colliding with the walls of the container. This force depends not only on the mass of the molecules, but also on their speed, that is, upon their momenta.

Chapter 4

Condensed Phases and Solutions

Phase Equilibria and Phase Diagrams

A **phase** is a sample of matter that is uniform throughout in both chemical composition and physical state. **Phase transitions** are the six possible changes linking the three physical states of matter. They are:

$$\text{solid} \rightleftharpoons \text{liquid, solid} \rightleftharpoons \text{gas and liquid} \rightleftharpoons \text{gas}$$

Some of the names for the six phase transitions are familiar, but other are not. Melting and freezing refer to the two directions of the solid-liquid transition; **sublimation** and condensation refer to solid-gas transitions; vaporization and condensation refer to gas-liquid transitions. Many substances have more than one solid phase (such as white tin and gray tin, see **problem 4-10**), but none has more than one liquid or gaseous phase. Depending on conditions two or even three phases of a pure substance may coexist indefinitely.

Consider a sample of a volatile substance (such as water) confined in a vessel but not completely filling the available volume. Substance evaporates until the pressure of the vapors in the space above the liquid reaches a characteristic value, the **vapor pressure** of that substance. The vapor pressure depends only on the temperature, and not on the size of the container See Figure 4-3 (text page 127). A **phase equilibrium** exists when the system under study persists with no further net transport of matter from one phase to another.

Phase equilibria have these important characteristics:

1. A phase equilibrium is a dynamic process on the molecular level; individual molecules continue to move from one phase to the other despite the absence of any visible sign of change.

2. The properties of the equilibrium are independent of the direction from which they are reached.

A **phase diagram** of a pure substance is a plot with T on the horizontal axis and P on the vertical axis. Lines (often curved) on this plot define combinations of temperature and pressure at which two phases coexist at equilibrium. If two phases are at equilibrium, then no further macroscopic changes are detectable, although microscopic exchange of molecules (or atoms) between the phases continues.

A generic phase diagram:

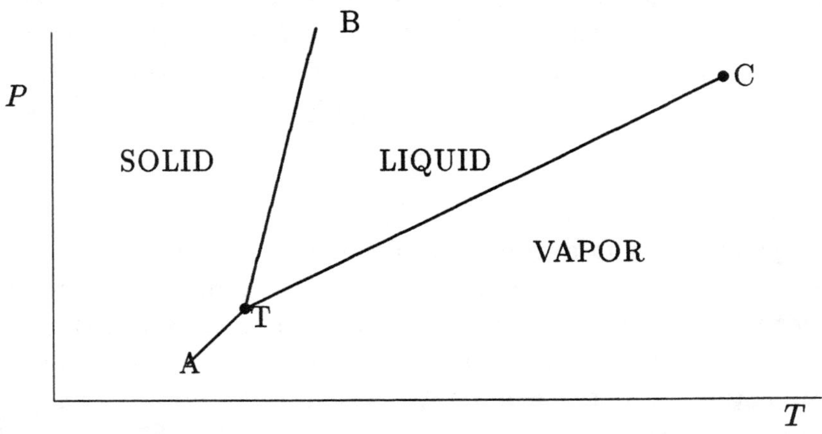

This typical diagram displays the relationships among three phases of a pure substance. It has three different coexistence lines radiating from a central point (T, the **triple point.** The liquid-gas coexistence line terminates at the **critical point** (labeled C). The other two coexistence lines have no sharp point of termination. Use this diagram as a model to solve **problem 4-64.** Other types of phase diagrams with variables other than P and T can also be plotted (see **problem 4-67).**

Freezing, Boiling and Sublimation Points

The **normal freezing point** of a pure substance is the temperature at which solid and liquid are in equilibrium when the pressure is one atmosphere. Similarly, the **normal boiling point** is the temperature at which liquid and gas are in equilibrium when the pressure is one atmosphere. Finally, **the normal sublimation point** of a pure substance is the temperature at which solid and gas are in equilibrium at one atmosphere pressure. All three, in diagrammatic terms, are intersections of the $P = 1$ atm line and the appropriate coexistence line on a phase diagram. No pure substance can have all three of these points. After all, a solid heated slowly at a constant pressure of 1 atm either turns to a liquid or a gas, but not both (see **problem 4-13).** Some substances have *none* of these points. They simply decompose chemically before the temperature is high enough for them to melt, boil or sublime.

Phase Equilibria and Triple Points

Consider just the three phases: solid, liquid, and gas. There are exactly three possible equilibrium lines (solid-liquid, solid-gas and liquid-gas) along which two of these three phases coexist. Regions between these lines are sets of P-T values at which only *one* phase exists at equilibrium. If two phase coexistence lines intersect, then, at the point of intersection, *three* phases coexist. This means that the third phase coexistence line must pass through the intersection of the first two. In general the coexistence lines on the P versus T diagram do intersect. The resulting specific combination of pressure and temperature at which three phases coexist is a **triple point** of a pure substance.

Critical Point

The essence of a phase transition is an abrupt discontinuous change in physical properties (*e.g.* density or viscosity) as the transition occurs. The distinction between liquid and gas *no longer exists* at temperatures and pressures that exceed the T and P of the **critical point,** a certain P-T combination on the phase diagram of every substance. The experimental terms, the **meniscus**, which is the boundary separating the more-dense liquid from the less-dense gas, disappears. A substance held at conditions beyond its critical point is a **supercritical fluid.** Distinguishable, abrupt liquid-to-gas transitions can no longer take point in the supercritical region of the phase diagram of a substance.

At temperatures above its **critical temperature,** a gas cannot be liquefied no matter how great the pressure. The **critical pressure** is the pressure just necessary to liquefy a gas at its critical temperature. The critical pressure exceeds normal atmospheric pressure for all common substances. For this reason the "normal" pattern of phase change is the melting of a solid to a liquid followed by boiling to a gas. The observed behavior of substances at and beyond their critical points seems paradoxical.

Many interesting and thought-provoking problems are constructed around the facts presented on the phase diagrams of pure substances.

- Description of the state of a substance in a condition at certain conditions **(problem 4-11)** or the events within a container of a pure substance when the temperature is changed at constant pressure or when the pressure is changed at constant temperature **(problem 4-12** and **4-13b).**

  A pressure change at constant temperature corresponds to a vertical line on a phase diagram. A temperature change at constant pressure corresponds to a horizontal line. When such lines cut through phase coexistence lines then phase transitions occur. Keep in mind that heating, for example, a sample of a pure liquid substance at constant pressure would mean coping with enormous changes in volume when vaporization occurred.

- Construction of a phase diagram given the temperature and pressure at the triple point and critical point and perhaps the densities of the solid and liquid phases. **Problem 4-67** involves the construction of another type of diagram.

- Determination of the equation defining one or more of the phase coexistence lines. Although these lines are in general *not* straight lines, segments of them (over limited T and P ranges) are often approximated as straight. The determination of the slopes of the coëxistence lines on phase diagrams requires use of the Clapeyron equation (see text page 699-700).

The solid-liquid coexistence line on a P-T phase diagram usually goes almost straight up, because freezing points are rarely strongly dependent on pressure. The slight slope that it has is generally to the right (a positive slope) because, for most substances, the density of the solid phase exceeds the density of the liquid phase. Water is an exception. Ice floats in liquid water because it is less dense than the liquid. The solid-liquid coexistence line for H_2O has a slight negative slope. For another exception see **problem 4-9.**

The liquid-gas coexistence line of all substances slopes to the right on the P-T phase diagram because the gas phase is always less dense than the liquid phase. Increases in pressure favor the more dense liquid phase. This line terminates at the critical point. The solid-vapor coexistence line also always has a positive slope and rises from somewhere in the lowest reaches of temperature and pressure to terminate at the triple point.

Qualitative discussions of phase diagrams often deal with tidy-looking diagrams that attain their apparent symmetry by virtue of severe distortion of the scales on the T and P axes. When phase diagrams are drawn on axes with regularly spaced increments of T and P, areas of interest (triple points, critical point, etc.) are commonly forced onto an edge of the diagram. The text deals with this issue wisely in Figure 4-6 (text page 130) by using a logarithmic scale for the pressure.

Composition of Solutions

Homogeneous systems that contain two or more substances are **solutions.** Solutions have ranges of composition. The major component in a solution is called the **solvent.** The minor component is the **solute.** A solution may have several solutes. For each additional solute, another variable (in addition to pressure, temperature and volume) is needed fully to describe the system.

The following measures are used to describe the composition of solutions:

Mass Percentage. The mass percentage (also called weight percentage) of a substance in a solution is the mass of the substance divided by the total mass of the solution and then multiplied by 100%. It is also possible to speak of the mass fraction of a substance by skipping the conversion to (or from) the basis of 100.

Mole fraction. The mole fraction of a substance in a mixture is its number of moles divided by the total number of moles of all the different substances present. Mole fraction is symbolized with a capital X: $X_1 = n_1/n_{\text{tot}}$.

The sum of the mole fractions of all of the components of a solution must equal 1. Two component systems are common in problems and the relationship $X_1 + X_2 = 1$ is often crucial in solving these problems. See **problems 4-17** and **4-19.**

Molality. The molality of a solution is the number of moles of solute divided by the number of kilograms of solvent. When more than one solute is present then the solution has a molality in the first, another molality in the second and so forth. The solvent is the component of the solution that is present in the greatest chemical amount. The symbol for molality is m, and the unit is mol kg^{-1}.

The only facts needed to convert freely between mole fractions and molalities are the molar masses of the components of the solution. See **problem 4-17.** Conversely, if the composition of a solution is expressed both in terms of molalities and mole fractions, then molar masses can be calculated.

The case of knowing the molality of a two-component system and its composition by mass is particularly common. These facts allow computation of the molar mass of a solute. See **problem 4-41.**

Molarity. The molarity of a solution is the number of moles of the solvent per liter of solution. The symbol for molarity is M and its unit is mol L^{-1}. See **problems 4-15** and **4-23**. Because this molarity has a volume in the denominator, conversion between it and either molality or mole fraction requires a knowledge of the density of the *solution* (not of the pure components). It follows that if the molarity and molality of a two-component solution are both known then the density of the solution can be computed. Such a problem is well-known on examinations. Also, if the mass percentage and molarity are known, the density can be calculated (see **problem 4-21**). Densities vary with temperature and consequently so do molarities. In *dilute* aqueous solutions the molarity is approximately the same as the molality because the density of the solution is approximately 1 g cm^{-3}. See **problem 4-15b.**

Many problems involve the conversion among different units of concentration. It is not wise to memorize specific formulas for these conversions. Instead learn the definitions of the different units of concentration and apply them (see **problem 4-17**).

The units of concentration fall into two categories, those having a *mass* in the denominator (mole fraction, molality, mass percent) and those having a *volume* in the denominator (molarity). To convert between these two categories requires knowledge of the density of the solution.

Preparation of Solutions

The preparation and mixing of solutions have enormous practical importance. To prepare a solution that is for example 1.20 M in a certain solute (contains 1.20 mol of solute per liter of solution), one takes 1.20 mol of the solute and adds enough solvent to make the total volume exactly 1 L. If only 517 mL of such a solution is required, then one takes (0.517×1.20) mol of the solute and adds enough solvent to bring the total volume exactly to 517 mL. This is not the same as adding 517 mL of solvent. The difference stems from the volume that the solute occupies. See **problem 4-17.** Special flasks called **volumetric flasks** are calibrated to hold precisely known volumes at specified temperatures. Volumetric flasks are used in preparing solutions of precisely known concentration.

Doubling the total volume of a solution by adding more solvent obviously cuts the concentrations of all solutes in half. It does not change the chemical amounts of the solutes that are present. For the dilution of a solution:

$$c_f = \frac{\text{moles solvent}}{\text{final solution volume}} = \frac{c_i V_i}{V_f}$$

where c and V stand for concentration and volume respectively and the subscripts refer to the final and initial values.

The preparation and dilution of solutions are the subjects of many exercises. Mixing of solutions of differing concentrations of a single solute are also possible **(problem 4-23).** A potentially puzzling variation of such dilution problems involves *concentrating* a solution from a lower to a higher concentration. The same formula would apply except that V_f would be smaller than V_i instead of larger.

The Stoichiometry of Reactions in Solutions

The molar concentration of a solution furnishes a new type of conversion factor to use in stoichiometric calculations. The new type relates the chemical amount of a substances with the volume of the solution. For example, if a solution of alcohol (C_2H_5OH in water has a concentration of 0.575 mol L^{-1}, the factors:

$$\left(\frac{0.575\text{ mol } C_2H_5OH}{1\text{ L solution}}\right) \quad \text{and} \quad \left(\frac{1\text{ L solution}}{0.575\text{mol } C_2H_5OH}\right)$$

can immediately be written. **Problems 4-27, 4-29, 4-33,** and **4-35** show the technique, which should be studied as an extension of the methods of Chapter 1. **Problem 4-71** is a good example of the combined use several types of conversion factors.

Titration

A **titration** is the addition of a measured volume of a solution of known concentration in one reactant to a solution of unknown concentration in a second reactant. The object is to determine the unknown concentration. The equation for the reaction taking place between the two substances must be known, and the reaction must go to complete. Also, there must be some provision for determining when the second reactant is used up. An **indicator** shows when the **endpoint** of a titration is reached. If reactants 1 and 2 are in solutions of concentration c_1 and c_2 and combine in a 1-to-1 molar ratio, then at the endpoint:

$$V_1c_1 = n_1 = n_2 = V_2c_2$$

This relationship applies to **problem 4-35.** If the two reactants do not react in a 1-to-1 molar ratio, then the ratio they do follow must be taken into account **(problem 4-33).**

A **acid-base titration** employs the neutralization reaction to determine the concentration of an unknown acid or base in a solution. To titrate an acid solution of unknown concentration, the chemical analyst carefully adds measured portions of a base of known (standardized) concentration to a quantity of the acid. Neutralization proceeds. Finally, some indication (such as the change in color of a sensitive dye that has been added in small amounts to the reaction) shows that the neutralization is just complete. The analyst notes the volume of base added to reach this point, which is the endpoint of the titration. If aqueous HCl for example is titrated with aqueous NaOH then according to the preceding equation:

$$V_{\mathrm{HCl}}c_{\mathrm{HCl}} = V_{\mathrm{NaOH}}c_{\mathrm{NaOH}}$$

at the endpoint. Even if the volume of the HCl solution is not known, the chemical amount of HCl that was originally present can be computed.

Titrating a base with an acid differs only in reversing the order of addition of the reactants.

Colligative Properties of Solutions

There are four colligative properties of solutions. **Colligative** properties depend only on the effect of the number of solute particles and not on the identity of the solute.

To understand these properties one must first understand **Raoult's law.** This law states that the vapor pressure of a component in a solution depends on its mole fraction times the vapor pressure exerted by the pure component:

$$\boldsymbol{P_1 = X_1 P_1^\circ}$$

If P_1° equals zero, then the component has no vapor pressure and is **non-volatile.**

Solutions that follow Raoult's law are **ideal solutions.** Compare them with ideal gases. In an ideal gas, there are negligible forces of attraction among the molecules of the gas. In an ideal solution, the intermolecular attractions are not negligible. Instead the solvent-to-solute attractions are the *same* as the solvent-solvent interactions.

Vapor Pressure Lowering

The case of a volatile solvent and a non-volatile solute is common. The presence of the solute in the solvent reduces of the total vapor pressure of the solution. The reduction is easily measured and is proportional to the mole fraction of the solute. Thus measurements of vapor pressure can give the composition of solutions:

$$\Delta P = P_2 - P_1 = \text{vapor pressure change} = -P^\circ_{(\textbf{solvent})} X_2$$

See **problem 4-37.** An observed vapor-pressure lowering, taken in combination with the molar mass of the solvent and the masses of the solute and solvent, gives the molar mass of the solute **(problems 4-37, 4-73).** An obvious variation is to give the molar mass of the *solute,* the masses of solute and solvent, and ask the molar mass of the solvent. The approach is essentially unchanged in such a case.

Boiling Point Elevation

The boiling point of a volatile solvent is raised by the presence of a non-volatile solute. The approximate equation representing this change is:

$$\Delta T_b = K_b m$$

where K_b is a constant that depends only on the solvent, m is the molality of the solute, and ΔT is the final boiling temperature minus the original boiling temperature. The quantity ΔT_b is always positive, by the nature of the phenomenon. Tables of K_b values are available (Table 4-1, text page 143). The values in such tables were established by performing boiling-point elevation experiments on solutions of known concentration of known solutes. The units of K_b are K kg mol^{-1} where it is understood that the kg refers to the mass of the solvent and the mol refers to the chemical amount of the solute.

The simple form of the boiling point elevation formula lends itself to many practical problems involving the determination of the molar mass of unknown non-volatile solutes. All kinds of solutes, not just water, have their boiling points raised by the presence of non-volatile solutes (**problem 4-40**).

Freezing Point Depression

Freezing point depression is analogous to boiling point elevation. The formula for the amount of change in the freezing point of a volatile solvent occasioned by the presence of a non-volatile solute is:

$$\Delta T = -K_f m$$

where K_f is the solvent's freezing point depression constant. The constants K_f and K_b differ from solvent to solvent. The negative sign appears because T_2 is always less than T_1 in a temperature lowering and K_f is positive (Table 4-1, text page 143). Typical applications again involve the determination of the molar masses of unknown solutes (**problem 4-41**).

Osmotic Pressure

The colligative property that is most used in biochemical applications is the osmotic pressure. All solutions have an osmotic pressure. In the case of dilute solutions:

$$\pi = cRT$$

where π is the osmotic pressure, c is the concentration of the solution, T is its absolute temperature and R is the gas constant. Osmotic pressures are measured in set-ups using semi-permeable membranes, materials that allow passage of molecules of solvent but not of solute. As is the case with the other colligative properties, observations of osmotic pressure are useful in determinations of molar mass (**problems 4-47** and **4-49**).

The colligative properties can closely related. This makes possible the prediction of one property based on the observation of another (**problem 4-77** and **4-81**).

Solutes That Dissociate

A complication arises in the case of solutions (nearly always aqueous) of solutes that dissociate. Such solutes give rise to two or more particles in solution for every particle added. This affects the magnitude of the several colligative properties. If a colligative property is enhanced in this way the apparent or effective molality, not actual molality, is increased **problem 4-45**). The theoretical number of particles is often evident from the formula and name of the solute: Na^+Cl^- has 2; $Ca^{2+}(Cl^-)_2$

has 3; $(K^+)_2(SO_4^{2-})$ has 3; and so forth, but this is no always so (see **problem 4-79).** Dissociation of the solute affects all four colligative properties **(problem 4-81).**

Mixtures and Distillation

Solutions of two (or more) components have phase diagrams, just as do pure substances. A full phase diagram of a two-component solution would require three axes, one for pressure, one for temperature, and the third for composition. A three-component solution would require *four* axes because two different composition variables require specification. Evidently, full phase diagrams for many-component solutions are at best difficult and often impossible to draw. Instead *cross-sections* of phase diagrams are sketched. Text Figure 4-16 (page 153) is a constant temperature cross-section of the phase diagram of a two-component ideal solution. Pressure and composition vary on the vertical and horizontal axes respectively. Text Figure 4-17 (page 153), shows composition on the horizontal axis but temperature on the vertical axis. The pressure is constant.

Phase diagrams of real solutions are complex and differ sharply from the simple predictions of Raoult's law. Nevertheless, at a sufficiently low concentration of solute (even in nonideal solutions) there is still some simplicity:

$$P_{\text{solute}} = kX_{\text{solute}}$$

This is **Henry's law**. The constant k is different for every solute-solvent combination. Henry's law often occurs in problems involving solutions of gases in liquids. See **problems 4-51, 4-53,** and **4-83.**

Distillation

When a solution of volatile components is heated, the total vapor pressure above the solution increases. The vapors in equilibrium with the liquid are in general richer in the more volatile components than the liquid. This enrichment is the basis for the separation of mixtures by **distillation.** If the solution is ideal, the degree of enrichment can be computed. Raoult's law gives the partial vapor pressures of the components, and Dalton's law gives the mole fractions of the components in the vapor. See **problems 4-55, 4-57,** and **4-84.** A solution of two or more volatile components can be separated by **fractional distillation,** in which components are repeatedly vaporized and recondensed. A Nonideal solution deviates from Raoult's law. Such solutions have azeotropic compositions. An **azeotrope** is a liquid mixture of that boils to give vapors having the same composition as the liquid. Azeotropic mixtures cannot be separated by distillation.

Colloidal Suspensions

Colloids are mixtures of two or more substances in which one phase is suspended as small particles in a second. The particles of the **dispersed phase** are larger than single molecules **(see problem 4-85)**. They are from 10^{-9} to 10^{-6} m in diameter and too small to be distinguished by eye. The particles are in a state of constant motion called **Brownian motion.** In principle, the dispersed phase in a colloid will settle out—eventually. In practice, this sedimentation is *slow.* When colloids are **flocculated** by adding soluble salts to the dispersing phase, **aggregation** of the particles occurs, and sedimentation is speeded up.

Detailed Solutions to Odd-Numbered Problems

4-1 Use the ideal-gas law to calculate the molar volume of the vapors of mercury in the space above the top of the column of liquid mercury. This should *not* be expected to turn out to equal 22.4 L mol^{-1} because the mercury vapor is nowhere near being at STP:

$$\frac{n}{V} = \frac{P}{RT} = \frac{2.87 \times 10^{-6} \text{ atm}}{(0.08206 \text{ L atm mol}^{-1}\text{K}^{-1})(300.0 \text{ K})} = 1.17 \times 10^{-7} \text{ mol L}^{-1}$$

Then convert from moles per liter to atoms per cubic centimeter:

$$\frac{1.17 \times 10^{-7} \text{ mol}}{1 \text{ L}} \times \left(\frac{1 \text{ L}}{1000 \text{ cm}^3}\right) \times \left(\frac{6.022 \times 10^{23} \text{ atom}}{1 \text{ mol}}\right) = 7.02 \times 10^{13} \frac{\text{atom}}{\text{cm}^3}$$

This answer is a number density (see text page 110).

4-3 The total pressure that is measured is the sum of the pressure of the C_2H_2 produced in the reaction and the pressure of the vapors of H_2O also present. Subtracting out the pressure of $H_2O(g)$ gives the pressure that the C_2H_2 would exert if it were by itself:

$$P_{C_2H_2} = P_{\text{total}} - P_{H_2O} = 0.9950 - 0.0728 = 0.9222 \text{ atm}$$

The chemical amount of C_2H_2 per liter of "wet" acetylene can now be computed by using the ideal-gas equation:

$$\frac{n_{C_2H_2}}{V} = \frac{P_{C_2H_2}}{RT} = \frac{0.9222 \text{ atm}}{(0.08206 \text{ L atm mol}^{-1}\text{K}^{-1})(313.15 \text{ K})} = 0.03589 \text{ mol L}^{-1}$$

This is changed to grams per liter by multiplying it by the molar mass of C_2H_2:

$$0.03589 \frac{\text{mol } C_2H_2}{\text{L}} \times \left(\frac{26.038 \text{ g } C_2H_2}{1 \text{ mol } C_2H_2}\right) = 0.9345 \text{ g } C_2H_2 \text{ L}^{-1}$$

4-5 The approach is very similar to the approach in problem 4-3 except now a specific volume is given, so that the calculation is no longer on a per-unit-volume basis

$$P_{CO_2} = P_{total} - P_{H_2O} = 0.9963 - 0.0231 = 0.9732 \text{ atm}$$

$$n_{CO_2} = \frac{P_{CO_2}V}{RT} = \frac{(0.9732 \text{ atm})(0.722 \text{ L})}{(0.08206 \text{ L atm mol}^{-1}\text{K}^{-1})(293.15 \text{ K})} = 0.0292 \text{ mol}$$

$$0.0292 \text{ mol CO}_2 \times \left(\frac{1 \text{ mol CaCO}_3}{1 \text{ mol CO}_2}\right) \times \left(\frac{100.1 \text{ g CaCO}_3}{1 \text{ mol CaCO}_3}\right) = 2.92 \text{ g CaCO}_3$$

4-7 The tea boils at 194°F. This temperature must be converted to the Celsius scale:

$$t_C = \frac{9}{5}(t_F - 32) = \frac{9}{5}(194 - 32) = 90.0°\text{C}$$

Figure 4-3 shows that at 90.0°C the vapor pressure of water is 0.69 atm. If the vapor pressure of water at the camp in the Andes is 0.69 atm, then the fraction of the earth's atmosphere that lies below the altitude of the camp is $1 - 0.69 = 0.31$.

4-9 At a given temperature, an increase in pressure leads to the formation of the phase that has the higher density. Since the density of liquid Pu is larger than the density of solid Pu, even very strong compression will not tend to produce the solid. No phase change is expected.

4-11 Each of the combinations of temperature and pressure defines a point on the phase diagram of argon. The physical state of argon at that P-T combination is given by the label for the area into which the point falls in Figure 4-6:
a) At a pressure of 50 atm and a temperature of 100 K, Ar is a liquid.
b) At a pressure of 8 atm and a temperature of 150 K, Ar is a gas.
c) At a pressure of 1.5 atm and a temperature of 25 K, Ar is a solid.
d) At a pressure of 0.25 atm and a temperature of 120 K, Ar is a gas.

4-13 It is wise to sketch and use a crude phase diagram.
a) The line in a phase diagram that defines the conditions for the equilibrium co-existence of a solid and gas always lies below the triple point in both temperature and pressure. It rises to terminate at the triple point. The triple-point temperature therefore exceeds −84.0°C.
b) The problem states that acetylene sublimes at 1 atm (760 torr) at 189 K. Heating acetylene from 10 K to 300 K at a pressure *less* than 1 atm will cause the acetylene to sublime at a temperature *below* 189 K. Liquid acetylene never forms in this process.

4-15 a) The problem is one of conversion of units. There are 10 dL per L and 1000 mg per gram. Hence:

$$\frac{214 \text{ mg cholesterol}}{1 \text{ dL}} \times \frac{1 \text{ g}}{1000 \text{ mg}} \times \frac{10 \text{ dL}}{1 \text{ L}} \times \frac{1 \text{ mol cholesterol}}{386.64 \text{ g cholesterol}} = \frac{0.00553 \text{ mol}}{1 \text{ L}}$$

b) Assume blood to have a density of 1.00 g mL^{-1}. One liter (1000 mL) of blood then has a mass of 1000 g, or 1.00 kg, of which 2.14 g is cholesterol. This leaves 0.998 kg) of solvent. The molality of the cholesterol is therefore:

$$\frac{0.00553 \text{ mol cholesterol}}{0.998 \text{ kg solvent}} = 0.00554 \text{ mol kg}^{-1}$$

c) If there is 2.14 g of cholesterol per liter of blood, then one liter of blood contains 2.14 g of cholesterol. This simple turn-about gives the second factor in the following expression:

$$8.10 \text{ g cholesterol} \times \left(\frac{1 \text{ L blood}}{2.14 \text{ g cholesterol}}\right) = 3.79 \text{ L blood}$$

4-17 To compute the various quantities, we need the masses and chemical amounts of the hydrochloric acid and of the water in some set quantity of solution. Then it is a matter of using definitions. Suppose there is exactly 100.0 g of the solution. There then must be 38.00 g of HCl and 62.00 g of H_2O. The volume of the sample is:

$$100 \text{ g solution} \times \frac{1 \text{ mL solution}}{1.1886 \text{ g solution}} = 84.133 \text{ mL solution}$$

Use the respectively molar masses of HCl and H_2O to convert the masses of the components to chemical amounts:

$$\frac{38.00 \text{ g HCl}}{36.4606 \text{ g mol}^{-1}} = 1.0422 \text{ mol HCl}: \quad \frac{62.00 \text{ g H}_2\text{O}}{18.0153 \text{ g mol}^{-1}} = 3.4415 \text{ mol H}_2\text{O}$$

By definition, the molarity of the HCl is the number of moles of HCl divided by the number of liters of solution. The 84.133 mL (0.084133 L) sample of solution contains 1.0422 mol of HCl, for a molarity of 12.39 mol L^{-1}.
The molality of the HCl is the number of moles of HCl divided by the number of kilograms of solvent. There is 1.0422 mol of HCl dissolved in 62.00 g, (0.06200 kg) of water. The molality of the HCl is therefore 16.81 mol kg^{-1}.
The mole fraction of water is the number of moles of water divided by the total number of moles of all components of the solution:

$$\frac{3.4415 \text{ mol}}{1.0422 + 3.4415 \text{ mol}} = 0.7676$$

Because there are only two components, the mole fraction of the HCl is simply 1 − 0.7676, which equals 0.2324.

4-19 From the definition of mole fraction in this two-component solution:

$$X_{H_2O} = 1.00 \times 10^{-5} = \frac{n_{H_2O}}{n_{N_2} + n_{H_2O}} = \frac{n_{H_2O}}{35.6972 + n_{H_2O}}$$

where the 35.6972 is the number of moles of N_2 in 1.00 kg of N_2 (non-significant figures are carried along deliberately). Solving for n_{H_2O} requires straightforward algebra. It turns out there is 3.5697×10^{-4} mol of H_2O per kilogram of nitrogen. This amounts to 0.00643 g of H_2O.

4-21 a) According to the problem, 100.00 g of commercial $H_3PO_4(aq)$ contains 90.00 g of pure H_3PO_4 and 10.00 g of H_2O. This much H_3PO_4 is 0.9184 mol of H_3PO_4 ($\mathcal{M} = 97.995$ g mol^{-1}). Then:

$$\frac{100.00 \text{ g solution}}{0.9184 \text{ mol } H_3PO_4} \times \left(\frac{12.2 \text{ mol } H_3PO_4}{1 \text{ L solution}}\right) = \frac{1.33 \times 10^3 \text{ g solution}}{1 \text{ L solution}}$$

This is a correct answer, but densities are usually given in gram per milliliter. In those units, the answer is 1.33 g mL^{-1}.
b) The 2.00 L of 1.00 M $H_3PO_4(aq)$ contains 2.00 mol of H_3PO_4. The volume of the 12.2 M $H_3PO_4(aq)$ solution containing 2.00 moles of H_3PO_4 is:

$$2.00 \text{ mol } H_3PO_4 \times \left(\frac{1 \text{ L solution}}{12.2 \text{ mol } H_3PO_4}\right) = 0.164 \text{ L solution}$$

To make 2.00 L of a 1.00 M H_3PO_4(aq) solution, one would add 0.164 L (164 mL) of 12.2 M H_3PO_4 to a 2.00 L volumetric flask and then add enough water to bring the total volume up to 2.00 L.

4-23 First calculate how many moles of NaOH were in the solution before any solid NaOH was added:

$$1.50 \text{ L solution} \times \left(\frac{2.40 \text{ mol NaOH}}{1 \text{ L solution}}\right) = 3.60 \text{ mol NaOH}$$

The molar mass of NaOH is 40.00 g mol^{-1}. The added 25.0 g of NaOH is therefore 0.625 mol of NaOH. After the addition there is a total of $3.60 + 0.625 = 4.23$ mol of NaOH in the container. Meanwhile, the final volume of the solution is 4.00 L. The molarity of NaOH is 4.23 mol/4.00 L = 1.06 mol L^{-1}.

4-25 The volatile water evaporates when the container is left uncovered, but the sodium sulfate stays. First, determine the mass of the 1.000 mol of Na_2SO_4 that the student put into the container:

$$1.000 \text{ mol } Na_2SO_4 \times \frac{142.05 \text{ g } Na_2SO_4}{1 \text{ mol } Na_2SO_4} = 142.1 \text{ g } Na_2SO_4$$

After evaporation the solid residue has the sharply larger mass of 322.2 g, which is a gain of 180.1 g. The gain must be from water chemically bound with the Na_2SO_4 because water was the only other substance that was in the flask (the walls of the flask are assumed to be inert). This mass is 9.997 mol of H_2O ($\mathcal{M}$ = 18.0153 g mol^{-1}). Almost exactly 10 mol of water has been bound by the 1 mol of sodium sulfate. The solid has the molecular formula $Na_2SO_4{\cdot}10H_2O$.

4-27 The chemical equation indicates a 4-to-2 molar relationship between HNO_3 and PbO_2 in the reaction. This fact can be used to construct a conversion factor. The given molarity can also be used as a conversion factor, and the molar mass of PbO_2 (239.2 g mol^{-1}) furnishes another conversion factor. The computation using these three factors is set up as

$$15.9 \text{ g PbO}_2 \times \left(\frac{1 \text{ mol PbO}_2}{239.2 \text{ g PbO}_2}\right) \times \left(\frac{4 \text{ mol HNO}_3}{2 \text{ mol PbO}_2}\right) \times \left(\frac{1 \text{ L HNO}_3 \text{ sol'n}}{7.91 \text{ mol HNO}_3}\right)$$

The answer is 0.0168 L of HNO_3 solution.

4-29 The carbon dioxide in this problem is a gas (with volume measured in liters), and the potassium carbonate is in aqueous solution (with volume measured in liters). The use of the same unit must not be allowed to cause confusion between these two! In the balanced equation the $CO_2(g)$ and $K_2CO_3(aq)$ react in a 1-to-1 molar ratio. The following uses this fact together with the concentration of the $K_2CO_3(aq)$ to determine the chemical amount of $CO_2(g)$ that reacts:

$$187 \text{ L solution} \times \left(\frac{1.36 \text{ mol K}_2\text{CO}_3}{1 \text{ L solution}}\right) \times \left(\frac{1 \text{ mol CO}_2}{1 \text{ mol K}_2\text{CO}_3}\right) = 254.3 \text{ mol CO}_2$$

The volume occupied by this much $CO_2(g)$ depends on the conditions under which it is held. Using the stated temperature and pressure in the ideal-gas equation gives:

$$V = \frac{nRT}{P} = \frac{(254.3 \text{ mol})(0.08206 \text{ L atm mol}^{-1}\text{K}^{-1})(323.15 \text{ K})}{1.00 \text{ atm}} = 6.74 \times 10^3 \text{ L}$$

4-31 a) Phosphorus trifluoride is PF_3, phosphorous acid is H_3PO_3, and hydrofluoric acid is HF. The unbalanced equation is $PF_3 + H_2O \rightarrow H_3PO_3 + HF$.
This is easily balanced by inspection. All of the fluorine ends up in HF, requiring a 3 as coefficient for the HF. All of the oxygen ends up in H_3PO_3, requiring a 3 as coefficient for the H_2O: $PF_3 + 3\,H_2O \rightarrow H_3PO_3 + 3\,HF$.
b) First determine the chemical amount of $PF_3(g)$ in 1.94 L of gaseous PF_3 at 25° C (298 K) and 0.970 atm:

$$n_{\text{PF}_3} = \frac{PV}{RT} = \frac{(0.970 \text{ atm})(1.94 \text{ L})}{(0.08206 \text{ L atm mol}^{-1}\text{K}^{-1})(298 \text{ K})} = 0.07695 \text{ mol}$$

For each mole of PF_3 reacted, 1 mol of H_3PO_3 and 3 mol of HF are produced. This means 0.07695 mol of H_3PO_3 and 0.2309 mol of HF are produced from the 0.07695 mol of PF_3. Both acids dissolve as they are formed. There is enough water to give a final volume of 872 mL (0.872 L). The acids are mixed with each other, but their respective concentrations are computed by *separately* dividing chemical amount by final volume:

$$[H_3PO_3] = \frac{0.07695 \text{ mol}}{0.872 \text{ L}} = 0.0882 \text{ M} \quad \text{and} \quad [HF] = \frac{0.2309 \text{ mol}}{0.872 \text{ L}} = 0.265 \text{ M}$$

4-33 The solution of potassium dichromate contains 5.134 g of the solute per 1000 mL, and 34.26 mL of it brings the titration to the endpoint. The amount of $K_2Cr_2O_7$ that reacts, expressed in moles, is computed as follows:

$$34.26 \text{ mL sol'n} \times \left(\frac{5.134 \text{ g } K_2Cr_2O_7}{1000 \text{ mL sol'n}}\right) \times \left(\frac{1 \text{ mol } K_2Cr_2O_7}{294.18 \text{ g } K_2Cr_2O_7}\right)$$

Giving an answer of 5.979×10^{-4} mol $K_2Cr_2O_7$. In aqueous solution, 1 mol of $Cr_2O_7^{2-}(aq)$ ion exists for every 1 mol of $K_2Cr_2O_7$ that was dissolved. Also, in the balanced equation (which is a net ionic equation) 1 mol of $Cr_2O_7^{2-}(aq)$ reacts with 6 mol of $Fe^{2+}(aq)$. Combine these facts:

$$5.979 \times 10^{-4} \text{ mol } K_2Cr_2O_7 \times \left(\frac{1 \text{ mol } Cr_2O_7^{2-}}{1 \text{ mol } K_2Cr_2O_7}\right) \times \left(\frac{6 \text{ mol } Fe^{2+}}{1 \text{ mol } Cr_2O_7^{2-}}\right)$$

The answer is 0.003587 mol Fe^{2+}. This is the chemical amount of Fe^{2+} in a 500.0 mL portion of solution. The amount per liter (1000.0 mL) is twice as much. The concentration of Fe^{2+} in the sample is 0.007175 mol L^{-1} of Fe^{2+}.

4-35 The problem is very similar to Example 4-6 (text page 139). Each mole of KOH dissolves in water to give one mole of $K^+(aq)$ ion and one mole of $OH^-(aq)$ ion. The chemical amount of $OH^-(aq)$ in 37.85 mL of 0.1279 M aqueous KOH equals the volume multiplied by the concentration:

$$37.85 \text{ mL} \times \left(\frac{0.1279 \text{ mol}}{1 \text{ L}}\right) = 4.841 \text{ mmol}$$

Note the use of the convenient unit, the millimole (mmol). Nitric acid furnishes one mole of $H^+(aq)$ ions per mole dissolved. Therefore, there is a 1-to-1 molar ratio in the acid-base reaction between HNO_3 and KOH. Thus, the chemical amount of HNO_3 in the 100.0 mL sample before the reaction was also 4.841 mmol. The concentration of HNO_3 in the original solution was:

$$[HNO_3] = \frac{4.841 \text{ mmol}}{100.0 \text{ mL}} = \frac{0.04841 \text{ mmol}}{1 \text{ mL}} = 0.04841 \text{ mol L}^{-1}$$

It is easy to confirm that a millimole per milliliter is the same as a mole per liter.

4-37 The change in the vapor pressure ΔP of the solvent is $0.411 - 0.437 = -0.026$ atm. Raoult's law can be rewritten in the case of a two-component system (consisting of solute and solvent) as $\Delta P_{\text{solvent}} = -X_{\text{solute}} P^{\circ}_{\text{solvent}}$. The solvent in this problem is CCl_4. The mole fraction of solute is therefore

$$X_{\text{solute}} = -\frac{\Delta P}{P^{\circ}_{CCl_4}} = -\frac{-0.026 \text{ atm}}{0.437 \text{ atm}} = 0.05950$$

There is 0.1000 kg of CCl_4, which is 0.6502 mol of CCl_4 ($\mathcal{M} = 153.8$ g mol^{-1}). Hence,

$$X_{\text{solute}} = \frac{n_{\text{solute}}}{n_{\text{solute}} + n_{CCl_4}} = 0.05950 = \frac{n_{\text{solute}}}{n_{\text{solute}} + 0.6502}$$

Solving for the chemical amount of the unknown solute gives 0.04114 mol. Because this amount of the solute is simultaneously 7.42 g, the molar mass of the solute is

$$\mathcal{M} = \frac{7.42 \text{ g}}{0.04114 \text{ mol}} = 180 \text{ g mol}^{-1}$$

4-39 The boiling point elevation of a solvent is proportional to the molal concentration of a single nonvolatile solute:

$$\Delta T_b = K_b m \quad \text{hence} \quad K_b = \frac{\Delta T_b}{m}$$

To determine m in this formula, use the definition of molality. The number of moles of anthracene is its mass divided by 178.2 g mol^{-1}, the molar mass of anthracene:

$$n_{\text{anthracene}} = \frac{7.80 \text{ g}}{178.2 \text{ g mol}^{-1}} = 0.04376 \text{ mol}$$

The molality of the anthracene in the toluene solution is this chemical amount (in moles) divided by the mass of the toluene (in kilograms):

$$\frac{0.4376 \text{ mol}}{0.1000 \text{ kg}} = 0.4376 \text{ mol kg}^{-1}$$

The change in the boiling temperature is clearly $112.06 - 110.60 = 1.46$°C. The elevation constant can now be computed by simple substitution:

$$K_b = \frac{\Delta T_b}{m} = \frac{1.46°\text{C}}{0.4376 \text{ mol kg}^{-1}} = 3.34°\text{C mol kg}^{-1}$$

This can also be expressed as 3.34 K mol kg^{-1}. Convert both temperatures in the problem from Celsius to absolute and then subtract the initial from the final to verify this. A Celsius degree and a kelvin are the same size.

4-41 Assume that the unknown solute is nonvolatile. The change in the freezing point of the solvent is then proportional to the molality of the solute:

$$\Delta T_f = -K_f m$$

We can use this equation to calculate the molality of the unknown in its solution in camphor:

$$m = -\frac{\Delta T_f}{K_f} = -\frac{(170.8 - 178.4)°\text{C}}{37.7°\text{C kg mol}^{-1}} = 0.20 \text{ mol kg}^{-1}$$

Note the change in the temperature units in the freezing-point depression constant from K to °C. This is legitimate since the problem concerns a *change* in temperature. See problem 4-39. There is 0.20 mol of unknown per kilogram of camphor, but the problem deals with 25.0 g (0.0250 kg) of camphor. The amount of unknown in the 25.0 g of camphor is:

$$0.20 \text{ mol kg}^{-1} \times 0.0250 \text{ kg} = 0.0050 \text{ mol}$$

The 0.840 g of unknown that was put in the 25.0 g of camphor was also 0.0050 mol of unknown. The molar mass of a substance is equal to its mass divided by its chemical amount: 0.840 g/0.0050 mol = 1.7×10^2 g mol^{-1}.

4-43 Imagine 1.0 g of ice cream mixture. Such a mixture contains 0.34 g of sucrose and 0.66 g of water. A solution containing 1000 g of water would contain:

$$1000 \text{ g water} \times \left(\frac{0.34 \text{ g sucrose}}{0.66 \text{ g water}}\right) = 515 \text{ g sucrose}$$

Dividing this mass by 342.3 g mol^{-1}, the molar mass of sucrose, converts the amount to moles. It is 1.50 mol. Since there is 1.50 mol of sucrose per 1000 g of the water, the molality of the 34% sucrose solution is 1.50 mol kg^{-1}. The change in the freezing point depression is:

$$\Delta T = -K_f m = (-1.86 \text{ K mol kg}^{-1})(1.50 \text{ kg mol}^{-1}) = -2.8 \text{ K}$$

The freezing point of the mixture is this change added to the freezing point of the pure solvent (0°C). It is −2.8°C. As ice freezes out, the solution becomes more concentrated in sucrose and the freezing point is depressed further.

4-45 Calculate the effective molality from the amount of freezing-point depression:

$$m = -\frac{\Delta T}{K_f} = -\frac{(-4.218 \text{ K})}{1.86 \text{ K kg mol}^{-1}} = 2.268 \text{ mol kg}^{-1}$$

The ratio of the effective molality to the actual molality is 2.268/0.8402 = 2.70. Thus the each Na_2SO_4 unit dissociates effectively into 2.70 particles. This is somewhat less than the theoretical value of 3.00 (corresponding to two Na^+ and one SO_4^{2-}) because the plus and minus ions in this rather concentrated solution associate to a degree, which reduces the effective number of free ions.

4-47 The osmotic pressure π of a solution is related to the concentration of the solute by the equation:

$$\pi = cRT$$

The measurement of the osmotic pressure therefore allows easy calculation of the concentration:

$$c = \frac{\pi}{RT} = \frac{0.0105 \text{ atm}}{(0.08206 \text{ L atm mol}^{-1}\text{K}^{-1})(300 \text{ K})} = 4.265 \times 10^{-4} \text{ mol L}^{-1}$$

This concentration was obtained by dissolving 200 mg (0.200 g) of the solute in 25.0 mL of solution, a procedure that gives the same concentration as dissolving 8.00 g solute in 1.00 L of water. Thus 8.00 g of solute is 4.265×10^{-4} mol of solute. The molar mass is:

$$\mathcal{M} = \frac{8.00 \text{ g}}{4.265 \times 10^{-4} \text{ mol}} = 1.88 \times 10^4 \text{ g mol}^{-1}$$

4-49 Text Figure 4-14 shows the experimental set-up. The difference h between the level of the solution in the tube and the level of the water outside is proportional to the osmotic pressure of the solution in the tube. The problem gives h as 15.2 cm (0.152 m) of water. To get the osmotic pressure, use the density ρ of the solution and the acceleration g due to gravity in the formula $\pi = \rho g h$ (text page 148). The density of the solution is 1.00 g cm^{-3}. This is equivalent to 1.00×10^3 kg m^{-3}. The latter unit must be used to make the pressure come out in the SI unit of pressure, which is the pascal:

$$\pi = \rho g h = (1.00 \times 10^3 \text{ kg m}^{-3})(9.807 \text{ m s}^{-2})(0.152 \text{ m}) = 1.49 \times 10^3 \text{ kg m}^{-1}\text{s}^{-2}$$

This also equals 1.49×10^3 Pa (see Table B-2 in text Appendix B). Converting to atm:

$$\pi = 1.49 \times 10^3 \text{ Pa} \times \left(\frac{1 \text{atm}}{101\,325 \text{ Pa}}\right) = 0.0147 \text{ atm}$$

Now, calculate the concentration of the polymer solution:

$$c = \frac{\pi}{RT} = \frac{0.0147 \text{ atm}}{(0.08206 \text{ L atm mol}^{-1}\text{K}^{-1})(288.15 \text{ K})} = 6.22 \times 10^{-4} \text{ mol L}^{-1}$$

It is possible to compute the concentration of the polymer without changing the pressure to atmospheres:

$$c = \frac{\pi}{RT} = \frac{1.49 \times 10^3 \text{ Pa}}{(8.315 \text{ J K}^{-1}\text{mol}^{-1})(288.15 \text{ K})} = 0.622 \text{ mol m}^{-3}$$

This answer is the same because there are 1000 L in a cubic meter. The solution holds 6.22×10^{-4} mol of polymer per liter of solution. There are also 4.64 g of polymer per liter of solution. Therefore:

$$\mathcal{M} = \frac{4.64 \text{ g}}{6.22 \times 10^{-4} \text{ mol}} = 7.46 \times 10^3 \text{ g mol}^{-1}$$

4-51 a) Henry's law relates the mole fraction of a solute (called component B) to the partial pressure of the vapors of that solute above the solution: $P_B = k_B X_B$. In this case the solute is CO_2 and its partial pressure above an aqueous solution is 5.0 atm. Then:

$$X_{CO_2} = \frac{P_{CO_2}}{k_{CO_2}} = \frac{5.00 \text{ atm}}{(1.65 \times 10^3 \text{ atm})} = 3.0 \times 10^{-3}$$

This fraction means that there is 3.0×10^{-3} mol of CO_2 in solution for every 0.997 mol of water. But there is 55.5 mol of water per liter of solution if the dilute solution is taken to have the same density as water. Hence:

$$\frac{3.0 \times 10^{-3} \text{ mol } CO_2}{0.997 \text{ mol } H_2O} \times \frac{55.5 \text{ mol } H_2O}{1.00 \text{ L solution}} = 0.17 \text{ mol } CO_2 \text{ L}^{-1}$$

The amount of CO_2 dissolved in a liter of water is therefore 0.17 mol.

b) Before the cap is removed, molecules of CO_2 are confined above the liquid and are in equilibrium with the dissolved molecules. This means that CO_2 molecules are constantly moving from the gas phase to the dissolved phase and back. The rate of movement of CO_2 out of solution equals the rate of movement into solution. When the cap is removed gaseous CO_2 escapes from the bottle. The rate of solution of the gas plummets but the rate of loss from the liquid stays the same. Equilibrium is re-established with a far smaller concentration of CO_2 in the solution.

4-53 We can determine the chemical amount of the methane that was dissolved in the 1.00 kg of solution from the conditions at which the gas is held after expulsion from solution by boiling. The assumptions are that *all* of the methane was expelled and that the gas is ideal:

$$n_{CH_4} = \frac{PV}{RT} = \frac{(1.00 \text{ atm})(3.01 \text{ L})}{(0.08206 \text{ L atm mol}^{-1}\text{K}^{-1})(273.15 \text{ K})} = 0.1343 \text{ mol}$$

The molar mass of CH_4 is 16.04 g mol^{-1}, so the methane that was expelled has a mass of 2.154 g. The 1.00 kg of solution contained only water and methane. The mass of the water left after removal of methane was then 0.9978 kg, the mass of the solution minus the mass of the methane. This mass of H_2O ($\mathcal{M} = 18.0153$ g mol^{-1}) is 55.39 mol of water. The mole fraction of CH_4 in the solution can now be calculated:

$$X_{CH_4} = \frac{0.1343 \text{ mol}}{0.1343 + 55.39 \text{ mol}} = 0.002419$$

This mole fraction of CH_4 was present in solution with 1.00 atm of CH_4 above it. Henry's law for this solution is:

$$P_{CH_4} = k_{CH_4} X_{CH_4}$$

Both the mole fraction of methane in the solution and the partial pressure of methane are known, so the Henry's law k is easily calculated:

$$k_{CH_4} = \frac{P_{CH_4}}{X_{CH_4}} = \frac{1.00 \text{ atm}}{0.002419} = 413 \text{ atm}$$

The Henry's law constant for methane in water at 25°C is 413 atm.

4-55 Write Raoult's law for both the benzene and toluene

$$P_{\text{benz}} = X_{\text{benz}} P^{\circ}_{\text{benz}} \quad \text{and} \quad P_{\text{tol}} = X_{\text{tol}} P^{\circ}_{\text{tol}}$$

Since equal numbers of moles of benzene and toluene were mixed, $X_{\text{benz}} = X_{\text{tol}} = 0.500$. Substitute these mole fractions and the vapor pressures of the pure substances to compute the partial pressure of each substance above the mixture:

$$P_{\text{benz}} = 0.500(0.0987 \text{ atm}) = 0.04935 \text{ atm} \quad P_{\text{tol}} = 0.500(0.0289 \text{ atm}) = 0.01445 \text{ atm}$$

From the ideal-gas law, the number of moles of a particular gas in a gaseous mixture at constant temperature is directly proportional to its partial pressure:

$$n_{\text{benz}} = P_{\text{benz}} \frac{V}{RT} \quad \text{and} \quad n_{\text{tol}} = P_{\text{tol}} \frac{V}{RT}$$

The mole fraction of benzene in the vapor is

$$X_{\text{benz,vap}} = \frac{n_{\text{benz}}}{n_{\text{benz}} + n_{\text{tol}}} = \frac{P_{\text{benz}}(V/RT)}{P_{\text{benz}}(V/RT) + P_{\text{tol}}(V/RT)} = \frac{P_{\text{benz}}}{P_{\text{benz}} + P_{\text{tol}}}$$

$$X_{\text{benz,vap}} = \frac{0.04935 \text{ atm}}{(0.04935 + 0.01445) \text{ atm}} = 0.774$$

Exactly half of the molecules in the liquid are benzene molecules, but well over three quarter of the molecules in the vapors above the liquid are benzene molecules. Such enrichment in the more volatile component is the basis for separation by distillation.

4-57 a) The molar masses of the CCl_4 and $C_2H_4Cl_2$ are respectively 153.82 g mol^{-1} and 98.96 g mol^{-1}. The masses of the two components of the solution are converted to chemical amount by dividing by these numbers with the results:

$$n_{CCl_4} = 0.1950 \text{ mol}; \quad n_{C_2H_4Cl_2} = 0.2021 \text{ mol}$$

From these values, the mole fraction of the CCl_4 in the solution is readily found:

$$X_{CCl_4} = \frac{0.1950 \text{ mol}}{0.1950 + 0.2021 \text{ mol}} = 0.491$$

b) The total vapor pressure above the solution is the sum of the partial pressures of the two components in the mixture. These partial pressures are given by Raoult's law. Hence:

$$P_{tot} = P_{CCl_4} + P_{C_2H_4Cl_2} = X_{CCl_4}P^{\circ}_{CCl_4} + X_{C_2H_4Cl_2}P^{\circ}_{C_2H_4Cl_2}$$

$$P_{tot} = X_{CCl_4}P^{\circ}_{CCl_4} + (1 - X_{CCl_4})P^{\circ}_{C_2H_4Cl_2}$$

The two mole fractions must add up to one because there are only two components in the solution. The mole fraction of CCl_4 in the solution was computed in part a). The mole and the vapor pressures of the pure components are given in the problem. Substitution of the numbers gives:

$$P_{tot} = (0.491)(0.293 \text{ atm}) + (1 - 0.493)(0.209 \text{ atm}) = 0.250 \text{ atm}$$

c) The mole fraction of a component in a gaseous mixture equals its partial pressure divided by the total pressure of the gaseous mixture:

$$X_{CCl_4,vap} = \frac{P_{CCl_4}}{P_{tot}}$$

See problem 4-55. According to Raoult's law, the partial pressure of $CCl_4(g)$ in the vapor above the solution is its mole fraction in the solution times its vapor pressure when pure: Hence

$$X_{CCl_4,vap} = \frac{P_{CCl_4}}{P_{tot}} = \frac{X_{CCl_4}P^{\circ}_{CCl_4}}{P_{tot}}$$

But $P^{\circ}_{CCl_4}$ is given in the problem and X_{CCl_4} was found in part a. Substituting them gives:

$$X_{CCl_4,vap} = \frac{0.491(0.293 \text{ atm})}{0.250 \text{ atm}} = 0.575$$

4-59 "Saturation" means that the vapor pressure of the water in the room has reached its maximum value. Water is condensing in droplets on the walls and ceiling. The partial pressure of water vapor in the room at saturation (also called 100% relative humidity) at this temperature is 0.03126 atm. The rest of the pressure in the room is furnished by the oxygen and nitrogen that are the normal components of air. Assume that Dalton's law holds so that the partial pressure of the water vapor equals the pressure the water vapor would have if there were no other gases in the room. Then, use the ideal gas law to calculate the number of moles of H_2O in the room:

$$n_{H_2O} = \frac{P_{H_2O}V}{RT} = \frac{(0.03126 \text{ atm})(1.00 \times 10^5 \text{ L})}{(0.08206 \text{ L atm mol}^{-1}\text{K}^{-1})(298.15 \text{ K})} = 127.8 \text{ mol}$$

The volume of the room has been converted from cubic meters to liters because liters appears in the units of the R that is used. The mass of this chemical amount of water is:

$$127.8 \text{ mol H}_2\text{O} \times \left(\frac{18.015 \text{ g H}_2\text{O}}{1 \text{ mol H}_2\text{O}}\right) = 2300 \text{ g H}_2\text{O}$$

4-61 Compute the chemical amount of air that was present in the 6.00-L portion of air saturated with the vapors of the unknown. This is possible because the physical state of the air after purification is fully described and the ideal-gas law can be assumed to apply:

$$n_{\text{air}} = \frac{PV}{RT} = \frac{(1.000 \text{ atm})(3.75 \text{ L})}{(0.08206 \text{ L atm mol}^{-1}\text{K}^{-1})(223.15 \text{ K})} = 0.2048 \text{ mol}$$

Next, use the ideal-gas law in the form $P = nRT/V$ to compute the pressure that this chemical amount of air exerted as part of the 6.00-L mixture:

$$P_{\text{air}} = \frac{(0.2048 \text{ mol})(0.08206 \text{ L atm mol}^{-1}\text{K}^{-1})(298.15 \text{ K})}{6.00 \text{ L}} = 0.835 \text{ atm}$$

But the total pressure above the unknown was 0.980 atm. By Dalton's law $P_{\text{unknown}} = 0.980 - 0.835 = 0.145$ atm. This is the desired vapor pressure of the unknown at 298.15 K (25°).

4-63 When chunks of solid CO_2 are added to room-temperature ethanol in an open beaker, portions of the solid will sublime to gaseous CO_2. Much bubbling and roiling will accompany the escape of this gas. The process will in theory chill the ethanol to −78.5°C, the sublimation temperature of the solid CO_2, but no lower. Since ethanol requires a temperature below −114.5°C to freeze, it stays liquid in this experiment. Once the ethanol is good and cold, small amounts of further sublimation will take place to counteract heat flowing in from the surroundings; the mixture will always fizz a little.

4-65 Substitution of the van der Waals a's and b's in the three equations given in the problem give the results in the fourth through sixth columns of the following table. These columns include the observed values (in parentheses) for comparison:

Gas	a (atm $L^{-2}mol^{-2}$)	b (L mol^{-1})	T_c (K)	P_c (atm)	$(V/n)_c$ (L mol^{-1})
O_2	1.360	0.03183	154.3(154.6)	49.72(49.8)	0.09549(0.0734)
CO_2	3.592	0.04267	304.0(304.2)	73.07(72.9)	0.1280(0.0940)
H_2O	5.464	0.03049	647.1(647.1)	217.7(217.6)	0.09147(0.0567)

The T_c and P_c for the gases are taken from experimental reports and are more precise than can be obtained by reading the scale in Figure 4-6. Molar volumes cannot be read from P-T phase diagrams at all.

4-67 a)

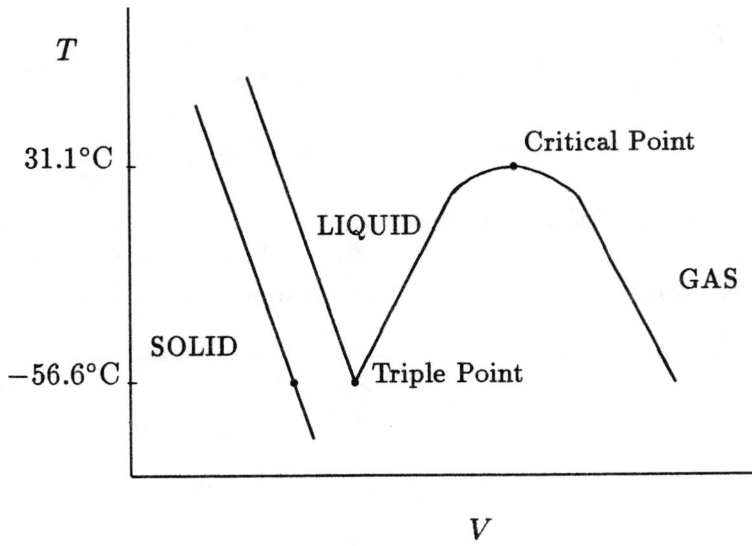

b) Such a diagram cannot be drawn because two different phases (liquid and solid) can have the same temperature and molar volume.

4-69 a) The element iodine is distributed in the Donovan's solution in different chemical forms. However, the total amount of iodine in the solution, which is prepared by mixing pure compounds that dissolve completely, depends only on the amounts of the compounds and the fraction of each comprised by iodine. The total mass of iodine in 100 mL of Donovan's solution is thus the fraction by mass of elemental iodine in AsI_3 ($\mathcal{M} = 455.6$ g mol^{-1}) multiplied by 1.00 g of AsI_3 plus the fraction by mass of iodine in HgI_2 ($\mathcal{M} = 454.4$ g mol^{-1}) multiplied by 1.00 g of HgI_2:

$$m_{\rm I} = \left(\frac{(3)(126.9)\text{ g I}}{455.6\text{ g AsI}_3}\right) 1.00\text{ g AsI}_3 + \left(\frac{(2)(126.9\text{ g I})}{454.4\text{ g HgI}_2}\right) 1.00\text{ g HgI}_2 = 1.39\text{ g}$$

Note that 126.9 is the atomic mass of iodine. The mass of iodine per liter (which is 10 times 100 mL) of solution is 10 times this answer or 13.9 g L^{-1}.
b) The 0.100 M AsI_3 solution contains 45.56 g of AsI_3 per liter and therefore furnishes 4.556 g of AsI_3 per 100 mL. To make 3.50 L of Donovan's solution, 35.0 g of AsI_3 is needed. Therefore, measure out $(35.0/4.556) \times 100$ mL $= 768$ mL of the AsI_3 solution. Add to it 35.0 g of $HgI_2(s)$ and 31.5 g of $NaHCO_3(s)$. Then add enough water to bring the total volume to 3.50 L.

4-71 The careful wording of the problem assures the reader that no Cl is lost at any point during the transformation $NaCl \rightarrow Cl_2 \rightarrow HCl$. For every mole of chlorine that was originally present in the 150 mL of 10.00% aqueous NaCl, there is one mole of Cl in the 250 mL of HCl(*aq*) that ultimately is formed. Compute this number of moles:

$$150 \text{ mL sol'n} \times \frac{1.0726 \text{ g sol'n}}{1 \text{ mL sol'n}} \times \frac{10.0 \text{ g NaCl}}{100 \text{ g sol'n}} \times \frac{1 \text{ mol NaCl}}{58.44 \text{ g NaCl}} = 0.2753 \text{ mol}$$

Note the third term, in which the mass percentage of NaCl in the solution is used as a conversion factor. If there is 0.2753 mol of NaCl, there is also 0.2753 mol of Cl, because each mole of NaCl contains one mole of Cl. All of the Cl ends up in the form of HCl, so there automatically must be 0.2753 mol of HCl in the 250 mL of solution that is formed. The concentration of the HCl accordingly is 0.2753 mol/0.250 L, which equals 1.10 mol L^{-1}.

4-73 The change in the vapor pressure ΔP is $0.3868 - 0.3914 = -0.0046$ atm. The mole fraction of the sulfur present in the sulfur-CS_2 system is therefore:

$$X_{\text{sulfur}} = -\frac{\Delta P}{P^\circ_{CS_2}} = -\frac{-0.0046 \text{ atm}}{0.3914 \text{ atm}} = 0.0117$$

There is 1.00 kg of CS_2 ($\mathcal{M} = 76.14$ g mol^{-1}), which is 13.13 mol of CS_2. Hence:

$$X_{\text{sulfur}} = 0.0117 = \frac{n_{\text{sulfur}}}{n_{\text{sulfur}} + n_{CS_2}} = \frac{n_{\text{sulfur}}}{n_{\text{sulfur}} + 13.13}$$

Solving for the chemical amount of sulfur gives 0.155 mol. Because this amount of sulfur is simultaneously 40.0 g of sulfur, the molar mass of sulfur as it exists in this solution is 40.0 g/0.155 mol = 257 g mol^{-1}. This is almost exactly eight times larger than the molar mass of atomic sulfur (see the periodic table). The molecular formula of the sulfur in the solution must be S_8.

4-75 The soft drink is a solution of (among other things) CO_2 in water. When the cap is on, gaseous CO_2 in the space above the fluid is held at a pressure exceeding 1 atm. In accord with Henry's law there is a higher concentration of CO_2 in solution than if the pressure of CO_2 were only 1 atm. The presence of this dissolved CO_2 depresses the freezing point of the solution. When the cap is popped, the pressure of CO_2 over the solution suddenly drops to far less than 1 atm. Gaseous CO_2 bubbles out of solution. The mole fraction (and also the molality) of all solutes in the soft drink taken as a group becomes less. The freezing point of the soft drink rises. If the temperature of the soft drink after the cap is removed is less than the higher freezing point that comes from loss of the gaseous solute, the solution will freeze.

4-77 Raoult's law allows calculation of an effective mole fraction of $CaCl_2$ in the solution at 25°C from vapor-pressure lowering:

$$X_2 = -\frac{P_1 - P_1^0}{P_1^0} = -\frac{0.02970 - 0.03126 \text{ atm}}{0.03126 \text{ atm}} = 0.0499$$

The $CaCl_2$ dissociates in water to form one mole of $Ca^{2+}(aq)$ cations and two moles of $Cl^-(aq)$ anions per mole dissolved. The X_2 just calculated is therefore not the true mole fraction of the solute, but is an *effective* mole fraction that is larger that the true mole fraction. Now, calculate the effective molality of the $CaCl_2$:

$$\frac{0.0499 \text{ CaCl}_2}{(1.000 - 0.0499) \text{ mol H}_2\text{O}} \times \left(\frac{1 \text{ mol H}_2\text{O}}{0.018015 \text{ kg H}_2\text{O}}\right) = 2.92 \text{ mol kg}^{-1}$$

Put this effective molality into the standard formula for freezing-point depression:

$$\Delta T = -K_f m = -(1.86 \text{ K kg mol}^{-1})(2.92 \text{ mol kg}^{-1}) = -5.43 \text{ K} = -5.43°\text{C}$$

Recall that the size the kelvin and degree Celsius are the same. The freezing point of the solution is the original freezing point plus the change:

$$0.00°\text{C} + (-5.43°\text{C}) = -5.43°\text{C}$$

where it has been assumed that the effective number of particles of solute is unchanged by the drop from 25°C where the vapor pressure was recorded, to −5.43°C.

4-79 The salt $GaCl_2$ would be expected to dissociate in water according to:

$$\text{GaCl}_2(aq) \rightarrow \text{Ga}^{2+}(aq) + 2\,\text{Cl}^-(aq)$$

If the "$GaCl_2$"were actually $Ga[GaCl_4]$, then the dissociation would be:

$$\text{Ga[GaCl}_4](aq) \rightarrow \text{Ga}^+(aq) + [\text{GaCl}_4]^-(aq)$$

In the former case, dissociation gives three ions; in the latter case, it gives only two ions. We can therefore measure a colligative property to distinguish between the two cases. For example, imagine that a sufficient mass of the compound were dissolved in water to make a solution that was 0.0100 mol kg^{-1} in $GaCl_2$. This solution would have a freezing point of −0.056°C if the formula $GaCl_2$ were correct. This freezing point is calculated by using an effective molality of 0.0300 mol kg^{-1} in the usual formula for freezing-point depression. The molality is tripled because three moles of ions are formed by dissociation of one mole of $GaCl_2$. Now try the formula $Ga[GaCl_4]$. The identical aqueous solution has a true molality of only 0.00500 mol kg^{-1}, because the molar mass of the solute is now twice as large; the solution has an effective molality of 0.0100 mol kg^{-1}, because two ions are formed upon dissociation. The predicted freezing point of −0.0186°C is measurably different.

4-81 First determine the effective molality of the solution of NaCl by comparing its freezing point to the freezing point of pure water:

$$m = -\frac{\Delta T_f}{K_f} = -\frac{-0.406\ \text{K}}{1.86\ \text{K kg mol}^{-1}} = 0.218\ \text{mol kg}^{-1}$$

If the molality and molarity are equal, then this number can immediately be substituted into the formula for the osmotic pressure of a solution:

$$\pi = cRT = (0.218\ \text{mol L}^{-1})(0.08206\ \text{L atm mol}^{-1}\text{K}^{-1})(298.15\ \text{K}) = 5.33\ \text{atm}$$

4-83 Write Henry's law for a solution of benzene in water:

$$P_{\text{benz}} = k_{\text{benz}} X_{\text{benz}} = (301\ \text{atm}) X_{\text{benz}}$$

The chemical amount of benzene ($\mathcal{M} = 78.11$ g mol^{-1}) in the solution described in the problem is 2.0 g/78.11 g mol^{-1} = 0.0256 mol = 0.026 mol. The mole fraction of benzene is:

$$X_{\text{benz}} = \frac{0.0256\ \text{mol}}{0.0256 + (55.5 \times 10^3)\ \text{mol}} = 4.6 \times 10^{-7}$$

The large amount of water (the 1000 L has a mass of one million grams) completely drowns out the contribution of the benzene to the denominator of this fraction. Combining the mole fraction of benzene and the Henry's law constant cited in the problem in Henry's law gives into Henry's law gives:

$$P_{\text{benz}} = k_{\text{benz}} X_{\text{benz}} = (301\ \text{atm})(4.6 \times 10^{-7}) = 1.4 \times 10^{-4}\ \text{atm}$$

This pressure and the temperature in kelvins are then inserted into the rearranged ideal-gas equation

$$\frac{n_{\text{benz}}}{V} = \frac{P_{\text{benz}}}{RT} = \frac{1.4 \times 10^{-4}\ \text{atm}}{(0.082\ \text{L atm K}^{-1}\text{mol}^{-1})(298.15\ \text{K})} = 5.7 \times 10^{-6} \frac{\text{mol}}{\text{L}}$$

The answer is the concentration of the benzene at equilibrium in the vapor above the solution. This concentration is converted to molecules per cubic centimeter as follows:

$$\frac{5.7 \times 10^{-6}\ \text{mol}}{1\ \text{L}} \times \left(\frac{1\ \text{L}}{1000\ \text{cm}^3}\right) \times \left(\frac{6.022 \times 10^{23}\ \text{molecule}}{1\ \text{mol}}\right) = 3.4 \times 10^{15} \frac{\text{molecule}}{\text{cm}^3}$$

4-85 The difference between a solution and a colloidal suspension is with the size of the dispersed particles. In a solution, the solute is dispersed at the molecular (or ionic) level. Each small particle is surrounded by a cage of several solvent particles. Examples are solutions of NaCl or alcohol in water. In a colloidal suspension, the dispersed particles are aggregates of hundreds to thousands of solute molecules. The aggregates are frequently surrounded by interacting solvent molecules that prevent them from sticking together to form a visible precipitate. The particles do not settle on the bottom of the container because the agitation caused by collisions of neighboring molecules is strong enough to keep them up. An example of a colloid is homogenized milk. The white opacity of milk is caused by tiny particles of fat that are too small to settle out or to be filtered. In some cases, it may be difficult to classify a mixture as a solution or a suspension. If the particles are aggregates of only small numbers of molecules, the properties of the mixture will be similar to those of a solution, but deviate somewhat toward those of a colloid.

Chapter 5

Chemical Equilibrium

Chemical Reactions and Equilibrium

A chemical system left to itself ultimately comes to a state of **equilibrium.** At equilibrium the pressures or concentrations of the reactants and products no longer change, and all tendency for observable change has been exhausted. Mechanical analogies are easy to find: the unwinding of a spring, the slump of a pile of gravel to a final angle of repose. Such analogies have some merit but also fail in some respects. True chemical equilibria have *all* of the following characteristics:

1. They display no macroscopic evidence of change.

2. They are reached through spontaneous processes.

3. They are a dynamic balance of forward and reverse processes.

4. They are the same regardless of direction of approach.

Most analogies fail on the third point. Although there is never a visible sign of change in a chemical system at equilibrium, there is nonetheless continuous exchange between the reactants and products on the molecular level. Dynamic balance is absent in mechanical equilibria.

The Law of Mass Action

Experiment shows that the partial pressures (or concentrations) of the reactants and products present in a chemical system at equilibrium are interrelated. Consider a general chemical change in which all of the reactants and products are gases:

$$a\,\mathrm{A}(g) + b\,\mathrm{B}(g) \rightleftharpoons c\,\mathrm{C}(g) + d\,\mathrm{D}(g)$$

The **reaction quotient** (Q) for this reaction is:

$$\frac{P_A^a P_B^b}{P_C^c P_D^d} = Q$$

The reaction tends to occur in one direction or the other (note the double arrow in the equation) to adjust the value of Q to equal a constant K that is in principle different for every reaction. When Q equals K, the reaction is at equilibrium:

$$\frac{\boldsymbol{P_A^a P_B^b}}{\boldsymbol{P_C^c P_D^d}} = \boldsymbol{K(T)} \quad \text{(at equilibrium)}$$

The constant $K(T)$ is the **equilibrium constant** of the reaction; K is written as an explicit function of T because K depends quite strongly upon the temperature. This **law of mass action** can be extended to cover solutions and equilibria involving any combination of solids, liquids and gases.

Large values of K correspond to nearly complete equilibrium conversion of reactants to products. Small values of K correspond to slight conversion of reactants to products.

Many problems give an initial set of partial pressures of reactants and products. Substitution of such a set of value into the mass-action expression gives the initial value of Q. Initial Q's are sometimes written Q_0 to distinguish them (see **problem 5-31**). If Q exceeds K, then the reaction proceeds from right to left, the reverse of the direction in which it is written. If K exceeds Q, then the reaction proceeds as written, from left to right. If by some chance $Q_0 = K$, then the system is at equilibrium as originally mixed and no observable change will ever occur. Thus it is easy to predict the direction of reaction given a set of initial pressures and a value of K. See **problems 5-31, 5-35, 5-51,** and **5-67.**

Identifying the *tendency* to react in one direction or the other does not tell how *fast* the reaction will be. Chemical systems can remain for years in states that are far from equilibrium without apparent change.

The Concept of Activity

Both K and Q are *dimensionless* quantities, that is, K and Q are pure numbers that have no units. The law of mass action is strictly true only when the partial pressures of gases (given in atm) are replaced by the unit-less **activities** of the gases. The activity of a dilute gas is the ratio of its partial pressure to some reference pressure P_{ref}.

The impact of the term "activity" is that the greater the partial pressure of a gas taking part in a reaction, the more influential it is in determining the exact nature of the eventual equilibrium. It is convenient to choose the reference pressure to be exactly 1 atm. Then, as long as partial pressures are in atmospheres their conversion to activities requires nothing more than the dropping of the unit. Numerical values of K obviously depend on the P_{ref}. Exactly 1 atm is a common choice for a reference pressure, and is the exclusive choice in this text. It is not the only choice.

Calculating Equilibrium Constants

Relationships Among Equilibrium Expressions

If the coefficients of a balanced chemical equation are all multiplied by a constant, the equation is still balanced. The corresponding mass-action expression is raised to a power equal to the multiplying constant, even if the multiplying constant is negative or fractional. Note that multiplying a chemical equation by a negative number corresponds to reversing its direction. Consider for example the chemical equation and mass-action expression:

$$\mathrm{A}(g) \rightleftharpoons \mathrm{B}(g) + 2\,\mathrm{C}(g) \qquad \frac{P_{\mathrm{B}}P_{\mathrm{C}}^2}{P\mathrm{A}} = K_1$$

Compare it to:

$$1/2\mathrm{A}(g) \rightleftharpoons 1/2\mathrm{B}(g) + \mathrm{C}(g) \qquad \frac{P_{\mathrm{B}}^{1/2}P_{\mathrm{C}}}{P_{\mathrm{A}}^{1/2}} = K_2$$

The second chemical equation is the first multiplied through by one-half. Hence, K_2 is the square root of K_1. because raising to the 1/2 power means taking the square root.

- If two balanced chemical equations are added then the corresponding mass action expressions are multiplied. See **problem 5-13, 5-15,** and **5-59.**

- If one chemical equation is subtracted from a second then the mass action expression of the resulting equation is the mass action expression of the second *divided* by the mass action expression of the first.

Whenever an equilibrium constant is quoted, it must be accompanied by a chemical equation. For example, if **problem 5-27**, stated "K for the synthesis of ammonia at 25° is 6.78×10^5" the meaning would not be clear. Is the K for $\mathrm{N_2}(g) + 3\,\mathrm{H_2}(g) \rightleftharpoons 2\,\mathrm{NH_3}(g)$ or perhaps $1/2\,\mathrm{N_2}(g) + 3/2\,\mathrm{H_2}(g) \rightleftharpoons \mathrm{NH_3}(g)$?

Calculation of Gas-Phase Equilibria

The law of mass action provides a mathematical equation relating the equilibrium partial pressures of the several gases taking part in a reaction. It is however only one relationship, and many different gases may be present at equilibrium. It is often desired to calculate the partial pressures of *all* the gases in an equilibrium mixture. To do so requires finding additional relationships among the partial pressures.

- The partial pressures substituted into a mass-action expression must be *equilibrium* partial pressures. Problems often quote initial partial pressures; final (equilibrium) partial pressures may differ sharply. Use initial values only to compute an initial reaction quotient. See **problem 5-31a.**

- The units of all partial pressures must be the same. Pressure is measured in an assortment of units. If a problem is presented in mixed units, convert to a single unit, preferably atmospheres, before going on.

- The initial partial pressures in a reaction mixture are determined by the person setting up the experiment. They may be given specifically in a problem as in **problems 5-21, 5-35,** and **5-69.** When they are not specified they are often equal to zero **(problem 5-23, 5-53)** or their values are given indirectly by the wording of the problem **(problem 5-27).**

- The stoichiometry of the equation determines the *changes* in the partial pressures of reactants and products as a chemical reaction moves toward equilibrium.

- The sum of the partial pressures of all of the gases in a system equals the total pressure. In practical reactions the total pressure is usually quite easy to measure. Knowing it provides an additional relationship among the several partial pressures. See **problems 5-73** and **9-75.**

To solve gas-phase equilibrium problems:

1. Write down the balanced reaction and the corresponding mass action expression.

2. Write down the initial partial pressures of all reactants and products. Include zero partial pressures for gases not initially present.

3. Determine the changes required in the partial pressures to reach equilibrium. If a change is not known, symbolize it with x or y. Changes made by the reaction as it goes to equilibrium are related by the coefficients in the balanced equation. A very effective device to keep track of this is a three-line table of the form:

	reactants	$\rightleftharpoons$	products
Initial pressure (atm)	–		–
Change in pressure (atm)	–		–
Equilibrium pressure (atm)	–		–

For each product and reactant, the entries on the third line equal the sum of the entries on the first and second lines. This format is used repeatedly in the text and in this Guide, starting with the solution to **problem 5-11.**

4. Substitute the expressions for the equilibrium partial pressures into the mass action expression and solve for the unknowns, which are related to the changes. Use the quadratic formula (text page A-16) if quadratic equations arise (they often do). Higher-order equations also develop. Numerical considerations can often lead to big simplifications **(problem 5-25)** although analytical solution is always an option **problems 5-27** and **5-65.**

5. Calculate the equilibrium partial pressures by adding the changes to the initial partial pressures.

6. Examine the answer critically. Negative equilibrium pressures are impossible and show that an error has been made. Negative *changes* in pressure correspond to a substance being used up in the reaction. They are quite possible, but getting them can be upsetting. In **problem 5-25,** a negative change in partial pressure is carefully avoided by a suitable definition of x.

Concentrations of Gases in Equilibrium Calculations

The partial pressure of a gas is directly proportional to its concentration, assuming the gas is ideal. The concentration in mol L^{-1} of a substance in a mixture is indicated by a set of brackets around its formula. The notation "$[CO_2]$" thus stands for a number whereas "CO_2" stands for a compound. This is source of frequent confusion.

The concentration and partial pressure of a gas are related. For the gas A:

$$[\mathrm{A}] = \frac{n_\mathrm{A}}{V} = \frac{P_\mathrm{A}}{RT}$$

as long as the assumption of ideal-gas behavior is valid. See **problems 5-59, 5-61,** and **5-63.** In view of this equation, a mass-action expression in terms of partial pressures is easy to recast to employ the concentrations of the various gases. This can often simplify calculations. See **problem 5-19.** Doing it obviously can never exert an influence on events in a reaction system which is coming to equilibrium. For the general reaction:

$$a\mathrm{A} + b\mathrm{B} \rightleftharpoons c\mathrm{C} + d\mathrm{D}$$

the law of mass action in terms of concentrations of the gases is:

$$\frac{[\mathrm{C}]^c[\mathrm{D}]^d}{[\mathrm{A}]^a[\mathrm{B}]^b} = K\left(\frac{RT}{P_\mathrm{ref}}\right)^{a+b-c-d} = K\left(\frac{RT}{P_\mathrm{ref}}\right)^{-\Delta n_\mathrm{g}}$$

The expression on the left is just the customary form of the equilibrium expression but using concentrations instead of partial pressures. The quantity $(c + d - a - b)$ is also symbolized Δn_g because it is the change in the chemical amount of gas between the two sides of the reaction.

External Effects and LeChatelier's Principle

LeChatelier's principle states that if a stress is applied to a system at equilibrium, then the position of the equilibrium will shift in the direction that counteracts the stress. For example:

- A product is taken out of a system at equilibrium. The equilibrium shifts to the right and partially remedies the loss.
- A reactant is removed from an equilibrium system. The equilibrium shifts to the left to minimize the loss, producing more reactants.
- A product is added to an equilibrium system. The equilibrium shifts to the left, consuming some of that product.
- A reactant is added to an equilibrium system. The equilibrium re-establishes itself after shifting to the right.
- The volume of an equilibrium chemical system is changed. A reduction in volume causes an increase in the pressure and the system responds by shifting toward the side of the reaction having the fewer moles of gas. Fewer moles of gas require smaller volume. Increasing the volume has the reverse effect.

LeChatelier's principle is a handy qualitative guide in grappling with equilibrium problems. Its use avoids a troublesome difficulty. In problems, it is often possible inadvertently to set up the algebraic sense of the changes required to come to equilibrium so that they work out to be negative. If one expects that an equilibrium will shift to the right, getting a negative sign for changes in partial pressure simply means that the equilibrium in fact shifts to the left. In practice, however, negative changes are not expected, especially if they come after some mathematical effort. LeChatelier's principle allows one to form a correct qualitative idea of the direction of prospective shifts. See **problem 5-25.**

Heterogeneous Equilibria

Chemical equilibria may involve solids, liquids and solutions in all combinations in addition to gases. The mass-action expression in a general case is constructed in the same way as in a case involving only gases. As before, gases enter the expression as partial pressures, in atmospheres. Dissolved species enter as concentrations, in moles per liter. In addition:

- Pure solids and pure liquids are omitted from mass action expressions. The activity of a pure solid or liquid is equal to 1. Even if it were raised to some large power in a mass action expression, an activity of 1 would make no numerical difference. See **problems 5-41** and **5-43.** Although the activities of pure solids and liquids does not appear explicitly in mass action expressions, the presence of the solids or liquids is still essential to the existence of the equilibrium. See **problem 5-49.**

- In dilute solutions the activity of the solvent is quite close to 1. For this reason the solvent can be omitted from mass action expressions. See **problem 5-43c.**

Extraction and Separation Processes

The separation of mixtures is important in research and industry. A type of equilibrium much used in separations is the partition of a substance between two immiscible (mutually insoluble) liquids. Carbon tetrachloride and water quickly separate into two layers after being shaken together in a flask. A third substance, like I_2, partitions itself between the layers according to the ratio of its solubilities in the two solvents. A heterogeneous equilibrium is established:

$$I_2(aq) \rightleftharpoons I_2(CCl_4) \quad \text{for which } K = \frac{[I_2]_{(CCl_4)}}{[I_2]_{(aq)}}$$

The above K is a **partition coefficient.** An impurity in the iodine will partition itself between the two liquids, too, but with some different K. Suppose the impurity has a smaller K than the I_2. Then it tends to concentrate in the water relative to the I_2. The I_2 tends to concentrate in the CCl_4 relative to the impurity. Even small differences in K can lead to effective separations if the equilibrium is repeatedly established in a cycled operation. See **problems 5-55, 5-57,** and **5-77. Extraction** takes advantage of the partitioning of a solute between two immiscible solvents to remove the solute from one solvent into the other. In the above example, I_2 is extracted from the water into the CCl_4.

Various methods of **chromatography** rely on the continuous extraction of a solute from one phase to another. In all forms of chromatography, a **mobile phase** is passed over a **stationary phase.** Solutes are partitioned between the phases with different partition coefficients, and separations are thereby achieved.

Detailed Solutions to Odd-Numbered Problems

5-1 The equilibrium expression for reactions like these consists of the partial pressures of the products raised to powers equal to their respective coefficients in the balanced equation divided by the partial pressures of the reactants raised to powers equal to their respective coefficients. See page 105 of this Guide. Apply this to the three specific examples in the problem:

$$\text{a)}\ \frac{P^2_{H_2O}}{P^2_{H_2}P_{O_2}} = K \qquad \text{b)}\ \frac{P_{XeF_6}}{P_{Xe}P^3_{F_2}} = K \qquad \text{c)}\ \frac{P^{12}_{CO_2}P^6_{H_2O}}{P^2_{C_6H_6}P^{15}_{O_2}} = K$$

5-3 One balanced equation is $P_4(g) + 2\,O_2(g) + 6\,Cl_2(g) \rightleftharpoons 4\,POCl_3(g)$ for which the equilibrium expression is:

$$\frac{P^4_{POCl_3}}{P_{P_4}P^2_{O_2}P^6_{Cl_2}} = K$$

Other answers are possible. If the equation is written with doubled coefficients, then the equilibrium expression has doubled exponents, and the value of K is squared.

5-5 a) Apply the law of mass action to the given reaction:

$$\frac{P_{CO_2}P_{H_2}}{P_{CO}P_{H_2O}} = K$$

b) The problem gives K and three of the four equilibrium partial pressures. Substitution in the equilibrium expression and solution of the resulting equation to find the equilibrium partial pressure of $H_2(g)$ is straightforward. First:

$$\frac{(0.70)(P_{H_2})}{(0.10)(0.10)} = 3.9$$

Note that each partial pressure was divided by the reference pressure of one atmosphere before insertion in the equilibrium expression. There are no units in equilibrium expressions once a reference pressure (or concentration) has been defined. Solving for the unknown in the preceding gives 0.056; the equilibrium pressure of gaseous hydrogen is 0.056 atm.

5-7 First, write the equilibrium expression. Then calculate its value based on the partial pressures given. For the reaction $3\,Al_2Cl_6(g) \rightleftharpoons 2\,Al_3Cl_9(g)$, the outcome is:

$$\frac{P^2_{Al_3Cl_9}}{P^3_{Al_2Cl_6}} = K = \frac{(1.02 \times 10^{-2})^2}{(1.00)^3} = 1.04 \times 10^{-4}$$

5-9 In a collection of, say, 10 000 molecules, 642 would be in the chair form and 9358 would be in the boat form at equilibrium. The partial pressure of gases are proportional to the number of molecules present, assuming ideality. That is, $P_{\text{gas}} = kN$ where k is a constant of proportionality. Therefore:

$$K = \frac{P_{\text{boat}}}{P_{\text{chair}}} = \frac{kN_{\text{boat}}}{kN_{\text{chair}}} = \frac{9358}{642} = 14.6$$

5-11 a) The chemical amount (in moles) of SO_2Cl_2 put into the flask can be calculated from the given mass:

$$3.174 \text{ g } SO_2Cl_2 \times \left(\frac{1 \text{ mol } SO_2Cl_2}{135.0 \text{ g } SO_2Cl_2}\right) = 0.02351 \text{ mol } SO_2Cl_2$$

Imagine that the SO_2Cl_2 vaporizes in one step and then reacts in a second distinct step. The partial pressure of the $SO_2Cl_2(g)$ after it fills the flask at 100°C but before it has a chance to react can be computed using the ideal-gas law:

$$P_{SO_2Cl_2} = n_{SO_2Cl_2}\left(\frac{RT}{V}\right) = 0.02351\left(\frac{(0.08206 \text{ L atm mol}^{-1}\text{K}^{-1})(373.15 \text{ K})}{1.000 \text{ L}}\right)$$

The use of an absolute temperature is of course obligatory. The partial pressure of the SO_2Cl_2 comes out to 0.7199 atm. As the reaction starts toward equilibrium, the partial pressures of the three compounds in the mixture change. The SO_2Cl_2 decomposes to generate SO_2 and Cl_2 in equal chemical amounts:

	$SO_2Cl_2(g) \rightleftharpoons$	$SO_2(g)\ +$	$Cl_2(g)$
Init. pressure (atm)	0.7199	0	0
Change in pressure (atm)	$-x$	$+x$	$+x$
Equil. pressure (atm)	$0.7199 - x$	x	x

The total pressure in the flask at equilibrium is the sum of the three equilibrium partial pressures:

$$P_{\text{tot}} = 1.30 \text{ atm} = P_{SO_2Cl_2} + P_{Cl_2} + P_{SO_2} = (0.7199 - x) + x + x$$

Solving gives x equal to 0.5801. The equilibrium partial pressures of the two products are accordingly both 0.58 atm, and the equilibrium partial pressure of the reactant is 0.14 atm.

b) The equilibrium constant is computed by substituting in the appropriate equilibrium expression:

$$K = \frac{P_{SO_2}P_{Cl_2}}{P_{SO_2Cl_2}} = \frac{(0.58)(0.58)}{(0.14)} = 2.4$$

5-13 The second reaction (reaction 2) is the same as reaction 1 but with all of the coefficients divided by three. The equilibrium expression for reaction 2 therefore is just like that for reaction 1 except with all of the exponents divided by 3. Dividing the exponents by three corresponds to taking the cube root: $K_2 = \sqrt[3]{K_1}$.

5-15 When two chemical equations are added to obtain a third equation, the equilibrium constant associated with third equation is the product of the equilibrium constants associated with the first two. Also if a equation is written in reverse, the new equilibrium constant is the reciprocal of that of the forward reaction. Writing the first equation in this problem in reverse and adding the second equation gives the equation of interest. Thus, K for the reaction of interest equals $K_2 \times (1/K_1)$ or K_2/K_1.

5-17 a) First, calculate the chemical amount n of the $C_6H_5CH_2OH$. Then use the ideal-gas law to calculate its initial partial pressure:

$$n = 1.20 \text{ g } C_6H_5CH_2OH \times \left(\frac{1 \text{ mol } C_6H_5CH_2OH}{108 \text{ g } C_6H_5CH_2OH}\right) = 1.11 \times 10^{-2} \text{ mol}$$

$$P = \frac{nRT}{V} = \frac{(1.11 \times 10^{-2} \text{ mol})(0.08206 \text{ L atm mol}^{-1}\text{K}^{-1})(523 \text{ K})}{2.00 \text{ L}} = 0.238 \text{ atm}$$

The following table shows how the partial pressures of the reactant and its products then change as equilibrium is approached:

	$C_6H_5CH_2OH(g) \rightleftharpoons$	$C_6H_5CHO(g)\ +$	$H_2(g)$
Init. pressure (atm)	0.238	0	0
Change in pressure (atm)	$-x$	$+x$	$+x$
Equil. pressure (atm)	$0.238 - x$	x	x

Substituting in the equilibrium expression gives:

$$\frac{P_{C_6H_5CHO}P_{H_2}}{P_{C_6H_5CH_2OH}} = 0.558 = \frac{(x)(x)}{(0.238 - x)} \text{from which} x^2 + 0.558x - 0.133 = 0$$

Use the quadratic formula (text page A-16) to solve for x:

$$x = \frac{-(0.558) \pm \sqrt{(0.558)^2 - 4(1)(0.133)}}{2(1)} = \frac{-0.558 \pm 0.918}{2}$$

$$x = 0.180 \quad \text{and} \quad x = -0.738$$

Disregard the solution $x = -0.738$ because it implies a negative partial pressure for C_6H_5CHO and H_2. The answer is $P_{C_6H_5CHO} = 0.180$ atm.

b) The fraction of the benzyl alcohol dissociated at equilibrium is the amount dissociated divided by the initial amount present. These amounts are proportional to the change in partial pressure of the benzyl alcohol and the initial partial pressure respectively. Hence:

$$f = \frac{\text{amt. } C_6H_5CH_2OH \text{ dissoc.}}{\text{orig. amt. } C_6H_5CH_2OH} = \frac{0.180 \text{ atm}}{0.238 \text{ atm}} = 0.756$$

5-19 a) At equilibrium, the contents of the bulb are a mixture of three gases: the two products and whatever reactant is left. The total pressure of this mixture is given as 0.895 atm. If this mixture follows Dalton's law, then its total pressure is equal to the sum of the equilibrium partial pressures of the components:

$$P_{\text{tot}} = P_{PCl_3} + P_{Cl_2} + P_{PCl_5} = 0.895 \text{ atm}$$

The balanced chemical equation given in the problem shows that the partial pressure of PCl_3 and the partial pressure of Cl_2 must be equal. Represent each of these two pressures by x. Then the partial pressure of PCl_5 equals $(0.895 - 2x)$ atm. The contents of the bulb are at equilibrium, so the equilibrium expression, which involves the three partial pressures, must equal K:

$$K = 2.15 = \frac{P_{Cl_2} P_{PCl_3}}{P_{PCl_5}} = \frac{(x)(x)}{0.895 - 2x}$$

Solving (using the quadratic formula) gives $x = 0.40866$ and a physically meaningless root ($x = -4.7087$) that is discarded. The equilibrium partial pressures of the Cl_2 and the PCl_3 are both 0.409 atm, and the partial pressure of the PCl_5 is 0.078 atm.
b) The temperature of the bulb is 523.15 K. The concentrations of the component gases in the mixture is calculated using this fact, the partial pressures from part a), and the ideal-gas law:

$$[Cl_2] = \frac{n_{Cl_2}}{V} = \frac{P_{Cl_2}}{RT} = \frac{0.409 \text{ atm}}{0.08206 \text{ L atm mol}^{-1}\text{K}^{-1}(523.15 \text{ K})} = 0.00952 \text{ mol L}^{-1}$$

The concentrations of Cl_2 and PCl_3 are both 0.00952 mol L^{-1}. The concentration of PCl_5 is 0.0018 mol L^{-1}.
c) At equilibrium, the bulb holds 0.0018 mol L^{-1} of PCl_5 and 0.00952 mol L^{-1} of PCl_3. For every mole per liter of PCl_3 present in the bulb at equilibrium there was one mole per liter of PCl_5 in the original charge. Also, for every mole per liter of PCl_5 left unchanged at equilibrium there was one mole of liter of PCl_5 in the original charge. The combined concentration of the two phosphorus-containing substances therefore is the original number of moles per liter of PCl_5 put into the bulb. The

combined concentration is 0.01132 mol L^{-1}. The bulb has a volume of 0.100 L, so the original amount of PCl_5 was 1.132×10^{-3} mol. Multiplying by 208.2 g mol^{-1} (the molar mass of PCl_5) gives 0.236 g of PCl_5 as the original charge.

5-21 Write the chemical equilibrium and the given data:

	$Br_2(g)$	$+I_2(g) \rightleftharpoons$	$2\,IBr(g)$
Init. pressure (atm)	0.0500	0.0400	0
Change in pressure (atm)	$-x$	$-x$	$+2x$
Equil. pressure (atm)	$0.0500 - x$	$0.0400 - x$	$2x$

$$\frac{P_{IBr}^2}{P_{Br_2} P_{I_2}} = 322 = \frac{(2x)^2}{(0.0500 - x)(0.0400 - x)}$$

Rearrangement leads to the equation: $x^2 - 0.09113x + 2.025 \times 10^{-3} = 0$. Using the quadratic formula to solve this equation gives $x = 0.0384$ and $x = 0.0527$. The second root is "unphysical" because there was only 0.0400 atm of I_2 at the start. Going down by 0.0527 atm is impossible. The correct partial pressures come from the first root. The partial pressure of IBr is 0.0768 atm, that of I_2 is 0.0016 atm, and that of Br_2 is 0.0116 atm.

5-23 In the reaction $2\,A_2B(g) \rightleftharpoons 2\,A_2(g) + B_2(g)$, the pressure in the container must increase as the reaction goes to the right and the fraction of A_2B that is dissociated becomes larger. Imagine arbitrarily that the original pressure of A_2B in the vessel is 10 atm. The original partial pressures of both products are zero because the only way they form is by means of the reaction. Now imagine that various fractions of the A_2B molecules undergo dissociation. Tabulate the partial pressures of all components:

Frac. Dissoc.	P_{A_2B}	P_{A_2}	P_{B_2}	P_{total}
0.0	10.0 atm	0.0 atm	0.0 atm	10.0 atm
0.10	9.0	1.0	0.5	10.5
0.30	7.0	3.0	1.5	11.5
0.50	5.0	5.0	2.5	12.5
0.60	4.0	6.0	3.0	13.0
1.00	0.0	10.0	5.0	15.0

The table is constructed on the principle that the pressure of a component of a gas mixture is proportional to its chemical amount. (This follows from the ideal-gas law and Dalton's law.) As the table shows, the reaction of A_2B causes its own partial pressure to drop but forms enough products to make the total pressure rise. If *all* of the A_2B reacts then the total pressure is $1\frac{1}{2}$ times its original value. In mathematical form, with α standing for the fraction dissociated and P_0 for the original pressure of A_2B:

$$P_{tot} = P_{A_2B} + P_{A_2} + P_{B_2}$$

$$P_{\text{tot}} = P_0(1-\alpha) + P_0\alpha + P_0\left(\frac{\alpha}{2}\right) = P_0\left(1+\frac{\alpha}{2}\right)$$

Having derived this formula, apply it to the situation described in the problem. In the problem, the final pressure is 30.0 percent more than the original pressure:

$$\frac{P_{\text{tot}}}{P_0} = 1.30 = \left(1+\frac{\alpha}{2}\right)$$

Solving gives α equal to 0.600. P_0 is the pressure of the A_2B before any reaction. The original chemical amount of A_2B is 75.0 g/150 g mol^{-1} = 0.500 mol. Substitution in the ideal-gas equation with V = 22.4 L and T = 323.15 K gives P_0 = 0.5919 atm. The volume of the vessel is 22.4 L, the famous STP molar volume, but this is just an accident and should not make one think that there is 1.00 mol of gas present or that P_0 is 1.00 atm.
To calculate the equilibrium constant, substitute the equilibrium partial pressures in the equilibrium expression:

$$K = \frac{P_{A_2}^2 P_{B_2}}{P_{A_2B}^2} = \frac{P_0^2\alpha^2 P_0(\alpha/2)}{P_0^2(1-\alpha)^2} = P_0\left(\frac{\alpha^3/2}{(1-\alpha)^2}\right)$$

Both P_0 and α are known so K can be computed readily. It is 0.400. The total pressure, P_{tot} in the vessel is 30 % higher than 0.5919 atm (P_0), which is 0.5919×1.30 or 0.7695 atm. This becomes 0.769 atm when rounded to 3 significant digits.

5-25 Write the equation, the initial partial pressures, the change in the pressures required to reach equilibrium, and the final partial pressures:

	$N_2(g)+$	$O_2(g) \rightleftharpoons$	$2\,NO(g)$
Init. pressure (atm)	0.41	0.59	0.21
Change in pressure (atm)	$+x$	$+x$	$-2x$
Equil. pressure (atm)	$0.41+x$	$0.59+x$	$0.21-2x$

Note the use of x. A positive x in this set-up corresponds to the loss of product and the formation of reactants as the reaction comes to equilibrium. The reaction quotient Q in the original mixture far exceeds K, so the drive toward equilibrium will occur right-to-left in the equation as written. The equilibrium constant expression relates the equilibrium partial pressures:

$$K = 4.2\times10^{-31} = \frac{P_{NO}^2}{P_{N_2}P_{O_2}} = \frac{(0.21-2x)^2}{(0.41+x)(0.59+x)}$$

Before attempting to solve this equation for x, consider the sizes of the numbers involved. This saves getting bogged down with algebra. The equilibrium constant is very small. Hence, the numerator of the equilibrium expression must be very small at equilibrium—almost all of the $NO(g)$ is consumed at equilibrium. Suppose that *all* of the $NO(g)$ reacts. Then, $2x = 0.21$ and $x = 0.105$. Using this value of x gives an equilibrium pressure of N_2 of $0.41 + 0.105 = 0.52$ atm. The equilibrium pressure of O_2 is, by a similar computation, 0.70 atm. To get the true (non-zero) equilibrium partial pressure of NO, substitute these two pressures back into the original expression:

$$4.2 \times 10^{-31} = \frac{P_{NO}^2}{P_{N_2} P_{O_2}} = \frac{(P_{NO})^2}{(0.52)(0.70)}$$

Solving for P_{NO} gives 3.9×10^{-16} atm. The reaction lies indeed quite far toward the reactants at equilibrium!

5-27 The equilibrium-constant expression for the synthesis of ammonia according to the equation in the problem is:

$$\frac{P_{NH_3}^2}{P_{H_2}^3 P_{N_2}} = K = 6.78 \times 10^5$$

This equation can be written effortlessly. The difficulty is in translating the other statements in the problem into mathematical terms. If the ratio of H to N atoms in the system is 3 to 1 then:

$$P_{H_2} = 3P_{N_2}$$

because the third component, NH_3, maintains, within itself, the required 3 to 1 ratio of atoms. The fact that the total pressure is 1.00 atm means:

$$P_{N_2} + P_{H_2} + P_{NH_3} = 1.00 \text{ atm}$$

Let x the equilibrium partial pressure of N_2 (P_{N_2}) and substitute in the mass action expression:

$$6.78 \times 10^5 = \frac{(1.00 - 4x)^2}{(3x)^3 x} = \frac{(1.00 - 4x)^2}{27x^4}$$

This equation is not too hard to solve analytically (see below), but is it necessary to even try? The x is expected to be small because at equilibrium the mixture is mostly NH_3 (K is big). Using this idea simplifies the algebra. Suppose $4x << 1.00$. Then:

$$6.78 \times 10^5 \approx \frac{1.00}{27x^4} \quad \text{from which} \quad x \approx 0.0153$$

Rapid improvement on this answer is gained by doing successive approximations. The idea is to guess values of x and compute Q, the right-hand side of the equation, for each. Revise each succeeding guess based on the results of the previous guess until the computed value becomes sufficiently close to 6.78×10^5. The following table is a map of such a process. It starts with the x from the rough solution:

x	$(1.00 - 4x)^2$	$27x^4$	Q
0.0153	0.8813	1.480×10^{-6}	5.96×10^5
0.0145	0.8874	1.194×10^{-6}	7.43×10^5
0.0149	0.8843	1.331×10^{-6}	6.64×10^5
0.0147	0.8859	1.261×10^{-6}	7.03×10^5
0.0148	0.8851	1.295×10^{-6}	6.83×10^5

The last row in the table gives a Q that is closer to K than in either preceding row. Therefore x equals 0.0148 to three significant digits. Then:

$$P_{N_2} = 0.0148 \text{ atm} \quad P_{H_2} = 0.0444 \text{ atm} \quad P_{NH_3} = 0.941 \text{ atm}$$

Analytical solution of the equation:

$$6.78 \times 10^5 = \frac{(1.00 - 4x)^2}{27x^4}$$

proceeds by multiplying through by 27 and then taking the square root of both sides to obtain:

$$4.278 \times 10^3 = \frac{1.00 - 4x}{x^2} \quad \text{which gives} \quad (4.278 \times 10^3)x^2 + 4x - 1.00 = 0$$

Substitution in the quadratic formula gives:

$$x = \frac{-b \pm \sqrt{b^2 - 4ac}}{2a} = \frac{-4 \pm \sqrt{16 + 17122}}{8556} = +0.0148 \text{ and } -0.0158$$

The positive root is the same answer obtained by the other procedure; the negative root is meaningless.

5-29 The reaction and corresponding equilibrium expression are:

$$CO(g) + Cl_2(g) \rightleftharpoons COCl_2(g) \qquad \frac{P_{COCl_2}}{P_{CO}P_{Cl_2}} = K = 0.20$$

where the 0.20 comes from problem 5-6. Write the equilibrium expression in terms of concentrations. This affects K according to the general pattern shown on text page 178:

$$\frac{[COCl_2]}{[CO][Cl_2]} = 0.20\left(\frac{RT}{P_{ref}}\right)^{-\Delta n}$$

Here, Δn is the change in the number of moles of gas from left to right in the reaction (it equals -1 in this case because there is 1 mol of gas on the right and 2 mol on the left of the equation); P_{ref} is the reference pressure used to compute the K quoted in problem 5-6 (it equals 1.00 atm in this case). Insert these values along with the gas constant R and the absolute temperature T:

$$\frac{[COCl_2]}{[CO][Cl_2]} = 0.20\left(\frac{(0.08206 \text{ L atm mol}^{-1}\text{K}^{-1})(873.15 \text{ K})}{1 \text{ atm}}\right)^{-(-1)} = 14.3$$

Finally, substitute the equilibrium concentrations of the CO and Cl_2:

$$\frac{[COCl_2]}{[2.3 \times 10^{-4}][1.7 \times 10^{-2}]} = 14.3$$

and solve for the equilibrium concentration of $COCl_2$. The answer is 5.6×10^{-5} mol L^{-1}.

5-31 a) The reaction quotient Q has the same form as the equilibrium-constant expression. It is given by the partial pressures of the products raised to their coefficients in the balanced equation divided by the partial pressures of the reactants raised to their respective coefficients. A true K is computed only by substitution of true *equilibrium* partial pressures into the form of the mass-action law. Computations of Q on the other hand may employ whatever partial pressures might temporarily happen to prevail, even pressures measured at points quite far from equilibrium. In the following such a point is signified by the subscript zero:

$$Q_0 = \frac{(P_{Al_3Cl_9})_0^2}{(P_{Al_2Cl_6})_0^3} = \frac{(1.02 \times 10^{-2})^2}{(0.473)^3} = 9.83 \times 10^{-4}$$

b) From problem 5-7, $K = 1.04 \times 10^{-4}$. The initial reaction quotient Q_0 exceeds K so the reaction approaches equilibrium by shifting to the left. The process consumes Al_3Cl_9 and produces Al_2Cl_6.

5-33 The fading of the color in the reaction flask as equilibrium is approached means that the reaction must be consuming $Br_2(g)$ as it goes toward equilibrium. The change is left to right in the equation as written. The initial reaction quotient Q_0 is accordingly less than K. The data given in the problem allow computation of Q_0:

$$Q_0 = \frac{(P_{HBr})_0^2}{(P_{H_2})_0(P_{Br_2})_0} = \frac{(0.90 \text{ atm})^2}{(0.40 \text{ atm})(0.40 \text{ atm})} = 5.1$$

Thus K must exceed 5.1.

5-35 a) Let P_{di} stand for the partial pressure of gaseous diphosphorus $P_2(g)$ and P_{tet} for the partial pressure of gaseous tetraphosphorus $P_4(g)$. For the process:

$$\frac{P_{di}^2}{P_{tet}} = Q$$

Initially $P_{\text{di}} = 2.00$ atm and $P_{\text{tet}} = 5.00$ atm making $Q_0 = 0.800$. Because $Q_0 > K$ the equilibrium will tend to shift to the left (reducing the numerator and increasing the denominator in the above expression) until Q equals K.

b) Let x equal the increase in the pressure of $P_4(g)$ during the change. Then:

$$K = \frac{(2.00 - 2x)^2}{(5.00 + x)} = 0.612$$

The equation can be solved with the quadratic formula. It is instructive however to get x numerically (See Appendix C of text). Construct a table:

x	$(2.00 - 2x)^2$	$5.00 + x$	Q
0.00	4.00	5.00	0.800
0.100	3.24	5.10	0.635
0.120	3.10	5.12	0.605
0.115	3.13	5.115	0.612

As x increases Q decreases from 0.800. As the table shows, progress to a good answer is quick. Electronic calculators make the progress painless as well. At x equal to 0.120, Q only slightly exceeds the true value of K. If $x = 0.115$, then $P_{\text{di}} = 1.77$ atm and $P_{\text{tet}} = 5.12$ atm.

c) If the volume of the system is increased, then, by LeChatelier's principle, there will be net *dissociation* of P_4. The system responds to its forced rarefaction by producing more molecules to fill the larger volume.

5-37 Chemical systems always tend toward equilibrium. If a stress is applied to a system that is at equilibrium, the system will react to that stress in such a way as to minimize the stress. This allows the system to get back to equilibrium. Examples of a stress are: removal or addition of a reactant or product, changing temperature, changing pressure.

a) The applied stress is the addition of $N_2O(g)$. The system will react in such a way as to decrease the concentration of N_2O. It does this by proceeding from right to left until a new equilibrium is reached.

b) The applied stress is the reduction in volume. The partial pressures of all the compounds will momentarily rise. The equilibrium will then shift in such a way as to reduce the number of molecules of gas (chemical amount of gas) in the container and reduce the total pressure. There are three moles of gas on the reactant side of the equation and two moles of gas on the product side. The equilibrium will thus shift from left to right.

c) In order to maintain a constant pressure, the volume of the system must have been increased. Thus, the reaction will shift from right to left.

d) The partial pressures of the gases are unchanged by the addition of an inert gas, and the equilibrium law is independent of total pressure. Consequently, there is no effect on the position of the equilibrium.

5-39 Reducing the volume shifts this particular equilibrium to the right. Such a stress always favors the side of the reaction with fewer moles of gas. Hence, there is a net *decrease* in the number of gas molecules from left to right in the reaction.

5-41 Pure solids and pure liquids are omitted from the expressions:

$$\text{a)}\ \frac{P^8_{H_2S}}{P^8_{H_2}} = K \qquad \text{b)}\ \frac{P_{COCl_2}P_{H_2}}{P_{Cl_2}} = K \qquad \text{c)}\ P_{CO_2} = K \quad \text{d)}\ \frac{1}{P^3_{C_2H_2}} = K$$

5-43 Pure solids and liquids are omitted from the expressions; concentrations of solutes and partial pressures of gases are used:

$$\text{a)}\ \frac{[Zn^{2+}]}{[Ag^+]^2} = K \quad \text{b)}\ \frac{[VO_3(OH)^{2-}][OH^-]}{[VO_4^{3-}]} = K \quad \text{c)}\ \frac{[HCO_3^-]^6}{[As(OH)_6^{3-}]^2 P^6_{CO_2}} = K$$

5-45 a) The graph consists of the values in the second column of the following table plotted (on the y-axis) versus the values in the first column (on the x-axis). The values in the third column are those of the second divided by those of the first:

$[N_2O_4]$	$[NO_2]^2$	$[NO_2]^2/[N_2O_4]$
0.190×10^{-3} M	0.784×10^{-5} M^2	4.12×10^{-2} M
0.686	2.70	3.94
1.54	5.27	3.42
2.55	10.8	4.23
3.75	13.7	3.65
7.86	29.9	3.80
11.9	44.1	3.71

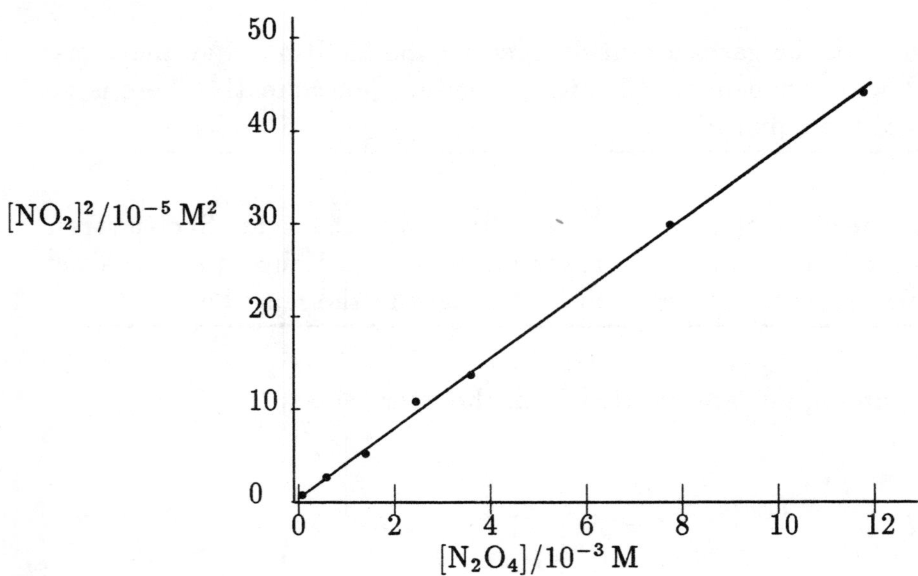

From the equilibrium-constant expression for this reaction it follows that:

$$[NO_2]^2 = K[N_2O_4]$$

The equation has the form of the equation for a straight line $y = mx + b$, with $y = [NO_2]^2$, $m = K$, and $x = [N_2O_4]$ and $b = 0$. Thus, the equilibrium constant is the slope of the line just plotted. See page A-15 of the text for a discussion of the determination of the slope of a straight-line graph.

b) The mean of the values in the last column of the table in the preceding part is the mean experimental K. It is 3.83×10^{-2}.

5-47 Write the equilibrium-constant expressions for the two reactions given in the problem:

$$\frac{1}{P_{H_2}} = K_1 = 4.0 \times 10^{-6} \quad \text{and} \quad \frac{P_{CO}}{P_{CO_2} P_{H_2}} = K_2 = 3.2 \times 10^{-4}$$

If both reactions are simultaneously at equilibrium in the same container then both equations must be fulfilled simultaneous. Compute the partial pressure of the hydrogen from the first equation:

$$P_{H_2} = \sqrt[3]{\frac{1}{K_1}} = \sqrt[3]{\frac{1}{4.0 \times 10^{-6}}} = 63 \text{ atm}$$

Insert this partial pressure into the (slightly rearranged) second equation:

$$\frac{P_{CO}}{P_{CO_2}} = K_2 P_{H_2} = (3.2 \times 10^{-4})63 = 0.020$$

5-49 a) The equilibrium expression for the reaction is:

$$\frac{1}{P_{NH_3}P_{HCl}} = K = 4.0 \quad \text{at } 340°\text{C}$$

If P_{NH_3} is 0.80 atm at equilibrium, then, by substitution, the equilibrium partial pressure of HCl(g) is 0.31 atm.

b) Obviously, the equilibrium cannot occur before addition of $NH_4Cl(s)$ to the container filled with ammonia—the ammonium chloride is the only source of HCl(g). For every mole of HCl(g) that is produced one mole of $NH_3(g)$ is added to the quantity responsible for the original 1.50 atm. The partial pressures of the $NH_3(g)$ and of HCl(g) are directly proportional to their respective chemical amounts (assuming ideal-gas behavior). This means that for every atmosphere of HCl(g) produced one additional atmosphere of $NH_3(g)$ is also produced. Let x equal the equilibrium partial pressure of HCl. Then the equilibrium partial pressure of NH_3 is $1.50 + x$ and:

$$P_{NH_3}P_{HCl} = (1.50 + x)x = 0.25 \quad \text{which gives} \quad x^2 + 1.50x - 0.25 = 0$$

Solving the quadratic equation gives x equal to 0.151 (the negative root is rejected). Therefore $P_{HCl} = 0.15$ atm, and $P_{NH_3} = 1.50 + 0.15 = 1.65$ atm.

5-51 The "water gas" reaction: $C(s) + H_2O(g) \rightleftharpoons CO(g) + H_2(g)$ has a reaction quotient of the form:

$$\frac{P_{CO}P_{H_2}}{PH_2O} = Q$$

If Q at any moment is less than K, the reaction tends to proceed ("shifts") from left to right; if Q at any moment greater than K, the reaction tends to proceed from right to left.

a) The value of Q is computed by substitution in the above expression:

$$Q = \frac{P_{CO}P_{H_2}}{P_{H_2O}} = \frac{(1.525)(0.805)}{0.600} = 2.05$$

This Q is less than K (which is 2.6) so the reaction shifts from left to right.

b) All three partial pressures are higher than in part a, and:

$$Q = \frac{P_{CO}P_{H_2}}{P_{H_2O}} = \frac{(1.714)(1.383)}{0.724} = 3.27$$

The Q is now more than K so the reaction shifts from right to left to reach equilibrium.

5-53 a) The container holds no gas until some of the $NH_4HSe(s)$ decomposes. Breakdown of the solid is the only source of the two gases than eventually fill the container at equilibrium. The stoichiometry of the decomposition requires that the partial pressure of the $H_2Se(g)$ always equal the partial pressure of the $NH_3(g)$ (assuming ideality). The total pressure is the sum of the partial pressures of the two gases: Therefore:

$$P_{H_2Se} = P_{NH_3} \text{ and } P_{H_2Se} + P_{NH_3} = 0.0184 \text{ atm}$$

Clearly both partial pressures equal 0.00920 atm. The equilibrium constant accordingly is computed by:

$$K = P_{H_2Se}P_{NH_3} = (0.00920)^2 = 8.46 \times 10^{-5}$$

b) As long as the temperature is constant, the size of the container has no effect on the K. Both partial pressures are different, but their product is still K:

$$8.46 \times 10^{-5} = P_{H_2Se}P_{NH_3} = P_{H_2Se}(0.0252) \text{ hence } P_{H_2Se} = 0.00336 \text{ atm}$$

5-55 Assume that there is exactly 1 L each of H_2O and CCl_4. Then at equilibrium there is 1.30×10^{-4} mol of I_2 in the aqueous phase. The remaining I_2 must be in the CCl_4 layer:

$$n_{I_2\,(CCl_4)} = 1.00 \times 10^{-2} - 1.30 \times 10^{-4} = 9.9 \times 10^{-3} \text{ mol}$$

The concentration of I_2 in the CCl_4 layer is its amount divided by the volume:

$$[I_2]_{(CCl_4)} = \frac{9.9 \times 10^{-3} \text{ mol}}{1 \text{ L}} = 9.9 \times 10^{-3} \text{ mol L}^{-1}$$

The equilibrium constant equals the ratio of the two concentrations:

$$K = \frac{[I_2]_{(CCl_4)}}{[I_2]_{(aq)}} = \frac{9.9 \times 10^{-3}}{1.30 \times 10^{-4}} = 76$$

5-57 a) The equilibrium-constant expressions for the dissolution of benzoic acid in water (K_1) and the dissolution of benzoic acid in ether (K_2) are both quite simple:

$$K_1 = [C_6H_5COOH]_{(aq)} \text{ and } K_2 = [C_6H_5COOH]_{(ether)}$$

The equilibrium concentration of C_6H_5COOH in the water is:

$$\frac{2.00 \text{ g } C_6H_5COOH}{1 \text{ L}} \times \left(\frac{1 \text{ mol } C_6H_5COOH}{122 \text{ g } C_6H_5COOH}\right) = 0.0164 \text{ M}$$

In the ether the equilibrium concentration is much higher:

$$\frac{660 \text{ g } C_6H_5COOH}{1 \text{ L}} \times \left(\frac{1 \text{ mol } C_6H_5COOH}{122 \text{ g } C_6H_5COOH}\right) = 5.4 \text{ M}$$

Hence K_1 is 0.164, and K_2 is 5.4.
b) The desired partition reaction equals reaction 1 in the previous part subtracted from reaction 2. Subtracting a reaction is the same as reversing it and adding it. The equilibrium constant for the partition reaction (reaction 3) is accordingly:

$$K_3 = K_2\left(\frac{1}{K_1}\right) = \frac{5.4}{0.0164} = 330$$

5-59 a) The partial pressures of the gases are in direct proportion to their chemical amounts (assuming Dalton's law and the ideal-gas law prevail):

$$P_{\text{gas}} = n_{\text{gas}}\left(\frac{RT}{V}\right)$$

Suppose that the volume of the container of gas is such that the actual chemical amounts of the four gases are 90 mol, 470 mol, 200 mol, and 45 mol. Then, the partial pressure of the $BCl_3(g)$ is $P_{BCl_3} = 90(RT/V)$; expressions for the partial pressures of the other gases are similar in form. The equilibrium law for the first reaction is:

$$\frac{P^3_{BFCl_2}}{P^2_{BCl_3}P_{BF_3}} = K$$

Substituting the three partial pressures gives

$$\frac{(45RT/V)^3}{(90RT/V)^2(470RT/V)} = K = \frac{(45)^3}{(90)^2(470)} = 0.024$$

The RT/V terms all cancelled out! The cancellation means the volume of the container has no effect on the position of this particular equilibrium. We were therefore fully justified in making any assumption we wanted about the volume. A similar procedure with the equilibrium expression for the second reaction gives $K = 0.40$.
b) The equation given in this part of the problem is just the sum of the two equations in part a) divided by three. Compute the required equilibrium constant either by going back to the original partial pressures and substituting in the equilibrium expression or by taking the cube root of the product of two constants computed in part a). By both methods, the desired K comes out to 0.21. There is no new information in this result because it derives from the previous results.

5-61 a) If we assume that 1 mol of the *cis* form is initially present, then the equilibrium chemical amounts of each species are:

	cis	*trans*
Initial amount	1 mol	0 mol
Change	$-x$	$+x$
Equilibrium amount	$1-x$	x

At equilibrium, 73.6% of the *cis* has been converted to the *trans* form. Therefore x is 0.736 mol. The equilibrium chemical amounts are:

$$n_{\text{cis}} = 1 - x = 1 - 0.736 = 0.264 \text{ mol} \quad \text{and} \quad n_{\text{trans}} = x = 0.736 \text{ mol}$$

Assuming ideal-gas behavior, the equilibrium partial pressures of the two forms are

$$P_{\text{cis}} = \frac{n_{\text{cis}}RT}{V} = \frac{(0.264)RT}{V} \text{ and } P_{\text{trans}} = \frac{n_{\text{trans}}RT}{V} = \frac{(0.736)RT}{V}$$

Substitution in the equilibrium-constant expression gives K:

$$\frac{P_{\text{trans}}}{P_{\text{cis}}} = \frac{(0.736)RT/V}{(0.264)RT/V} = 2.788 \approx 2.79$$

b) From the work in part a) it is clear that:

$$P_{\text{trans}} = 2.788(P_{\text{cis}}) \quad \text{and} \quad n_{\text{trans}} = 2.788(n_{\text{cis}})$$

at equilibrium. Also, the combined chemical amount of the two forms of the compound is 0.525 mol because one mole of the *cis* is consumed for every mole of the *trans* created:

$$n_{\text{trans}} + n_{\text{cis}} = 0.525$$

Solving the two equations in two unknowns gives n_{trans} equal to 0.386 mol at equilibrium. The partial pressure of the *trans* form is now computed using the ideal-gas equation:

$$P_{\text{trans}} = n_{\text{trans}}\left(\frac{RT}{V}\right) = 0.386 \text{ mol}\left(\frac{(0.08206 \text{ L atm mol}^{-1}\text{K}^{-1})(698.75 \text{ K})}{15.00 \text{ L}}\right)$$

Completion of the arithmetic gives a partial pressure of 1.48 atm.

5-63 Equilibrium constants are strongly dependent on temperature. The equation in the problem gives the experimentally determined dependence of K for the hydrogenation reaction upon T.

a) Substitute $T = 500$ K into the expression:

$$\log_{10} K = -20.281 + \frac{10560}{T} = -20.281 + \frac{10560}{500} = 0.839$$

Take the antilog of both sides: $K = 10^{0.839} = 6.90$.

b) Assume that the hydrogenation reaction is at equilibrium. Then the expression:

$$\frac{P_{\text{C}_5\text{H}_{11}\text{N}}}{P_{\text{C}_5\text{H}_5\text{N}}P^3_{\text{H}_2}} = K = 6.90$$

must be satisfied. Inserting the partial pressure of the hydrogen (1.00 atm) gives:

$$P_{C_5H_{11}N} = 6.90 P_{C_5H_5N}$$

Suppose that the pressure of the piperidine ($C_5H_{11}N$) is y atm. The pressure of the pyridine (C_5H_5N) is then, by the preceding equation, $6.90y$ atm. The total pressure of two nitrogen-containing gases is $7.90y$ atm. The fraction of this pressure that comes from the pyridine is $1.00y/7.90y = 0.127$. Assuming ideality, the number of moles of nitrogen present in each compound is equal to some constant times the partial pressure of that compound, and the constant is the same for the two compounds. Hence, the fraction of nitrogen in the form of pyridine is 0.127.

5-65 The setup of the problem is exactly as given in problem 5-27 except that now the K is different, and the total pressure is 100 atm. The result is:

$$3.19 \times 10^{-4} = \frac{(100 - 4x)^2}{(3x)^3 x}$$

where x is P_{N_2}, the equilibrium partial pressure of nitrogen. This equation can be simplified as follows:

$$3.19 \times 10^{-4} = \frac{16(25 - x)^2}{27x^4} \quad \text{which gives} \quad 5.383 \times 10^{-4} = \frac{(25 - x)^2}{x^4}$$

Taking the square root of both sides of this equation leads to a quadratic equation in x that can be solved routinely. It is also quite quick to solve by successive approximations. First, observe that x must be less than 25 because if P_{N_2} exceeded 25 atm the pressure of just two (the nitrogen and hydrogen) out of the three gases alone would exceed the total pressure in the container. Now, guess values of x in the range 0 to 25 and compute the right-hand side of the equation for each. Revise the guesses until the computed value becomes sufficiently close to 5.383×10^{-4}. The following table shows the process. The first guess is near the middle of the range:

x	$(25.0 - 4x)^2/x^4$	Comment
15	1.97×10^{-3}	x is too small
20	1.56×10^{-4}	x is too large
18	4.67×10^{-4}	x is too large
17.5	6.00×10^{-4}	x is too large
17.7	5.43×10^{-4}	x is a bit too small
17.72	5.375×10^{-4}	x acceptable

The answer is $x = 17.72$. If $x = 17.72$ then:

$$P_{N_2} = 17.7 \text{ atm} \quad P_{H_2} = 53.2 \text{ atm} \quad P_{NH_3} = 29.1 \text{ atm}$$

Check the answers by noting that the sum of these three partial pressures is 100 atm and that:

$$\frac{P_{NH_3}^2}{P_{H_2}^3 P_{N_2}} = \frac{(29.1)^2}{(53.2)^3(17.7)} = 3.18 \times 10^{-4} = K$$

5-67 a) The reaction quotient for this reaction has the form:

$$Q = \frac{P_{Cl_2} P_{PCl_3}}{P_{PCl_5}}$$

All of the partial pressures are given. By substitution, $Q = 120$.
b) As originally mixed, the system has $Q = 120$ and $K = 11.5$. The system will tend to react from right to left, adjusting partial pressures until $Q = K$. Let y equal the partial pressure of $Cl_2(g)$ that is removed by this reaction. An equal pressure of $PCl_3(g)$ is removed, and an equal pressure of $PCl_5(g)$ is produced. This follows from the stoichiometry of the equation. Therefore:

$$K = 11.5 = \frac{(6.0 - y)(2.0 - y)}{(0.10 + y)} \quad \text{from which} \quad y^2 - 19.5y + 10.85 = 0$$

The roots of the quadratic equation are 0.573 and 18.93. Only the first makes physical sense. The second corresponds to negative partial pressures. Complete the computation by finding the various partial pressures:

$$P_{PCl_3} = 2.0 - 0.573 = 1.4 \text{ atm}$$
$$P_{PCl_2} = 6.0 - 0.573 = 5.4 \text{ atm}$$
$$P_{PCl_5} = 0.10 + 0.573 = 0.67 \text{ atm}$$

c) By LeChatelier's principle, an increase in volume causes the reaction to shift to the side having *more* moles of gas. The amount of $PCl_5(g)$ will decrease.

5-69 The first reaction and its equilibrium constant expression are:

$$Cl_2(g) \rightleftharpoons Cl_2(aq) \qquad \frac{[Cl_2]}{P_{Cl_2}} = K$$

Both the equilibrium concentration of the dissolved chlorine and the partial pressure of the gaseous chlorine are given in the problem. Hence:

$$K = \frac{[Cl_2]}{P_{Cl_2}} = \frac{0.061}{1.00} = 0.061$$

The equilibrium constant for: $Cl_2(aq) + H_2O(l) \rightleftharpoons H^+(aq) + Cl^-(aq) + HOCl(aq)$ is also easily calculated from the information given. The $[H^+]$ must be equal to $[Cl^-]$ and $[HOCl]$ because of the 1-to-1-to-1 stoichiometry of the second reaction. Therefore:

$$K_2 = \frac{[HOCl][H^+][Cl^-]}{[Cl_2]} = \frac{(0.030)(0.030)(0.030)}{(0.061)} = 4.4 \times 10^{-4}$$

5-71 a) The decomposition reaction is $NH_4HS(s) \rightleftharpoons NH_3(g) + H_2S(g)$. One mole of gaseous ammonia forms for every mole of gaseous hydrogen sulfide. Assuming that the product mixture follows Dalton's law, then the equilibrium partial pressure of each gas is half the total pressure at equilibrium:

$$P_{NH_3} = P_{H_2S} = \frac{0.659 \text{ atm}}{2}$$

Put these values in the equilibrium constant expression and compute K:

$$K = P_{NH_3} P_{H_2S} = \left(\frac{0.659}{2}\right)^2 = 0.109$$

b) Bringing the equilibrium partial pressure of $NH_3(g)$ to 0.750 atm will by LeChatelier's principle reduce the partial pressure of $H_2S(g)$. It will not change the equilibrium constant. Hence:

$$P_{H_2S} = \frac{K}{P_{NH_3}} = \frac{0.109}{0.750} = 0.145 \text{ atm}$$

5-73 The equilibrium constant for this reduction reaction relates the partial pressures of the $CO(g)$ and the $CO_2(g)$. The nickel and nickel(II) oxide are pure solids and do not enter the equilibrium expression. Thus:

$$\frac{P_{CO_2}}{P_{CO}} = K = 255.4$$

The total pressure of the system is the sum of these two partial pressures:

$$P_{CO_2} + P_{CO} = 2.50 \text{ atm}$$

Once the system is at equilibrium, the two equations relating the two partial pressures must be satisfied simultaneously. Therefore, at equilibrium:

$$\frac{P_{CO_2}}{(2.50 - P_{CO_2})} = 255.4$$

Solving gives P_{CO_2} equal to 2.49 atm. Substitution back into the equilibrium-constant expression gives P_{CO} equal to 9.75×10^{-3} atm.

5-75 There are equal amounts of PCB-2 and PCB-11 in some volume V of water. Then, a like volume V of octanol is added. The treatment with octanol does nothing to alter the total chemical amount of either PCB but redistributes both. Once the redistribution comes to equilibrium:

$$n_{PCB-2} = [2]_{(aq)} V + [2]_{(oct)} V \quad \text{and} \quad n_{PCB-11} = [11]_{(aq)} V + [11]_{(oct)} V$$

where the bracketed numbers stand for the concentrations of the PCB-2 and PCB-11 and the "(aq)" and "(oct)" refer to water and octanol solutions. Setting the two chemical amounts equal to each other and dividing through by V gives:

$$[2]_{(\mathrm{aq})} + [2]_{(\mathrm{oct})} = [11]_{(\mathrm{aq})} + [11]_{(\mathrm{oct})}$$

The equilibrium-constant expressions for the partition of the two PCB's between the solvents are:

$$K_2 = \frac{[2]_{(\mathrm{oct})}}{[2]_{(\mathrm{aq})}} \quad \text{and} \quad K_{11} = \frac{[11]_{(\mathrm{oct})}}{[11]_{(\mathrm{aq})}}$$

from which it follows that:

$$[2]_{(\mathrm{oct})} = K_2[2]_{(\mathrm{aq})} \quad \text{and} \quad [11]_{(\mathrm{oct})} = K_{11}[11]_{(\mathrm{aq})}$$

Substitution of these expressions into the first equation yields:

$$[2]_{(\mathrm{aq})} + K_2[2]_{(\mathrm{aq})} = [11]_{(\mathrm{aq})} + K_{11}[11]_{(\mathrm{aq})}$$

Solve this equation for the ratio of the concentration of the two PCB's in the water phase:

$$\frac{[2]_{(\mathrm{aq})}}{[11]_{(\mathrm{aq})}} = \frac{1 + K_{11}}{1 + K_2} = \frac{1 + 1.26 \times 10^5}{1 + 3.98 \times 10^4} = 3.17$$

The PCB-2 predominates over the PCB-11 by a factor of 3.17 in the water. This result is interesting, but is *not* the ratio desired in the problem, which is instead the ratio of the amounts of the two PCB's in the octanol. To find this ratio, divide the equation for $[2]_{(\mathrm{oct})}$ in terms of K_2 by the similar expression for $[11]_{(\mathrm{oct})}$ in terms of K_{11}:

$$\frac{[2]_{(\mathrm{oct})}}{[11]_{(\mathrm{oct})}} = \frac{K_2[2]_{(\mathrm{aq})}}{K_{11}[11]_{(\mathrm{aq})}} = \frac{K_2}{K_{11}}\left(\frac{[2]_{(\mathrm{aq})}}{[11]_{(\mathrm{aq})}}\right)$$

The quantity in parentheses is the "interesting result" previously determined. Substitute for it to give:

$$\frac{[2]_{(\mathrm{oct})}}{[11]_{(\mathrm{oct})}} = \frac{K_2}{K_{11}}\left(\frac{1 + K_{11}}{1 + K_2}\right)$$

Insertion of the numerical values of the two equilibrium constants gives:

$$\frac{[2]_{(\mathrm{oct})}}{[11]_{(\mathrm{oct})}} = \frac{3.98 \times 10^4}{1.26 \times 10^5}\left(\frac{1 + 1.26 \times 10^5}{1 + 3.98 \times 10^4}\right) = 1.00$$

5-77 a) The chemical amount of I_2 initially present in aqueous solution is:

$$0.100 \text{ L} \times \left(\frac{2 \times 10^{-3} \text{ mol } I_2}{1 \text{ L}}\right) = 2 \times 10^{-4} \text{ mol } I_2$$

Addition of the 0.025 L of CCl_4 allows the I_2 to distribute itself between the two phases. If y is the amount of I_2 that dissolves into the CCl_4 at equilibrium, then the amount of I_2 left in the aqueous phase is $2 \times 10^{-4} - y$ mol. At equilibrium:

$$[I_2]_{(aq)} = \left(\frac{2 \times 10^{-4} - y}{0.100}\right) \text{ mol L}^{-1} \quad \text{and} \quad [I_2]_{(CCl_4)} = \left(\frac{y}{0.025}\right) \text{ mol L}^{-1}$$

The law of mass action for this system is:

$$\frac{[I_2]_{(CCl_4)}}{[I_2]_{(aq)}} = K = 85 \quad \text{so that} \quad \frac{(2 \times 10^{-4} - y)/0.100}{y/0.025} = 85$$

Rearrangement gives $y = 4.25 \times 10^{-3} - 21.25y$. This is easily solved for y, which equals 1.91×10^{-4} mol. Remember that y is the amount of I_2 that transfers to the CCl_4. By simple subtraction, the amount remaining is 0.09×10^{-4} mol. The fraction remaining is the amount remaining divided by the original amount:

$$f = \frac{0.09 \times 10^{-4}}{2 \times 10^{-4}} = 0.045 = 0.04$$

b) The first extraction with 0.025 L of CCl_4 leaves only the fraction 0.045 of the I_2 in the water. Another extraction with a like amount of CCl_4 will leave only 0.045 of that 0.045. The fraction remaining after these successive treatments is:

$$f = 0.045 \times 0.045 = 0.0020$$

c) From example 5-11 (text page 187), the fraction of I_2 remaining in the water after one 0.050 L extraction is 0.023, which is substantially larger than 0.0020. It is actually about 11 times more efficient to extract the iodine with two half-sized portions of CCl_4 rather than one big portion. In general, it is more efficient to use several small portions of solvent, rather than one or two big ones in performing separations by extraction.

Chapter 6

Acid-Base Equilibria

The concepts of chemical equilibrium (chapter 5) apply to acid-base reactions just as to the other two types of types of chemical reaction identified by the text. This chapter starts a systematic discussion of equilibria in each kind of reaction. In aqueous solution, acid-base reactions generally come to a condition of equilibrium as rapidly as the reacting solutions can be mixed.

Properties of Acids and Bases

A **Brønsted-Lowry acid** is a hydrogen-ion donor; a **Brønsted-Lowry base** is a hydrogen-ion acceptor. The chemical equation:

$$\text{acid} \rightleftharpoons \text{base} + H^+$$

summarizes the definition. The acid and base in this equation make a **conjugate acid-base pair:** the acid is the conjugate acid of the base; the base is the conjugate base of the acid.

A common error with the Brønsted-Lowry definition is to reverse acid and base. To avoid this, become familiar with a few important acids and bases and use them as touchstones in recollection. There are just six common **strong acids** in water:

$$HClO_4,\ HNO_3,\ HCl,\ HBr,\ HI,\ \text{and}\ H_2SO_4$$

Each contains hydrogen ion and donates it with essentially complete efficiency in water solution. All acids not in this group may be assumed to be **weak acids** (see below). The common **strong bases** in water are NaOH, KOH, RbOH, CsOH, $Ca(OH)_2$, $Sr(OH)_2$ and $Ba(OH)_2$. All dissolve completely in water to give $OH^-(aq)$, an excellent hydrogen-ion acceptor. Bases not in this group may be assumed to be **weak bases** (see below). Acidic hydrogen atoms are often segregated to the extreme left or extreme right when chemical formulas are written. Thus, C_6H_5COOH is an acid having the conjugate base $C_6H_5COO^-$.

A standard exercise is to write the formulas of the conjugate acids or conjugate bases of a list of chemical species. See **problems 6-1 and 6-2.** The conjugate acid is obtained by adding H^+ to the formula of the species; the conjugate base is obtained by subtracting H^+ from the formula. If there is no H in a compound's formula, then that compound has no Brønsted-Lowry conjugate base.

Brønsted-Lowry acids and bases are *not* necessarily electrically neutral. Hydrogen ions can be donated by a positively charged species (*e.g.* NH_4^+) and a neutral species (*e.g.* HCN) or a negatively charged species (*e.g.* HCO_3^-). Each of these species can serve as a acid. Similarly, bases can be negatively charged (like CN^-), neutral (like NH_3) and positively charged (like $NH_2NH_3^+$). Probably the most common positively charged acid in problems is the NH_4^+ ion (the ammonium ion). Ammonium ions are put into aqueous solution by dissolving a salt like NH_4Cl or NH_4NO_3. Many positively charged acids can be viewed as derived from the NH_4^+ ion. For example thiamine hydrochloride, $C_{12}H_{17}ON_4SCl{\cdot}HCl$, **(problem 6-61)** dissolves by dissociation into $ThiH^+$ (where "Thi" equals $C_{12}H_{17}ON_4SCl$) and Cl^- ions. The $ThiH^+$ ion is analogous to the "NH_3H^+" ion.

Amphoterism

A species that can act as both acid and base is **amphoteric.** Under the Brønsted-Lowry definition, *every* species containing hydrogen is in principle amphoteric because formulas of both a conjugate acid and base can be written. The idea is that even a strong base could be forced to act as an acid by a super-strong base wresting away a hydrogen ion, and that even a strong acid could be forced by a super-strong acid. to accept hydrogen ions.

Common amphoteric species are H_2O, HCO_3^- (see **problem 6-3**) HS^-, and HSO_4^-. The way in which an amphoteric species reacts depends on its surroundings. In the presence of stronger acids an amphoteric molecule or ion *accepts* hydrogen ions. In the presence of stronger bases the same species *donates* hydrogen ions.

Under the Brønsted-Lowry definition, NaOH is a base in water. It dissolves by dissociation to $Na^+(aq)$ and $OH^-(aq)$ ions. The $OH^-(aq)$ ions accept hydrogen ions from water molecules. Sodium hydrogen carbonate ($NaHCO_3$) behaves similarly. It dissolves in water to give $Na^+(aq)$ and $HCO_3^-(aq)$ ions. The $HCO_3^-(aq)$ ions accept hydrogen ions from water to form their conjugate acid, $H_2CO_3(aq)$:

$$HCO_3^-(aq) + H_2O(l) \rightleftharpoons H_2CO_3(aq) + OH^-(aq)$$

Sodium acetate does the same kind of thing in water as sodium hydroxide. It just does it to a lesser extent.

Autoionization of Water

The conjugate acid and base of H_2O are the **hydronium ion** (H_3O^+) and the **hydroxide ion** (OH^-) respectively. Water is amphoteric. Remarkably, it establishes an acid-base equilibrium with itself:

$$2\,H_2O(l) \rightleftharpoons H_3O^+(aq) + OH^-(aq)$$

This reaction is called the **autoionization** of water. Even more remarkably, water is not the only substance that autoionizes. See **problem 6-67.**

In pure water, $[H_3O^+] = [OH^-]$, that is, there is one H_3O^+ ion for every OH^- ion. When a solute that donates or accepts hydrogen ions interacts with water this equality is destroyed. Autoionization is an equilibrium, so that at equilibrium in water and all aqueous solutions:

$$[H_3O^+][OH^-] = K_w$$

This means that the *product* of the hydronium ion and the hydroxide ion concentrations in water is a constant. As one goes up, the other comes down.

The strongest acid that may exist in water is the H_3O^+ ion; the strongest base is the OH^- ion. All acids stronger than hydronium ion donates effectively all available hydrogen ions to H_2O molecules when placed in water. Hydronium ion (H_3O^+) results. Two strong acids (such as HNO_3 and HCl) therefore have the same apparent strength when dissolved in water. This is called the **leveling effect.** All bases stronger than hydroxide ion accept hydrogen ions from H_2O when placed in water. Hydroxide ion (OH^-) results. The leveling effect applies to bases as well as acids.

The autoionization constant K_w is equal to 1.0×10^{-14} at 25°C. Like all equilibrium constants it depends on the temperature. Many calculations emphasize other aspects of acid-base equilibria, assuming tacitly that the temperature is 25°C. **Problems 6-37 and 6-71** consider the temperature dependence of K_w.

The pH Function

The pH of an aqueous solution is defined as:

$$\mathbf{pH = -\log_{10}\,[H_3O^+]}$$

Logarithms of numbers less than 1 are negative. The minus sign in the definition makes the pH of commonly-encountered solutions, in which $[H_3O^+]$ is less than one, positive. Unfortunately, it also makes the pH go *down* as the concentration of hydronium ion goes *up,* a source of confusion for beginners.

Negative pH's are uncommon but not impossible. A standard pH problem asks the hydronium ion concentration in an aqueous solution with a pH of 0 at 25°C (Answer: 1 M).

The "p" operator (it says: "take the negative logarithm of what follows") is frequently applied to quantities other than $[H_3O^+]$. Thus, pOH is the negative logarithm of the concentration of hydroxide ion and pK_w is the negative logarithm of the autoionization constant of water. Taking the logarithm of both sides of $[H_3O^+] \times [OH^-] = K_w$, multiplying by -1 and using the "p" operator gives:

$$\text{pH} + \text{pOH} = \text{p}K_w$$

At 25°C, pK_w is 14.0. If the pH of an exceedingly acidic aqueous solution at 25°C is for instance -1, then its pOH is 15. If the pH of an aqueous solution at 25°C is 7, then the pOH is also 7. Such a solution is **neutral.**

The computation of pH is as easy as pushing the "log" button on an electronic calculator. Getting from pH to $[H_3O^+]$ requires taking an antilogarithm, also called an inverse logarithm. There are two kinds of logarithm in ordinary use. The pH function uses "$\log_{10}$", logarithm to the base 10, not "ln", logarithm to the base e.

Acid and Base Strength

To be donated, a hydrogen ion must also be accepted. Free hydrogen ions do not exist in solution, and it is therefore impossible to determine the intrinsic tendencies of acids to donate H^+ (or bases to accept H^+). Instead, in considering acid strength or base strength the H_3O^+ and OH^- ions must be used as references. An equilibrium constant (symbolized by K_a, for **acid ionization constant**) for the ionization of an acid in water in reality measures the outcome of a *competition* between two acids in donating a hydrogen ion. Consider the equilibrium that results from putting the acid BH^+ in water:

$$BH^+(aq) + H_2O(l) \rightleftharpoons B(aq) + H_3O^+(aq) \quad K_a = \frac{[B]H_3O^+}{[BH^+]}$$

The competition is between BH^+ and H_3O^+. A large K_a means that BH^+ wins. The equilibrium lies well to the right (and the BH^+ ion is nearly all gone). The BH^+ is a strong acid. A small K_a means that H_3O^+ wins. The equilibrium then lies well to the left, most of the BH^+ remains, and little H_3O^+ ion is formed. The BH^+ is a **weak acid.** There are naturally gradations of strength within the category of weak acids. The larger the K_a, then the stronger the acid (see **problem 6-15).**

What about the base strength of B, the conjugate base of BH^+? The strength of B relative to OH^- is judged in the equilibrium:

$$B(aq) + H_2O(l) \rightleftharpoons BH^+(aq) + OH^-(aq) \quad K_b = \frac{[BH^+][OH^-]}{[B]}$$

A common error is to perceive this reaction to be the reverse of the acid-ionization reaction of BH^+. *It is not.* It shows B acting as a base to seize hydrogen ions from H_2O and generate OH^-. Its equilibrium constant is, quite logically, symbolized K_b.

- K_b's refer to equilibria that form OH^- from a base plus water.
- K_a's refer to equilibria that form H_3O^+ from an acid plus water.

A **weak base** has a small K_b so that its base-ionization reaction lies well to the left.

Compare the K_b equilibrium of B(aq) and the K_a equilibrium of BH^+:

$$B(aq) + H_2O(l) \rightleftharpoons BH^+(aq) + OH^-(aq) \quad K_b$$

$$BH^+(aq) + H_2O(l) \rightleftharpoons B(aq) + H_3O^+(aq) \quad K_a$$

Adding these two reactions gives:

$$2\,H_2O(l) \rightleftharpoons OH^-(aq) + H_3O^+(aq) \quad K_w$$

which is the autoionization of water. Adding two reactions entails multiplication of their equilibrium constants. Therefore, for a conjugate pair:

$$K_aK_b = K_w = 1.0 \times 10^{-14} \quad \text{(at 25°C)}$$

This relationship is very useful (as in **problems 6-9, 6-13b,** and **6-85).** It shows that weaker acids (smaller K_a's) have stronger conjugate bases (larger K_b's) and vice versa. Thanks to this relationship, Table 6-2 (text page 206) provides the K_b's of all the respective conjugate bases as well as the K_a's of a set of weak acids. Thus, the nitrite ion (NO_2^-) has, because it is the conjugate base of HNO_2, a K_b equal to K_w divided by 4.6×10^{-4}, which is the K_a of HNO_2. Nitrous acid is a weak acid and nitrite ion is a weak base.

A strong acid or base is not necessarily either concentrated or hazardous. The term "strong" refers only to the relative ability of the acid or base to donate or accept hydrogen ions. Organic acids are prominent weak acids. Examples are acetic acid, CH_3COOH, and formic acid, HCOOH. Acetic acid is typical weak acid. It is so common that the special formula "HOAc" to often used to represent it Important weak bases are NH_3 ammonia and $C_2H_3O_2^-$ the acetate ion.

Indicators

An **indicator** is a weak conjugate acid-base pair in which the acid and base forms have intense and different colors. It is added to a solution in such small molar amount that its effect on the pH of the solution is negligible. Its effect on the color of the solution is on the other hand profound. The acid-ionization equilibrium of the indicator and the (slightly rearranged) mass-action expression for it are:

$$HIn(aq) + H_2O(l) \rightleftharpoons H_3O^+(aq) + In^-(aq) \qquad [H_3O^+] = K_a\left(\frac{[HIn]}{[In^-]}\right)$$

At high concentrations of hydronium ion (low pH's), most of the indicator is in the HIn form. The corresponds to a large numerator and small denominator in the fraction in the preceding equation. The solution takes on the color of the predominant acid form of the indicator. As the hydronium ion concentration decreases, the ratio $[HIn]/[In^-]$ gets small, in order to maintain the equation. Eventually, the base form of the indicator predominates, and the solution has a different color.

If $[H_3O^+]$ is about equal to the indicator's K_a, then the acid and base forms of the indicator have about the same concentrations. The color of the solution is a mixture of the colors of the two forms. Experience shows that this intermediate range of color exists when the [HIn] to $[In^-]$ ratio is between 1/10 and 10/1. In other words, a complete color change requires a hundred-fold change in $[H_3O^+]$, or a change of two pH units.

There are numerous structurally complex organic acids and bases having different colors in the acid and base forms. These indicators have different K_a (or K_b) values and different colors. See Figure 6-5, text page 210.

One can get an estimate of the pH of an unknown solution by adding a different indicator to each of several small portions and noting the colors. Finding that a solution is acidic to one indicator and basic to another brackets the solution's pH. See **problem 6-17b**.

Calculating the pH of Solutions of Acid and Base

Suppose c_0 mol L^{-1} of the acid HA is placed in water. It interacts with H_2O, which serves as a base and accepts hydrogen ions. Equilibrium is quickly established:

$$HA(aq) + H_2O(l) \rightleftharpoons A^-(aq) + H_3O^+(aq)$$

For every $A^-(aq)$ that forms one H_3O^+ also forms, and one $HA(aq)$ is consumed. If the reaction goes all the way to the right, then the final concentration of HA is zero, and the final concentrations of A^- and H_3O^+ both equal c_0. A large value of K_a corresponds to the acid ionization lying, at equilibrium, *almost* all the way to the right. A large K_a makes HA a strong acid. Therefore, if HA is strong, the $[H_3O^+]$ of the solution equals c_0.

If K_a is less than 1, then HA is scarcely ionized at equilibrium. It is a weak acid. If y mol L^{-1} of it actually does ionize, then the equilibrium concentrations of A^- and H_3O^+ are both y, and the concentration of the $HA(aq)$ that is left unreacted is $c_0 - y$. The mass-action expression becomes:

$$K_a = \frac{y^2}{(c_0 - y)}$$

Many pH calculations (such as **problems 6-21a, 6-25,** and **6-27**) require nothing more than setting up and solving equations just like this one. It is always possible to solve for y, either by using the quadratic formula or by successive approximations. The analysis for aqueous solutions of bases is exactly similar, except that K_b replaces K_a and y represents the concentration of OH^-. See **problem 6-19.**

The drawback in focussing exclusively on the dissolved acid or base as *the* source of H_3O^+ (or OH^-) is that every aqueous solution has a second, simultaneous equilibrium going on. This is the autoionization of water. Most of the time, the concentration of H_3O^+ (or OH^-) coming from autoionization is negligible compared to the concentration coming from the weak acid (or base). But sometimes it is not.

In Section 6-7, the text derives an *exact* equation for the H_3O^+ concentration of an aqueous solution of a monoprotic weak acid having an original concentration of weak acid equal to c_a. The weak acid donates hydrogen ions, changing the hydrogen-ion concentration such that:

$$[H_3O^+]^3 + K_a[H_3O^+]^2 - (K_w + c_aK_a)[H_3O^+] - K_aK_w = 0$$

The derivation includes no approximations.[1] In principle, the pH of any aqueous solutions prepared by dissolving a single weak acid can be computed using this equation.

Memorizing the equation is *not* recommended. For one thing, relying on a memorized equation leaves one powerless to deal with obvious variations, such as a mixture of two different weak acids. What counts is chemical insight. Therefore, study the equation for insight into acid-base equilibria. Start a series of what-if questions.

- What if K_a gets smaller and smaller? This corresponds to making the acid weaker and weaker. Clearly all of the terms in which K_a multiplies another quantity go toward zero. The cubic equation transforms to:

  $$[H_3O^+]^2 = K_w$$

  The hydronium ion concentration becomes equal to $\sqrt{K_w}$, which is its concentration in pure water. An infinitely weak acid has no effect on pH.

- What if K_a in the cubic equation gets very large? This corresponds to making the weak acid strong. The terms containing K_a then completely overshadow the terms that do not contain K_a. Neglect the small terms. The cubic equation becomes:

  $$K_a[H_3O^+]^2 - c_aK_a[H_3O^+] - K_aK_w = 0 \quad K_a \text{ large}$$

[1] Actually, the derivation is for a more general case, the mixture of a weak acid with its conjugate base. The cubic equation given here is a rearrangement of equation c′ on text page 232 with c_b set to zero.

But K_a appears in each term and can be divided out:

$$[H_3O^+]^2 - c_a[H_3O^+] - K_w = 0 \quad K_a \text{ large}$$

The chemical insight here is that once K_a is big it does not matter how big. When an acid is already effectively at 100 percent efficiency in donating hydrogen ions, further increases in K_a achieve no more. Two very strong acids cannot be distinguished in strength. The two are leveled to the same strength. In this equation, K_w is on the order of 10^{-14} (see Table 6-1, text page 202). Unless c_a is very small the third term is much smaller than the second and can be neglected. Doing so gives:

$$[H_3O^+]^2 - c_a[H_3O^+] = 0 \qquad K_a \text{ large, } c_a \text{ not tiny}$$

This means $[H_3O^+]$ equals c_a when K_a is large, and c_a is anything but tiny. In this context, "tiny" is less than about 10^{-5} M. In other words, the concentration of hydronium ion in a solution of a strong acid in water is (nearly always) simply equal to the concentration of the acid.

- What if c_a becomes very small? Chemically, this corresponds to the removal of the weak acid from the solution. If c_a goes to zero, the cubic equation becomes:

$$[H_3O^+]^3 + K_a[H_3O^+]^2 - K_w[H_3O^+] - K_aK_w = 0$$

This equation has only one physically meaningful root *regardless* of the value of K_a. That root is $+\sqrt{K_w} = 10^{-7}$. The solution becomes neutral.

- What if K_a and c_a are typical practical values? The relative importance of the terms changes with dilution. Unless the solution is quite dilute (c_a tiny) the fourth term, K_aK_w, is much smaller than the third, $(K_w + c_aK_a)[H_3O^+]$. Furthermore, K_w is much smaller than c_aK_a. Neglecting the small quantities gives the tractable quadratic equation:

$$[H_3O^+]^2 + K_a[H_3O^+] - c_aK_a = 0$$

The absence of K_w in this equation means that the autoionization of water has been neglected. The equation is equivalent to the quadratic (in y) that worked out when attention was restricted to the acid-ionization reaction. Because K_w is so small, it is usually (but not always) safe to ignore it (set it to zero) in practical cases. Remember to consider K_w in problems in which the acid is very weak (K_a less than about 10^{-10}), and problems in which the acid is very dilute (c_a less than about 10^{-5}).

Example: Calculate the pH of a 1×10^{-7} M solution of HCl at 25° C. **Solution:** The naive answer is 7. But making pure water (pH 7) even a little bit acidic will inevitably lower the pH. The correct answer is 6.79. Although HCl is a strong acid, this solution of it is so dilute that the autoionization of water makes a non-negligible contribution to the H_3O^+ concentration.

Hydrolysis

In general, **hydrolysis** refers to the splitting apart ("lysis") of a substance by reaction with water. Every acid ionization and base ionization reaction in water has H_2O among the reactants and is in the general sense an hydrolysis. In a more restricted sense, hydrolysis refers to those acid-base reactions that occur when a salt dissolves in water to give ions with large enough K_a or K_b to change the pH of the solution (see **problem 6-73).** For example, sodium acetate dissolves in water to give sodium ions and acetate ions. Acetate ion, the conjugate base of the weak acid acetic acid has a K_b of about 10^{-10}. Placing sodium acetate in water raises the pH because the reaction:

$$CH_3COO^-(aq) + H_2O(l) \rightleftharpoons CH_3COOH(aq) + OH^-(aq)$$

provides OH^- ion to a measurable extent. **Problem 6-23** treats the hydrolysis of the very weak base IO_3^-.

If NH_4Cl is dissolved in water, the NH_4^+ ion hydrolyzes to give H_3O^+ and NH_3:

$$NH_4^+(aq) + H_2O(l) \rightleftharpoons NH_3(aq) + H_3O^+(aq)$$

The equilibrium constant is the K_a of the NH_4^+ ion. This hydrolysis lowers the pH. Meanwhile, the chloride ion arising in the dissolution has a K_b so small that hydrolysis is effectively non-existent. However, some the cation and anion can in principle hydrolyze at the same time. If some ammonium acetate, $NH_4^+C_2H_3O_2^-$, is dissolved in water, then two ions hydrolyze simultaneously and the net effect on the pH is a compromise between the separate effects of the two.

Buffer Solutions

Buffer solutions resist changes in pH when either a strong acid or a strong base is added. A typical buffer solution is made by mixing a weak acid with its conjugate base (see **problem 6-35).** A solution of a weak base and its conjugate acid also is a buffer solution **problem 6-33**. The weak base tends to neutralize any acid added to the buffer, and the weak acid tends to neutralize any added base. Buffers are not magic. Acids or bases do affect the pH of buffer solutions, even if only slightly. Also, enough acid or base can always overwhelm the **buffer capacity** of a buffer solution and lower or raise the pH substantially.

Calculations of the pH of Buffer Solutions

Consider a typical buffer solution. Suppose that a weak acid HA is mixed in water with a source of its conjugate base A^-. The salt NaA, which dissociates completely when it dissolves, is such a source. The problem is to compute the concentration of H_3O^+ (and then the pH). Let the initial concentration of HA be c_a and the initial concentration of the conjugate base be c_b. In Section 6-7, it is shown that:

$$[H_3O^+] = K_a \frac{c_a - [H_3O^+] + [OH^-]}{c_b + [H_3O^+] - [OH^-]}$$

This equation is actually a rearrangement of equation **c**$'$ on text page 232. Do not memorize this equation but rather study for chemical insight.

- Suppose K_a is neither very big nor very small but is a number like 1×10^{-5}:

  1. As long as the concentrations of the weak acid and its conjugate base are big they drown out the terms added to and subtracted from them. The equation becomes:

     $$[H_3O^+] \approx K_a \frac{c_a}{c_b}$$

     Under these circumstances, the concentration *ratio,* not the actual values, determines the hydrogen ion concentration. Adding a little acid from an outside source slightly reduces c_b and slightly increases c_a. The ratio, c_a/c_b, hardly changes. Adding a little base from outside also causes only a slight change in the ratio. This is the crux of the solution's buffer action.

     Taking the negative logarithm of both sides of the preceding equation and using the definition of the "p" operation gives:

     $$\text{pH} \approx \text{p}K_a - \log \frac{c_a}{c_b}$$

     This useful equation works to solve a vast majority of practical buffer problems, such as **problem 6-33,** but its indiscriminate use can lead to trouble (see **problem 6-75).**

  2. If either c_a or c_b goes to zero, their ratio goes either to zero or infinity. Either way, the ratio no longer has any meaning in determining $[H_3O^+]$. Buffer action no longer occurs. The insight is that both acid and conjugate base must be present for buffer action. Moreover, the buffer is most effective when c_a equals c_b (sometimes called the **buffer point).** When c_a or c_b goes to zero the approximate equation gives impossible answers, but the original (exact) equation transforms to describe a weak base by itself in water or a weak acid by itself in water.

3. Rewriting the equation with "a" for "b" (and "b" for "a"), OH^- for H_3O^+ (and the reverse) and K_b for K_a, describes a buffer solution of a weak base and its conjugate acid. The symmetry of this transformation echoes the fundamental chemical symmetry between acid and base.

- Taking K_a to zero transforms the exact equation to $[OH^-] = c_b$. This is an expression for the OH^- concentration in a solution of a strong base of concentration c_b. The chemical insight is that as K_a gets very small the acid A becomes no real acid at all and the conjugate base B automatically becomes a strong base.

- Making K_a very large transforms the equation to $[H_3O^+] = c_a$. The is the hydronium-ion concentration in a solution of a strong acid of concentration c_a. As the acid becomes a strong acid its conjugate base becomes so weak that it no longer exerts any effect on the pH of the solution.

Acid-Base Titration Curves

A acid-base titration is an analytical procedure for determining the concentration of an acid or base in a solution. During a titration, acid and base are allowed to neutralize each other in a controlled fashion. The progress of the neutralization is followed by measuring the pH of the solution as a function of the volume of titrant added. A **titration curve** is a plot of pH versus volume of added titrant. Figures 6-9 and 6-10 in the text are typical titration curves.

Calculation of Points on a Titration Curve

During the course of a titration two effects occur to change the pH of the solution:

1. The acid and base neutralize each other. In water, the neutralization reaction is:

$$H_3O^+(aq) + OH^-(aq) \rightleftharpoons 2\,H_2O(l)$$

Naturally, the pH goes up when an acid is titrated with a base and down when a base is titrated with acid.

2. The two solutions dilute each other. Simple dilution affects pH. For example, 10 mL of 0.1 M HCl has a pH of 1. Diluting it to 100 mL changes the concentration to 0.01 M HCl and raises the pH to 2.

In calculations concerning titrations consider the two effects separately. A general procedure is:

1. Imagine that no reaction occurs until mixing is complete. Compute the effect of dilution on the concentrations of all species of importance.

2. Allow the neutralization reaction to proceed 100 percent to the right until the limiting reactant, whether acid or base is entirely consumed.

3. Identify the important equilibria and use them with appropriate K's to compute $[H_3O^+]$.

Suppose a strong base is being added to a strong acid. The course of the titration has four regions.

The Starting Point. No titrant has been added. The hydronium-ion concentration is just the molar concentration of the acid.

Approach to Equivalence. Some base has been added. The reaction:

$$H_3O^+(aq) + OH^-(aq) \rightarrow 2\,H_2O(l)$$

may be assumed to go to completion. The amount (not concentration) of H_3O^+ that remains in solution equals the amount originally present minus the amount of OH^- added. The concentration of H_3O^+ is this answer divided by the volume of the solution. The volume of the solution is the sum of the original volume of acid solution and the volume of titrant added.

Equivalence Point. The amount of base that has been added exactly equals the amount of H_3O^+ originally present. The $[H_3O^+]$ in the solution comes entirely from the autoionization of water. The pH is 7.

Beyond the Equivalence Point. The excess strong base builds up in solution. The concentration of OH^- is the molar amount of this excess divided by the total volume of the solution.

The titration of a weak acid with a strong base has the same four region. **Problem 6-45** and **problem 6-63** show the calculation of points on the titration curve when a weak acid is titrated with a strong base. **Problem 6-43** covers the conjugate case—the titration of a weak base with a strong acid.

Polyprotic Acids

Acids that contain two or more ionizable H atoms are **polyprotic** acids. Sulfuric acid (H_2SO_4) is a diprotic acid; phosphoric acid (H_3PO_4) is a triprotic acid. Although acetic acid (CH_3COOH) contains four hydrogen atoms per molecule, it is only a monoprotic acid. The three H atoms bonded to the C atom are not ionizable.

Polyprotic acids ionize in two or more stages. Typically the equilibrium constant for the first stage (K_{a1}) is about 10^5 times larger than for the second stage (K_{a2}). A crucial point comes up in working with equilibria of polyprotic acids (or their conjugate bases): *when two or more acid-ionization reactions go on simultaneously each contributes hydronium ion, but the hydronium ion concentration used in all the equilibrium constant expressions is one and the same.* The different stages of ionization of a polyprotic acid interact with each other.

Exact calculations of pH of solutions of polyprotic acid therefore in general require solving sets of several simultaneous equations. They are hard. In practical problems, especially those concerned with inorganic polyprotic acids, K_{a1}, K_{a2} and other K_a's almost always differ so much in magnitude that the equilibria can be treated as if they were separate. Thus, in **problem 6-53** the second ionization of arsenic acid adds only negligibly to the hydronium ion concentration coming from the first.

Because the several stages in the ionization of a polyprotic acid interact, changing the pH of a solution of a polyprotic acid must shift all of the acid-base equilibrium at once. At high pH, most of the concentration of a polyprotic acid will be in the form of the most negative conjugate base; at low pH most of the concentration will be in the form of the most acidic form. **Problems 6-59** and **6-81** show how to compute the fraction of each of the forms of a polyprotic acid that is present at a given pH.

Exact Treatment of Acid-Base Equilibria

Principle of Electrical Neutrality

Every solution must be electrically neutral. When a solution contains ions the total amount of positive charge equals the total amount of negative charge. This fact furnishes an important mathematical relationship among the concentrations of the ions in a solution. For example, in an aqueous solution of NaCl:

$$[Na^+] + [H_3O^+] = [Cl^-] + [OH^-]$$

The principle of electrical neutrality applies to electrical charges and not to ions. In a solution of sulfuric acid in water:

$$[H_3O^+] = [OH^-] + [HSO_4^-] + 2\,[SO_4^{2-}]$$

where the coefficient of 2 for the sulfate concentration reflects the contribution by each sulfate of two negative charges to the total negative charge in the solution. See **problem 6-41** for the use of this principle in analyzing a problem.

Material Balance

Substances often, when dissolved in water, end up partially converted to new forms. A substance may distribute itself in a number of ways among different forms. See

problem 6-61.

• The sum of the concentrations of the derivative forms and the unconverted original form always equals the number of moles per liter originally added.

For example, if 0.1 mol of H_2SO_4 is placed in a liter of solution it interacts with the solvent to produce HSO_4^- and SO_4^{2-} ions. The material balance condition is:

$$0.1 = [H_2SO_4] + [HSO_4^-] + [SO_4^{2-}]$$

The Exact Treatment

The exact treatment of acid-base equilibria involves first identifying all the species in the equilibrium solution and then writing equations to relate their concentrations. Specifically:

1. List all of the species that can be present in solution at equilibrium. In aqueous solutions this includes H_3O^+, OH^- and H_2O.

2. Employ the principle of electrical neutrality to write a single mathematical equation relating the concentrations of ions in the solution.

3. Apply the principle of material balance to each substance that was placed in solution. If, for example, five weak acids are mixed in the solution there are five material balance equations, one for each of the five.

4. Write a mass-action expression for every equilibrium taking place in the solution, including the autoionization of water.

5. Look up the equilibrium constants (at the proper temperature) for all of the equilibria that have been identified.

6. Verify that the number of mathematical equations equals the number of different chemical species in solution. Follow the procedure properly and it always will.

7. Solve the system of independent equations.

An exact solution to such a set of equations usually involves considerable algebraic effort. The best course is to use chemical insight to simplify the equations. This is done by neglecting terms in the equations which make only tiny contributions when added to or subtracted from other terms. Never neglect terms that are multipliers or divisors even if they are small.

Amphoteric Equilibria

One class of problem does require consideration of simultaneous equations involving K_{a1} and K_{a2}. The two-step ionization of a diprotic acid gives an amphoteric intermediate. This species is ready to donate H^+ to form its conjugate base (according to K_{a2}) and also ready to accept H^+ to give back the original diprotic acid (according to $K_{b2} = K_w/K_{a1}$). When such an amphoteric species is put in water the hydrogen-ion concentration is related to the K's as follows:

$$[H_3O^+] \approx \sqrt{\frac{K_{a1}K_{a2}c_0 + K_{a1}K_w}{K_{a1} + c_0}}$$

where c_0 is the original concentration of the amphoteric species. If K_{a1} is negligible compared to c_0, and if K_w is negligible compared to $K_{a2}c_0$ then this equation becomes:

$$[H_3O^+] \approx \sqrt{K_{a1}K_{a2}}$$

These equations are used in **problem 6-85** and discussed further there.

Detailed Solutions to Odd-Numbered Problems

6-1 a) The chloride ion Cl^- cannot act as a Brønsted-Lowry acid because it does not have any hydrogen.
b) The hydrogen sulfate ion HSO_4^- can act as a Brønsted-Lowry acid; its conjugate base is SO_4^{2-} (the sulfate ion).
c) The ammonium ion NH_4^+ can act as a Brønsted-Lowry acid; its conjugate base is NH_3 (ammonia).
d) Ammonia NH_3 can act as a Brønsted-Lowry acid; its conjugate base is NH_2^- (the amide ion).
e) Water H_2O can act as a Brønsted-Lowry acid; its conjugate base is OH^- (hydroxide ion).

6-3 The reaction is $H_3C_5H_5O_7(aq) + HCO_3^-(aq) \rightleftharpoons H_2C_5O_7^-(aq) + H_2CO_3(aq)$. Citric acid donates a hydrogen ion to the hydrogen carbonate ion, which serves as a base. The H_2CO_3 formed thereby very soon decomposes to water and gaseous carbon dioxide, which causes the cookies to rise. A preliminary step is the dissolution of the solid sodium hydrogen carbonate in the lemon juice.

6-5 The pH of an aqueous solution is by definition the negative logarithm of the hydronium-ion concentration: $\text{pH} = -\log[H_3O^+] = -\log(2.0 \times 10^{-4}) = 3.70$.

6-7 The two ends of the pH range for urine each give a H_3O^+ concentration:

$$[H_3O^+] = 10^{-5.5} = 3 \times 10^{-6}\text{ M}; \quad \text{and} \quad [H_3O^+] = 10^{-6.5} = 3 \times 10^{-7}\text{ M}$$

The pOH can be determined from the pH in the two cases:

$$\text{pOH} = \text{p}K_w - \text{pH} = 14.0 - 5.5 = 8.5 \quad \text{pOH} = \text{p}K_w - \text{pH} = 14.0 - 6.5 = 7.5$$

where the use of $\text{p}K_w = 14.0$ assumes a temperature of 25°C. Then:

$$[\text{OH}^-] = 10^{-\text{pOH}} = 10^{-8.5} = 3 \times 10^{-9}\ \text{M} \quad \text{and} \quad [\text{OH}^-] = 10^{-7.5} = 3 \times 10^{-8}\ \text{M}$$

6-9 Use 13.776 instead of 14.00 as $\text{p}K_w$ when calculating pOH:

$$[\text{H}_3\text{O}^+] = 10^{-8.00} = 1.0\times10^{-8}\ \text{M and } [\text{OH}^-] = 10^{-(13.776-8.00)} = 10^{-5.776} = 1.7\times10^{-6}\ \text{M}$$

6-11 The first equation $2\,\text{K}(s) + 2\,\text{H}_2\text{O}(l) \rightarrow 2\,\text{KOH}(aq) + \text{H}_2(g)$ is the better representation of what really happens. The reaction is instantaneous and continues with great vigor. If the second equation, the one involving H_3O^+, were the case, one might expect a slow process since H_3O^+ is present in only 10^{-7} M concentration in pure water. Even if the low concentration of H_3O^+ sufficed to react with the great vigor, the progress of the reaction would generate $\text{OH}^-(aq)$, raising the pH and thereby rapidly lowering the concentration of H_3O^+. Therefore, the reaction according to the second equation would be expected to grow progressively slower as it proceeded. On the other hand, the first equation represents a direct interaction of $\text{K}(s)$ with $\text{H}_2\text{O}(l)$. As this reaction proceeds, the concentration of H_2O remains relatively constant and high (about 56 M), assuming the water is in big excess.

6-13 a) A base (Brønsted-Lowry) is a hydrogen-ion acceptor. Ephedrine and most other organic bases, have a nitrogen atom as the site of attachment of the hydrogen ion:

$$\text{C}_{10}\text{H}_{15}\text{ON}(aq) + \text{H}_2\text{O}(l) \rightleftharpoons \text{C}_{10}\text{H}_{15}\text{ONH}^+(aq) + \text{OH}^-(aq)$$

b) Use the equation $K_aK_b = K_w$, which relates the acid-ionization constant and base-ionization constant of an aqueous acid and its conjugate base. Substitution gives:

$$K_a = \frac{K_w}{K_b} = \frac{1.0 \times 10^{-14}}{1.4 \times 10^{-4}} = 7.1 \times 10^{-11}$$

c) The K_a for ammonium ion, NH_4^+, is 5.6×10^{-10} (see Table 6-2). The K_b for ammonia can accordingly be calculated:

$$K_b = \frac{K_w}{K_a} = \frac{1.0 \times 10^{-14}}{5.6 \times 10^{-10}} = 1.8 \times 10^{-5}$$

A K_b is an equilibrium constant for the general reaction:

$$\text{B}(aq) + \text{H}_2\text{O}(l) \rightleftharpoons \text{BH}^+(aq) + \text{OH}^-(aq)$$

The larger K_b corresponds to the stronger base. The stronger base is ephedrine.

6-15 We can arrive at this reaction by combining two equations appearing in Table 6-2:

$$H_2O(l) + HClO_2(aq) \rightleftharpoons H_3O^+(aq) + ClO_2^-(aq) \quad K_a = 1.1 \times 10^{-2}$$

$$H_2O(l) + HNO_2(aq) \rightleftharpoons H_3O^+(aq) + NO_2^-(aq) \quad K_a = 4.6 \times 10^{-4}$$

If the second is subtracted from the first the result is:

$$HClO_2(aq) + NO_2^-(aq) \rightleftharpoons HNO_2(aq) + ClO_2^-(aq) \qquad K = 24$$

When one chemical equation is subtracted from another to give a third equation, the equilibrium constant of the one being subtracted is divided into the equilibrium constant of the other to give the constant of the final equation. This has been done in the preceding. This constant, which is greater than 1, means that the reaction "tilts" toward the products. The relatively small concentration of $HClO_2$ present at equilibrium means that $HClO_2$ is a stronger acid than HNO_2; the comparatively large concentration of HNO_2 at equilibrium means NO_2^- is a stronger base than ClO_2^-.

6-17 a) The pH at which an indicator changes color is roughly equal to pK_a of the indicator. The color changes of indicators occur over ranges of 1 to 1.9 pH units (see text Figure 6-5). Take the middle of these range to be the pK_a's:

bromocresol green	p$K_a = 4.6$	$K_a = 2.51 \times 10^{-5}$
methyl orange	p$K_a = 3.8$	$K_a = 1.58 \times 10^{-4}$

The acid form of methyl orange is the stronger acid because it has the larger K_a.
b) Both of these indicators are in their transition ranges (see text Figure 6-6). These means that the pH of the solution must simultaneously lie between pH 3.8 and pH 5.4 (bromocresol green) and 3.2 and 4.4 (methyl orange) Therefore, the pH of the solution is in the range 3.8 to 4.4.

6-19 The K_b applies to the reaction $M(aq) + H_2O(l) \rightleftharpoons MH^+(aq) + OH^-(aq)$ where "M" stands for morphine. The initial concentration (after its dissolution in water but before the preceding reaction with water) of the morphine is:

$$[M]_0 = 0.0400 \text{ mol}/0.600 \text{ L} = 0.0667 \text{ mol L}^{-1} = 0.0667 \text{ M}$$

Represent the amount of MH^+ formed at equilibrium as x:

	$M(aq)$	$+H_2O(aq) \rightleftharpoons$	$MH^+(aq)$ +	$OH^-(aq)$
Init. Conc. (M)	0.0667	—	0	small
Change in Conc. (M)	$-x$	—	$+x$	$+x$
Equil. Conc. (M)	$0.0667 - x$	—	x	x

Substitute the equilibrium concentrations into the equilibrium expression:

$$\frac{[MH^+][OH^-]}{[M]} = K_b \quad \text{or} \quad \frac{x^2}{0.0667 - x} = 8 \times 10^{-7}$$

This equation can be solved using the quadratic formula. A better method however is to note that x must be quite small compared to 0.0667. Neglecting the x in the denominator allows very quick computation of $x = 2.301 \times 10^{-4}$. The equilibrium concentration of OH^- is thus 2.3×10^{-4} M. The pOH is $-\log[OH^-] = 3.64$, and the pH is this number subtracted from 14.0, which is 10.4. Use of one significant figure (the 4) in the final answer reflects the fact that K_b has only one significant figure.

6-21 a) The formula of benzoic acid is C_6H_5COOH. Represent it as "HOBz." As the 0.20 M HOBz comes to equilibrium in water, it generates a concentration x of H_3O^+:

	$HOBz(aq)$	$+H_2O(l) \rightleftharpoons$	$OBz^-(aq)\ +$	$H_3O^+(aq)$
Init. Conc. (M)	0.20	–	0	small
Change in Conc. (M)	$-x$	–	$+x$	$+x$
Equil. Conc. (M)	$0.20 - x$	–	x	x

The concentration of H_3O^+ arising from the autoionization of water is very small compared to that arising from the ionization of the benzoic acid, so x is the concentration of H_3O^+ at equilibrium:

$$\frac{[OBz^-][H_3O^+]}{[HOBz]} = K_a \quad \text{or} \quad \frac{x^2}{0.20 - x} = 6.46 \times 10^{-5}$$

This can be rearranged to the equation: $x^2 + 6.46 \times 10^{-5}x - 1.29 \times 10^{-5} = 0$. Applying the quadratic formula gives:

$$x = \frac{-6.46 \times 10^{-5} \pm \sqrt{4.17 \times 10^{-9} + 5.16 \times 10^{-5}}}{2}$$

$$x = \frac{-6.46 \times 10^{-5} \pm 7.18 \times 10^{-3}}{2} = -0.00352 \text{ and } 0.00356$$

The negative root has no physical meaning. Hence $[H_3O^+] = 0.00356$ M. The pH is $-\log[H_3O^+] = 2.45$. Note that if x is neglected in comparison to 0.020 in the equation, then the very simple approximate equation $x = \sqrt{(0.20)(6.46 \times 10^{-5})}$ results. The positive root of this equation is $x = 0.0036$. This is, to two significant figures, the same answer obtained by exact solution of the quadratic equation.

b) The equilibrium concentration of $[H_3O^+]$ must be 3.56×10^{-3} M. Start as in part a) but now x is known and the aim is to solve for c_0, the initial concentration of acetic acid:

$$\frac{x^2}{c_0 - x} = K_a = 1.76 \times 10^{-5} = \frac{(3.54 \times 10^{-3})^2}{c_0 - 3.56 \times 10^{-3}} = 1.76 \times 10^{-5}$$

Solving gives $c_0 = 0.72$ M. This means 0.72 mol of acetic acid must be dissolved per liter of solution.

6-23 The problem as originally devised asked for the pH of a 0.100 M of HIO_3 (*not* KIO_3, which is what appears in the text). Both versions are solved here.
The answer to the intended problem uses the same approach as problem 6-21a:

	$HIO_3(aq)$	$+\ H_2O(aq) \rightleftharpoons$	$IO_3^-(aq)\ +$	$H_3O^+(aq)$
Init. Conc. (M)	0.100	–	0	small
Change in Conc. (M)	$-x$	–	$+x$	$+x$
Equil. Conc. (M)	$0.100 - x$	–	x	x

$$\frac{[IO_3^-][H_3O^+]}{[HIO_3]} = K_a = \frac{x^2}{0.100 - x} = 0.16$$

This yields: $x^2 + 0.16x - 0.016 = 0$. Using the quadratic formula:

$$x = \frac{-0.16 \pm \sqrt{0.0256 + 0.064}}{2} = 0.0697 \text{ and } -0.230$$

The negative solution of this quadratic has no physical meaning. Hence $[H_3O^+] = 0.0697$ M, and the pH is $-\log[H_3O^+] = 1.16$. The final pH is well on the acid side because HIO_3 is a rather strong weak acid. Note that getting a quick solution by neglecting x in comparison to 0.100 gives an hydronium-ion concentration of 0.126 M, which is wrong.
The pH of a 0.100 M solution of KIO_3 is completely different. Such a solution contains K^+ ions, which have a K_a that is equal to zero, and IO_3^- ions, which are very weakly basic. The pH of the solution should lie just over 7.00. Because IO_3^- is such a weak base, the contribution of the autoionization of water to the pH will *not* be negligible. Therefore, start with the highlighted equation developed on text page 233 for the exact treatment of acid-base equilibria:

$$[H_3O^+]^3 + (c_b + K_a)[H_3O^+]^2 - (K_w + c_aK_a)[H_3O^+] - K_aK_w = 0$$

where c_a is the concentration of a weak acid put into a solution and c_b is the concentration of its conjugate base. In this problem, c_a is zero, because no HIO_3 was put into the solution, and c_b is 0.100 M. Also, K_w is 1.0×10^{-14}, and K_a is 0.16. Substitution of these numbers gives:

$$[H_3O^+]^3 + (0.100 + 0.16)[H_3O^+]^2 - (0.0 + 1.0 \times 10^{-14})[H_3O^+] - (0.16)(1.0 \times 10^{-14}) = 0$$

This cubic equation is fairly easily solved by successive approximation, starting with the guess that $[H_3O^+]$ equals 1.0×10^{-7} M. Clearly, the answer must be somewhat less than this starting value. The answer comes out to 0.7844×10^{-7} M, which rounds to 7.8×10^{-8} M. The pH is therefore 7.11.

6-25 The $papH^+Cl^-$, a salt, dissolves completely in water to give $papH^+$ ion and Cl^- ion. The Cl^- ion does not react significantly with the water. The $papH^+$ ion reacts as a weak acid:

$$\text{papH}^+(aq) + \text{H}_2\text{O}(l) \rightleftharpoons \text{pap}(aq) + \text{H}_3\text{O}^+(aq)$$

Assume that this reaction is the predominant source of H_3O^+ in the solution. The concentration of H_3O^+ is 4.90×10^{-4} M (calculated from the pH of 3.31), and the concentration of "pap" is also 4.90×10^{-4} M. The concentration of $papH^+$ is its original value minus the portion converted into pap. This is $0.205 - 4.90 \times 10^{-4} \approx 0.205$ M. Substitute these equilibrium values into the K_a expression to compute K_a:

$$K_a = \frac{(4.90 \times 10^{-4})(4.90 \times 10^{-4})}{(0.205)} = 1.2 \times 10^{-6}$$

6-27 Hydrofluoric acid is a weak acid in water:

$$\text{HF}(aq) + \text{H}_2\text{O}(l) \rightleftharpoons \text{F}^-(aq) + \text{H}_3\text{O}^+(aq) \qquad K_a = 6.6 \times 10^{-4}$$

Because the pH of the HF solution at 25°C is 2.13:

$$[\text{H}_3\text{O}^+] = \text{antilog}(-2.13) = 10^{-2.13} = 7.41 \times 10^{-3}\text{ M}$$

If the hydrofluoric acid is the only source of H_3O^+, then $[F^-]$ is also 7.41×10^{-3} M. The operation of the equilibrium guarantees that:

$$K_a = \frac{[\text{H}_3\text{O}^+][\text{F}^-]}{[\text{HF}]} = 6.6 \times 10^{-4}$$

In this expression, all of the quantities expect [HF] are known. Solving for [HF]:

$$[\text{HF}] = \frac{(7.41 \times 10^{-3})^2}{6.6 \times 10^{-4}} = 0.083\text{ M}$$

6-29 Aqueous NaOH solutions contain Na^+ ions and OH^- ions. The latter react with acetic acid to produce water and acetate ions. However, acetate ion is a weak base in its own right. It causes the solution of sodium acetate formed by the treatment of aqueous acetic acid with an equal chemical amount of NaOH to be basic. Whenever a weak acid is titrated with a strong base, the pH at the equivalence point exceeds 7. Similarly, whenever a weak base is titrated with a strong acid, the pH at the equivalence point is less than 7.

6-31 The ammonium bromide dissolves to give the NH_4^+ ion, a weak acid, and the Br^- ion, which is essentially non-existent as a base. This solution is therefore acidic. The HCl, a strong acid, ionizes completely in water, and gives an even more acidic solution than the NH_4Br. Meanwhile, the NaOH, a strong base, gives a highly basic solution, and the $NaCH_3COO$, the salt of a strong base and a weak acid, gives a somewhat basic solution. The KI, a salt of a strong base and a strong acid, gives a neutral solution. The order is therefore: HCl < NH_4Br < KI < $NaCH_3COO$ < NaOH.

6-33 The pK_a of the conjugate acid of "tris" can be computed from the pK_b of tris itself:

$$pK_a = 14.00 - pK_b = 14.00 - 5.92 = 8.08$$

The addition of the HCl converts some of the tris to its conjugate acid:

$$\text{tris}(aq) + \text{HCl}(aq) \rightleftharpoons \text{trisH}^+(aq) + \text{Cl}^-(aq)$$

The resulting solution is a mixture of a weak acid (trisH^+) and its conjugate base (tris). It is a buffer by virtue of the equilibrium:

$$\text{trisH}^+(aq) + \text{H}_2\text{O}(l) \rightleftharpoons \text{tris}(aq) + \text{H}_3\text{O}^+(aq) \quad K_a = \frac{[\text{tris}][\text{H}_3\text{O}^+]}{[\text{trisH}^+]}$$

Compute the concentrations of the tris and its trisH^+ after complete reaction with the HCl but before the preceding equilibrium is established:
$[\text{tris}]_0 = (0.050 - 0.025 \text{ mol})/2.00 \text{ L} = 0.0125 \text{ M}$
$[\text{trisH}^+]_0 = 0.025 \text{ mol}/2.00 \text{ L} = 0.0125 \text{ M}$
The equilibrium reaction now reduces the concentration of the weak acid as it forms H_3O^+ and tris in equal amounts. If x is the equilibrium concentration of $H_3O^+(aq)$ ion, then:

	$\text{trisH}^+(aq)$	$+\text{H}_2\text{O}(aq) \rightleftharpoons$	$\text{tris}(aq) +$	$\text{H}_3\text{O}^+(aq)$
Init. Conc. (M)	0.0125	—	0.0125	small
Change in Conc. (M)	$-x$	—	$+x$	$+x$
Equil. Conc. (M)	$0.0125 - x$	—	$0.0125 + x$	x

$$K_a = \frac{[\text{tris}][\text{H}_3\text{O}^+]}{[\text{trisH}^+]} = \frac{(0.0125 + x)x}{(0.0125 - x)}$$

Assume that x is small compared to 0.0125. Then the 0.0125's cancel out and

$$[\text{H}_3\text{O}^+] = K_a \quad \text{so that} \quad \text{pH} = pK_a = 8.08$$

Clearly x is less than 10^{-7}, so the assumption was justified. Note that the pH is exactly equal to the pK_a in this solution. This is always the case in a buffer solution in which the acid and conjugate base concentrations are equal.

6-35 a) The problem is exactly like Example 6-6 (text page 217). After mixing and dissolution but before any other chemical change, the solution is 0.10 M in acetic acid and 0.040 M in acetate ion (from the sodium acetate). Then the acid-ionization equilibrium comes into play. Let x equal the equilibrium concentration of H_3O^+:

	$HOAc(aq)$	$+H_2O(aq) \rightleftharpoons$	$OAc^-(aq)\ +$	$H_3O^+(aq)$
Init. Conc. (M)	0.10	–	0.040	small
Change in Conc. (M)	$-x$	–	$+x$	$+x$
Equil. Conc. (M)	$0.10 - x$	–	$0.040 + x$	x

$$K_a = 1.76 \times 10^{-5} = \frac{[OAc^-][H_3O^+]}{[HOAc]} = \frac{(0.040 + x)x}{(0.10 - x)}$$

Assume that x is small compared to 0.10. The x's that are added and subtracted can be neglected in the above expression to give:

$$x = 1.76 \times 10^{-5} \left(\frac{0.10}{0.040}\right) = 4.4 \times 10^{-5}$$

The assumption is clearly justified; the pH is the negative logarithm of x or 4.36.
b) The addition of 0.010 mol of $OH^-(aq)$ ion (in the form of NaOH) immediately converts 0.010 mol of acetic acid HOAc to 0.010 mol of acetate ion OAc^-. The concentrations of HOAc and OAc^- right after the conversion but before the action of the HOAc as a weak acid are:

$$[HOAc] = \frac{0.050 - 0.010 \text{ mol}}{0.500 \text{ L}} = 0.080 \text{ M};\ [OAc^-] = \frac{0.020 - 0.010 \text{ mol}}{0.500 \text{ L}} = 0.060 \text{ M}$$

Now the acid-ionization equilibrium acts. Let y equal the equilibrium concentration of H_3O^+:

	$HOAc(aq)$	$+H_2O(aq) \rightleftharpoons$	$OAc^-(aq)\ +$	$H_3O^+(aq)$
Init. Conc. (M)	0.080	–	0.060	small
Change in Conc. (M)	$-y$	–	$+y$	$+y$
Equil. Conc. (M)	$0.080 - y$	–	$0.060 + y$	y

$$K_a = 1.76 \times 10^{-5} = \frac{[OAc^-][H_3O^+]}{[HOAc]} = \frac{(0.060 + y)y}{(0.080 - y)}$$

Assume that y is small compared to 0.060 and 0.080. Then:

$$y = 1.76 \times 10^{-5} \left(\frac{0.080}{0.060}\right) = 2.34 \times 10^{-5}$$

The assumption is clearly justified. The pH is the negative logarithm of y or 4.63. Note that the pH rises only from 4.36 to 4.63 from part a) to part b) even though the amount of strong base added was substantial (20% of the amount of weak acid present). This resistance to a change in pH is characteristic of buffered solutions.

6-37 Buffer solutions are most efficient at resisting changes in pH at their "buffer points." A buffer solution is at its buffer point if the concentration of the weak acid in the solution equals the concentration of the conjugate base. At the buffer point, the pH of the buffer solution equals the pK_a of the weak acid. The physician should therefore select a weak acid having a pK_a as close as possible to the desired pH. The acid of choice on the list is *m*-chlorobenzoic acid, $pK_a = 3.98$.

6-39 In this problem, 500 mL of 0.100 M formic acid is titrated with 0.0500 M NaOH. Before any NaOH is added, the solution contains $0.500 \text{ L} \times 0.100 \text{ mol L}^{-1}$ or 0.0500 mol of formate-containing species. Most of this is $HCOOH(aq)$, that is, most is un-ionized formic acid. There is also a small quantity of $HCOO^-(aq)$, the formate ion. Adding strong base to this solution does nothing to change the total *amount* of formate-containing species. What it does do is to convert $HCOOH(aq)$ to $HCOO^-(aq)$ according to the reaction:

$$HCOOH(aq) + NaOH(aq) \rightleftharpoons Na^+(aq) + HCOO^-(aq) + H_2O(l)$$

and thereby raise the pH. Suppose that exactly one mole of strong base is added for every mole of formic acid. To reach this point, the *equivalence point*, requires 0.0500 mol of NaOH, which is furnished by 1000 mL of 0.0500 M NaOH. At the equivalence point, the solution is simply dilute sodium formate. Sodium formate is the salt of a weak acid and strong base, so the solution at the equivalence point is basic—its pH exceeds 7 (see problem 6-29).

To raise the pH to only 4.00 clearly requires less than 1000 mL of 0.0500 M NaOH. At pH = 4.00, $[H_3O^+] = 1.0 \times 10^{-4}$ M. Write the acid-ionization equilibrium expression:

$$1.77 \times 10^{-4} = \frac{[H_3O^+][HCOO^-]}{[HCOOH]} \quad \text{which means} \quad \frac{[HCOO^-]}{[HCOOH]} = 1.77$$

Suppose that V L of 0.0500 M NaOH brings the 0.100 M formic acid up to a pH of 4.00. The total volume of the titration mixture when the pH hits 4.00 is $(0.500 + V)$ L. Each mole of NaOH converts one mole of $HCOOH(aq)$ to $HCOO^-(aq)$. Assume that this reaction is the only source of $HCOO^-(aq)$. This assumption neglects the fact that at equilibrium additional $HCOOH(aq)$ indeed does ionize to give $H_3O^+(aq)$ and more $HCOO^-(aq)$ ion. However, if only a small amount of $HCOOH(aq)$ undergoes the ionization, then the assumption is *nearly* correct. The question of how near must be answered later. Meanwhile, because $0.0500V$ mol of NaOH has been added:

$$[HCOO^-] = \frac{0.0500V}{(0.500 + V)}$$

The numerator of this expression is the chemical amount of $HCOO^-(aq)$ produced by the acid-base reaction. The denominator is the total volume of the solution in liters. Dilution alone reduces the total concentration of formate-containing species from 0.100 M to $(0.500/0.500 + V) \times 0.100$ M. There are only two formate-containing species, HCOOH(aq) and $HCOO^-(aq)$. Therefore:

$$[\mathrm{HCOOH}] = \frac{(0.500)V}{(0.500 + V)}(0.100) - \frac{(0.050)V)V}{(0.500 + V)}$$

Inserting the expressions for [HCOOH] and $[HCOO^-]$ into the expression for their ratio allows cancellation of the two denominators. The result is:

$$1.77 = \frac{0.0500V}{(0.0500 - 0.0500V)}$$

It is easy to solve this equation for V, which equals 0.639 L or 639 mL. The total volume of the solution when the pH reaches 4.00 is 1.139 L. The final concentrations of HCOOH(aq) and $HCOO^-(aq)$ are 0.0158 and 0.0280 M respectively. Both values are large in comparison to the hydronium ion concentration, 1.0×10^{-4} M.

6-41 The substance $Ba(OH)_2$ is a strong base. It completely dissociates in solution producing two moles of OH^- for every mole dissolved. Before any acid is added, $[OH^-] = 2 \times 0.3750 = 0.7500$ M. The pOH, the negative logarithm of this number, is 0.125; pH= 14.000− pOH $= 14.000 - 0.125 = 13.875$.
The number of moles of OH^- ion in the 100.0 mL of solution under titration is the molarity of the OH^- ion multiplied by the volume in liters of the solution. It equals 0.07500 mol. To reach the equivalence point in a titration with the strong acid $HClO_4$ requires 0.07500 mol of $HClO_4$. The volume of 0.4540 M $HClO_4$ needed to provide this $HClO_4$ is:

$$\frac{1\ \mathrm{L}}{0.4540\ \mathrm{mol\ HClO_4}} \times 0.0750\ \mathrm{mol\ HClO_4} = 0.1652\ \mathrm{L} = 165.2\ \mathrm{mL}$$

When the titration is 1.00 mL short of the equivalence point, 164.2 mL of 0.4540 M $HClO_4$, for a total of only 0.074547 mol of $HClO_4$, has been added. Some OH^- ion remains not neutralized. Its amount is the difference between the amount of OH^- originally present and the amount of $HClO_4$ added so far. This difference equals 4.5×10^{-4} mol OH^-. Note the correct use of only two significant figures in this difference. The concentration of OH^- is:

$$[\mathrm{OH^-}] = \frac{4.5 \times 10^{-4}\ \mathrm{mol}}{0.1000 + 0.1642\ \mathrm{L}} = 0.0017\ \mathrm{M}$$

The pOH is 2.77, and the pH is therefore 11.23.
At the equivalence point, the pH will be 7.000, since this is the titration of a strong base with a strong acid.

When the titration is 1.00 mL past the equivalence point all of the OH^- from $Ba(OH)_2$ has been reacted and an excess of $HClO_4$ is present. The amount of the excess $HClO_4$ is:

$$\left(\frac{0.4540 \text{ mol}}{1 \text{ L}} \times 0.1662 \text{ L}\right) - 0.07500 \text{ mol} = 4.5 \times 10^{-4} \text{ mol}$$

Since $HClO_4$ is a strong acid, it is completely ionized producing 4.5×10^{-4} mol of H_3O^+. The concentration of H_3O^+ is

$$\frac{4.5 \times 10^{-4} \text{ mol}}{0.1000 + 0.1662 \text{ L}} = 0.0017 \text{ M}$$

Hence, pH $= -\log(0.0017) = 2.77$. In working this problem, the autoionization of water is ignored. Even very small amounts of strong acid or base completely overshadow water as a source of H_3O^+ or OH^- ion. Note the way the pH plummets (from 11.23 to 2.77) upon addition of only 2 mL of titrant in the range of the equivalence point.

6-43 The problem describes the titration of a weak base, ethylamine, with a strong acid, HCl. The titration falls into four ranges: *before* the addition of acid; *between* the first addition of acid and the equivalence point; *at* the equivalence point; *beyond* the equivalence point. The pH of the starting 40.00 mL of 0.1000 M ethylamine exceeds 7 because ethylamine is a base. As 0.1000 M HCl is added to this solution, the pH falls.

• *Before Addition of Acid.* Ethylamine raises the pH of pure water by the equilibrium reaction:

$$C_2H_5NH_2(aq) + H_2O(l) \rightleftharpoons C_2H_5NH_3^+(aq) + OH^-(aq)$$

The corresponding mass-action expression is:

$$K_b = 6.41 \times 10^{-4} = \frac{[C_2H_5NH_3^+][OH^-]}{[C_2H_5NH_2]}$$

Let $[OH^-] = y$, and assume that the concentration of hydroxide ion from the autoionization of water is small. Then, because no HCl has been added:

$$[OH^-] = [C_2H_5NH_3^+] = y \quad \text{and} \quad [C_2H_5NH_2] = 0.100 - y$$

$$6.41 \times 10^{-4} = \frac{y^2}{0.1000 - y}$$

This expression is identical in form to the one developed in the text for the titration of acetic acid with NaOH. The difference is that y is now $[OH^-]$. Rearranging it and solving gives $y = [OH^-] = 0.00769$ M. The pOH is 2.11, and the pH is $14.00 - 2.114 = 11.89$. Formulas exactly analogous to the ones developed in the text can be used to compute the pH in the other three ranges. The only complication is that, as just shown, the natural choice for an unknown is $[OH^-]$ rather than $[H_3O^+]$.

• *After First Addition of Acid, Before Equivalence Point.* In this range of the titration:

$$[C_2H_5NH_3{}^+] = \frac{c_tV}{V_0 + V} + y$$

where c_t is the concentration of the titrant (0.1000 M), V_0 is 0.0400 L, the original volume of ethylamine solution, V is the volume of titrant added and y is the concentration of OH^-. The quantity c_tV is the chemical amount of $C_2H_5NH_3^+$ generated by the 1-to-1 reaction between the titrant and the $C_2H_5NH_2(aq)$. Note that $V_0 + V$ is the total volume of the solution. Their quotient is the concentration of $C_2H_5NH_3^+(aq)$ from the neutralization reaction alone. The equilibrium that produces $OH^-(aq)$ is an additional source of $C_2H_5NH_3^+(aq)$. The addition of y on the right-hand side of the above equation takes it into account. Similarly:

$$[C_2H_5NH_2] = \frac{c_0V_0 - c_tV}{V_0 + V} - y$$

where the new symbol, c_0 stands for the original ethylamine concentration (0.1000 M). After 5.00 mL of HCl has been added, $V = 0.00500$ L. Hence:

$$[C_2H_5NH_3^+] = 0.01111 + y \quad \text{and} \quad [C_2H_5NH_2] = 0.07777 - y$$

Putting the concentrations in the base-ionization expression:

$$6.41 \times 10^{-4} = \frac{y(0.01111 + y)}{(0.07777 - y)}$$

In this equation, y is not negligible compared to 0.01111 or 0.07777. Omitting it from the two terms on the right-hand side gives $y = 0.00487$, which is 43% of 0.01111! Solve the equation instead by rearranging it and using the quadratic formula. Solution for y gives $[OH^-] = 0.00331$ M; pH = 11.52.

At 20.00 mL, similar substitution and solution give $[OH^-] = 6.18 \times 10^{-4}$ M, and a pH of 10.79. At 39.90 mL, the same sequence of operations gives a pH of 8.20.

• *At the Equivalence Point.* At equivalence, the titration has produced what is nothing other than 80.00 mL of 0.05 M ethylammonium chloride, $C_2H_5NH_3^+Cl^-$. The cation of this salt is a weak acid:

$$C_2H_5NH_3^+(aq) + H_2O(l) \rightleftharpoons C_2H_5NH_2(aq) + H_3O^+(aq)$$

The equilibrium constant for its acid ionization is: $K_a = K_w/K_b$. Let $x = [H_3O^+]$. Then:

$$K_a = 1.56 \times 10^{-11} = \frac{x^2}{(0.05000 - x)}$$

Solution gives $x = 8.83 \times 10^{-7}$ so the pH $= -\log(8.83 \times 10^{-7}) = 6.05$
Using the formula that works in the range before the equivalence point gives a deceptive result at this point:

$$[C_2H_5NH_2] = 0 - [OH^-]$$

which cannot be right since $[OH^-]$ and $[C_2H_5NH_2]$ in fact must both be positive.
• *Beyond the Equivalence Point.* In this range the solution behaves like a simple solution of HCl. In comparison to the strong acid HCl, the weakly acidic ethylammonium ion contributes essentially nothing to the $H_3O^+(aq)$ concentration.
When 40.10 mL of HCl has been added, the first 40.00 mL has been consumed in producing $C_2H_5NH_3^+(aq)$ ion by neutralization of all the $C_2H_5NH_2(aq)$. The remaining 0.10 mL is free to act as a strong acid. Thus, 0.10 mL of 0.1000 M HCl is diluted to 80.10 mL. Every HCl gives one H_3O^+ so:

$$[H_3O^+] = \frac{0.10}{80.10} \times 0.1000 = 1.25 \times 10^{-4}\ \text{M} \quad \text{The pH is 3.90.}$$

At 50.00 mL, $[H_3O^+] = (10.00/90.00) \times 0.1000 = 1.111 \times 10^{-2}$ M; pH = 1.95.

6-45 Hydrazoic acid, a weak acid, lowers the pH of water by the reaction:

$$HN_3(aq) + H_2O(l) \rightleftharpoons N_3^-(aq) + H_3O^+(aq) \quad K_a = 1.9 \times 10^{-5}$$

Before any base is added, the initial concentration of the HN_3 is 0.1000 M. Let x equal the concentration of H_3O^+ present at equilibrium in the solution. Then:

$$\frac{[N_3^-][H_3O^+]}{[HN_3]} = \frac{x^2}{0.1000 - x} = 1.9 \times 10^{-5}$$

where it has been assumed that the acid ionization of hydrazoic acid is the only significant source of H_3O^+ in the solution. Rearranging gives the quadratic equation $x^2 + (1.9 \times 10^{-5})x - 1.9 \times 10^{-6} = 0$. for which the positive root is $x = 1.37 \times 10^{-3}$. Therefore, before the titration:

$$[H_3O^+] = 1.37 \times 10^{-3}\ \text{M}; \quad \text{pH} = 2.86$$

Sodium hydroxide is a strong base. Each added mole of NaOH converts one mole of $HN_3(aq)$ to one mole of $N_3^-(aq)$. The 25.00 mL of solution that is 0.1000 M in NaOH furnishes:

$$\frac{0.100\ \text{mol OH}^-}{1\ \text{L}} \times 0.250\ \text{L} = 2.500 \times 10^{-3}\ \text{mol OH}^-$$

Assume that the added OH^- reacts completely with the HN_3. The reaction produces 2.500×10^{-3} mol of N_3^- and leaves an excess of 2.500×10^{-3} mol of HN_3. In fact, exactly half of the hydrazoic acid is neutralized, which explains why this is sometimes called the half-equivalence point in a titration. The total volume of the solution is 0.0750 L, so the "original" concentrations of the weak acid and its conjugate base are both 0.0333 M. "Original" is in quotation marks because these concentrations refer to a state after the mixing of the solutions but before the acid-base equilibrium has a chance to get established. When it does go, this equilibrium generates H_3O^+ and changes both concentrations slightly. The changes are so slight that the approximate equation:

$$\text{pH} \approx \text{p}K_a - \log \frac{[\text{HN}_3]_0}{[\text{N}_3^-]_0}$$

is valid at this point in the titration. This equation is developed on page 220 of the text. Also, the calculation of the pH in this stage of a titration is just like the calculation shown in Example 6-8 (text page 219). Substitution give pH = 4.72. Note that the pH equals pK_a of the weak acid being titrated. This is always so at the half-equivalence point in the titration of a weak acid.

The addition of 50.00 mL of the NaOH solution brings the titration exactly to its equivalence point—the number of moles of OH^- added equals the number of moles of NH_3 originally present. Presume that there is no hydrazoic acid present at all. Then the concentration of N_3^- ion is its number of moles, which is 0.00500 mol, divided by the volume of the mixture, which is 0.1000 L. The answer is 0.0500 M. Now, some of the N_3^- ion reacts with water because the ion is a weak base:

	$N_3^-(aq)$	$+H_2O(aq) \rightleftharpoons$	$HN_3(aq) +$	$OH^-(aq)$
Init. Conc. (M)	0.0500	—	0	small
Change in Conc. (M)	$-x$	—	$+x$	$+x$
Equil. Conc. (M)	$0.0500 - x$	—	x	x

The K_b for this reaction is 5.26×10^{-10}, obtained by dividing K_w by K_a for hydrazoic acid Use the base-ionization expression:

$$\frac{[\text{OH}^-][\text{HN}_3]}{[\text{N}_3^-]} = \frac{x^2}{0.0500 - x} = K_b = 5.26 \times 10^{-10}$$

Solving for x gives $[OH^-] = 5.13 \times 10^{-6}$. This corresponds to a pOH of 5.29 and therefore a pH of 8.71.

After 51.00 mL of NaOH(aq) is added, all of the acid has been reacted and there is some OH^- left over. The concentration of OH^- remaining is the amount of OH^- added minus the amount reacted divided by the volume of the solution:

$$[\text{OH}^-] = \frac{(0.05100 \text{ L} \times 0.100 \text{ M}) - 5.000 \times 10^{-3} \text{ mol}}{0.1010 \text{ L}} = 9.910 \times 10^{-4} \text{ M}$$

With this much OH^- remaining in solution, the base ionization to convert N_3^- to HN_3 will not affect the concentration of OH^- significantly:

$$\text{pOH} = -\log(9.901 \times 10^{-4}) = 3.00 \quad \text{hence} \quad \text{pH} = 11.00$$

6-47 Addition of 46.50 mL of the 0.393 M NaOH(aq) solution brings the solution very near to an equivalence point. This is concluded from the fact that at the equivalence point in a titration small additions of base or base cause large changes in pH—in this case that one more drop raises the pH by over one unit. Assume that 46.51 mL of the base corresponds exactly to the equivalence point. The number of moles of NaOH added at the this point is

$$n_{\text{NaOH}} = 0.04651 \text{ L} \times \left(\frac{0.393 \text{ mol OH}^-}{1 \text{ L}}\right) = 0.01828 \text{ mol}$$

This strong base finishes off the neutralization of the HCl, but some of the HCl was neutralized by the benzoate ion ($C_6H_5COO^-$), also a base, present in solution from the original sample. That is,

$$n_{\text{HCl}} = n_{\text{NaOH}} + n_{\text{C}_6\text{H}_5\text{COO}^-}$$

The number of moles of HCl is:

$$n_{\text{HCl}} = 0.0500 \text{ L} \times \left(\frac{0.500 \text{ mol HCl}}{1 \text{ L}}\right) = 0.0250 \text{ mol}$$

Substitution gives:

$$0.0250 = 0.01828 + n_{\text{C}_6\text{H}_5\text{COO}^-} \qquad \text{hence } n_{\text{C}_6\text{H}_5\text{COO}^-} = 6.722 \times 10^{-3} \text{ mol}$$

The mass of C_6H_5COONa in the sample is

$$6.722 \times 10^{-3} \text{ mol C}_6\text{H}_5\text{COO}^- \times \left(\frac{1 \text{ mol C}_6\text{H}_5\text{COONa}}{1 \text{ mol C}_6\text{H}_5\text{COO}^-}\right) \times$$

$$\left(\frac{144.11 \text{ g C}_6\text{H}_5\text{COONa}}{1 \text{ mol C}_6\text{H}_5\text{COONa}}\right) = 0.969 \text{ g C}_6\text{H}_5\text{COONa}$$

6-49 Diethylamine and hydrochloric acid react in a 1-to-1 molar ratio. Therefore, the chemical amount of HCl necessary to reach the equivalence point equals the chemical amount of diethylamine originally present. The chemical amount of HCl used equals the volume of the HCl solution that was added times the molarity of that solution:

$$n_{\text{HCl}} = (15.90 \text{ mL}) \times \left(\frac{0.0750 \text{ mmol HCl}}{1 \text{ mL}}\right) = 1.1925 \text{ mmol}$$

Thus there was originally 1.1925 mmol of diethylamine. Convert to in grams:

$$1.1925 \times 10^{-3} \text{ mol} \times \left(\frac{73.14 \text{ g (C}_2\text{H}_5)_2\text{NH}}{1 \text{ mol (C}_2\text{H}_5)_2\text{NH}}\right) = 0.0872 \text{ g (C}_2\text{H}_5)_2\text{NH}$$

Imagine that at the equivalence point, all of the diethylamine is converted to its conjugate acid, the diethylammonium ion, $(C_2H_5)_2NH_2^+$. If so, the concentration of diethylammonium ion equals its chemical amount, 1.1925 mmol, divided by the volume of the solution (115.90 mL); the concentration would be 0.0103 M. In actuality, some of the diethylammonium ion is reacted away as it donates H^+ ions to increase the H_3O^+ concentration in the solution:

$$(C_2H_5)_2NH_2^+(aq) + H_2O(l) \rightleftharpoons (C_2H_5)_2NH + H_3O^+(aq)$$

The equilibrium expression for this reaction is:

$$K_a = \frac{[H_3O^+][(C_2H_5)_2NH]}{[(C_2H_5)_2NH^+]}$$

where K_a is K_w divided by the K_b of diethylamine. Let x stand for the concentration of H_3O^+, and assume that all H_3O^+ in the solution comes from the acid ionization of the diethylammonium ion. Then:

$$K_a = \frac{K_w}{K_b} = \frac{1.0 \times 10^{-14}}{3.09 \times 10^{-4}} = 3.236 \times 10^{-11} = \frac{x^2}{0.0103 - x}$$

Solving give $x = 5.77 \times 10^{-7}$ for a pH of 6.24.
This concentration of H_3O^+ is only about six times larger than the concentration furnished by autoionization in pure water. How valid then is the assumption that the acid ionization of the diethylammonium ion furnishes all of the H_3O^+? One way to check is to use an expression (page 233 of the text) that takes the autoionization of water into account in solutions of weak acids:

$$[H_3O^+]^3 + K_a[H_3O^+]^2 - (K_w + c_a)[H_3O^+] - K_aK_w = 0$$

Inserting $K_w = 1.0 \times 10^{-14}$, $c_a = 0.0103$, and $K_a = 3.24 \times 10^{-11}$ gives the cubic equation:

$$[H_3O^+]^3 + 3.24 \times 10^{-11}[H_3O^+]^2 - 3.433 \times 10^{-13}[H_3O^+] - 3.24 \times 10^{-25} = 0$$

Solving this equation analytically is fairly time-consuming. A better way is to reason that the last term is certainly much smaller than any of the other three because $[H_3O^+]$ can only be larger than 5.77×10^{-7}. After all, a second source of hydronium ion can only raise the concentration of that ion, not lower it. Omitting the last term on this basis and dividing through by $[H_3O^+]$ gives:

$$[H_3O^+]^2 + 3.24 \times 10^{-11}[H_3O^+] - 3.433 \times 10^{-13} = 0$$

The applicable root of this quadratic equation is $[H_3O^+] = 5.86 \times 10^{-7}$ M, for a pH of 6.23. Including the contribution of water autoionization makes only a minimal

difference in the pH at the equivalence point on this titration, despite the fact that the diethylammonium ion is a very weak acid. It is *wrong* to take autoionization into consideration by naively adding 1.0×10^{-7} to the answer 5.77×10^{-7} M! Water autoionization contributes far less hydronium ion to this solution than the 1.0×10^{-7} M it furnishes in pure water. The reason is that the autoionization reaction is shifted to the left by the hydronium ion from the diethylammonium ion, in accord with LeChatelier's principle.
A suitable indicator for the titration is bromothymol blue, which changes color in the pH range surrounding 6.24 (see Figure 6-5 on text page 210).

6-51 This buffer solution contains $C_6H_{13}NO$ (call it "morph") and its conjugate acid $C_6H_{13}NOH^+$ (morphH^+), which forms in considerable quantity from the reaction of the morph with HCl. The two species are in chemical equilibrium

$$\text{morph}(aq) + H_2O(l) \rightleftharpoons \text{morphH}^+(aq) + OH^-(aq) \quad K_b = \frac{[OH^-][\text{morphH}^+]}{[\text{morph}]}$$

This expression can be used to compute K_b if equilibrium values of all of the concentrations on the right can determined. The concentration of OH^- is easy because the pH of the solution is given: $[OH^-] = 1.0 \times 10^{-7}$. There was 10.00 mmol of morph in the solution before the addition of the HCl. The added HCl amounts to 8.00 mmol. These two chemical amounts were figured by multiplying the concentrations of the solutions by their respective volumes. If the neutralization reaction goes to completion, it forms 8.00 mmol of morphH^+ and leaves 2.00 mmol of morph unreacted (in excess). The volume of the solution rises to 58.00 mL upon addition of the acid; additional water is added to make the volume of the solution up to 100.0 mL. The equilibrium concentrations of morph and morphH^+ are:

$$[\text{morph}] = \frac{2.00 \text{ mmol}}{100.0 \text{ mL}} - 1.0 \times 10^{-7} \text{ M}; \ [\text{morphH}^+] = \frac{8.00 \text{ mmol}}{100.0 \text{ mL}} + 1.0 \times 10^{-7} \text{ M}$$

The subtraction and addition of the 1.0×10^{-7} accounts for the slight amount of conversion of morph to morphH^+ by the action of the equilibrium. Substitution into the base-ionization equilibrium expression now gives K_b:

$$K_b = \frac{[OH^-][\text{morphH}^+]}{[\text{morph}]} = \frac{(1.0 \times 10^{-7})(0.0800)}{(0.0200)} = 4 \times 10^{-7}$$

Note that 1.0×10^{-7} is so small compared to 0.0200 or 0.0800 that actually subtracting or adding it is not worth the trouble.

6-53 Arsenic acid in aqueous solution donates hydrogen ions in three steps. Each step has a successively smaller K_a:

$$H_3AsO_4(aq) + H_2O(l) \rightleftharpoons H_2AsO_4^-(aq) + H_3O^+(aq) \quad K_{a1} = 5.0 \times 10^{-3}$$
$$H_2AsO_4^-(aq) + H_2O(l) \rightleftharpoons HAsO_4^{2-}(aq) + H_3O^+(aq) \quad K_{a2} = 9.3 \times 10^{-8}$$
$$HAsO_4^{2-}(aq) + H_2O(l) \rightleftharpoons AsO_4^{3-}(aq) + H_3O^+(aq) \quad K_{a3} = 3.0 \times 10^{-12}$$

The K_{a2} is many thousands of times smaller than K_{a1}, and K_{a3} is thousands of times smaller yet. This suggests that the first ionization alone is significant in producing H_3O^+ and that the contributions of the two subsequent stages of ionization are negligible by comparison. The plan of attack then is to compute the hydronium-ion concentration as if the first step occurred separately and then to get the other concentrations by substitution into the equilibrium expressions for the following steps. Ignoring the interaction of the equilibria means not having to solve complicated systems of simultaneous equations. The proper approach for the first step considered separately is just the approach used in problem 6-21a. Let x be the equilibrium H_3O^+ concentration, which equals the equilibrium $H_2AsO_4^-$ concentration. Writing the equilibrium expression then gives:

$$\frac{x^2}{(0.1000 - x)} = K_a = 5.0 \times 10^{-3}$$

After rearrangement and substitution into the quadratic formula this becomes:

$$x = \frac{-5.0 \times 10^{-3} \pm \sqrt{2.50 \times 10^{-5} + 2.0 \times 10^{-3}}}{2}$$

The positive root of the equation is 0.0200. Thus: $[H_3AsO_4]$ is 0.080 M; $[H_2AsO_4^-]$ is 0.020 M; $[H_3O^+]$ is 0.020 M. Now consider the second step in the step-wise ionization. Let y equal the concentration of $HAsO_4^{2-}$ produced at equilibrium:

	$H_2AsO_4^-(aq)$	$+H_2O(l) \rightleftharpoons$	$HAsO_4^{2-}(aq) +$	$H_3O^+(aq)$
Init. Conc. (M)	0.0200	—	0	0.0200
Change in Conc. (M)	$-y$	—	$+y$	$+y$
Equil. Conc. (M)	$0.020 - y$	—	y	y

Use of the equilibrium expression for K_{a2} gives the equation

$$9.3 \times 10^{-8} = \frac{y(0.0200 + y)}{0.0200 - y}$$

This equation is very easily solved once it is realized that y must be small compared to 0.0200. Then $y = 9.3 \times 10^{-8}$ M $= [HAsO_4^{2-}]$. Note that $[HAsO_4^{2-}]$ is equal to K_{a2}. Finally, consider the third ionization. Let z equal the concentration of AsO_4^{3-} produced at equilibrium:

	$HAsO_4^{2-}(aq)$	$+H_2O(l) \rightleftharpoons$	$AsO_4^{3-}(aq) +$	$H_3O^+(aq)$
Init. Conc. (M)	9.3×10^{-8}	—	0	0.0200
Change in Conc. (M)	$-z$	—	$+z$	$+z$
Equil. Conc. (M)	$9.8 \times 10^{-8} - z$	—	z	$0.0200 + z$

Use of the equilibrium expression for K_{a3} gives the equation:

$$3.0 \times 10^{-12} = \frac{z(0.0200 + z)}{9.8 \times 10^{-8} - z} \quad \text{from which} \quad z = 1.4 \times 10^{-17}\ \text{M} = [\text{AsO}_4^{3-}]$$

This is a very small concentration. One liter of the arsenic acid solution fewer than ten million AsO_4^{3-} ions!

6-55 The phosphate ion gains hydrogen ions from water in three stages:

$$PO_4^{3-}(aq) + H_2O(l) \rightleftharpoons HPO_4^{2-}(aq) + OH^-(aq) \qquad K_{b1} = K_w/K_{a3} = 4.55 \times 10^{-2}$$
$$HPO_4^{2-}(aq) + H_2O(l) \rightleftharpoons H_2PO_4^-(aq) + OH^-(aq) \qquad K_{b2} = K_w/K_{a2} = 1.61 \times 10^{-7}$$
$$H_2PO_4^-(aq) + H_2O(l) \rightleftharpoons H_3PO_4(aq) + OH^-(aq) \qquad K_{b3} = K_w/K_{a1} = 1.33 \times 10^{-12}$$

The same kind of simplifying approximation that worked in problem 6-53 will work in this problem. Each successive equilibrium may be treated independently. Consider the first stage of the reaction of the PO_4^{3-} ion as a base:

	$PO_4^{3-}(aq)$	$+H_2O(l) \rightleftharpoons$	$HPO_4^{2-}(aq)+$	$OH^-(aq)$
Init. Conc. (M)	0.050	—	0	small
Change in Conc. (M)	$-x$	—	$+x$	$+x$
Equil. Conc. (M)	$0.050 - x$	—	x	x

Let x be the equilibrium OH^- concentration, which equals the equilibrium HPO_4^{2-} concentration. Writing the equilibrium expression then gives:

$$\frac{x^2}{(0.050 - x)} = K_b = 4.55 \times 10^{-2}$$

After rearrangement and substitution into the quadratic formula:

$$x^2 + (4.55 \times 10^{-2})x - 2.27 \times 10^{-3} = 0 \quad \text{which gives} \quad x = 3.02 \times 10^{-2}$$

Hence: $[OH^-] = 0.030$ M; $[HPO_4^{2-}] = 0.030$ M; $[PO_4^{3-}] = 0.050 - 0.0302 = 0.020$ M. Consider the second stage as a completely separate reaction:

	$HPO_4^{2-}(aq)$	$+H_2O(l) \rightleftharpoons$	$H_2PO_4^-(aq)+$	$OH^-(aq)$
Init. Conc. (M)	0.0302	—	0	0.0302
Change in Conc. (M)	$-y$	—	$+y$	$+y$
Equil. Conc. (M)	$0.0302 - y$	—	y	$0.0302 + y$

Use of the equilibrium expression for K_{b2} gives the equation

$$1.61 \times 10^{-7} = \frac{y(0.0302 + y)}{0.0302 - y}$$

This equation is very easily solved once it is realized that y must be small compared to 0.0302. Then $y = 1.61 \times 10^{-7} = [H_2PO_4^-]$. Finally, consider the third stage of the base ionization. Let z equal the concentration of H_3PO_4 produced at equilibrium:

	$H_2PO_4^-(aq)$	$+H_2O(l) \rightleftharpoons$	$H_3PO_4(aq) +$	$OH^-(aq)$
Init. Conc. (M)	1.61×10^{-7}	—	0	0.0302
Change in Conc. (M)	$-z$	—	$+z$	$+z$
Equil. Conc. (M)	$1.61 \times 10^{-7} - z$	—	z	$0.0302 + z$

Use of the equilibrium expression for K_{b3} gives the equation

$$1.33 \times 10^{-12} = \frac{z(0.0302 + z)}{1.61 \times 10^{-7} - z}$$

This equation is easily solved because z is small compared to 1.61×10^{-7}. The result is $z = 7.1 \times 10^{-18}$ M $= [H_3PO_4]$.

6-57 Henry's Law states that the solubility of a gas is directly proportional to its partial pressure above the solvent:

$$P_{gas} = kX$$

where X is the mole fraction of the gas and k is a constant. The first task is to determine k for carbon dioxide in water. The problem states that 1.00 g of $H_2O(l)$ dissolves 0.759 cm^3 of $CO_2(g)$ at 25°C and 1 atm pressure. Assuming that the $CO_2(g)$ behaves ideally, this is 3.10×10^{-5} mol of $CO_2(g)$. The mole fraction of CO_2 in the solution is the number of moles of CO_2 divided by the total number of moles present. Using 18.015 g mol^{-1} as the molar mass of water, and 1.00 g cm^{-3} as the density, there is 55.51×10^{-3} mol of $H_2O(l)$ in a cm^3 of water. Hence, the mole fraction: of CO_2 is

$$X_{CO_2} = \frac{3.102 \times 10^{-5} \text{ mol}}{55.51 \times 10^{-3} \text{ mol} + 3.102 \times 10^{-5} \text{ mol}} = 5.58 \times 10^{-4}$$

Substituting into the Henry's law expression gives:

$$1.000 \text{ atm} = k(5.588 \times 10^{-4}) \quad \text{and} \quad k = 1.79 \times 10^2 \text{ atm}$$

At Denver, the partial pressure of CO_2 is only 0.833 atm, but k is of course the same as everywhere else. Applying Henry's law gives the mole fraction of CO_2 in saturated water at Denver as:

$$X_{CO_2} = \frac{0.833 \text{ atm}}{1.79 \times 10^2 \text{ atm}} = 4.65 \times 10^{-4}$$

In a liter of saturated water (with density 1.00 g cm^{-3}) at Denver there is 55.51 mol of water. There must be 2.58×10^{-2} mol of CO_2 in this liter of water to make the mole fraction of CO_2 come out to 4.65×10^{-4}. Thus, the molarity of CO_2 in the water at Denver is 2.58×10^{-2} M.

Now go on to find the pH of the solution. Dissolved CO_2 decreases the pH of the water because it is a diprotic weak acid:

$$CO_2(g) + 2H_2O(l) \rightleftharpoons HCO_3^-(aq) + H_3O^+(aq) \quad K_{a1} = 4.3 \times 10^{-7}$$
$$HCO_3^-(aq) + H_2O(l) \rightleftharpoons CO_3^{2-}(aq) + H_3O^+(aq) \quad K_{a2} = 4.8 \times 10^{-11}$$

The difference between K_{a1} and K_{a2} is about four orders of magnitude. This large difference justifies ignoring the second equilibrium as a source of $H_3O^+(aq)$. Working from the first equilibrium and letting $x = [H_3O^+]$:

$$4.3 \times 10^{-7} = \frac{x^2}{2.58 \times 10^{-2} - x} \qquad \text{so that } x = 1.05 \times 10^{-4}$$

If $[H_3O^+] = 1.05 \times 10^{-4}$ M, then the pH is 3.98.

6-59 The major natural contributor to the acidity of rainwater is dissolved CO_2, which reacts to form carbonic acid: $CO_2(g)+H_2O(l) \rightleftharpoons H_2CO_3(aq)$. In recent times, $SO_3(g)$ and $NO_2(g)$ pollutants have also become significant contributors to the acidity of rain. The two steps in the ionization of carbonic acid are:

$$H_2CO_3(aq) + H_2O(l) \rightleftharpoons HCO_3^-(aq) + H_3O^+(aq) \quad K_{a1} = 4.3 \times 10^{-7}$$
$$HCO_3^-(aq) + H_2O(l) \rightleftharpoons CO_3^{2-}(aq) + H_3O^+(aq) \quad K_{a2} = 4.8 \times 10^{-11}$$

At equilibrium, the following two mathematical equations are satisfied, based on the preceding two chemical equations:

$$K_{a1} = \frac{[H_3O^+][HCO_3^-]}{[H_2CO_3]}; \quad K_{a2} = \frac{[H_3O^+][CO_3^{2-}]}{[HCO_3^-]}$$

The pH of the raindrop is 5.6. It follows that $[H_3O^+] = 2.51 \times 10^{-6}$ M. Substituting the known value of $[H_3O^+]$ and the two K_a's gives:

$$0.1731 = \frac{[HCO_3^-]}{[H_2CO_3]}; \quad 1.912 \times 10^{-5} = \frac{[CO_3^{2-}]}{[HCO_3^-]}$$

It is convenient to recast these equations so that the concentration of the same species, say HCO_3^-, is in the denominator of each of the fractions:

$$5.777 = \frac{[H_2CO_3]}{[HCO_3^-]}; \quad 1.912 \times 10^{-5} = \frac{[CO_3^{2-}]}{[HCO_3^-]}$$

The first of this pair of equations is the reciprocal of the first of the preceding pair. The second is identical to the second of the preceding pair. The fraction f of any of the three species present equals its concentration divided by the sum of the concentrations of all three. For example:

$$f_{H_2CO_3} = \frac{[H_2CO_3]}{[H_2CO_3] + [HCO_3^-] + [CO_3^{2-}]}$$

This expression can be simplified by dividing numerator and denominator by $[HCO_3^-]$ and inserting the ratios just calculated:

$$f_{H_2CO_3} = \frac{[H_2CO_3]/[HCO_3^-]}{[H_2CO_3]/[HCO_3^-] + 1 + [CO_3^{2-}]/[HCO_3^-]}$$

$$f_{H_2CO_3} = \frac{5.777}{5.777 + 1 + 1.912 \times 10^{-5}} = \frac{5.777}{6.777} = 0.852$$

Similar expressions are generated to compute the fractions of the other three species by changing the numerator as required. The resulting fractions are 0.1476 for HCO_3^-, and 2.82×10^{-6} for CO_3^{2-}. The total concentration of all three forms of the carbon-containing species is 1.0×10^{-5} M. The desired concentrations are the respective fractions times this total:

$$[H_2CO_3] = 8.5 \times 10^{-6}\text{ M}; \quad [HCO_3^-] = 1.5 \times 10^{-6}\text{ M}; \quad [CO_3^{2-}] = 2.8 \times 10^{-11}\text{ M}.$$

6-61 The molar mass of thiamine hydrochloride is 337.27 g mol^{-1}. Placing 3.0×10^{-5} g of this substance in 1.00 L of water makes a solution that is 8.89×10^{-8} M in $ThiH^+Cl^-$, (analogous to $NH_4^+Cl^-$). The $ThiH^+(aq)$ cation is a weak acid:

$$ThiH^+(aq) + H_2O(l) \rightleftharpoons Thi(aq) + H_3O^+(aq) \qquad K_a = 3.4 \times 10^{-7}$$

This equilibrium produces only small amounts of $H_3O^+(aq)$ because K_a is small and the original concentration of $ThiH^+(aq)$ is quite small. The simultaneous equilibrium:

$$2\,H_2O(l) \rightleftharpoons H_3O^+(aq) + OH^-(aq)$$

must be considered as a source of $H_3O^+(aq)$.

The following four mathematical relationships always hold in this solution:

$$3.4 \times 10^{-7} = \frac{[Thi][H_3O^+]}{[ThiH^+]} = K_a \qquad 1.0 \times 10^{-14} = [H_3O^+][OH^-] = K_w$$

$$8.89 \times 10^{-8} = [Thi] + [ThiH^+] = c_a \qquad [H_3O^+] + [ThiH^+] = [OH^-] + [Cl^-]$$

The last equation follows from the principle of electrical neutrality: for every positive charge in the solution there must be a negative charge. The second-to-last equation represents a material balance. Whatever the distribution between its two possible forms, the *total* concentration of thiamine-material is known. The first two equations are the usual mass-action expressions.

The $[Cl^-]$ is 8.89×10^{-8} M because $Cl^-(aq)$ does not react to any extent with other species. The set of four equations therefore involves four unknowns. It is a question of careful algebra to solve for $[H_3O^+]$. The details of the algebra are given in the text for a similar case (See Section 6-7). The result is a cubic equation in $[H_3O^+]$:

$$[H_3O^+]^3 + K_a[H_3O^+]^2 - (K_w + cK_a)[H_3O^+] - K_aK_w = 0$$

When the numbers specific to this case are substituted:

$$[H_3O^+]^3 + (3.4 \times 10^{-7})[H_3O^+]^2 - (4.03 \times 10^{-14})[H_3O^+] - 3.4 \times 10^{-21} = 0$$

Direct solution of the original cubic equation gives two meaningless negative roots and $[H_3O^+] = 1.367 \times 10^{-7}$ M for a pH of 6.86. A good alternate route to this solution is to guess a value of $[H_3O^+]$ near 7 and proceed by successive approximations.

If the autoionization of water is (wrongly!) neglected, the result is the quadratic equation: $[H_3O^+]^2 + (3.4 \times 10^{-7})[H_3O^+] - 3.02 \times 10^{-14} = 0$. Solving gives $[H_3O^+] = 7.31 \times 10^{-8}$ M for a pH of 7.14. This pH is on the basic side of 7, which is of course impossible when an acid is dissolved in water.

6-63 Maleic acid is a diprotic acid, but the two ionization constants differ by about four orders of magnitude. The two ionization steps can therefore be treated separately. Before any 0.1000 M NaOH is added, the predominant source of $H_3O^+(aq)$ in the solution is the equilibrium:

$$H_2mal(aq) + H_2O(l) \rightleftharpoons Hmal^-(aq) + H_3O^+(aq)$$

Let y equal $[H_3O^+]$. Then:

$$K_{a1} = 1.42 \times 10^{-2} = \frac{[Hmal^-][H_3O^+]}{[H_2mal]} = \frac{y^2}{0.1000 - y}$$

Solving this equation for y gives 0.0313. This is the concentration of $[H_3O^+]$ in mol L^{-1}. The pH equals 1.51.

As the first 5.00 mL of 0.1000 M NaOH is added to the aqueous maleic acid, it neutralizes $H_2mal(aq)$ to $Hmal^-(aq)$. The 5.00×10^{-3} L of the 0.1000 M NaOH contributes 5.00×10^{-4} mol of NaOH. There is originally 50.00×10^{-4} mol of H_2mal. The base is the limiting reagent, and the maximum yield of $Hmal^-(aq)$ is 5.00×10^{-4} mol. The reaction leaves 45.00×10^{-4} mol of $H_2mal(aq)$ unreacted. Adding 5.00 mL of dilute aqueous solution raises the volume of the solution to 55.00 mL so, still assuming complete neutralization:

$$[Hmal^-] = 9.091 \times 10^{-3} \text{ M} \quad \text{and} \quad [H_2mal] = 8.182 \times 10^{-2} \text{ M}$$

The equilibrium now acts to alter these concentrations slightly. It generates hydronium ions. For every $H_3O^+(aq)$ that the equilibrium produces, it uses up one $H_2mal(aq)$ and makes one additional $Hmal^-(aq)$. Thus:

$$1.42 \times 10^{-2} = y\frac{(9.091 \times 10^{-3}) + y}{(8.182 \times 10^{-2}) - y}$$

Rearranging and solving this quadratic equation for y gives 0.0244 M. This is the concentration of $H_3O^+(aq)$. The pH equals 1.61.

At the next point in the titration, 25.00 mL of titrant has been added, and the total volume is 75.00 mL. The equation develops:

$$1.42 \times 10^{-2} = y\frac{0.0333 + y}{0.0333 - y}$$

Solving gives $y = 8.45 \times 10^{-3}$ and pH $= 2.07$.

When 50.00 mL of 0.1000 M NaOH has been added the titration has exactly attained the *first* equivalence point. The solution consists of 100.00 mL of 0.0500 M Na^+Hmal^- (sodium hydrogen maleate). The $Hmal^-(aq)$ ion is amphoteric. It behaves as an acid:

$$Hmal^-(aq) + H_2O(l) \rightleftharpoons H_3O^+(aq) + mal^{2-}(aq) \quad K_{a2} = 8.57 \times 10^{-7}$$

and it behaves as a base:

$$Hmal^-(aq) + H_2O(l) \rightleftharpoons OH^-(aq) + H_2mal(aq) \quad K_b = \frac{K_w}{K_{a1}} = 7.04 \times 10^{-13}$$

$Hmal^-(aq)$ is just like $HCO_3^-(aq)$ in this respect. By an analysis just like the one presented for $HCO_3^-(aq)$ in Section 6-7 (text page 234):

$$[H_3O^+] \approx \sqrt{\frac{K_{a1}K_{a2}[Hmal^-]_0 + K_{a1}K_w}{K_{a1} + [Hmal^-]_0}}$$

where $[Hmal^-]_0$ is the "original" concentration of $Hmal^-(aq)$, 0.0500 M. Solving gives $[H_3O^+] = 9.73 \times 10^{-5}$ M for a pH of 4.01. Notice that the approximate formula:

$$[H_3O^+] \approx \sqrt{K_{a1}K_{a2}}$$

gives a wrong pH of 3.96. It fails because the formula depends on the assumption that K_{a1} is negligible compared to $[Hmal^-]$. But K_{a1} is in fact equal to 0.0142, which is 28 percent of 0.050.

After 75.00 mL of 0.1000 M NaOH has been added, the titration is half-way to the *second* equivalence point. In this range, the main source of $H_3O^+(aq)$ is the second ionization:

$$Hmal^-(aq) + H_2O(l) \rightleftharpoons mal^{2-}(aq) + H_3O^+(aq) \quad K_{a2} = 8.57 \times 10^{-7}$$

The concentration of $H_2mal(aq)$ is quite close to zero because so much base has been added. The first ionization accordingly now has only a negligible effect on the pH. The titration can be viewed as the addition of 25.00 mL of 0.1000 NaOH to 100.00 mL of 0.0500 NaHmal, making a total volume of 125.00 mL. Assuming complete neutralization, this much NaOH converts exactly half of the $Hmal^-(aq)$ to $mal^{2-}(aq)$. Dilution then reduces the concentrations of both species by the factor 100/125. Thus *after* neutralization but *before* the action of any equilibrium:

$$[Hmal^-] = 0.025 \times \frac{100}{125} = 0.020 \text{ M}; \quad [mal^{2-}] = 0.025 \times \frac{100}{125} = 0.020 \text{ M}$$

The K_{a2} acid-ionization equilibrium now acts to change these concentrations slightly. It adds y to the concentration of $mal^{2-}(aq)$ and removes y from the concentration of $Hmal^-(aq)$, where y is the concentration of hydronium ion that it produces. The mass-action expression is:

$$K_{a2} = 8.57 \times 10^{-7} = y\frac{0.020 - y}{0.020 + y}$$

Solving gives $y = 8.57 \times 10^{-7}$ so the pH is 6.07. By similar computations, the pH after the addition of 99.90 mL of 0.1000 M NaOH is 8.77.

At the *second* equivalence point the solution is really 150.00 mL of 0.0333 M Na_2mal. Consider the *hydrolysis* of $mal^{2-}(aq)$ as the principal equilibrium at this point:

$$mal^{2-}(aq) + H_2O(l) \rightleftharpoons Hmal^-(aq) + OH^-(aq)$$

The K_b for this equilibrium is $(K_w/8.57 \times 10^{-7})$ or 1.17×10^{-8}. Let x be the equilibrium concentration of $OH^-(aq)$. Then, based on the mass-action expression for the K_b equilibrium:

$$K_{b1} = 1.17 \times 10^{-8} = \frac{x^2}{0.0333 - x}$$

Solving for x gives 1.97×10^{-5} so the pOH is 4.71 and the pH is 9.29.

After 105 mL of 0.1000 M NaOH has been added, the above hydrolysis is completely overshadowed as a source of $OH^-(aq)$ by the excess strong base NaOH. The first 100.00 mL of NaOH was used up neutralizing the diprotic acid. The remaining 5.00 mL of 0.100 NaOH makes a solution that is $(5.00/155.00) \times 0.100$ or 3.23×10^{-3} M in $OH^-(aq)$. This corresponds to pH = 11.51. The calculation requires no use of equilibrium constants. NaOH is a strong base in water. This means that its K_b is large.

6-65 Use the relations: pOH = 14.00 − pH; $[H_3O^+] = 10^{-pH}$; $[OH^-] = 10^{-pOH}$.

a) $[H_3O^+] = 1.58 \times 10^{-3}$ M; pOH = 11.2 $\Longrightarrow$ $[OH^-] = 6.31 \times 10^{-12}$ M

b) $[H_3O^+] = 7.94 \times 10^{-5}$ M; pOH = 9.90 $\Longrightarrow$ $[OH^-] = 1.26 \times 10^{-10}$ M

c) $[H_3O^+] = 7.94 \times 10^{-5}$ M; pOH $= 9.90 \Longrightarrow [OH^-] = 1.26 \times 10^{-10}$ M
d) $[H_3O^+] = 3.16 \times 10^{-9}$ M; pOH $= 5.50 \Longrightarrow [OH^-] = 3.16 \times 10^{-6}$ M
e) $[H_3O^+] = 1.26 \times 10^{-12}$ M; pOH $= 2.10 \Longrightarrow [OH^-] = 7.9 \times 10^{-3}$ M

6-67 a) The density of the nitric acid is 1.503 g cm^{-3}, which is 1.503 kg L^{-1}. The nitrate ion has a molality of 0.25 mol kg^{-1} in this solution. The molarity of the nitrate ion is then:

$$\left(1.503\ \frac{\text{kg}}{\text{L}}\right) \times \left(0.25\ \frac{\text{mol NO}_3^-}{\text{kg}}\right) = 0.38 \text{ mol L}^{-1} = 0.38 \text{ M}$$

b) In this case, $HNO_3(l)$ is the solvent, and H_2O is one of the products of the autoionization. The HNO_3 is therefore omitted from the autoionization expression on the grounds it is a pure solvent, but the H_2O is included:

$$[NO_3^-][NO_2^+][H_2O] = K_{HNO_3}$$

If the concentration of nitrate ion is 0.38 M, then the concentrations of nitronium ion and of water are also 0.38 M—the reaction is the only source of any of the three species. $K_{HNO_3} = (0.38)^3 = 5.5 \times 10^{-2}$.

6-69 The pH should be low. LeChatelier's principle indicates that increasing $[H_3O^+]$ shifts the equilibrium in the problem to the left, favoring $Cl_2(aq)$ at the expense of $Cl^-(aq)$.

6-71 Write down the acid-ionization equilibria equation and solve for the equilibrium concentration of H_3O^+ at both temperatures. At 25°C let $[H_3O^+]$ be y. Then:

$$K_a = 1.76 \times 10^{-5} = \frac{y^2}{0.10 - y} \quad \text{from which} \quad y^2 + (1.76 \times 10^{-5})y - 1.76 \times 10^{-6} = 0$$

The applicable root of the quadratic equation is 1.318×10^{-3}, so at 25°C the pH is 2.88. Repeat the calculation using the conditions at 50°C. Let x represent the concentration of H_3O^+. Some attention must immediately be paid to the change in the initial molarity of the acetic acid caused by the expansion of the solution going from 25 to 50°. The volume at 50°C is 1.012 times the volume at 25°C (this is calculated from the fact that the density is 0.9881 times the density at 25°C: 1/0.9881 = 1.0121). The $[HOAc]_0$ is therefore 0.0988 M at 50°C. The rest is routine:

$$1.63 \times 10^{-5} = \frac{x^2}{0.0988 - x} \quad \text{from which} \quad x^2 + (1.63 \times 10^{-5})x - 1.61 \times 10^{-6} = 0$$

The applicable root is 0.00126 so at 50°C the pH is 2.90. The pH is higher at the higher temperature because the acid weakens and the solution is more dilute. The extent of the autoionization of water, which also depends on temperature, is not considered in either calculation. I is a completely negligible source of hydronium ion at both temperatures.

6-73 The 0.100 M solutions of HCl and NH_4Cl are acidic. The 0.100 M solutions of Na_3PO_4 and $NaCH_3COO$ are basic. The 0.100 M solution of KNO_3 is neutral. The compounds are, in order, a strong acid, the salt of a strong acid and a weak base, the salt of a weak acid and a strong base, the salt of a weak acid and a strong base, and the salt of a strong acid and a strong base.

6-75 The initial concentrations of C_6F_5COOH and $C_6F_5COO^-$ are:

$$[C_6F_5COOH]_0 = \frac{0.050 \text{ mol}}{2.00 \text{ L}} = 0.025 \text{ M}; \quad [C_6F_5COO^-]_0 = \frac{0.060 \text{ mol}}{2.00 \text{ L}} = 0.030 \text{ M}$$

These are the concentrations *after* the complete mixing of the two solutions, but before any reactions involving the pentafluorobenzoic acid and its conjugate base the pentafluorobenzoate ion have a chance to occur. Now set up the problem as in Example 6-6 (text page 217), or use the general expression derived and discussed on text page 233-4. Let y equal the equilibrium concentration of hydronium ion. Then:

$$K_a = \frac{[H_3O^+][C_6F_5COO^-]}{[C_6F_5COOH]} = \frac{y(0.030 + y)}{(0.025 - y)} = 0.033$$

If y is neglected as small in comparison to 0.025 and 0.030, then the equation is very easy to solve, and y equals 0.0275. This is obviously *wrong* because 0.0275 is actually larger than 0.025 rather than being a great deal smaller. When such things happen, it is necessary to start over and set up the equation in y without neglecting anything. The result is: $y^2 + 0.063y - 0.000825 = 0$. Use of the quadratic formula gives the positive root $y = 0.01113$. It follows that the pH of the buffer is 1.95.

6-77 Procedure d) would not make an effective buffer. A good buffer exists when a weak acid and its conjugate base are mixed, both in substantial amounts, in a solution. Procedure d) yields a solution that is 1.77×10^{-5} M in HCl mixed with some NaCl. The solution would change pH greatly upon addition of either acid or base. All of the other solutions are acetate buffers.

6-79 a) The 0.100 M solution of the weak acid HA must have a volume of 50.00 mL because it is brought to its equivalence point by 50.00 mL of the 0.1000 M base. The addition of 40.00 mL of the base to the 50.00 mL of acid converts 40.00/50.00 of the acid to its conjugate base and creates a solution with a volume of 90.00 mL. The pH of this 90.00 mL of solution is 4.50:

$$[H_3O^+] = 10^{-4.50} = 3.16 \times 10^{-5} \text{ M}$$

The solution is at equilibrium. At equilibrium, the concentration of HA is the amount that remains not neutralized divided by the volume of the solution minus the small concentration lost by the donation of H^+ to water. Similarly, the concentration of A^- is the amount formed by the neutralization of the HA divided by the volume of the solution plus the small concentration gained by the action of the equilibrium to produce A^-. These values are all known. Hence:

$$K_a = \frac{[H_3O^+][A^-]}{[HA]} = \frac{(3.16\times10^{-5})\left(\frac{40.00(0.1000)}{90.00} + [H_3O^+]\right)}{\left(\frac{10.00(0.1000)}{90.00} - [H_3O^+]\right)} = 1.26\times10^{-4}$$

b) At the equivalence point, the solution is the same as if 5.000 mmol of A^- ion were dissolved in 100.00 mL of water. In such a solution, the initial concentration of A^- ion is 0.05000 M. To determine the pH consider the reaction of this A^- ion, which is a base, with water to give OH^-:

$$A^-(aq) + H_2O(l) \rightleftharpoons HA(aq) + OH^-(aq)$$

The equilibrium expression for this reaction is:

$$\frac{[HA][OH^-]}{[A^-]} = \frac{x^2}{0.05000 - x} = K_b = \frac{K_w}{K_a} = 7.91\times10^{-11}$$

Solving for x gives the hydroxide-ion concentration of 1.99×10^{-6} M. The pOH is accordingly 5.70, and the pH is 8.30.

6-81 Abbreviate phosphonocarboxylic acid as H_3Pho. The three steps in the ionization of this acid are:

$$H_3Pho(aq) + H_2O(l) \rightleftharpoons H_2Pho^-(aq) + H_3O^+(aq) \quad K_{a1} = 1.0\times10^{-2}$$
$$H_2Pho^-(aq) + H_2O(l) \rightleftharpoons HPho^{2-}(aq) + H_3O^+(aq) \quad K_{a2} = 7.8\times10^{-6}$$
$$HPho^{2-}(aq) + H_2O(l) \rightleftharpoons Pho^{3-}(aq) + H_3O^+(aq) \quad K_{a3} = 2.0\times10^{-9}$$

At equilibrium, the following three relationships hold based on the preceding three chemical equations:

$$K_{a1} = \frac{[H_3O^+][H_2Pho^-]}{[H_3Pho]}; \quad K_{a2} = \frac{[H_3O^+][HPho^{2-}]}{[H_2Pho^-]}; \quad K_{a3} = \frac{[H_3O^+][Pho^{3-}]}{[HPho^{2-}]}$$

The pH of the blood is 7.40, and the buffering action of the blood maintains this pH despite the addition of the drug. It follows that $[H_3O^+] = 3.98\times10^{-8}$ M. Substituting the known value of $[H_3O^+]$ and the three K's gives:

$$2.510\times10^5 = \frac{[H_2Pho^-]}{[H_3Pho]}; \quad 195.927 = \frac{[HPho^{2-}]}{[H_2Pho^-]}; \quad 0.05024 = \frac{[Pho^{3-}]}{[HPho^{2-}]}$$

It is convenient to recast these equations so that the concentration of the same species, say H_2Pho^-, is in the denominator of each of the fractions:

$$3.984 \times 10^{-6} = \frac{[H_3Pho]}{[H_2Pho^-]}; \quad 195.927 = \frac{[HPho^{2-}]}{[H_2Pho^-]}; \quad 9.8430 = \frac{[Pho^{3-}]}{[H_2Pho^-]}$$

The first of this new triad of equations is the reciprocal of the first of the preceding triad. The third equation comes by multiplying the second and third equations in the first set. The fraction f of any of the four species present equals its concentration divided by the sum of the concentrations of all four. For example:

$$f_{H_3Pho} = \frac{[H_3Pho]}{[H_3Pho] + [H_2Pho^-] + [HPho^{2-}] + [Pho^{3-}]}$$

This expression can be simplified by dividing numerator and denominator by $[H_2Pho^-]$ and inserting the ratios just calculated

$$f_{H_3Pho} = \frac{[H_3Pho]/[H_2Pho^-]}{[H_3Pho]/[H_2Pho^-] + 1 + [HPho^{2-}]/[H_2Pho^-] + [Pho^{3-}]/[H_2Pho^-]}$$

$$f_{H_3Pho} = \frac{3.98 \times 10^{-6}}{3.98 \times 10^{-6} + 1 + 195.927 + 9.8430} = \frac{3.98 \times 10^{-6}}{206.77} = 1.92 \times 10^{-8}$$

Similar expressions are generated to compute the fractions of the other three species by changing the numerator as required. The resulting fractions are 0.004836 for H_2Pho^-, 0.94756 for $HPho^{2-}$, and 0.0476 for Pho^{3-}. The total concentration of all four forms of the drug is 1.0×10^{-5} M. The desired concentrations are this total multiplied by the respective fractions. The answers are: $[H_3Pho] = 1.92 \times 10^{-13}$ M; $[H_2Pho^-] = 4.83 \times 10^{-8}$ M; $[HPho^{2-}] = 9.48 \times 10^{-6}$ M; $[Pho^{3-}] = 4.76 \times 10^{-7}$ M.

6-83 a) $HC_3H_5O_3(aq) + NaHCO_3(aq) \rightarrow NaC_3H_5O_3(aq) + H_2O(l) + CO_2(g)$

b) Compute the chemical amount of sodium bicarbonate that is neutralized:

$$\frac{1}{2}\,\text{t} \times \left(\frac{236.6\text{ mL}}{48\text{ t}}\right) \times \left(\frac{1\text{ cm}^3}{1\text{ mL}}\right) \times \left(\frac{2.46\text{ g NaHCO}_3}{1\text{ mL}}\right) \times \frac{1\text{ mol NaHCO}_3}{84.01\text{ g NaHCO}_3}$$

$$= 0.0633\text{ mol NaHCO}_3$$

One mole of lactic acid neutralizes one mole of sodium bicarbonate. Therefore one cup of sour milk contains 0.0633 mol of lactic acid. The concentration of lactic acid is 0.0633 mol/0.2366 L = 0.268 M.

c) The same chemical amount (0.0633 mol) of $CO_2(g)$ will be produced. Its volume can be computed with the ideal-gas law:

$$V = \frac{nRT}{P} = \frac{(0.0633\text{ mol})(0.08206\text{ J K}^{-1}\text{mol}^{-1})(450\text{ K})}{1\text{ atm}} = 2.34\text{ L}$$

6-85 Represent the amino acid glycine as glyH. The conjugate acid is then $glyH_2^+$, and the conjugate base is gly^-.
Copy the two equilibria given in the problem:

$$\text{glyH}(aq) + \text{H}_2\text{O}(l) \rightleftharpoons \text{gly}^-(aq) + \text{H}_3\text{O}^+(aq) \quad K_a = 1.7 \times 10^{-10}$$
$$\text{glyH}(aq) + \text{H}_2\text{O}(l) \rightleftharpoons \text{glyH}_2^+(aq) + \text{OH}^-(aq) \quad K_b = 2.2 \times 10^{-12}$$

The following equations convey the *same* chemical relationships:

$$\text{glyH}_2^+(aq) + \text{H}_2\text{O}(l) \rightleftharpoons \text{glyH}(aq) + \text{H}_3\text{O}^+(aq) \quad K_1 = K_w/K_b$$
$$\text{glyH}(aq) + \text{H}_2\text{O}(l) \rightleftharpoons \text{gly}^-(aq) + \text{H}_3\text{O}^+(aq) \quad K_2 = K_a$$

In this latter pair of equilibria the second equation given in the problem has been *reversed* and *added* to the water autoionization equation. Now, the emphasis is on $glyH_2^+$ ($^+H_3N—CH_2—COOH$) as a diprotic acid rather than on self-neutralization by the glycine This presentation clearly shows the similarity of a solution of glycine to solutions of other amphoteric species. The 0.10 M aqueous glycine is like 0.10 M $HCO_3^-(aq)$, for example. The case of the pH of an aqueous solution of $HCO_3^-(aq)$ is extensively treated in the text (Section 6-7, text page 233-4) where the approximate formula $[H_3O^+]^2 \approx K_1K_2$ is derived. According to this formula the $[H_3O^+]$ in the solution is independent of the concentration of the glycine! Substitution in the formula gives:

$$[\text{H}_3\text{O}^+]^2 = K_1K_2 = \left(\frac{K_w}{K_b}\right)K_a = \left(\frac{(1.0 \times 10^{-14})}{2.2 \times 10^{-12}}\right)(1.7 \times 10^{-10})$$

$$[\text{H}_3\text{O}^+] = 8.8 \times 10^{-7}\ \text{M} \quad \text{and} \quad \text{pH} = 6.06$$

A more exact analysis (text page 234) gives:

$$[\text{H}_3\text{O}^+] \approx \sqrt{\frac{K_1K_2c_0 + K_1K_w}{K_1 + c_0}}$$

The more exact treatment leading to *this* formula includes only one approximation: that the equilibrium concentration of glycine is close to c_0, its original concentration. Substitution of $c_0 = 0.10$ M and the three constants gives $[H_3O^+] = 8.6 \times 10^{-7}$ M and pH $= 6.07$, not much different from 6.06.
If c_0 were 0.01 M, one-tenth of the value given in the problem, the pH of the glycine solution would be 6.14. This deviates substantially from 6.06, the first answer. Thus, as c_0 goes down it becomes *more* important in determining the pH, at least at first. As c_0 becomes exceedingly small the solution approximates pure water with the pH equal to 7.00.

Chapter 7

Dissolution and Precipitation Equilibria

The Nature of Solubility Equilibria

The dissolution and subsequent re-precipitation of a soluble substance from a solvent is **recrystallization**. Recrystallization is a powerful method of purification. Successful recrystallizations require adroit manipulation of a dissolution-precipitation equilibria. In **dissolution** a solvent attacks a solute and brings it into solution by **solvating** it at the level of individual molecules or ions. The solute is seen to "melt away." In **precipitation,** the reverse occurs. Solid substance is seen to deposit from a solution.

Solvent of crystallization may be incorporated into a re-precipitated solid. This leads to formulas like $Li_2SO_4{\cdot}H_2O(s)$ (text page 244), $CoCl_2{\cdot}6H_2O(s)$ (mentioned on page 27 of the text), and $Cu(NO_3)_2{\cdot}6H_2O(s)$, used on page 6 of the text.

- To have equilibrium in a dissolution-precipitation reaction, there must be both solid and solution. Complete dissolution of a solid give an **unsaturated** solution. Dissolution has occurred, but the reverse reaction cannot be occurring (or the solid would be visible). These equilibria resemble those between pure liquid and its vapor (see text page 168).

- Dissolution-precipitation reactions are slow to come to equilibrium. **Supersaturation**, in which the concentration of the solute exceeds its equilibria value and persist for a fairly long time in this non-equilibrium condition, is common.

Solubility of Salts

Ionic solids (salts) typically dissolve in water to give independent separately aquated (hydrated) anions and cations. A bf saturated solution of a salt is a solution in which equilibrium exists between the solid salt and its dissolved ions. The **solubility** of a substance is the number of grams or moles of it present at equilibrium in a given quantity of solvent or a given quantity of solution. The amount of a salt that can be dissolved before saturation varies hugely depending on the identity of the cation and anion. **It is important to know which combinations of ions are insoluble, which are slightly soluble, and which are soluble.** The generalizations of Table 7-1 (text page 247) have practical importance because they allow one to predict what happens when solutions of common chemicals are mixed.

Ionic Equilibria

In a saturated aqueous solution of a salt, equilibrium exists between the solid salt and its aquated ions, and not between the solid salt and dissolved molecules of salt. For example:

$$PbSO_4(s) \rightleftharpoons Pb^{2+}(aq) + SO_4^{2-}(aq)$$

This heterogeneous equilibria is described by the mass-action expression:

$$K_{sp} = [Pb^{2+}][SO_4^{2-}]$$

The K_{sp} is a **solubility-product** constant. It does not differ fundamentally from other equilibrium constants. The special name merely emphasizes that the mass-action expression has the form of a product and not the more usual form of a quotient. See **problems 7-5** and **7-59** for examples of writing a K_{sp} expression. Note that:

- K_{sp} expressions are not written for highly soluble salts because such solutions are too nonideal. Use K_{sp}'s only for slightly soluble salts.

- K_{sp} values come from experiment (solubility measurements, see below). Compilations of the K_{sp}'s for most slightly soluble salts are widely available, but disagreements among different compilations of K_{sp}'s may be found, owing to experiment difficulties.

- Numerical values of K_{sp} are never good to more than two significant figures and often good to only one significant figure.

- K_{sp}'s depend strongly on temperature; they more often increase with increasing T, but sometimes diminish.

The solubility and the K_{sp} of a salt *not* the same constant, although there is a relation between solubility and K_{sp}. Simple relationships between K_{sp} and S (solubility in mol L^{-1}) of a salt exist if:

- the solution is ideal or near-ideal;

- no side-reactions reduce concentrations of ions once the salt is in solution. For examples of dealing with the complication of side-reactions, see **problems 7-49** and **7-71.**

Some examples of simple K_{sp} to S relationships are: For AgCl or $BaSO_4$, $K_{sp} = S^2$; for $AgCrO_4$, $K_{sp} = S(2S)^2 = 4S^3$.

Solubilities that are expressed in units other than molarity (such as g L^{-1}, or g/100mL) must be converted to mol L^{-1} for use in K_{sp} expressions. Solubilities depend strongly on the temperature **(problem 7-3).** Assume room temperature (25°C) if a solubility is quoted but the temperature is not explicitly given in a problem.

Typical Problems Involving K_{sp}

- The calculation of K_{sp} from the solubility of a solid and the reverse. See **See problems 7-7, 7-9, 7-11** and **7-67.**

- The prediction of the concentrations of ions required in the solution for precipitation of a given solid upon mixing its constituent ions. Precipitation occurs whenever the product of the proper powers of the concentrations of the ions composing a salt exceeds the value of K_{sp}. See **problems 7-17, 7-21,** and **7-65.**

- Evaluation of proposed selective precipitations in which, for example, a mixture of two cations is separated on the basis of the differing solubilities of their salts with an added anion. See **problems 7-25, 7-41,** and **7-43.**

Precipitation and the Solubility Product

A salt tends to precipitate from aqueous solution if the concentrations of its constituent ions are such that Q_{sp} exceeds K_{sp}; otherwise, it stays in solution. The concentrations of the constituent ions need not be in the molar ratio found in the salt. Unrelated ratios of constituent ions are in fact the general case when solutions are mixed. To compute equilibrium concentrations of ions after mixing two solutions:

- Realize that mixing two solutions dilutes both: "initial" concentration (in three-line format used in previous chapters) is concentration after dilution, before reaction

- Assume that precipitation goes to completion (uses up the ion that is the limiting reactant). Subsequent dissolution then restores some of that ion. See **problem 7-23. Problem 7-47** applies the same theme to a non-dissolution equilibrium.

The Common Ion Effect

If a salt is placed in water that is not pure but contains one of the ions produced by the dissolution of the salt (a common ion) then the solubility of the salt is *reduced.* This is a straightforward conclusion from LeChatelier's principle. Although the common ion effect drastically changes the solubilities of salts, it does not change their K_{sp}'s, which are more fundamental quantities. See **problem 7-29.**

The Effects of pH on Solubility

The pH of the solvent strongly influences solubility of many salts. For hydroxide salts (which are numerous and important), the pH of the solvent affects solubility directly. Metal hydroxides have enhanced solubility at low pH, but depressed solubility at high pH. See **problems 7-31** and **7-67.** When either the cation or the anion in a salt is an acid or a base, the pH of the solvent affects the solubility of the salt indirectly because H_3O^+ or OH^- reacts with the anion or cation. For example, $AgCH_3COO$ more soluble at low pH than at moderate pH because the acetate ion, a base, accepts H^+ and is removed causing the dissolution equilibrium to shift to the right partially to make up the lost acetate.

Selective Precipitation of Ions

Metal sulfides are an important group of salts in which the anion is a base. The sulfide ion (S^{2-}) does not in fact exist to any real extent in aqueous solution because it is too strong a base. It accepts H^+ from the water to give HS^- ion. The dissolution of for example $CdS(s)$ is therefore best represented: $CdS(s) + H_2O(l) \rightleftharpoons Cd^{2+}(aq) + HS^-(aq) + OH^-(aq)$ and has the equilibrium expression:

$$K_{sp} = [Cd^{2+}][HS^-][OH^-]$$

Such equilibria are very easily manipulated by changing the pH (using buffers). In computations describing such experiments, remember:

- The concentration of HS^- ion in a solution is linked by an acid-ionization equilibrium to the concentration of the acid H_2S. See **problem 7-41** and **7-43.** Hydrogen sulfide H_2S is a somewhat soluble gas. A saturated aqueous solution of H_2S has a concentration of 0.10 M at room temperature. See **problem 7-71.**

Complex Ions and Solubility

In a **coordination complex**, a central metal ion is bound to one or more **ligand** molecules or ions. Ligands usually capable of independent existence. Examples are

Cl^-, NH_3, OH^-, and, most importantly, H_2O. The Interaction between metal ion and ligand is called coordination. A central metal ion may coordinate several ligands; two, four and six are common coordination numbers.

Not all ligands in a complex have to be same. Complex ions undergo reactions in which different ligands exchange for each other. The **instability constant** K_{inst} is the equilibrium constant for replacement of all ligands other than water by water. Step-wise dissociation constants of a complex are for the first replacement by water (K_{d1}), the second (K_{d2}), and so forth. The product of the several K_d's is K_{inst}. See **problem 7-45.** Formation of a complex ion can shift a solubility equilibrium to bring a salt into solution. An example is the dissolution of $AgCl(s)$ in aqueous ammonia as: $[Ag(NH_3)_2]^+(aq) + Cl^-(aq)$. See **problem 7-69.**

Hydrolysis and Amphoterism of Complex Ions

Solutions of many metal ions are acidic. A metal ion coordinated by water can be an effective H^+-donor. Example: $Fe(H_2O)_6^{3+}$ has a K_a of 7.7×10^{-5}. A metal ion coordinated by H_2O and by OH^- is amphoteric, can donate H^+ or accept H^+ depending on other species present. See **problems 7-51, 7-53,** and **7-55.**

Complex-Ion Equilibria

A complex will lose ligands one by one just as molecules of polyprotic acids (such as H_3PO_4) in water solution lose their protons one by one. If a metal ion has four ligands surrounding it, then the metal-ligand complex may lose the four in a series of four consecutive reactions. Each of these is an equilibrium and has an equilibrium constant. The constants are K_{d1}, K_{d2}, etc. where the "d" stands for dissociation.

• Complex-ion equilibria are fundamentally no different than any other equilibria in aqueous solution.

The product, $K_{d1} \times K_{d2} \times K_{d3} \times K_{d4}$, is the **instability constant** of the complex ion. It is written K_{inst}. Taking the product of the K's corresponds to *adding up* the four step-wise equilibria.

Consider the dissociation of the complex $Cd(NH_3)_4^{2+}$. The overall dissociation is:

$$Cd(NH_3)_4^{2+}(aq) \rightleftharpoons Cd^{2+}(aq) + 4\,NH_3(aq) \quad K_{inst} = 3 \times 10^{-7}$$

The K comes from Table 18-3, text p. 637. It is true at equilibrium that:

$$3 \times 10^{-7} = \frac{[Cd^{2+}][NH_3]^4}{[Cd(NH_3)_4^{2+}]}$$

It is *not* true that $[NH_3] = 4\,[Cd^{2+}]$. Much of the total cadmium concentration in the solution at equilibrium is tied up in the form of the $Cd(NH_3)_3^{2+}$ and the other intermediate complexes. The $Cd^{2+}(aq)$ concentration is much less than 1/4 the concentration of ammonia. Writing false relationships like the above is a major source of error in solving problems involving complex-ion equilibria. The wisest policy is always to use the principles of electrical neutrality and material balance to generate needed relationships among concentrations.

There is another source of difficulty. In text Table 18-3, the successive Kd values for many complex ions are close together in magnitude. For example, for the $Cd(NH_3)_4^{2+}$ complex, Kd1 is 0.16 and Kd2 is 0.05. Calculations for complex ion dissociations often become tedious because it is necessary to treat two or more stages of the equilibrium simultaneously. This is rarely the case with the ionization of polyprotic acids (Chapter 6).

Fortunately, there is a common special case in practical problems. *If the ligand is present in excess and the K_d's are small, then:*

1. Most of the metal ion will be tied up in the complex with the highest coordination number.

2. The concentration of the free ligand will approximately equal the concentration of *excess* ligand that is calculated assuming complete complexation.

These are exactly the conditions that prevail in text Example 7-8. They also are the conditions in **problem 7-45** and **7-73.** When these conditions do *not* prevail, the calculations become complicated. If there is uncertainty about what assumptions might be valid for a given system, then set up the problem in complete terms, using the principles of electrical neutrality of the solution and material balance for the different species present. See Chapter 6 of the text. Then apply chemical and mathematical insight to find out what terms (if any) can be neglected. **See problem 6-41** for an example.

The formation of complex ions can exert a significant effect on the solubilities of salts. In such cases, the complex-ion equilibria run simultaneously with the K_{sp} equilibria that govern the solubility of the salt. It is useful to think of these reactions in terms of a competition among different species in binding to the metal ion. In **problem 7-49,** Ag^+ ion is distributed among three competitive forms, $AgCl(s)$, $AgCl_2^-(aq)$ and $Ag^+(aq)$. Factors determining the outcome of the competition are the intrinsic stability of the different forms (as conveyed in the equilibrium constants for their formation or dissociation) and the relative concentrations of the $Cl^-(aq)$ and $Ag^+(aq)$.

The aqua and hydroxo complexes of the metal ions are of particular interest because they furnish species that affect the pH of their solution directly. A metal ion that is six-coordinated with water has 12 protons that could in theory be donated if a strong enough base were present to accept them. In fact, a typical aqua complex of a metal ion is a weak acid. The values of K_{a1}, K_{a2} etc. for its function as an acid are related to the dissociation constants of its *hydroxo* complexes containing, respectively,

1, 2, etc. OH^- ligands in place of coordinated water molecules. The relationship is:

$$K_d = K_w/K_a$$

This relationship is covered in **problem 7-75.**

Detailed Solutions to Odd-Numbered Problems

7-1 Comparison of the two formulas reveals that the hydration reaction uses 3/2 mol of water for every mole of plaster of paris. Thus:

$$25.0 \text{ kg CaSO}_4\cdot\frac{1}{2}\text{H}_2\text{O} \times \left(\frac{1 \text{ mol CaSO}_4\cdot\frac{1}{2}\text{H}_2\text{O}}{0.14515 \text{ kg CaSO}_4\cdot\frac{1}{2}\text{H}_2\text{O}}\right) \times \left(\frac{3 \text{ mol H}_2\text{O}}{2 \text{ mol CaSO}_4\cdot\frac{1}{2}\text{H}_2\text{O}}\right)$$

$$\times \left(\frac{0.01802 \text{ kg H}_2\text{O}}{1 \text{ mol H}_2\text{O}}\right) \times \left(\frac{1 \text{ L H}_2\text{O}}{1 \text{ kg H}_2\text{O}}\right) = 4.65 \text{ L H}_2\text{O}$$

7-3 On the graph, scan across from "80" on the vertical axis until the solubility curve is reached. Then drop down to the horizontal axis and read the temperature. A solubility of 80 g HBr per 100 g H_2O is reached at a temperature of approximately 48°C. The last of the KBr will dissolve at approximately 48°C.

7-5 The expressions for solubility-product constants (K_{sp}'s) follows the general rules for heterogeneous equilibria (see text page 184-185). Pure solids and H_2O do not appear in these expressions:

$$\text{Fe}_2(\text{SO}_4)_3(s) \rightleftharpoons 2\,\text{Fe}^{3+}(aq) + 3\,\text{SO}_4^{2-}(aq) \qquad K_{sp} = [\text{Fe}^{3+}]^2[\text{SO}_4^{2-}]^3$$

7-7 The dissolution of thallium(I) iodate is represented:

$$\text{TlIO}_3(s) \rightleftharpoons \text{Tl}^+(aq) + \text{IO}_3^-(aq) \text{ for which } K_{sp} = [\text{Tl}^+][\text{IO}_3^-] = 3.07 \times 10^{-6}$$

If S mol per liter of $TlIO_3$ dissolves, then there must be S mol per liter of Tl^+ and also S mol per liter of IO_3^- present at equilibrium, as long as neither ion undergoes any further reactions once in solution. Write the K_{sp} expression and substitute the unknown:

$$K_{sp} = 3.07 \times 10^{-8} = [\text{Tl}^+][\text{IO}_3^-] = S^2$$

Solving gives $S = 1.752 \times 10^{-3}$ mol L^{-1}. This means that 1.752×10^{-3} mol of thallium(I) iodate dissolves in a liter of solution at equilibrium. This equals the solubility of the thallium(I) iodate in 1000 mL of water because the addition of a small amount of solute does not make the volume of the solution measurably from the volume of the pure solvent. This chemical amount of thallium(I) iodate has a mass of 0.665 g (using $\mathcal{M}$ = 379.3 g mol^{-1} for thallium(I) iodate). The mass of the salt that can dissolve in 100 mL of water is naturally only 1/10 as great: 0.0665 g $TlIO_3$ dissolves per 100.0 mL H_2O at 25°C.

7-9 The equilibrium and K_{sp} expression are:

$$Hg_2I_2(s) \rightleftharpoons Hg_2^{2+}(aq) + 2\,I^-(aq) \qquad K_{sp} = [Hg_2^{2+}][I^-]^2$$

If S mol L^{-1} of Hg_2I_2 dissolves, then S mol L^{-1} Hg_2^{2+} and $2S$ mol L^{-1} of I^- are produced. Substituting these values and the K_{sp} from the problem gives:

$$[Hg_2^{2+}][I^-]^2 = S(2S)^2 = 1.2 \times 10^{-28} \text{ from which } 4S^3 = 1.2 \times 10^{-28}$$

Solving gives $S = 3.1 \times 10^{-10}$ M. Hence:

$$[Hg_2^{2+}] = S = 3.1 \times 10^{-10} \text{ M} \qquad [I^-] = 2S = 6.2 \times 10^{-10} \text{ M}$$

7-11 As seen in problems 7-7 and 7-9, solubility is not the same as the solubility-product constant, although there is often a simple relationship between the two. The equilibrium of interest is:

$$Ag_2CrO_4(s) \rightarrow 2\,Ag^+(aq) + CrO_4^{2-}(aq) \qquad K_{sp} = [Ag^+]^2[CrO_4^{2-}]$$

The problem states that 0.0129 g of silver chromate dissolves in 500 mL (0.500 L) of water. First, compute the chemical amount of silver ions in solution at equilibrium:

$$n_{Ag^+} = 0.0129 \text{ g} \times \frac{1 \text{ mol } Ag_2CrO_4}{331.7 \text{ g } Ag_2CrO_4} \times \frac{2 \text{ mol } Ag^+}{1 \text{ mol } Ag_2CrO_4} = 7.78 \times 10^{-5} \text{ mol}$$

The chemical amount of chromate ions in solution is exactly half of this, or 3.89×10^{-5} mol. The concentrations of the two ions are their chemical amounts divided by the 0.500 L, which is the volume of the solution:

$$[Ag^+] = 1.56 \times 10^{-4} \text{ mol L}^{-1}; \quad [CrO_4^{2-}] = 7.78 \times 10^{-5} \text{ mol L}^{-1}$$

Substitute these values in the K_{sp} expression and compute a numerical value:

$$K_{sp} = [Ag^+]^2[CrO_4^{2-}] = (1.56 \times 10^{-4})^2(7.78 \times 10^{-5}) = 1.9 \times 10^{-12}$$

7-13 The dissolution-precipitation equilibrium and K_{sp}-expression are:

$$AgCl(s) \rightleftharpoons Ag^+(aq) + Cl^-(aq) \qquad K_{sp} = [Ag^+][Cl^-]$$

At equilibrium in the solution:

$$n_{\mathrm{Ag^+}} = 0.018 \text{ g} \times \left(\frac{1 \text{ mol AgCl}}{143 \text{ g AgCl}}\right) \times \left(\frac{1 \text{ mol Ag}^+}{1 \text{ mol AgCl}}\right) = 1.26 \times 10^{-4} \text{ mol}$$

Since the volume of the solution is 1.00 L, the concentration of Ag^+ ion is 1.26×10^{-4} M. The concentration of Cl^- ion is the same—one mole per liter of chloride ion is produced in solution for every mole per liter of silver ion. Substitute the concentrations into the equilibrium-constant expression:

$$K_{sp} = [\mathrm{Ag^+}][\mathrm{Cl^-}] = (1.26 \times 10^{-4})(1.26 \times 10^{-4}) = 1.6 \times 10^{-8}$$

7-15 If 5.57% of 1.00 mol of NaCl in solution is associated as ion pairs, then 0.9443 mol NaCl dissolves to produce free ions. When one mole of NaCl dissolves, two moles of ions are produced, one of Na^+ and one of Cl^-. Therefore:

$$0.9443 \text{ mol NaCl} \times \frac{2 \text{ mol ion}}{1 \text{ mol NaCl}} \times \frac{6.022 \times 10^{23} \text{ ion}}{1 \text{ mol ion}} = 1.14 \times 10^{24} \text{ free ions}$$

7-17 Get the initial concentrations of Ba^{2+} ion and CrO_4^{2-} ion and use them to calculate the initial reaction quotient Q for the reaction:

$$\mathrm{BaCrO_4}(s) \rightleftharpoons \mathrm{Ba^{2+}}(aq) + \mathrm{CrO_4^{2-}}(aq)$$

"Initial" in this case means after the solution has reached 25°C, but before any reaction has had a chance to occur. Compare this number with K_{sp}. If Q exceeds K_{sp}, then a precipitate will form as the reaction proceeds from right to left, if Q is less than K_{sp}, then there will be no chance of a precipitate forming. The initial chemical amount of the barium ion in the solution is:

$$n_{\mathrm{Ba^{2+}}} = 0.0063 \text{ g} \times \left(\frac{1 \text{ mol BaCrO}_4}{253 \text{ g BaCrO}_4}\right) \times \left(\frac{1 \text{ mol Ba}^{2+}}{1 \text{ mol BaCrO}_4}\right) = 2.49 \times 10^{-5} \text{ mol}$$

The initial concentration of Ba^{2+} ion is 2.49×10^{-5} M, if the volume of the cooled solution is taken as 1.00 L. Doing this ignores the slight contraction of water when it is cooled from 100 to 25°C as well as the very slight change in volume caused by the dissolution of the barium chromate in hot water in the first place. The initial concentration of the CrO_4^{2-} ion is by a similar calculation equal to 2.49×10^{-5} M. Then:

$$Q_0 = [\mathrm{Ba^{2+}}]_0[\mathrm{CrO_4^{2-}}]_0 = (2.49 \times 10^{-5})(2.49 \times 10^{-5}) = 6.2 \times 10^{-10}$$

From text Table 7-2, K_{sp} for $BaCrO_4$ is 2.1×10^{-10}, which is smaller than this Q_0. A precipitate will form.

7-19 The dissolution equilibrium and equilibrium expression are:

$$CaF_2(s) \rightleftharpoons Ca^{2+}(aq) + 2\,F^-(aq) \quad [Ca^{2+}][F^-]^2 = K_{sp} = 3.9 \times 10^{-11}$$

The numerical value comes from text Table 7-2. Choose several concentrations for the calcium ion and calculate the concentrations of the fluoride ion that must be present at each:

$[F^-]$	$[Ca^{2+}]$	$[F^-]$	$[Ca^{2+}]$
0.050×10^{-3} M	15.6×10^{-3} M	0.10×10^{-3} M	3.9×10^{-3} M
0.14×10^{-3}	2.0×10^{-3}	0.194×10^{-3}	1.00×10^{-3}
0.50×10^{-3}	0.156×10^{-3}	1.0×10^{-3}	0.039×10^{-3}

Then plot the data with $[F^-]$ on the y-axis and $[Ca^{2+}]$ on the x-axis.

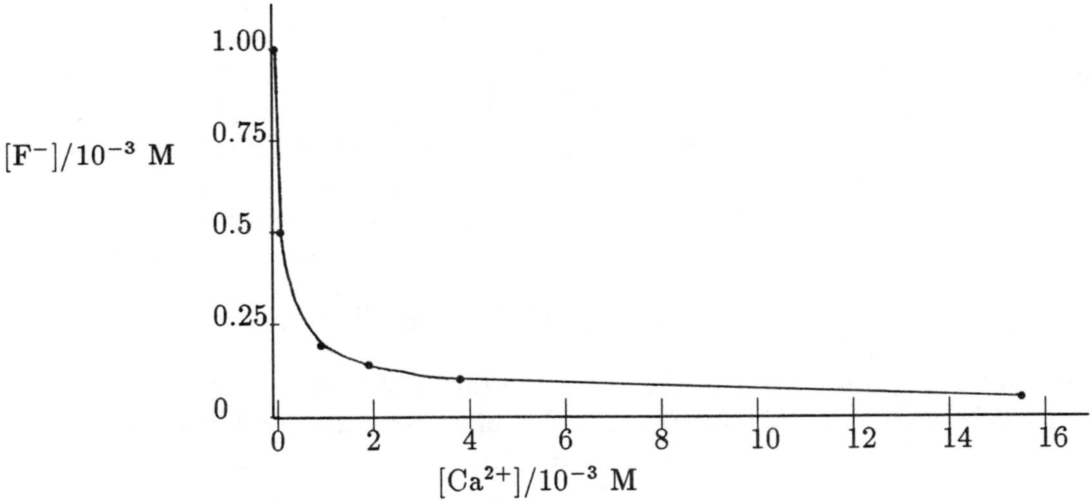

At all combinations of concentration of F^- and Ca^{2+} in the region above the curve ($Q > K_{sp}$) precipitation will occur. In the region below the curve ($Q < K_{sp}$), dissolution will occur. Along the curve Q equals K_{sp} and the system is at equilibrium.

7-21 Calculate the initial reaction quotient Q_0 and compare its value to K_{sp} for the reaction $Ce(IO_3)_3(s) \rightleftharpoons Ce^{3+}(aq) + 3\,IO_3^-(aq)$.
The expression for the reaction quotient is $Q_0 = [Ce^{3+}]_0[IO_3^-]_0^3$.
The "initial concentrations" to use in calculating Q_0 are those existing before the reaction occurs but after the solutions are mixed. After mixing, the volume of solution equals 250.0 mL (the volume of the cerium(III) nitrate) plus 150.0 mL (the volume of the potassium iodate) or 400.0 mL. Mixing dilutes both solutions. Before mixing,

the concentration of the Ce^{3+} ion is 0.0020 M. After mixing:

$$[Ce^{3+}]_0 = 0.0020 \text{ M} \times \left(\frac{250.0 \text{ mL}}{400 \text{ mL}}\right) = 1.25 \times 10^{-3} \text{ M}$$

Before mixing, the concentration of the IO_3^- ion is 0.0100 M. After mixing:

$$[IO_3^-]_0 = 0.0100 \text{ M} \times \left(\frac{150.0 \text{ mL}}{400 \text{ mL}}\right) = 3.75 \times 10^{-3} \text{ M}$$

$$Q = [Ce^{3+}]_0[IO_3^-]_0^3 = (1.25 \times 10^{-3})(3.75 \times 10^{-3})^3 = 6.6 \times 10^{-11}$$

This Q is less than K_{sp} so a precipitate does *not* form.

7-23 The 50.0 mL of 0.0500 M $Pb(NO_3)_2$ provides 2.50 mmol of Pb^{2+} ion in solution. The 40.0 mL of 0.200 Na IO_3 provides 8.00 mmol of IO_3^- ion in solution. The Pb^{2+} and IO_3^- ions react in a 1-to-2 ratio to give a precipitate of lead(II) iodate:

$$Pb^{2+}(aq) + 2\,IO_3^-(aq) \rightleftharpoons Pb(IO_3)_2(s)$$

Assume that this reaction goes to completion. Then no Pb^{2+} ion at all remains in solution (it is the limiting reactant), but $8.00 - 2(2.50) = 3.00$ mmol of IO_3^- remains in excess. The concentration of the excess IO_3^- is 3.00 mmol/90.00 mL = 0.0333 M. Actually, the reaction does not go to completion, but stops short at an equilibrium state. If it did go to completion, a small amount of the precipitate would soon redissolve according to:

$$Pb(IO_3^-)_2(s) \rightleftharpoons Pb^{2+}(aq) + 2\,IO_3^-(aq) \qquad K_{sp} = [Pb^{2+}][IO_3^-]^2 = 2.6 \times 10^{-13}$$

The equilibrium state is of course the same regardless of how it is attained. Let S equal the concentration of Pb^{2+} furnished in the solution by the preceding dissolution, which also adds $2S$ to the concentration of the IO_3^- ion. Thus, at equilibrium:

$$[Pb^{2+}] = S \quad \text{and} \quad [IO_3^-] = 0.0333 + 2S$$

Substitution in the K_{sp} expression gives:

$$(S)(0.0333 + 2S)^2 = 2.6 \times 10^{-13}$$

This cubic equation is simplified by assuming that $2S << 0.0333$. Then:

$$(0.0333)^2(S) \approx 2.6 \times 10^{-13} \quad \text{from which} \quad S \approx 2.3 \times 10^{-10}$$

$$[Pb^{2+}] = 2.3 \times 10^{-10} \text{ M} \quad \text{and} \quad [IO_3^-] = 0.0333 + 2(2.3 \times 10^{-10}) = 0.033 \text{ M}$$

The assumption that $2S$ is much smaller than 0.0333 is clearly valid.

7-25 Follow the procedure used in problem 7-23. There is 5.00 mmol of $AgNO_3$ and 1.8 mmol of Na_2CrO_4 in the solution "initially" (after mixing but before reaction). Assume that the precipitation reaction:

$$2\,Ag^+(aq) + CrO_4^{2-}(aq) \rightleftharpoons Ag_2CrO_4(s)$$

then goes to completion. As it does, it consumes 2 mmol of Ag^+ for 1 mmol of CrO_4^{2-} ion. Because the chemical amount of Ag^+ ion is more than twice that of CrO_4^{2-} ion, the CrO_4^{2-} ion is the limiting reactant. The chemical amount of Ag^+ left in excess is $5.00 - 2(1.80) = 1.40$ mmol. The concentration of the Ag^+ ion is this amount divided by the volume of the solution:

$$[Ag^+] = \frac{1.40 \text{ mmol}}{(50.0 + 30.0) \text{ mL}} = 0.01750 \text{ M}$$

Now suppose that some of the precipitate redissolves:

$$Ag_2CrO_4(s) \rightleftharpoons 2\,Ag^+(aq) + CrO_4^{2-}(aq) \quad K_{sp} = 1.9 \times 10^{-12} = [Ag^+]^2[CrO_4^{2-}]$$

Let S equal the concentration of CrO_4^{2-} in solution after this equilibrium is attained. Then the concentration of Ag^+ ion is $0.1750 + 2S$ and:

$$K_{sp} = (0.01750 + 2S)^2 S$$

Assume that $2S$ is much smaller than 0.0175. Then:

$$(0.01750)^2 S = 1.9 \times 10^{-12} \quad \text{and } S = 6.2 \times 10^{-9}$$

$$[CrO_4^{2-}] = 6.2 \times 10^{-9} \text{ M} \quad [Ag^+] = 0.01750 + 2(6.2 \times 10^{-9}) = 0.018 \text{ M}$$

7-27 The F^- ion present from the soluble NaF lowers the solubility of CaF_2 from its value in pure water. This is the common-ion effect. The CaF_2 dissolves according to the equation:

	$CaF_2(s) \rightleftharpoons$	$Ca^{2+}(aq)+$	$2\,F^-(aq)$
Init. Conc. (mol L^{-1})		0	0.040
Change in Conc. (mol L^{-1})		$+S$	$+2S$
Equil. Conc. (mol L^{-1})		S	$0.040 + 2S$

where the S is the solubility of the salt. For this equilibrium:

$$K_{sp} = [Ca^{2+}][F^-]^2 = 3.9 \times 10^{-11}$$

Substitute the equilibrium concentrations from the table and assume that $2S << 0.040$. Then:

$(S)(0.040 + 2S)^2 = 3.9 \times 10^{-11}$ which gives $S(0.040)^2 = 3.9 \times 10^{-8}$; $S = 2.4 \times 10^{-8}$

The computed value is S is indeed much smaller than 0.040, so the assumption is valid. The solubility of the CaF_2 is 2.4×10^{-8} mol L^{-1}.

7-29 a) Assume that there is 1.00 L of solution. For every S mol of $Ni(OH)_2$ that dissolves, S mol of Ni^{2+} and $2S$ mol of OH^- are produced in solution.

$$Ni(OH)_2(s) \rightleftharpoons Ni^{2+}(aq) + 2\,OH^-(aq) \quad K_{sp} = [Ni^{2+}][OH^-]^2 = 1.6 \times 10^{-16}$$

Substituting gives $(S)(2S)^2 = 1.6 \times 10^{-16}$ and $S = 3.4 \times 10^{-6}$ M.

b) The presence of a common ion (the OH^- ion) reduces the solubility of the nickel(II) hydroxide. Set up the usual three-line table:

	$Ni(OH)_2(s) \rightleftharpoons$	$Ni^{2+}(aq)+$	$2\,OH^-(aq)$
Init. Conc. (mol L^{-1})		0	0.100
Change in Conc. (mol L^{-1})		$+S$	$+2S$
Equil. Conc. (mol L^{-1})		S	$0.100 + 2S$

Substituting the equilibrium concentrations gives:

$$K_{sp} = [Ni^{2+}][OH^-]^2 = S((0.100 + 2S)^2 = 1.6 \times 10^{-16}$$

If $2S << 0.100$, then:

$$(0.100)^2(S) = 1.6 \times 10^{-16} \text{ so that } S = 1.6 \times 10^{-14}$$

The assumption is valid, and the solubility is therefore 1.6×10^{-14} M.

7-31 As long as the solution is in equilibrium with $Mg(OH)_2(s)$, the following equation is satisfied:

$$K_{sp} = 1.2 \times 10^{-11} = [Mg^{2+}][OH^-]^2$$

Let S represent the solubility of the $Mg(OH)_2(s)$ *before* any NaOH is added. At this stage:

$$[Mg^{2+}] = S \text{ and } [OH^-] = 2S$$

so that

$$1.2 \times 10^{-11} = [Mg^{2+}][OH^-]^2 = 4\,S^3$$

Solving gives $S = 1.44 \times 10^{-4}$ mol L^{-1}. Sodium hydroxide dissociates completely in water. Adding it to the solution is the same as the addition of $Na^+(aq)$ and $OH^-(aq)$ ions in a 1-to-1 ratio. The concentration of OH^- goes up with the addition of NaOH, and the concentration of Mg^{2+} must diminish to maintain the solubility product at a constant value. This is the common ion effect in action. Additional magnesium hydroxide precipitates as NaOH is added. The problem states that the solubility of $Mg(OH)_2$ is reduced to 0.0010 of its original value. This means that after the addition:

$$S = [Mg^{2+}] = 0.0010(1.44 \times 10^{-4}) = 1.44 \times 10^{-7} \text{ M}$$

Then

$$[OH^-] = \sqrt{\frac{K_{sp}}{1.44 \times 10^{-7}}} = \sqrt{\frac{1.2 \times 10^{-11}}{1.44 \times 10^{-7}}} = 9.1 \times 10^{-3} \text{ M}$$

7-33 Assume 1.00 L of solution. For every S mol of AgOH(s) that dissolves, S mol of Ag^+ and S mol of OH^- ions are produced. Assume that the concentration of OH^- ion from the dissolution is much larger than the concentration of OH^- ion in the water from the autoionization of water:

	$AgOH(s) \rightleftharpoons$	$Ag^+(aq)$ +	$OH^-(aq)$
Init. Conc. (mol L^{-1})	–	0.0	small
Change in Conc. (mol L^{-1})	–	$+S$	$+S$
Equil. Conc. (mol L^{-1})	–	S	S

At equilibrium in pure water:

$$K_{sp} = [Ag^+][OH^-] = S^2 = 1.5 \times 10^{-8} \quad \text{so that} \quad S = 1.2 \times 10^{-4}\ \text{M}$$

If the solution is buffered at pH 7.00, then the pH will remain at 7.00 even though the dissociation of AgOH produces OH^-. A pH of 7.00 means a pOH of 7.00. Hence:

$$[OH^-] = 10^{-7.00} = 1.0 \times 10^{-7}\ \text{M}$$

Put this concentration into the K_{sp}-expression to obtain the equilibrium concentration of $Ag^+(aq)$:

$$[Ag^+] = \frac{K_{sp}}{[OH^-]} = \frac{1.5 \times 10^{-8}}{1.0 \times 10^{-7}} = 0.15\ \text{M}$$

The solubility of the AgOH(s) equals the chemical amount of Ag^+ ion in solution per liter; it is therefore 0.15 mol L^{-1} in the solution buffered at pH 7.

7-35 a) The solubility of PbI_2 will remain unchanged as the pH of a solution at pH 7 is lowered. The anion is an exceedingly weak base and has little interaction with H_3O^+ even at a high concentration of H_3O^+.

b) The solubility of AgOH will increase greater as the pH of a solution is lowered. The additional H_3O^+ drives the dissolution by removing OH^- ion.

c) The solubility of $Ca_3(PO_4)_2$ will increase as the pH of a solution at pH 7 is lowered. Higher concentration of H_3O^+ drives the dissolution by removing PO_4^{3-} ion as HPO_4^{2-} ion.

7-37 a) Imagine slowly adding oxalate ion to the mixture of the magnesium and lead ions. Both oxalate salts will stay in solution until their Q_{sp}'s respectively exceed their K_{sp}'s. The K_{sp}'s:

$$PbC_2O_4(s) \rightleftharpoons Pb^{2+}(aq) + C_2O_4^{2-}(aq) \quad K_{sp} = [Pb^{2+}][C_2O_4^{2-}] = 2.7 \times 10^{-10}$$
$$MgC_2O_4(s) \rightleftharpoons Mg^{2+}(aq) + C_2O_4^{2-}(aq) \quad K_{sp} = [Mg^{2+}][C_2O_4^{2-}] = 8.6 \times 10^{-5}$$

The concentration of oxalate ion needed to precipitate magnesium oxalate from the 1.0 M Mg^{2+} solution is:

$$[C_2O_4^{2-}] = \frac{K_{sp}}{[Mg^{2+}]} = \frac{8.6 \times 10^{-5}}{0.10} = 8.6 \times 10^{-4}\ \text{M}$$

If the $[C_2O_4^{2-}]$ is kept at or below 8.6×10^{-4} M, then the magnesium salt cannot precipitate. Only the lead salt can precipitate. It precipitates before the magnesium salt because its K_{sp} is far smaller than that of the magnesium salt.
b) If the $[C_2O_4^{2-}]$ is held at 8.6×10^{-4} M, then:

$$[Pb^{2+}] = \frac{K_{sp}}{[C_2O_4^{2-}]} = \frac{2.7 \times 10^{-11}}{8.6 \times 10^{-4}} = 3.1 \times 10^{-8} \text{ M}$$

$$\text{fraction Pb}^{2+} \text{ remaining} = \frac{3.1 \times 10^{-8} \text{ M}}{0.10 \text{ M}} = 3.1 \times 10^{-7}$$

7-39 Calculate the $[I^-]$ that is just sufficient to bring about precipitation of each metal ion. The precipitation reaction are:

$$Hg_2I_2(s) \rightleftharpoons Hg_2^{2+}(aq) + 2\,I^-(aq) \quad K_{sp} = 1.2 \times 10^{-28} = [Hg_2^{2+}][I^-]^2$$
$$PbI_2(s) \rightleftharpoons Pb^{2+}(aq) + 2\,I^-(aq) \quad K_{sp} = 1.4 \times 10^{-8} = [Pb^{2+}][I^-]^2$$

The required concentration is quite small in the case of $Hg_2I_2(s)$ because the K_{sp} is very small. In the case of $PbI_2(s)$ the required concentration is:

$$[I^-] = \sqrt{\frac{1.4 \times 10^{-8}}{0.0500}} = 5.3 \times 10^{-4}$$

The optimum $[I^-]$ would be just below 5.3×10^{-4} M. If the concentration of I^- is raised to this value at equilibrium, then no $PbI_2(s)$ can yet precipitate but almost all of the $Hg_2^{2+}(aq)$ will have been precipitated as $Hg_2I_2(s)$.

7-41 Metal ions form sulfides of greatly varying but generally low solubility. Careful control of the pH, which strongly affects these solubilities, allows the separation of the metal ions by differential precipitation of the sulfides. The dissolution reaction for ZnS is:

$$ZnS(s) + H_2O(l) \rightleftharpoons Zn^{2+}(aq) + HS^-(aq) + OH^-(aq)$$

The corresponding solubility-product expression is a triple product, but is otherwise not exceptional. It is:

$$[Zn^{2+}][HS^-][OH^-] = K_{sp} = 2 \times 10^{-25}$$

The value of the constant was found on text page 258. Both the OH^- and HS^- concentrations depend strongly on the concentration of H_3O^+ according to the equations:

$$2\,H_2O(l) \rightleftharpoons H_3O^+(aq) + OH^-(aq) \qquad K_w = [H_3O^+][OH^-]$$
$$H_2S(aq) + H_2O(aq) \rightleftharpoons H_3O^+(aq) + HS^-(aq) \quad K_{a1} = [HS^-][H_3O^+]/[H_2S]$$

Solving the equilibrium expressions for $[OH^-]$ and $[HS^-]$ respectively and substituting into the solubility-product expression gives:

$$[Zn^{2+}]\left(\frac{K_{a1}[H_2S]}{[H_3O^+]}\right)\left(\frac{K_w}{[H_3O^+]}\right) = 2 \times 10^{-25}$$

The K_{a1} of H_2S equals 9.1×10^{-8}; K_w is 1.0×10^{-14}; the concentration of the H_2S is 0.10 M; the concentration of H_3O^+ is 0.0010 M. Insert these numbers and solve for the concentration of Zn^{2+}. It is 2×10^{-13} M.

7-43 Precipitation of FeS(s) can begin only when the pH is high enough to make the Q for the reaction:

$$FeS(s) + H_2O(l) \rightleftharpoons Fe^{2+}(aq) + HS^-(aq) + OH^-(aq)$$

exceed K_{sp}. The K_{sp} expression is:

$$[Fe^{2+}][HS^-][OH^-] = K_{sp} = 5 \times 10^{-19}$$

where the numerical K_{sp} comes from Table 7-3 on text page 258. By the method used in problem 7-41, this expression becomes:

$$[Fe^{2+}]\left(\frac{K_{a1}[H_2S]}{[H_3O^+]}\right)\left(\frac{K_w}{[H_3O^+]}\right) = 5 \times 10^{-19}$$

The K_{a1} of H_2S is 9.1×10^{-8}. Also, K_w is 1.0×10^{-14} for K_w; 0.10 is the concentration of Fe^{2+}; 0.10 is the concentration of H_2S. Substituting and solving gives $[H_3O^+] = 4.3 \times 10^{-3}$ M. This is the minimum concentration of H_3O^+ that keeps FeS in solution. The maximum pH is therefore 2.4. Higher pH, (implying lower $[H_3O^+]$ and higher $[OH^-]$) shifts the above reaction from right to left and favors precipitation of FeS(s). The dissolution reaction of PbS(s) is exactly similar to that of FeS(s):

$$PbS(s) + H_2O(l) \rightleftharpoons Pb^{2+}(aq) + HS^-(aq) + OH^-(aq)$$

and the expression:

$$[Pb^{2+}]\left(\frac{K_{a1}[H_2S]}{[H_3O^+]}\right)\left(\frac{K_w}{[H_3O^+]}\right) = K_{sp} = 3 \times 10^{-28}$$

can be derived as in problem 7-41. The K_{sp} is from Table 7-3 (text page 258). Substitution of 0.10 M for $[H_2S]$, 4.3×10^{-3} for $[H_3O^+]$, and 9.1×10^{-8} for K_{a1} gives $[Pb^{2+}]$ equal to 6×10^{-11} M. The point of the problem is that at a pH of 2.4, all of the Fe^{2+} ion but (essentially) no Pb^{2+} ion stays in solution.

7-45 Assume that all of the $Cu^{2+}(aq)$ initially present reacts with $NH_3(aq)$ (which is present in excess) to form $Cu(NH_3)_4^{2+}(aq)$:

$$Cu^{2+}(aq) + 4\,NH_3(aq) \rightleftharpoons Cu(NH_3)_4^{2+}(aq)$$

For every 1 mol of Cu^{2+} ion that reacts, 4 mol of NH_3 reacts and 1 mol $Cu(NH_3)_4^{2+}$ ion is formed. Then:

$$[Cu(NH_3)_4^{2+}] = 0.10\text{ M} \quad \text{and} \quad [NH_3] = 1.5 - 4(0.10) = 1.1\text{ M}$$

All free Cu^{2+} ion in solution is now thought of as coming from the dissociation of the complex ion $Cu(NH_3)_4^{2+}$. The dissociation of the complex proceeds through four steps. The sum of these steps is the exact reverse of the above formation reaction. If reactions are added together, then K for the resulting reaction is the product of the K's for the reactions being added. The steps are:

$$Cu(NH_3)_4^{2+}(aq) \rightleftharpoons Cu(NH_3)_3^{2+}(aq) + NH_3(aq) \quad K_{d1} = 1.1 \times 10^{-2}$$
$$Cu(NH_3)_3^{2+}(aq) \rightleftharpoons Cu(NH_3)_2^{2+}(aq) + NH_3(aq) \quad K_{d2} = 2 \times 10^{-3}$$
$$Cu(NH_3)_2^{2+}(aq) \rightleftharpoons Cu(NH_3)^{2+}(aq) + NH_3(aq) \quad K_{d3} = 5 \times 10^{-4}$$
$$Cu(NH_3)^{2+}(aq) \rightleftharpoons Cu^{2+}(aq) + NH_3(aq) \quad K_{d4} = 1.0 \times 10^{-4}$$

Therefore:

$$Cu(NH_3)_4^{2+}(aq) \rightleftharpoons Cu^{2+}(aq) + 4\,NH_3(aq) \quad K = K_{d1}K_{d2}K_{d3}K_{d4} = 1.0 \times 10^{-12}$$

where the four step-wise constants come from Table 7-4 (text page 261). This K is the instability constant (K_{inst}) of the complex. The K_{inst} given in the table is somewhat smaller (9×10^{-13}) because of round-off errors. Let x equal the equilibrium concentration of Cu^{2+} from the dissociation of the complex:

	$Cu(NH_3)_4^{2+}(aq) \rightleftharpoons$	$Cu^{2+}(aq)$	$+4\,NH_3(aq)$
Init. Conc. (mol L^{-1})	0.10	0	1.1
Change in Conc. (mol L^{-1})	$-x$	$+x$	$+4x$
Equil. Conc. (mol L^{-1})	$0.10 - x$	x	$1.1 + x$

$$\text{Then:} \quad \frac{[Cu^{2+}][NH_3]^4}{[Cu(NH_3)_4^{2+}]} = \frac{x(1.1 + 4x)^4}{0.10 - x} = K = 9 \times 10^{-13}$$

Assuming that x is small compared to 0.10 simplifies this to:

$$\frac{x(1.1)^4}{0.10} = 9 \times 10^{-13} \quad \text{so that } x = 6 \times 10^{-14}$$

Therefore $[Cu^{2+}] = 6 \times 10^{-14}$ M, and $[Cu(NH_3)_4^{2+}] = 0.10 - 6 \times 10^{-14} = 0.10$ M.

7-47 Assume that the reaction between $K^+(aq)$ and "crown" goes 100% to completion and that free K^+ then comes from back-dissociation of K-crown$^+$. The equation for this dissociation is the reverse of the equation in the problem:

	$\text{K crown}^+(aq) \rightleftharpoons$	$K^+(aq)$ +	$\text{crown}(aq)$
Init. Conc. (mol L^{-1})	0.0080	0	0
Change in Conc. (mol L^{-1})	$-x$	$+x$	$+x$
Equil. Conc. (mol L^{-1})	$0.0080 - x$	x	x

The reaction has the equilibrium-constant expression:

$$\frac{[K^+][\text{crown}]}{[\text{Kcrown}^+]} = \frac{1}{111.6} \quad \text{hence :} \quad \frac{1}{111.6} = \frac{x^2}{0.080 - x}$$

Rearranging gives the quadratic equation:

$$x^2 + (8.961 \times 10^{-3})x - 7.169 \times 10^{-5} = 0$$

Solving for x gives a root of 0.0051. The equilibrium concentration of $K^+(aq)$ is 0.0051 M. The calculation of the concentration of free Na^+ proceeds similarly. The quadratic equation is now:

$$y^2 + 0.152y - 0.00122 = 0$$

(where y is the concentration of free Na^+ ion) because the equilibrium constant of the reverse reaction is 1/6.6 instead of 1/111.6. Solving the new equation gives an equilibrium concentration of Na^+ ion of 0.0076 M.

7-49 The problem really asks for a calculation of the solubility of AgCl in 1.00 M NaCl solution. Let S equal this solubility. For every mole per liter of AgCl that dissolves either an Ag^+ or a $AgCl_2^-$ (dichloroargenate(I)) ion forms. In mathematical form:

$$S = [Ag^+] + [AgCl_2^-]$$

Obviously, this assumes that all of the dissolved silver is present as one of two ions. The solubility equilibrium for AgCl assures that:

$$K_{sp} = 1.6 \times 10^{-10} = [Ag^+][Cl^-]$$

as long as any solid silver chloride is present in the system. The dissociation equilibrium of the $AgCl_2^-$ complex ion leads to the expression:

$$6 \times 10^{-6} = \frac{[Ag^+][Cl^-]^2}{[AgCl_2^-]}$$

Solve these equations for $[Ag^+]$ and $[AgCl_2^-]$ and substitute the results into the expression for S:

$$S = \frac{1.6 \times 10^{-10}}{[Cl^-]} + \frac{(1.6 \times 10^{-10})[Cl^-]}{6 \times 10^{-6}}$$

Assume that the concentration of Cl^- from the complete dissociation of the NaCl is so large at 1.00 M that it is not substantially reduced by reaction with Ag^+. If $[Cl^-] = 1.00$ M, then:

$$S = \frac{1.6 \times 10^{-10}}{1.00} + \frac{1.6 \times 10^{-10}(1.00)}{6 \times 10^{-6}} = 3 \times 10^{-5}\ \text{M}$$

The assumption that $[Cl^-]$ is only negligibly reduced from its original value is vindicated by this low solubility—only about 6×10^{-5} M of Cl^- ion is tied up in the complex. The solubility of AgCl in this solution is more than double the solubility of AgCl in pure water, which is 1.3×10^{-5} M (by a computation like the one in problem 7-27). Hence, AgCl(s) dissolves to a *greater* extent in 1.00 M NaCl than in pure water. Note the remarkable reversal from the prediction of made by considering the common-ion effect only. Complexation plays a potent role in determining solubilities. In 0.100 M NaCl (in which $[Cl^-]$ is 0.100 M), the above solubility expression becomes:

$$S = \frac{1.6 \times 10^{-10}}{0.100} + \frac{(1.6 \times 10^{-10})0.100}{6 \times 10^{-6}} = 1.6 \times 10^{-9} + 2.7 \times 10^{-6} = 2.7 \times 10^{-6}\ \text{M}$$

This rounds off to 3.0×10^{-6} and is *less* than the solubility of AgCl(s) in pure water.

7-51 In aqueous solution, Cu^{2+} ion is coordinated by water to form the $Cu(H_2O)_4^{2+}$ ion. This complex ion acts as a Brønsted-Lowry acid according to the equation:

$$Cu(H_2O)_4^{2+}(aq) + H_2O(l) \rightleftharpoons H_3O^+(aq) + Cu(H_2O)_3OH^+(aq)$$

The solution is acidic because the concentration of H_3O^+ is increased by this reaction. An equivalent answer is the equation:

$$Cu^{2+}(aq) + H_2O(l) \rightleftharpoons H_3O^+(aq) + [CuOH]^+(aq)$$

7-53 The computation is just like other computations of the pH of solutions of weak acids, such as problem 6-21a. The coordinated cobalt(II) ion is acidic as follows:

	$Co(H_2O)_6^{2+}$	$+H_2O \rightleftharpoons$	$H_3O^+(aq)$ +	$Co(H_2O)_5OH^+$
Init. Conc. (mol L^{-1})	0.10		0	0
Change in Conc. (mol L^{-1})	$-x$		$+x$	$+x$
Equil. Conc. (mol L^{-1})	$0.10 - x$		x	x

The equilibrium expression is:

$$\frac{[H_3O^+][Co(H_2O)_5OH^+]}{[Co(H_2O)_6^{2+}]} = K_a = 3 \times 10^{-10} = \frac{x^2}{0.10 - x}$$

Neglect x in comparison to 0.10 and solve the equation:

$$\frac{x^2}{0.10} = 3 \times 10^{-10}$$

for x. The concentration of H_3O^+ is 5.5×10^{-6}, and the pH is therefore 5.3.

7-55 The reaction $Pt(NH_3)_4^{2+}(aq) + H_2O(l) \rightleftharpoons H_3O^+(aq) + Pt(NH_3)_3NH_2^+(aq)$ is responsible for the acidity of the solution. The equilibrium-constant expression for this reaction is:

$$\frac{[Pt(NH_3)_3NH_2^+][H_3O^+]}{[Pt(NH_3)_4^{2+}]} = K_a$$

The concentration of the $Pt(NH_3)_3NH_2^+$ ion equals the concentration of the H_3O^+ ion in solution if this reaction is the only significant source of either ion. The concentration of the $Pt(NH_3)_4^{2+}$ ion equals 0.15 M minus the concentration of H_3O^+ ion. The H_3O^+ concentration can be calculated from the pH given in the problem:

$$[H_3O^+] = 10^{-4.92} = 1.20 \times 10^{-5} \text{ M}$$

Substitute this and the other concentrations in the K_a-expression:

$$K_a = \frac{(1.20 \times 10^{-5})(1.20 \times 10^{-5})}{0.15 - 1.20 \times 10^{-5}} = 9.6 \times 10^{-10}$$

7-57 The problem concerns the solubility of $Pb^{2+}(aq)$ in a solution adjusted to pH 13 by the addition of NaOH. The reaction:

$$Pb^{2+}(aq) + 2\,OH^-(aq) \rightleftharpoons Pb(OH)_2(s)$$

acts to reduce the concentration of $Pb^{2+}(aq)$. The K for this reaction is the reciprocal of K_{sp} of $Pb(OH)_2(s)$. It is quite large (K_{sp} is 4.2×10^{-15} so $1/K_{sp}$ is 2.38×10^{14}). This means that 1 M Pb^{2+} ion would give a precipitate of $Pb(OH)_2(s)$ at pH 13. But there is a complication. The equilibrium:

$$Pb^{2+}(aq) + 3\,OH^-(aq) \rightleftharpoons Pb(OH)_3^-(aq)$$

ties up Pb^{2+} ion in a *soluble* form and acts to counter precipitation of $Pb(OH)_2(s)$. The K for this reaction is the reciprocal of K_{inst} of $Pb(OH)_3^-(aq)$ and is large; it equals 3.33×10^{14}. Whether solid $Pb(OH)_2$ forms depends on the resolution of the competition between these reactions. Suppose $Pb(OH)_2(s)$ *does* form. Then, at pH 13, where $[OH^-] = 0.10$ M, the concentration of Pb^{2+} must fulfill the equation:

$$4.2 \times 10^{-15} = [Pb^{2+}][OH^-]^2$$

This means $[Pb^{2+}]$ would equal 4.2×10^{-13} M. The mass-action expression for the second equilibrium is:

$$3.33 \times 10^{14} = \frac{[Pb(OH)_3^-]}{[Pb^{2+}][OH^-]^3}$$

If $[Pb^{2+}]$ were 4.2×10^{-13} M, then $[Pb(OH)_3^-]$ would be 0.14 M, assuming that there is solid $Pb(OH)_2$ at the bottom of the solution to maintain the K_{sp} equilibrium.
An initial concentration of 1.00 M $Pb^{2+}(aq)$, exceeds the critical value that was just identified. Therefore, $Pb(OH)_2(s)$ *does* precipitate until $[Pb^{2+}]$ equals 4.2×10^{-13} M. At this point, the concentration of $Pb(OH)_3^-$ equals 0.14 M.
An initial concentration of 0.0500 M Pb^{2+}, is less than the critical minimum of 0.14 M. There is no precipitate of $Pb(OH)_2(s)$. The K_{sp} equilibrium is *not* in effect. Essentially the entire concentration of the Pb^{2+} ion is tied up in the complex so that:

$$K_{inst} = 3 \times 10^{-15} = \frac{[Pb^{2+}][OH^-]^3}{[Pb(OH)_3^-]} \approx \frac{[Pb^{2+}](0.10)^3}{0.050}$$

Solving gives the concentration of free $Pb^{2+}(aq)$ as 1.5×10^{-13} M.

7-59 The mercurous ion exists as Hg_2^{2+} in aqueous solution, according to discussion on text page 55. The dissolution equilibrium is:

$$Hg_2Cl_2(s) \rightleftharpoons Hg_2^{2+}(aq) + 2\,Cl^-(aq) \qquad K_{sp} = [Hg_2^{2+}][Cl^-]^2$$

7-61 The dissolution of barium sulfate is represented:

$$BaSO_4(s) \rightleftharpoons Ba^{2+}(aq) + SO_4^{2-}(aq) \quad [Ba^{2+}][SO_4^{2-}] = K_{sp} = 1.1 \times 10^{-10}$$

If the equilibrium concentration of Ba^{2+} is S, then the concentration of SO_4^{2-} is also S, as long as there are no additional sources (or sinks) of either ion. Then $S^2 = 1.1 \times 10^{-10}$ so that $S = [Ba^{2+}] = 1.1 \times 10^{-5}$ M. This concentration is too low for any bad effects on patients drinking the suspension of $BaSO_4(s)$.

7-63 Two solubility equilibria are going on:

$$AgBr(s) \rightleftharpoons Ag^+(aq) + Br^-(aq) \quad K_{sp} = [Ag^+][Br^-] = 7.7 \times 10^{-13}$$
$$CuBr(s) \rightleftharpoons Cu^+(aq) + Br^-(aq) \quad K_{sp} = [Cu^+][Br^-] = 4.2 \times 10^{-8}$$

Divide the K_{sp} expression for the second by the K_{sp} for the first:

$$\frac{[Cu^+][Br^-]}{[Ag^+][Br^-]} = \frac{4.2 \times 10^{-8}}{7.7 \times 10^{-13}} = 5.45 \times 10^4$$

A single solution at equilibrium can have only one concentration of Br^- ion (or any other ion). Hence the $[Br^-]$ in the numerator equals the $[Br^-]$ in the denominator. Cancellation then gives:

$$\frac{[Cu^+]}{[Ag^+]} = 5.5 \times 10^4$$

7-65 When Q exceeds K_{sp}, precipitation begins. Calculate the concentration of Ag^+ that will just suffice to precipitate each salt by setting $Q = K_{sp}$. The equilibria and K_{sp}-expressions are:

$$AgCl(s) \rightleftharpoons Ag^+(aq) + Cl^-(aq) \qquad K_{sp} = [Ag^+][Cl^-] = 1.6 \times 10^{-10}$$
$$Ag_2CrO_4(s) \rightleftharpoons 2\,Ag^+(aq) + CrO_4^{2-}(aq) \qquad K_{sp} = [Ag^+]^2[CrO_4^{2-}] = 1.9 \times 10^{-12}$$

Inserting the given concentrations of Cl^- and CrO_4^{2-} ions shows that the concentration of Ag^+ ion required for the precipitation of $AgCl(s)$ is 1.6×10^{-9} M, and the concentration of Ag^+ for the precipitation of $AgCrO_4(s)$ is 2.8×10^{-5} M, which is much larger. Thus, $AgCl(s)$ will precipitate first.
Solid Ag_2CrO_4 just starts to precipitate when $[Ag^+] = 2.8 \times 10^{-5}$ M. Use the K_{sp} expression to calculate $[Cl^-]$ when $[Ag^+] = 2.8 \times 10^{-5}$ M:

$$[Cl^-] = \frac{K_{sp}}{[Ag^+]} = \frac{1.6 \times 10^{-10}}{2.8 \times 10^{-5}} = 5.7 \times 10^{-6}\text{ M}$$

The fraction of Cl^- remaining is:

$$f_{Cl^-} = \frac{[Ag^+]}{[Ag^+]_0} = \frac{5.7 \times 10^{-6}\text{ M}}{0.10\text{ M}} = 5.7 \times 10^{-5}$$

7-67 Magnesia dissolves in water and raises the pH by generating $OH^-(aq)$ ion:

$$MgO(s) + H_2O(l) \rightleftharpoons Mg^{2+}(aq) + 2\,OH^-(aq) \quad K = [Mg^{2+}][OH^-]^2$$

The pH of the solution at 25°C is 10.16. At this temperature, the sum of the pH and pOH of an aqueous solution is 14.00, making the pOH of the saturated solution of magnesia 3.84. By the definition of pOH:

$$[OH^-] = 10^{-3.84} = 1.44 \times 10^{-4}\text{ M}$$

Assume that dissolution of magnesia is the predominant source of OH^- in the solution, far outpacing the autoionization of water. Then:

$$2\,[Mg^{2+}] = [OH^-]$$

Substituting on the right-hand side of this equation gives the value of K:

$$K = [Mg^{2+}][OH^-]^2 = \frac{[OH^-]}{2}[OH^-]^2 = \frac{(1.44 \times 10^{-4})^3}{2} = 1.5 \times 10^{-12}$$

The solubility of the MgO equals the final concentration of $Mg^{2+}(aq)$:

$$[Mg^{2+}] = \frac{[OH^-]}{2} = 7.2 \times 10^{-5}\ \text{M}$$

7-69 Silver ion is complexed strongly by ammonia. The product is the $Ag(NH_3)_2^+$ ion:

$$Ag^+(aq) + 2\,NH_3(aq) \rightleftharpoons Ag(NH_3)_2^{2+}(aq) \quad K = 1.7 \times 10^7$$

The large K means that the equilibrium lies far to the right. When $AgBr(s)$ is placed in aqueous ammonia, a new dissolution reaction:

$$AgBr(s) + 2\,NH_3(aq) \rightleftharpoons Ag(NH_3)_2^{2+}(aq) + Br^-(aq)$$

replaces the usual reaction:

$$AgBr(s) \rightleftharpoons Ag^+(aq) + Br^-(aq)$$

The new dissolution reaction is the sum of this K_{sp} reaction and the complexation reaction. Its K is 1.7×10^7 times larger than the K_{sp}, which explains the increase in solubility of AgBr when ammonia is present.

7-71 **a)** The aqueous HCl reacts with $CdS(s)$ to give both $CdCl_4^{2-}(aq)$ and $H_2S(aq)$:

$$CdS(s) + 2\,H_3O^+(aq) + 4\,Cl^-(aq) \rightleftharpoons CdCl_4^{2-}(aq) + H_2S(aq) + 2\,H_2O(l)$$

The $HCl(aq)$ acts on the $CdS(s)$ both by removing sulfide ion (as H_2S) and cadmium ion (as the tetrachloro complex). The problem states that some $CdS(s)$ remains, so the above equation is an accurate description of the final equilibrium.

b) The equilibrium in the preceding part can be constructed as the *sum* of four reactions. The first represents the dissolution of $CdS(s)$ in pure water; the second and third account for the interactions of the basic ions HS^- and OH^- with water; the fourth is the formation of the complex between the Cd^{2+} and Cl^- ions:

$$\begin{array}{ll}
CdS(s) + H_2O(l) \rightleftharpoons Cd^{2+}(aq) + OH^-(aq) + HS^-(aq) & K_1 = K_{sp} \\
HS^-(aq) + H_2O(l) \rightleftharpoons H_2S(aq) + OH^-(aq) & K_2 = K_w/K_{a1} \\
2\,OH^-(aq) + 2\,H_3O^+(aq) \rightleftharpoons 4\,H_2O(l) & K_3 = (1/K_w)^2 \\
Cd^{2+}(aq) + 4\,Cl^-(aq) \rightleftharpoons CdCl_4^{2-}(aq) & K_4 = 1/K_{inst}
\end{array}$$

where K_{a1} is for the first stage of the acid ionization of $H_2S(aq)$. The equilibrium constants of the four reactions are numbered for identification. The desired constant is the product of the four because the equation of interest is the sum of the four equations in the list:

$$K = K_1K_2K_3K_4 = K_1\left(\frac{K_w}{K_a}\right)\left(\frac{1}{K_w}\right)^2\left(\frac{1}{K_{inst}}\right)$$

The K_1 is 7×10^{-28} (Table 7-3, text page 258); the K_{a1} is 9.1×10^{-8} (text Table 6-2, page 206); and K_{inst} is 1.3×10^{-3} (see the statement of the problem). Inserting these numbers and the well-known value of K_w into the preceding gives:

$$K = (7 \times 10^{-28})\left(\frac{1.0 \times 10^{-14}}{9.1 \times 10^{-8}}\right)\left(\frac{1}{1.0 \times 10^{-14}}\right)^2\left(\frac{1}{1.3 \times 10^{-3}}\right) = 5.9 \times 10^{-4}$$

Thus, at equilibrium:

$$\frac{[CdCl_4^{2-}][H_2S]}{[H_3O^+]^2[Cl^-]^4} = K = 5.9 \times 10^{-4}$$

c) Let S equal the molar solubility of the $CdS(s)$ in 6 M HCl. If all of the cadmium in solution is in the form of $CdCl_4^{2-}(aq)$, and if all of the sulfur in solution is in the form of $H_2S(aq)$, then:

$$S = [CdCl_4^{2-}] = [H_2S]$$

Also, every Cd^{2+} ion that goes into solution consumes 4 Cl^- ions, and every S^{2-} ion that goes into solution consumes 2 H_3O^+ ions. At equilibrium: $[H_3O^+] = 6 - 2S$ and $[Cl^-] = 6 - 4S$. The 6 comes from the original concentration of H_3O^+, which was 6 M. Substitution in the equilibrium-constant expression derived in the preceding parts gives:

$$5.9 \times 10^{-4} = \frac{S^2}{(6 - 2S)^2(6 - 4S)^4}$$

Trying to solve this equation by assuming that S is much less than 6 M does not work. It is clear however that S lies between 0 and 6. As a guess, suppose that S equals 1.0 M. Then the right side equals 3.9×10^{-3}, which exceeds 5.9×10^{-4}. Guess a smaller S, such as 0.6 M. Now, the right side equals 9.3×10^{-5}, which is less than 5.9×10^{-4}. Further approximations in the range between 0.6 and 1.0 M give improved values of S. The guess $S = 0.8$ fits the equation well. The solubility of $CdS(s)$ in 6 M HCl is thus apparently 0.8 M. Now check assumptions. It was assumed that all of the sulfur in solution is in the form of $H_2S(aq)$. Therefore, the concentration of $H_2S(aq)$ is also 0.8 M. But this exceeds the maximum molar solubility of H_2S at room temperature, which is only 0.1 M (see text page 259). Adding 6 M HCl to

$CdS(s)$ therefore forces gaseous H_2S out of solution. The gas bubbles out until the concentration of the residual $H_2S(aq)$ falls to 0.1 M. The previous equation is replaced by:

$$5.9 \times 10^{-4} = \frac{S\,(0.1)}{(6 - 2S)^2(6 - 4S)^4}$$

Solving this new equation (by approximations), gives S, the molar solubility of $CdS(s)$, equal to 1 M. This is a better answer than 0.8 M.

7-73 The solution is prepared by mixing 0.020 mol of $CuCl_2$ and 0.10 mol of NaCN in 1.0 L of water. The original concentration of Cu^{2+} ion is therefore 0.020 M, and the original concentration of CN^- is 0.100 M. These concentrations do not last long. The $Cu^{2+}(aq)$ ion and $CN^-(aq)$ quickly react to form a complex ion:

$$Cu^{2+}(aq) + 4\,CN^-(aq) \rightleftharpoons Cu(CN)_4^{2-}(aq) \qquad K = \frac{[Cu(CN)_4^{2-}]}{[Cu^{2+}][CN^-]^4}$$

The equilibrium constant for this process equals 2×10^{30}, the reciprocal of the K of the reverse reaction, which is given in the problem. The K for the formation of the complex is so large that effectively all of the Cu^{2+} ion is tied up as complex. At equilibrium then, $[Cu(CN)_4^{2-}] = 0.020$ M, and the concentration of uncomplexed Cu^{2+} is very small. Complexation reduces the concentration of CN^- ion, which is in excess relative to the Cu^{2+}, from its original 0.100 M to $0.100 - 4(0.020) = .020$ M because each mole of Cu^{2+} reacts essentially completely with 4 mol of $CN^-(aq)$. The values:

$$[CN^-] = 0.02 \text{ M} \quad \text{and} \quad [Cu(CN)_4^{2-}] = 0.020 \text{ M}$$

might now be substituted into the equilibrium expression for the formation of the complex and used to compute a equilibrium concentration of Cu^{2+} ion. Doing so would however overlook the fact that $CN^-(aq)$ reacts as a base with $H_2O(l)$ to give $HCN(aq)$:

$$CN^-(aq) + H_2O(l) \rightleftharpoons HCN(aq) + OH^-(aq) \qquad \frac{[HCN][OH^-]}{[CN^-]} = \frac{K_w}{K_a} = 2.03 \times 10^{-5}$$

This reaction lowers the concentration of $CN^-(aq)$ from 0.020 M. Let x equal the concentration of CN^- that reacts in this way. Then:

$$\frac{[HCN][OH^-]}{[CN^-]} = 2.03 \times 10^{-5} = \frac{x^2}{0.020 - x}$$

Solving gives x equal to 6.27×10^{-4} M. The correct equilibrium concentration of CN^- is therefore $0.020 - 6.27 \times 10^{-4} = 0.0194$ M. Put this value in the equilibrium expression for the complexation:

$$\frac{[Cu(CN)_4^{2-}]}{[Cu^{2+}][CN^-]^4} = 2 \times 10^{30} = \frac{(0.020)}{(0.0194)^4[Cu^{2+}]}$$

and solve for $[Cu^{2+}]$. The answer is 7×10^{-26} M.

7-75 a) From Example 7-10 (text page 264), K_{a1} for $Fe(H_2O)_6^{3+}$ is 7.7×10^{-3}. The problem gives K_{a2} as 2.0×10^{-5}. This K_{a2} is about 400 times smaller than K_{a1} the second stage of the acid ionization probably does *not* have a big effect on the pH. As shown in Example 7-10, the pH of the 0.100 M solution of $Fe(NO_3)_3$, is 1.62, based solely on the first ionization. This corresponds to $[H_3O^+] = 2.4 \times 10^{-2}$ M. Thus, considering only the first stage of acid ionization:

$$[H_3O^+] = [Fe(H_2O)_5(OH)^{2+}] = 2.4 \times 10^{-2}\ \text{M}$$

The equilibrium constant expression for the *second* ionization is:

$$K_{a2} = 2.0 \times 10^{-5} = \frac{[H_3O^+][Fe(H_2O)_4(OH)_2^+]}{[Fe(H_2O)_5OH^{2+}]}$$

Substitute the concentrations arrived at in the first-stage-only calculation into this expression. Solution of the resulting equation shows that the Fe-containing product of the second stage has a concentration of only 2.0×10^{-5} M. This concentration is negligible compared to 0.024 M, the concentration of $Fe(H_2O)_5OH^{2+}(aq)$. The $[H_3O^+]$ that arises in the second stage is also negligible compared to 0.024 M.

b) The question requires writing the two dissociation equations:

$$Fe(OH)_2^+(aq) \rightleftharpoons Fe(OH)^{2+}(aq) + OH^-(aq)$$
$$Fe(OH)^{2+}(aq) \rightleftharpoons Fe^{3+}(aq) + OH^-(aq)$$

These equilibria have dissociation constants K_{d1} and K_{d2}. The mass action expression for the first equation is:

$$K_{d1} = \frac{[Fe(OH)^{2+}][OH^-]}{[Fe(OH)_2^+]}$$

Divide this equation into the equilibrium expression for the autoionization of water, $K_w = [H_3O^+][OH^-]$. The result is:

$$\frac{K_w}{K_{d1}} = \frac{[H_3O^+][Fe(OH)_2^+]}{[Fe(OH)^{2+}]}$$

The right-hand side of this equation is identical to the mass-action expression for K_{a2} in the previous part except that the associated H_2O molecules are shown explicitly in the formulas in the K_{a2} expression. The crucial point is that "$Fe(OH)_2^+$" is really the same as "$Fe(H_2O)_4(OH)_2^+$". Hence, $K_{a2} = K_w/K_{d1}$. Similarly, $K_{a1} = K_w/K_{d2}$. Numerical values for K_{a1} and K_{a2} are available. Substitution gives $K_{d1} = 5.0 \times 10^{-10}$ and $K_{d2} = 1.3 \times 10^{-12}$. The K_{inst} of $Fe(OH)_2^+(aq)$ is the product of these numbers because dissociation of $Fe(OH)_2^+(aq)$ proceeds through $Fe(OH)^{2+}(aq)$ to $Fe^{3+}(aq)$. It is 6.5×10^{-22}.

Chapter 8

Thermodynamic Processes and Thermochemistry

Thermodynamics is the study of transfers of energy accompanying physical and chemical changes. The goals of thermodynamics are to predict which types of processes are possible and the conditions under which desired processes can occur.

Systems, States, and Processes

The basic terms in thermodynamics require careful definition:

System. A real or imaginary portion of the universe that is confined by physical boundaries or mathematical constraints. In essence, a thermodynamic system is that part of the universe which we decide to study. In problem-solving, a shrewd choice of system clarifies difficult situations. A useful tactic is to define a system as the sum of a set of sub-systems (see **problem 8-9).** One then concentrates attention on each of the sub-systems in turn. In complex problems the system should be fully defined *in writing* before starting any computations (see **problem 8-75).**

Closed System. A system that has boundaries that do not allow the transfer of matter. Closed systems are common in problems.

Open System. A system that *does* allow the transfer of matter across its boundaries. The typical chemical reaction, performed in an open flask in the laboratory, takes place in an open system.

Surroundings. The portion of the universe that lies outside the system.

Extensive Property. Extensive properties depend on the *extent,* or size of a system. The value of an extensive property for the whole system is the sum of the values for the individual sub-systems. A good example of an extensive property is the volume. Two sub-systems of volume 3.0 and 7.0 L make a system of volume 10.0 L. The energy is another important extensive property.

Intensive Property. A property of a system that does not depend on how big the system is. Examples are the pressure and temperature.

Thermodynamic State. A system is in a unique thermodynamic state when each of its properties (pressure, temperature, volume, energy, etc.) has a definite, time-independent value. Left to itself such a system will remain unchanged, in a state of *equilibrium,* indefinitely.

State Function. A state function is a property of a system which depends only on the thermodynamic state of the system. The value of a state function does not depend on the way in which the state was achieved. State functions are symbolized in thermodynamic equations by capital letters. A change in a state function is indicated by a Δ. Thus:

$$\Delta P = P_2 - P_1$$

means the difference in pressure between the final state (state 2) and the initial state (state 1). In many applications the *change* in a state function is much more important than the actual final and initial values of the function. Hence thermodynamic equations feature many Δ's. **In any process, the change in a state function depends only on the initial and final states and not on the path by which the change occurred.** In problems, it is best to adhere strictly to the convention that Δ applied to a function means the *final* value minus the *initial* value. Careless reversal of this convention leads to tiresome sign errors.

Thermodynamic Process. Such processes lead to changes in the thermodynamic states of systems. The **path** of a process is the sequence of intermediate conditions occurring in the system during a change.

An **irreversible** process passes through intermediate conditions that are not thermodynamic states. If the outside pressure on a sample of gas is suddenly relaxed, then the gas expands irreversibly. A **reversible** process proceeds through a continuous series of equilibrium states. At any moment during a reversible change, an infinitesimal alteration in external conditions is enough to reverse the direction of the process. Reversible compression of a sample of gas would require an infinite series of infinitely small reductions in volume. A reversible decrease in temperature would require a similar series of temperature reductions.

Energy, Work, and Heat

Work (w) and heat (q) are the two different means by which energy is transferred into or out of a system. The signs of w and q tells the direction of the transfer:

$+q$	the system gains heat
$-q$	the system loses heat
$+w$	work is done on the system
$-w$	the system performs work

Strict adherence to this sign convention reduces errors in problem solving. See the solutions to **problems 8-15** and **8-61.**

Work

Work is the product of an external force acting on a body times the distance through which the force acts. Work is thus the transfer of energy by oriented, non-random macroscopic motions. A transfer of work into a system goes to increase the kinetic energy of the system or the potential energy or both. Work is measured in joules, the unit used to measure energy, since work is a manifestation of energy. Force is measured in newtons (N), and distance in meters (m). Work is therefore also measured in newton meters (N m). This is all right because a newton is a kg m s^{-2}, which makes a newton-meter a kg m^2 s^{-2}, which is a joule. See Appendix B, text page A-11.

Of major concern in chemistry are electrical work (see Chapter 10) and mechanical work. The transfer of mechanical work requires mechanical contact between a system and its surroundings. In the absence of levers, wheels and other mechanical contrivances, the only kind of mechanical work possible in chemical processes is **pressure-volume work.** Pressure-volume work results from the change in volume (compression or expansion) of a system against a resisting pressure. If a system changes volume against an constant external pressure, then work flows:

$$w = -P_{\text{ext}}\Delta V$$

The negative sign must not be omitted. It maintains the convention that positive work is work done on the system. For an expansion, ΔV is positive because V_2 is larger than V_1. The system pushes back its surroundings; it does work on the surroundings. Under the sign convention this fact is stated as follows: negative work is done on the system. If the pressure is in atmospheres and the volume is in liters, then the unit of pressure-volume work is the L atm. A L atm is equal to 101.325 J.

Heat

Heat, like work, is a way in which energy is exchanged between a system and its surroundings. As a manifestation of energy it has the same units (joules) as energy. The transfer of heat requires thermal contact (rather than mechanical contact) across a system boundary and occurs because of a temperature difference. Heat involves random, non-directed motion instead of motion that has overall coherence and direction. Heat lacks the large-scale organization of motion that is characteristic of work.

Internal Energy

Two types of energy are distinguished. The **kinetic energy** of a system is its energy of motion:

$$KE = 1/2Mv^2$$

where M is its mass and v its velocity. An object has kinetic energy based on its own motion as a whole. A system has *internal* kinetic energy based on the relative motions of the atoms that compose it.

The **potential energy** of an object is its energy of position. An object has gravitational potential energy relative to the center of the earth. If its height h near the surface of the earth is changed, then the change in its potential energy is:

$$\Delta(PE) = Mg\Delta h$$

where g is the acceleration of gravity (9.81 m s^{-2}) and M is the mass of the object (see **problem 8-3**). *Internal* potential energy is potential energy that a system has due to the relative positions of the atoms that compose it. A pound of TNT for instance has much internal potential energy relative to the internal potential energy of the mixtures of gases that it forms when it explodes.

The total internal energy E of a system is the sum of its internal kinetic energy and internal potential energies. Changes in the internal energy of systems (symbolized ΔE) are very important in chemical thermodynamics.

Calorimetry and Heat Capacity

Calorimetry is the measurement of quantities of heat. The name of an alternative unit of heat, the *calorie,* reflects this. (One calorie is 4.184 J exactly.) The fundamental method in calorimetry is to let some heat flow into a system, checking the temperature of the system before and after. As the system gains heat its temperature increases. If the transfer is at constant pressure in a system that stays in the same state throughout the process, then the specific heat of the system can be measured: the **specific heat** c_s of a system is the amount of heat needed to raise the temperature of a one-gram mass by 1°C. The units for specific heat are J $g^{-1}(°C)^{-1}$. (or J $g^{-1}K^{-1}$, see **problem 8-9**). Once a specific heat is known, it can be used to measure flows of heat in practical experiments. In such cases:

$$q = Mc_s\Delta T$$

where M is the mass of the system in grams.

The **heat capacity** of a system is the quantity of heat necessary to raise its temperature 1 Kelvin. Heat capacity has the units J K^{-1}. In this definition, the focus is not on a one-gram portion of the system, but instead on the entire system. The heat capacity of a system is an extensive property.

The molar heat capacity is a heat capacity put on a per-mole basis. It is, like the specific heat, an intensive property. The **molar heat capacity** of a substance is the amount of heat necessary to raise the temperature of one mole of the substance by one kelvin. Accordingly, molar heat capacities have the unit J $mol^{-1}K^{-1}$. Molar heat capacities cannot be used unless the system is a substance or known mixture of substances (see **problem 8-75**).

Heat capacities depend on whether the heating is carried out at constant volume or constant pressure. The **constant-pressure molar heat capacity** is symbolized c_p; **constant-volume molar heat capacity** is symbolized c_v. The difference exists because a system at constant pressure may change volume as it absorbs heat. By changing volume, the system exchanges some work with its surroundings. This affects its internal energy and final temperature. At constant volume, there can be no such pressure-volume work.

At constant volume: $q_v = nc_v\Delta T$ at constant pressure: $q_p = nc_p\Delta T$

The difference between c_v and c_p is small for liquids and solids and is often ignored. Thus **problems 8-5** and **8-7** ask for molar heat capacities of several solids without stating whether volume or pressure was constant. The distinction between c_v and c_p becomes important with gases and figures strongly in many problems (see below).

• *Always check units particularly carefully in calorimetry problems.* Mixing up the heat capacity, an extensive variable, with the molar heat capacity or specific heat, intensive variables, is a common error. Although the absolute temperature scale has a zero different from the zero of the Celsius scale, 1 K is exactly the same size as 1°C. Values of heat capacities and specific heats quoted in J $(°C)^{-1}$ or J $g^{-1}(°C)^{-1}$ are for this reason numerically equal to values quoted in J K^{-1} or J $g^{-1}K^{-1}$, respectively. See **problem 8-9**.

The First Law of Thermodynamics

The first law of thermodynamics is a process-oriented re-statement of the principle of conservation of energy. Any change in the internal energy of a system during a process must equal the work *done on* the system plus the heat *absorbed* by the system:

$$\boldsymbol{\Delta E = w + q}$$

In a general process, both heat and work cross the boundaries of the system. The two are positive when they flow into the system. This explains the emphasis on *absorption* in the statement of the first law.

What if a system is carefully insulated so that heat cannot flow across its boundaries? Heat is neither absorbed nor lost so q equals 0. If the system now does some work, it does so entirely at the expense of its internal energy. When the internal energy diminishes in a process, E_2 is smaller than E_1; $\Delta E = E_2 - E_1$ is negative in such a case. According to the first law, if q is zero and ΔE is negative, then w has to be negative. See **problem 8-53.**

Although both q and w depend on the path along which a change occurs, their sum, ΔE, is a state function and does *not* depend on the path.

Heat Capacities of Ideal Gases

At constant volume the molar heat capacity of a monatomic ideal gas is $3/2R$. The gas constant R is equal to 8.315 J $K^{-1}mol^{-1}$ so this c_v is 12.472 J $mol^{-1}K^{-1}$. At constant pressure, the same monatomic ideal gas has a molar heat capacity of $5/2R$, which equals 20.788 J $mol^{-1}K^{-1}$. It is 1.666 times larger. **Problem 8-17b** requires these facts as do **problems 8-15, 8-19, 8-51,** and **8-53.**

In the first two of these problems, the solver must recall that argon and neon are monatomic. The qualification "monatomic" is important because diatomic and polyatomic gases have larger values of c_v and c_p than monatomic gases. The quantities c_p and c_v increase with the number of atoms per molecule. Gaseous C_8H_{18} has a c_p of 327 J $mol^{-1}K^{-1}$ compared to only 29.8 J $mol^{-1}K^{-1}$ for $N_2(g)$. See **problem 8-75.** For ideal gases, whether monatomic or polyatomic: $\boldsymbol{c_p = c_v + R}$.

Thermochemistry

Enthalpy

The enthalpy is an additional, very useful state function. It is defined in terms of other state functions:

$$\boldsymbol{H = E + PV}$$

The definition means that the enthalpy of a system is computed by determining its internal energy and adding the product of its pressure and volume. Clearly H has the same units as E (joules). Although PV ordinarily comes out in L atm, an easy conversion (1 L atm = 101.325 J) would allow adding it to E. In practice, one never actually performs such an addition. Absolute values of E are unattainable because of the impossibility of defining a zero for energy. The emphasis is on *changes* in E and H instead. For a process (change) in a system:

$$\Delta H = \Delta E + \Delta(PV)$$

The definition of enthalpy was deliberately and carefully chosen so that:

$$q_\text{p} = \Delta H$$

which means that the change in the enthalpy of a system during a process equals the amount of heat that the system would absorb if the process were carried out at constant pressure.

An change does *not* actually have to occur at constant pressure for the accompanying change in enthalpy to be meaningful. Enthalpy is a state function. However, the enthalpy is such a very useful function just because many processes *do* occur under constant (atmospheric) pressure. This circumstance makes q_p relatively easy to measure. The similarity of the relationships between q_p and ΔH to that between q_v and ΔE is not accidental:

$$\boldsymbol{q_\text{v} = \Delta E} \quad \text{and} \quad \boldsymbol{q_\text{p} = \Delta H}$$

These are key relationships in thermochemistry.

Enthalpies of Reaction

Large changes in the internal energy and in the enthalpy of systems occur during chemical reactions because reactions consist of the making and breaking of chemical bonds, events that greatly alter the internal potential energy of systems. A negative enthalpy change in a reaction means that H_2, the enthalpy of the products, is *less* than H_1, the enthalpy of the reactants:

$$\Delta H = H_2 - H_1 = q_\text{p} < 0$$

Because q_p is negative the reaction gives off heat. The reaction is **exothermic.** Reactions in which heat is absorbed are **endothermic** and have a positive ΔH.

The enthalpy change of a reaction is often written immediately after the reaction on the same line. In this position, ΔH refers to the enthalpy change of the balanced chemical reaction with which it is associated. For example, in **problem 8-71** the enthalpy change for the combustion of octane is computed:

$$C_8H_{18}(l) + 25/2\,O_2(g) \rightarrow 8\,CO_2(g) + 9\,H_2O(l) \quad \Delta H = -5520 \text{ kJ}$$

If all of the coefficients in the reaction are doubled, then ΔH is also doubled:

$$2\,C_8H_{18}(l) + 25\,O_2(g) \rightarrow 16\,CO_2(g) + 18\,H_2O(l) \quad \Delta H = -11040 \text{ kJ}$$

A ΔH is a change in an extensive property; the coefficients in the balanced equation give the extent (size) of the system. **Problems 8-23** and **8-71** show how the quantities of reactants are considered in figuring out how much heat a reaction evolves.

If the direction of a reaction is reversed the roles of products and reactants are interchanged. Such an interchange makes H_2 into H_1 and H_1 into H_2. The upshot is to change the sign of ΔH, but not its magnitude. Thus, for the reverse of the combustion of octane:

$$8\,CO_2(g) + 9\,H_2O(l) \rightarrow C_8H_{18}(l) + 12\,1/2\,O_2(g) \quad \Delta H = +5520 \text{ kJ}$$

The fact that H is a state function implies that ΔH is independent of the path followed during the conversion of reactants to products. Consequently, if two or more chemical equations are added to give some overall equation their enthalpy changes add up to the enthalpy change of the overall equation. The overall reaction can be imagined to proceed by stages consisting of the first reaction, then the second, then the third, and so on. **Hess's law** is a statement of this fact: **if two or more chemical equations are added to give another chemical equation, the corresponding enthalpies of reaction must be added.** This means that the ΔH for the conversion of a given set of reactants into a given set of products is the same regardless of whether the conversion occurs in one step or a series of steps.

Hess's law is important because it assures that whenever a chemical reaction can be constructed (on paper) as a combination of other reactions with known ΔH's, then the ΔH of the new reaction need not be measured. It can instead be calculated. **Problem 8-31** is an example showing the use of known ΔH's to get a ΔH for a new reaction.

To compute ΔE for a reaction when ΔH is known, use the relationship:

$$\Delta E = \Delta H - \Delta(PV)$$

For solids and liquids the term $\Delta(PV)$ is small. For this reason, for reactions involving exclusively solids and liquids, the difference between ΔE and ΔH can be safely neglected. For reactions involving gases, neglecting the difference is *not* safe. It is however nearly always acceptable to assume that the ideal-gas equation is followed by the gases. If so, then, at constant temperature:

$$\Delta E = \Delta H - \Delta n_g RT \quad \text{(at constant temperature)}$$

where Δn_g is the number of moles of gas among the products minus the number of moles of gas among the reactants. As **problems 8-43** and **8-69** show, the difference between ΔE and ΔH is usually only a small fraction of ΔH. A common mistake in applications is to compute $\Delta n_g RT$ in joules and then to subtract it from a ΔH value that is in kilojoules. Always check the units of these quantities before adding or subtracting.

Standard State Enthalpies

The standard state of a substance is its stable form at a temperature of 298.15 K and a pressure of 1 atm. The standard state of dissolved species is unit molarity (1 M). This last definition figures in **problem 8-39.** A superscript of zero ("naught") following a thermodynamic symbol refers to standard conditions.

Every chemical compound can, in principle, form directly from the elements that make it up. Reactions of this type are **formation** reactions. Equations representing formation reactions always have some assortment of pure elements on the left-hand side and a single compound, the compound being formed, on the right-hand side.

• **The standard enthalpy of formation ΔH_f° of a substance is the enthalpy change of the reaction that produces one mole of the substance from its constituent elements in their standard states.**

The units of standard enthalpies of formation are kJ mol^{-1}. For example, the formation of ammonium iodide from its constituent elements is represented:

$$1/2\,N_2(g) + 2\,H_2(g) + 1/2\,I_2(s) \rightarrow NH_4I(s) \quad \Delta H_f^\circ = 201.4 \text{ kJ}$$

To emphasize that this reaction is a formation reaction its ΔH is subscripted with an "f" (for formation). As written the reaction produces 1 mol of $NH_4I(s)$. Hence the ΔH_f° of ammonium iodide is 201.4 kJ mol^{-1}. This particular ΔH was measured at 298.15 K and 1 atm which are defined as the standard conditions. It is a standard enthalpy change, ΔH_f°.

The standard enthalpy of formation of pure elements is zero. For example, the equation:

$$O_2(g) \rightarrow O_2(g)$$

represents the formation of the element oxygen "from its constituent elements in their standard states." The two sides of the equation are the same, so there is no change in the enthalpy. The ΔH_f° of $O_2(g)$ is zero.

Standard enthalpies of formation of compounds are tabulated in long lists. See text Appendix D. Such tables are useful because a reaction, no matter how complicated, can be imagined to go by the decomposition of all of the reactants into the constituent elements followed by the formation of all of the products directly from those elements. As a consequence of Hess's law then:

$$\boldsymbol{\Delta H^\circ = \sum \Delta H_f^\circ\,(\text{products}) - \sum \Delta H_f^\circ\,(\text{reactants})}$$

This equation is heavily used in solving practical problems. **Problems 8-35**, **8-69** and **8-71** are examples. Confusion sometimes arises concerning the units. ΔH_f°'s are given in kJ mol^{-1}. To apply the above equation:

1. Multiply each compound's ΔH_f° by the number of moles of that compound shown in the balanced chemical equation.

2. Add these numbers together for all of the products and do the same for all of the reactants.

3. Subtract the latter from the former.

The result is $\Delta H°$ of the reaction and has the units of kJ. In those cases where ΔH of a reaction is given in kJ mol^{-1} (or J mol^{-1}), it refers to the enthalpy change *per mole of reaction as written.* If the reaction:

$$H_2(g) + 1/2\,O_2(g) \rightarrow H_2O(l)$$

is said to have a $\Delta H°$ of -285.83 kJ mol^{-1}, then it follows that the related reaction:

$$2\,H_2(g) + O_2(g) \rightarrow 2\,H_2O(l)$$

(derived by doubling all of the coefficients) has a $\Delta H°$ of -571.66 kJ mol^{-1}.

Bond Enthalpies

A **bond enthalpy** is the $\Delta H°$ of the reaction breaking a given type of bond in a substance in the gas phase. For example, the measured $\Delta H°$ for the reaction $Cl_2(g) \rightarrow 2\,Cl(g)$ is the bond enthalpy for one mole of Cl—Cl bonds. This particular reaction is the breakdown of a molecule into atoms. Bond enthalpies of this type are called **enthalpies of atomization.** Bond enthalpies, which are directly measurable, are always positive. Moreover, bond enthalpies are fairly reproducible from one compound to another. That is, the bond enthalpy for a C—H bond is about the same (within about 10 percent) for a four C—H bonds in the simple compound CH_4 and all 50 C—H bonds in the complicated compound $C_{30}H_{50}$.

There are sometimes big differences in the bond enthalpies between two types of atoms, but these are attributed to the existence of different categories of bonds (single, double, and triple bonds) between the two types. Thus Table 8-3 (text page 296) gives different enthalpies for C—N, C=N, and C≡N bonds.

Many average bond enthalpies have been tabulated. Tables of bond enthalpies (together with atomization enthalpies) allow approximate computation of $\Delta H_f°$'s of reactions in the gas phase **(problem 8-45)**; such tables also allow approximate computation of $\Delta H°$'s of reactions in the gas phase **(problem 8-47).** Bond enthalpies are discussed in Chapter 14 (text page 530-2) as well as in Chapter 8.

Reversible Gas Processes

Changes in state functions depend only on the initial and final state. They depend not at all on the path along which a process or reaction takes place. Sometimes, however, knowledge of the path of a process is available and can help in computing ΔH, ΔE, and other changes in state variables. To compute values for q and w, which *do* depend on the path a process follows, specification of the path is of course essential. Several technical terms describe paths:

Isothermal. An isothermal process goes on at constant temperature; ΔT is zero in an isothermal change. The internal energy of an ideal gas depends only on its temperature. If T is constant, $\Delta E = 0$ and:

$$q = -w \qquad \text{(isothermal process, ideal gas)}$$

In addition, ΔH is zero in an isothermal change of an ideal gas. For non-ideal gases, liquids, and solids, ΔE and ΔH may be non-zero even if the process is isothermal.

Isothermal and Reversible. An isothermal and reversible process goes on at constant temperature by a series of infinitesimally small steps. The system is always at equilibrium and at any time the direction of the process can be changed. For an ideal gas in such a process:

$$q = -w = nRT \ln\left(\frac{V_2}{V_1}\right) \qquad \text{(isothermal reversible process, ideal gas)}$$

Boyle's law states that $P_1V_1 = P_2V_2$ for an ideal gas at constant temperature. When solved for V_2 and substituted into the previous equation, this gives:

$$q = -w = nRT \ln\left(\frac{P_1}{P_2}\right) \qquad \text{(isothermal reversible process, ideal gas)}$$

In an isothermal *expansion*, V_2 is bigger than V_1. The gas expands and absorbs heat from the surroundings as it does so. It simultaneously performs an exactly equivalent amount of work on the surroundings. In that way ΔE remains at zero.

Adiabatic. In an adiabatic process there is no transfer of heat into or out of the system. This means that q equals zero in all adiabatic processes. In an adiabatic change, any work that the system performs comes at the expense of its internal energy; any work done on the system adds to its internal energy: $\Delta E = w$.

Adiabatic and Reversible. For an ideal gas it can be shown that:

$$P_1V_1^\gamma = P_2V_2^\gamma \quad \text{(adiabatic reversible process, ideal gas)}$$

where γ is the ratio of c_p to c_v. In addition:

$$T_1V_1^{\gamma-1} = T_2V_2^{\gamma-1} \quad \text{(adiabatic reversible process, ideal gas)}$$

ideal, adiabatic, reversible

The solutions to **problem 8-17** and **8-19** show a common pitfall with these equations. They do not apply unless an *ideal* gas undergoes an *adiabatic, reversible* change. All three qualifications must be met. **Problems 8-53, 8-75** and **8-77** show more applications of these equations to more complex cases.

Heat Engines and Refrigerators

The Carnot Cycle

A **cyclic process** takes a system through a series of thermodynamic states and returns it ultimately to the exact state where it started. In a **Carnot cycle,** an ideal gas undergoes a series of four reversible changes during an excursion out from and then back to its original state. These are:

1. An isothermal reversible expansion. The gas, at its original temperature, T_h, expands from V_A to V_B. It performs work on the surroundings. It simultaneously absorbs heat from the surroundings to keep the same internal energy.

2. An adiabatic reversible expansion. The gas continues to expand, going from V_B to V_C. Now, however, q is zero. The continued expansion of the gas performs more work on its surroundings. The energy to do the work comes from the internal energy of the gas. The temperature of the gas drops to T_ℓ.

3. An isothermal reversible compression. At T_ℓ the surroundings compress the gas, from V_C to V_D. Work is performed upon the gas. The temperature is constant and just enough heat flows from the gas to the surroundings so that ΔE for this step stays zero.

4. An adiabatic reversible compression. The surroundings continue to do work on the gas. Now however the gas is thermally insulated so that $q = 0$. All the work adds to the internal energy of the gas. The temperature rises toward T_h. The volume diminishes toward V_A. The compression continues until the gas reaches its original state.

In a Carnot cycle the system extracts heat from the surroundings at T_h, converts some of it into work, and dumps the rest to the surroundings at T_ℓ. The net work in one passage around the cycle is:

$$w_{net} = -nR(T_h - T_\ell)\ln\left(\frac{V_B}{V_A}\right)$$

The net work is negative, consistent with the convention that work performed *by* a system is negative. The net work depends on how much gas (n is the chemical amount) is taken around the cycle, on the two extremes of temperature and by how much the gas expands in the first step.

Heat Engines, Refrigerators, and Heat Pumps

A **heat engine** is a device which converts the natural heat flow from higher to lower temperature into useful work. A **refrigerator** does the reverse. It destroys work to produce a difference in temperature.

A heat engine operates between a high and low temperature, T_h and T_ℓ. The **efficiency** ϵ of a heat engine is the net work that it performs divided by the heat that it absorbs. Maximum efficiency is obtained only if when the engine operates reversibly, and is thus an unattainable ideal. This maximum efficiency depends only on the temperatures:

$$\epsilon = 1 - \frac{T_\ell}{T_h}$$

This equation is derived from consideration of the Carnot cycle. Why does this equation give the theoretical maximum efficiency? The proof consists of assuming the opposite and showing that an engine with greater efficiency would make it possible to transfer heat in a continuous cycle out of a low temperature reservoir into a high temperature reservoir without expending any work. In all human experience it is impossible to do this. Heat never spontaneously flows“ uphill” (against a temperature gradient). **It is impossible to construct a device that will transport heat from a cold reservoir to a hot reservoir in a continuous cycle without any net expenditure of work.**

The theoretical operation of a refrigerator or heat pump is modeled on a Carnot cycle running in reverse. The maximum heat q absorbed from the interior of a refrigerator is:

$$q = w_{\text{net}}\left(\frac{T_\ell}{T_h - T_\ell}\right)$$

where T_ℓ and T_h are the low and temperatures and w_{net} is the work during the refrigeration cycle. These equations are used in **problems 8-55** and **8-57.**

Detailed Solutions to Odd-Numbered Problems

8-1 The work done *on* a gas in a change of volume at constant pressure is given by $w = -P_{\text{ext}}\Delta V$. The problem states unmistakable values for the external pressure and for a final and initial volume. Substituting gives:

$$w = -P_{\text{ext}}\Delta V = -(50.0\text{ atm})(974\text{ L} - 542\text{ L}) = -2.16 \times 10^4\text{ L atm}$$

where, as ever, the change in a quantity (in this case the volume) is the final value minus the initial. To convert to joules, use the conversion factor from the text (page 278):

$$-2.16 \times 10^4\text{ L atm} \times \left(\frac{101.325\text{ J}}{1\text{ L atm}}\right) = -2.19 \times 10^6\text{ J}$$

This is the work done on the gas. The negative sign means that the gas does positive work on its surroundings. The difference is a matter of point of view. Do we picture ourselves with the gas looking at the surroundings, or in the surroundings looking at the gas?

8-3 A ball of mass M under the influence of gravity possesses potential energy relative to the ground. This potential energy equals Mgh where h is the height of the ball and g is the acceleration of gravity. A change in height Δh leads to a corresponding change in potential energy. If the ball falls, the potential energy is converted into kinetic energy. The total energy of the ball does not change during the fall. At impact, the (non-bouncing) ball stops. Its kinetic energy becomes suddenly zero. Simultaneously, its potential energy attains zero because h is zero. According to the problem, the total energy of ball still does not change. All of the energy becomes internal energy, and the temperature of the ball is thereby increased. This is expressed mathematically:

$$Mc_s\Delta T + Mg\Delta h = \Delta(\text{total energy}) = 0$$

where c_s is the heat capacity per unit mass (specific heat) of the ball. The M's cancel out, and it is easy to solve for Δh:

$$\Delta h = -\frac{c_s\Delta T}{g}$$

In this problem ΔT equals 1.00°C, which is the same as 1.00 K, and c_s is given as 0.850×10^3 J kg^{-1}K^{-1}. In the following, putting the specific heat of the ball on a per kilogram basis (it was given on a per gram basis) aids the cancellation of units. The acceleration of gravity at the earth's surface is 9.81 m s^{-2}. Substituting:

$$\Delta h = -\frac{(0.850 \times 10^3 \text{ J kg}^{-1}\text{K}^{-1})(1.00 \text{ K})}{9.81 \text{ m s}^{-2}} = -86.6 \text{ J kg}^{-1}\text{m}^{-1}\text{s}^2$$

By its definition a joule is equal to a kg m^{2}s^{-2}. Therefore, in the above cluster all units but meters cancel out: Δh is -86.6 m. The negative sign simply means that the final height of the ball is less than the initial height. The ball falls down (not up) a distance of 86.6 m.

8-5 The molar heat capacity of a substance equals its specific heat multiplied by its molar mass. Here is a sample calculation for lithium

$$c_p = c_s \times \mathcal{M} = 3.57 \text{ J g}^{-1}\text{K}^{-1} \times 6.94 \text{ g mol}^{-1} = 24.8 \text{ J K}^{-1}\text{mol}^{-1}$$

The full set of values in the group:

Li(s)	Na(s)	K(s)	Rb(s)	Cs(s)	
24.8	28.3	29.6	31.0	32.2	J K^{-1}mol^{-1}

Beyond sodium there is a steady increase of about 1.3 J $K^{-1}mol^{-1}$ for every element. Extrapolation of the trend gives francium a molar heat capacity of about 33.5 J $K^{-1}mol^{-1}$. Although there is a distinct trend in the values, it is small. Indeed, the molar heat capacities of the metallic elements are remarkably constant. This constancy is the law of Dulong and Petit (see problem 8-59).

8-7 The calculations proceed as in problem 8-5 with the results:

Ni(s)	Zn(s)	Rh(s)	W(s)	Au(s)	U(s)	
26.1	25.4	25.0	24.3	25.4	27.6	J $K^{-1}mol^{-1}$

8-9 Let the system under consideration consist of two sub-systems: the metal and the water. If the mixing of hot metal and cool water takes place in a well-insulated container (which prevents leaks of heat), then the heat absorbed by the system equals zero. The system is the sum of the two sub-systems. Therefore:

$$q_{\text{sys}} = 0 = q_{\text{m}} + q_{\text{w}}$$

For both sub-systems, the amount of heat gained equals the specific heat times the mass times the temperature change:

$$q_{\text{m}} + q_{\text{w}} = M_{\text{w}} c_{s,\text{w}} \Delta T_{\text{w}} + M_{\text{m}} c_{s,\text{m}} \Delta T_{\text{m}} = 0$$

Solving for the specific heat of the metal:

$$c_{\text{s,m}} = \frac{-M_{\text{w}} c_{s,\text{w}} \Delta T_{\text{w}}}{M_{\text{m}} \Delta T_{\text{m}}} = -\frac{(100.0 \text{ g})\, 4.18 \text{ J g}^{-1}\text{K}^{-1}(6.39°\text{C})}{(61.0 \text{ g})(-93.61°\text{C})} = 0.468 \text{ J g}^{-1}\text{K}^{-1}$$

Note that in setting up this equation we do not bother to convert °C to K. This is because a change of one degree Celsius is identical to a change of one kelvin. The Kelvin and Celsius scales have the same size increments. They differ only in the location of their zeros. Changes in temperature measured on the two scales are numerically equal.

8-11 Body 1 and body 2 are originally at different temperatures. They are brought into thermal contact with each other and held in thermal isolation from other objects. Then:

$$q_1 + q_2 = M_1 c_{s1} \Delta T_1 + M_2 c_{s2} \Delta T_2 = 0$$

If the masses of the two bodies are equal, then $M_1 = M_2$, and:

$$c_{s1} \Delta T_1 = -c_{s2} \Delta T_2$$

$$\frac{c_{s1}}{c_{s2}} = -\frac{\Delta T_2}{\Delta T_1}$$

This equation shows that the specific heats of the two bodies are inversely proportional to the temperature changes they undergo in this experiment.

if directly $\propto$ then
$C_{s1} = k\Delta T_1$ for example

8-13 The difference in temperature ΔT between water at its boiling point and melting point is 100°C. The amount of heat required to bring 1.00 g of water at 0°C to 100°C is:

$$q = Mc_s\Delta T = (1.00 \text{ g})(4.18 \text{ J g}^{-1}\,°\text{C}^{-1})(100°\text{C}) = 418 \text{ J}$$

The amount of heat needed to melt 1.00 g of ice is, according to the statement of Lavoisier and Laplace, 3/4 of this amount or 314 J. More recent experiments set the amount of heat to melt 1.00 g of ice at 333 J.

8-15 The 0.500 mol of neon expands against a constant pressure of 0.100 atm. Let the neon comprise the system. Before the expansion, the volume of the system is 11.207 L (calculated by applying the ideal gas equation to 0.500 mol of gas at STP). After the expansion, the volume is 43.08 L (calculated from the ideal-gas equation with $P = 0.200$ atm, $n = 0.500$ mol, and $T = 210$ K). The gas expands against a constant pressure (of 0.100 atm). Presumably this is possible because the pressure is suddenly raised to its final value of 0.200 atm just after the gas reaches its final volume. Therefore the work done on the system is:

$$w = -P_{\text{ext}}\Delta V = -0.100 \text{ atm}(43.08 - 11.207) \text{ L} = -3.19 \text{ L atm}$$

The gas cools from 273.15 to 210 K. Since it is an ideal monatomic gas, the change in its internal energy is directly proportional to the change in its temperature; the constant of proportionality is $n(\frac{3}{2})R$, the heat capacity at constant volume:

$$\Delta E = nc_v\Delta T = n\left(\frac{3}{2}R\right)\Delta T$$

Substituting gives:

$$\Delta E = 0.500 \text{ mol}\left(\frac{3}{2}\,0.08206 \text{ L atm mol}^{-1}\text{K}^{-1}\right)(-63.14 \text{ K}) = -3.89 \text{ L atm}$$

By the first law:

$$q = \Delta E - w = -3.89 \text{ L atm} - (-3.19 \text{ L atm}) = -0.70 \text{ L atm}$$

The units of all three answers can be converted to joules (1 L atm = 101.325 J): $w = -323$ J $\quad \Delta E = -394$ J $\quad q = -71$ J.

8-17 a) The statement of the problem gives the initial quantity (2.00 mol), pressure (3.00 atm), and temperature (350 K) of the ideal monatomic gas. The initial volume of the gas is $V = nRT/P = 19.15$ L. The final volume is *twice* this original volume or 38.30 L. The change in volume ΔV is $38.30 - 19.15 = 19.15$ L.

b) The adiabatic expansion occurs against a *constant* pressure of 1.00 atm. Under that circumstance, the work done on the gas is:

$$-P\Delta V = -1.00(19.15) \text{ L atm} \times \left(\frac{101.325 \text{ J}}{1 \text{ L atm}}\right) = -1.94 \times 10^3 \text{ J}$$

The expansion is adiabatic so $q = 0$ by definition, and:

$$\Delta E = q + w = 0 - 1.94 \times 10^3 \text{ J} = -1.94 \times 10^3 \text{ J}$$

c) Any change in the internal energy of an ideal gas causes a change in temperature in direct proportion:

$$\Delta E = nc_v \Delta T$$

Solving for ΔT and substituting the various values:

$$\Delta T = \frac{\Delta E}{nc_v} = \frac{-1.94 \times 10^3 \text{ J}}{2.00 \text{ mol}(3/2)8.315 \text{ J K}^{-1}\text{mol}^{-1}} = -77.8 \text{ K}$$

Thus, T_2, the final temperature, is $T_1 + \Delta T = 350 + (-77.8) = 272$ K.

8-19 The system consists of the 6.00 mol of argon. The change in internal energy of this monatomic gas (assuming ideality) is:

$$\Delta E = nc_v \Delta T = (6.00 \text{ mol}) \left(\frac{3}{2} 8.315 \text{ J K}^{-1}\text{mol}^{-1}\right) (150 \text{ K}) = +11.2 \times 10^3 \text{ J}$$

The change is adiabatic which means that $q = 0$. From the first law:

$$w = \Delta E - q = 11.2 \times 10^3 \text{ J} - 0 = +11.2 \times 10^3 \text{ J}$$

The work done on the argon is 11.2×10^3 J, *all* of which goes to increase its internal energy.

8-21 a) During the heating process, heat goes from the surroundings to the system. Therefore, q is positive. Since the container is rigid, it neither expands nor contracts. The rigidity means that ΔV is zero so that no pressure-volume work is performed on the system. No other type of work is possible, so $w = 0$. By the first law, ΔE is then positive.

b) During the cooling process, the heat absorbed by the system is negative ($q < 0$). No work can be done on the system ($w = 0$). The energy of the system is therefore lowered ($\Delta E < 0$).

c) Since no work was done in either step 1 or step 2, $w_1 + w_2 = 0$. Nothing in the problem suggests that the system is in the same thermodynamic state after it is cooled back to its original temperature. All that is stated is that the temperature is the same. Other variables, like the internal energy, might be greatly changed by the heating-cooling cycle. Hence $\Delta E_1 + \Delta E_2$ is *not* necessarily zero. The two traps here are the unjustified assumption that the system is an ideal gas (for which the internal energy depends only on the temperature) and the idea that any chemical changes brought on by the heating are exactly reversed by the cooling. All that can be said is $\Delta E_1 + \Delta E_2 = q_1 + q_2$. The two sides of this equation could be positive, negative or zero.

8-23 The balanced equation tells the enthalpy change taking place during the production or consumption of a specific number of moles of product or reactant. All that is necessary is to put these enthalpy changes on a basis of mass.

a)

$$\Delta H = \frac{-828 \text{ kJ}}{2 \text{ mol Na}_2\text{O}} \times \left(\frac{1.00 \text{ mol Na}_2\text{O}}{62.0 \text{ g Na}_2\text{O}}\right) = -6.68 \text{ kJ g}^{-1}$$

b)

$$\Delta H = \frac{302 \text{ kJ}}{1 \text{ mol MgO}} \times \left(\frac{1.00 \text{ mol MgO}}{40.31 \text{ g MgO}}\right) = +7.49 \text{ kJ g}^{-1}$$

c)

$$\Delta H = \frac{33.3 \text{ kJ}}{2 \text{ mol CO}} \times \left(\frac{1.00 \text{ mol CO}}{28.01 \text{ g CO}}\right) = +0.594 \text{ kJ g}^{-1}$$

8-25 Only 119.0 J of the measured 121.3 J of heat is due to the reaction of the 0.00288 mol of $Br_2(l)$. The rest of the heat (2.34 J) is added mechanically (see Figure 8-7, text page 280) by breaking the capsule and stirring the liquid. The amount of heat evolved from 1.00 mole of $Br_2(l)$ is:

$$\frac{119.0 \text{ J}}{2.88 \times 10^{-3} \text{ mol}} = 41.3 \times 10^3 \text{ J mol}^{-1}$$

8-27 The vaporization process is $CO(l) \rightarrow CO(g)$ for which ΔH_{vap} is 6.04 kJ mol^{-1}. The following series of conversions solves the problem:

$$2.38 \text{ g CO} \times \left(\frac{1 \text{ mol CO}}{28.01 \text{ g CO}}\right) \times \left(\frac{6.04 \text{ kJ}}{1 \text{ mol CO}}\right) = +0.513 \text{ kJ}$$

8-29 The ice cube is 2.00 mol in extent. It is put in contact with 20.0 mol of 20°C water. The ice is well below its melting point. It must be heated up before it can start to melt. Warming the ice cube from −10°C to 0°C would require:

$$q = nc_p\Delta T = (2.00 \text{ mol})(38 \text{ J mol}^{-1}\text{K}^{-1})(10 \text{ K}) = 760 \text{ J}$$

Melting the ice at 0°C to water at 0°C would then take:

$$q = n\Delta H_{\text{fus}} = (2.00 \text{ mol})(6007 \text{ J mol}^{-1}) = 12\,014 \text{ J}$$

On the other hand, cooling 20.0 mol of 20°C water down to 0°C would absorb:

$$q = nc_{\text{p}}(T_{\text{f}} - T_{\text{i}}) = (20.0 \text{ mol})(75 \text{ J mol}^{-1}\text{K}^{-1})(-20 \text{ K}) = -30\,000 \text{ J}$$

This result can be rephrased: cooling 20.0 mol of water from 20°C to 0°C would require *removal* of +30 000 J. Since the ice cube absorbs less than this by warming up and then melting to liquid water at 0°C, T_{f}, the final temperature of the mixture, must be above 0°C. No heat is lost to the surroundings. The heat absorbed in warming up and melting the ice, and then warming the melt-water to the actual T_{f} therefore can be added with the heat absorbed in cooling the 20.0° water to T_{f} to equal zero:

$$\underbrace{12\,774 \text{ J}}_{q \text{ for ice}} + \underbrace{(2.00 \text{ mol})(75 \text{ J mol}^{-1}\text{K}^{-1})(T_{\text{f}} - 0)}_{q \text{ for meltwater}}$$

$$+ \underbrace{(20.0 \text{ mol})(75 \text{ J mol}^{-1}\text{K}^{-1})(T_{\text{f}} - 20.0)}_{q \text{ for warm water}} = 0$$

Solving gives $T_{\text{f}} = 10.4$°C. A frequent source of difficulty in this problem is the wrong idea that ice must always be at 0°C. Like any other material held below its melting point, ice comes to the temperature of its surroundings.

8-31 Add the following two equations to obtain the equation for the desired reaction:

$$\begin{array}{ll} 2\,CH_4(g) + 4\,O_2(g) \rightarrow 2\,CO_2(g) + 4\,H_2O(g) & \Delta H = -1604.6 \text{ kJ} \\ 2\,CO_2(g) + H_2O(g) \rightarrow CH_2CO(g) + 2\,O_2(g) & \Delta H = 981.1 \text{ kJ} \\ \hline 2\,CH_4(g) + 2\,O_2(g) \rightarrow CH_2CO(g) + 3\,H_2O(g) & \Delta H = -623.5 \text{ kJ} \end{array}$$

The top equation has been multiplied through by 2 as has its ΔH. The second equation has been written as the reverse of what was given, and its ΔH accordingly multiplied by -1. Summing the two equations give the desired equation, and, by Hess's law, the enthalpy change associated with this reaction equals the sum of the enthalpy changes of the two equations being added.

8-33 The conversion $C(gr) \rightarrow C(dia)$ is endothermic (positive ΔH). Therefore, one pound of diamonds contains more enthalpy than one pound of graphite. Both diamond and graphite give the same product (carbon dioxide) when burned. When burned, the pound of diamonds will give off more heat.

8-35 A reaction enthalpy is calculated by summing the enthalpies of formation of the products and subtracting the enthalpies of formation of the reactants:

$$N_2H_4(l) + 3\,O_2(g) \rightarrow 2\,NO_2(g) + 2\,H_2O(l)$$

$$\Delta H^\circ = 2\underbrace{(33.18)}_{NO_2(g)} + 2\underbrace{(-285.83)}_{H_2O(l)} - 1\underbrace{(50.63)}_{N_2H_4(l)} - 3\underbrace{(0)}_{O_2(g)} = -555.93 \text{ kJ}$$

In the preceding equation, all of the ΔH_f°'s are in kJ mol^{-1}. All are multiplied by the number of moles of the substance represented in the balanced equation.

8-37 a) As in problem 8-35:

$$2\,ZnS(s) + 3\,O_2(g) \rightarrow 2\,ZnO(s) + 2\,SO_2(g)$$

$$\Delta H^\circ = 2\underbrace{(-348.28)}_{ZnO(s)} + 2\underbrace{(-296.83)}_{SO_2(g)} - 2\underbrace{(-205.98)}_{ZnS(s)} - 3\underbrace{(0)}_{O_2(g)} = -878.26 \text{ kJ}$$

b) The reaction of 2 mol of ZnS(s) has ΔH equal to -878.26 kJ. Therefore, ΔH of the reaction is -439.13 kJ per mole of ZnS(s). Compute the chemical amount of ZnS (in moles) and multiply it by this molar ΔH to get the amount of heat absorbed in the roasting of the 3.00 metric tons of ZnS:

$$3.00 \times 10^6 \text{ g ZnS} \times \left(\frac{1\text{mol ZnS}}{97.456 \text{ g ZnS}}\right) \times \left(\frac{-439.13 \text{ kJ}}{1 \text{ mol ZnS}}\right) = -1.35 \times 10^7 \text{ kJ}$$

8-39 a) The balanced equation is $CaCl_2(s) \rightarrow Ca^{2+}(aq) + 2\,Cl^-(aq)$. Combine the enthalpies of formation as follows:

$$\Delta H^\circ = 2\underbrace{(-167.16)}_{Cl^-(aq)} - 1\underbrace{(542.83)}_{Ca^{2+}(aq)} - 1\underbrace{(-795.8)}_{CaCl_2(s)} = -81.4 \text{ kJ}$$

b) Compute ΔH° for the dissolution of 20.0 g of $CaCl_2(s)$:

$$\Delta H^\circ = 20.0 \text{ g } CaCl_2 \times \left(\frac{1 \text{ mol } CaCl_2}{111.0 \text{ g } CaCl_2}\right) \times \left(\frac{-81.35 \text{ kJ}}{1 \text{ mol } CaCl_2}\right) = -14.66 \text{ kJ}$$

The process of dissolution absorbs -14.66 kJ. The immediate surroundings of the dissolution (the water) therefore must absorb $+14.66$ kJ. The temperature change of the water is the heat it absorbs divided by its heat capacity:

$$\Delta T = \frac{q}{c_p m} = \frac{14.66 \times 10^3 \text{ J}}{418 \text{ J K}^{-1}} = 35.1 \text{ K}$$

The final temperature will be $T_f = 20.0°C + 35.1°C = 55.1°C$. Remember that a change of one degree celsius is identical to a change of one kelvin.

8-41 The balanced equation is $C_6H_{12}(l) + 9\,O_2(g) \rightarrow 6\,CO_2(g) + 6\,H_2O(l)$. Set up the calculation of the standard enthalpy of the combustion reaction in terms of the standard enthalpies of formation of the products and reactants. Let x equal the unknown ΔH_f°:

$$\Delta H^\circ = -3923.7 = 6\underbrace{(-393.51)}_{CO_2(g)} + 6\underbrace{(-285.83)}_{H_2O(g)} - 1\underbrace{(x)}_{C_6H_{12}(g)} - 9\underbrace{(0)}_{O_2(g)}$$

The standard enthalpy of combustion and the standard enthalpies of formation were all in kJ mol^{-1}. All were therefore multiplied by the number of moles of each substance appearing in the balanced equation. Solving for x gives the ΔH_f° of liquid cyclohexane as -152.3 kJ mol^{-1}.

8-43 a) The equation is: $C_{10}H_8(s) + 12\,O_2(g) \rightarrow 10\,CO_2(g) + 4\,H_2O(l)$.
b) The amount of heat evolved $(-q)$ was observed to be 25.79 kJ. Since the combustion was performed at constant volume, no work was done on the system ($w = 0$). Therefore, $\Delta E = q + w = -25.79$ kJ $+ 0 = -25.79$ kJ. Put this on a molar basis:

$$\Delta E = \left(\frac{-25.79 \text{ kJ}}{0.6410 \text{ g } C_{10}H_8}\right) \times \left(\frac{128.17 \text{ g } C_{10}H_8}{1 \text{ mol } C_{10}H_8}\right) = -5157 \text{ kJ mol}^{-1}$$

The temperature is essentially 25°C both before and after the reaction. Hence, the ΔE° in the combustion of 1.000 mol of naphthalene is -5157 kJ.
c) To calculate ΔH use the definition: $\Delta H = \Delta E + \Delta(PV)$. If the gases behave ideally, then $\Delta(PV) = \Delta(nRT)$ so that:

$$\Delta H = \Delta E + \Delta(nRT) = \Delta E + (\Delta n)RT$$

for this change at constant temperature. The value of Δn is calculated by making the valid assumption that the change in the amount of gases is the only factor that contributes significantly to $\Delta(PV)$. The combustion of 1.000 mol of naphthalene produces 10.00 mol of gas, but requires 12.00 mol of gas. Hence:

$$(\Delta n_g)RT = (-2.00 \text{ mol})(8.315 \text{ J K}^{-1}\text{mol}^{-1})(298.15 \text{ K}) = -4.96 \text{ kJ}$$

$$\Delta H^\circ = \Delta E^\circ + (\Delta n)RT = -5157 - 4.96 = -5162 \text{ kJ}$$

d) The equation is still $C_{10}H_8(s) + 12\,O_2(g) \rightarrow 10\,CO_2(g) + 4\,H_2O(l)$.

$$\Delta H^\circ = -5162 \text{ kJ} = 10\underbrace{(-393.51 \text{ kJ})}_{CO_2(g)} + 4\underbrace{(-285.83 \text{ kJ})}_{H_2O(l)} - 12\underbrace{(0 \text{ kJ})}_{O_2(g)} - x$$

where x is the ΔH_f° of naphthalene. Solving gives x equal to $+84$ kJ mol^{-1}.

8-45 First, write an equation for the formation of $CCl_3F(g)$ from the "naked atoms":

$$C(g) + 3\,Cl(g) + F(g) \rightarrow CCl_3F(g)$$

From the average bond enthalpies tabulated in text Table 8-3, estimate the ΔH° corresponding to the preceding equation as:

$$\Delta H^\circ = 1\,\underbrace{(-441)}_{\text{C–F}} + 3\,\underbrace{(-328)}_{\text{C–Cl}} = -1425 \text{ kJ}$$

Next, write equations that show the preparation of the naked atoms from the elements in their standard states. Each of these equations has a corresponding atomization enthalpy computed from the data in Table 8-3:

$$\begin{array}{ll} C(s) \rightarrow C(g) & \Delta H^\circ = 716.7 \text{ kJ} \\ 3/2\,Cl_2(g) \rightarrow 3\,Cl(g) & \Delta H^\circ = 365.1 \text{ kJ} \\ 1/2\,F_2(g) \rightarrow F(g) & \Delta H^\circ = 79.0 \text{ kJ} \end{array}$$

The atomization enthalpies in the table are per mole of atom formed. Each ΔH° in the above has therefore been multiplied by the number of moles of atom formed. Combine the four reactions to arrive at the formation reaction of $CCl_3F(g)$ from the elements in their standard states:

$$\begin{array}{ll} C(g) + 3\,Cl(g) + F(g) \rightarrow CCl_3F(g) & \Delta H^\circ = -1425 \text{ kJ} \\ C(s) \rightarrow C(g) & \Delta H^\circ = 716.7 \text{ kJ} \\ 3/2\,Cl_2(g) \rightarrow 3\,Cl(g) & \Delta H^\circ = 365.1 \text{ kJ} \\ 1/2\,F_2(g) \rightarrow F(g) & \Delta H^\circ = 79.0 \text{ kJ} \\ \hline C(s) + 3/2\,Cl_2(g) + 1/2\,F_2(g) \rightarrow CCl_3F(g) & \Delta H^\circ = -264 \text{ kJ} \end{array}$$

8-47 a) The structure of the compound is

$$\begin{array}{ccccccc} & & H & & H & & H & \\ & & | & & | & & | & \\ H & — & C & — & C & — & C & — & H \\ & & | & & | & & | & \\ & & F & & H & & H & \end{array}$$

b) The reaction is the combustion of propane in oxygen:

$$C_3H_8(g) + 5\,O_2(g) \rightarrow 3\,CO_2(g) + 4\,H_2O(g)$$

As this reaction proceeds bonds are both broken and formed. Broken are: 2 mol of C—C bonds, 8 mol of C—H bonds, and 5 mol of O=O double bonds. Formed are: 6 mol of C=O double bonds and 8 mol of O—H bonds. The net enthalpy change is:

$$\Delta H \approx 6\,\underbrace{(728)}_{\text{C=O}} + 8\,\underbrace{(463)}_{\text{O–H}} - 5\,\underbrace{(498)}_{\text{O=O}} - 8\,\underbrace{(413)}_{\text{C–H}} - 2\,\underbrace{(348)}_{\text{C–C}} = -1.58 \times 10^3 \text{ kJ}$$

bonds formed − bonds broken

The result -1.582×10^3 kJ is correct according to the rules for significant digits (on text page A-6). In view of the approximate nature of average bond enthalpies (discussed on text page 295-296), -1.58×10^3 kJ is a better answer.

8-49 The Lewis structures are:

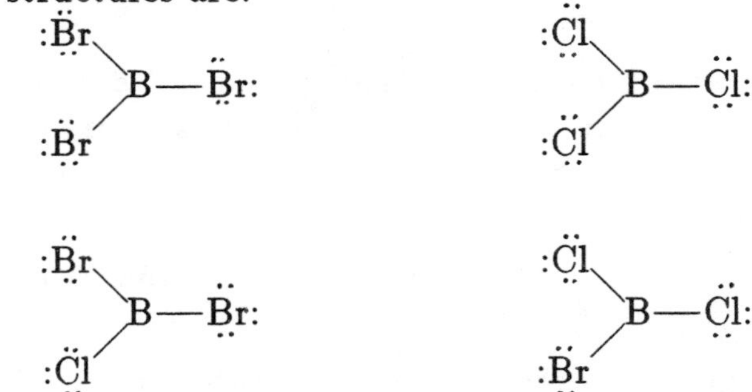

As the reaction proceeds, three boron-bromine bonds and three boron-chlorine bonds are broken, but the same number of boron-bromine and boron-chlorine bonds are formed in the products. Therefore, the sum of the bond enthalpies in the products equals the sum of the bond enthalpies in the reactants.

8-51 In an isothermal change, there is by definition no change of temperature ($\Delta T = 0$). The internal energy of an ideal gas depends only on its temperature which means that $\Delta E = 0$. As for the enthalpy:

$$\Delta H = \Delta E + \Delta(PV) = 0 + \Delta(nRT) = 0$$

This system, a non-reacting ideal gas undergoing a constant-temperature change also has ΔH equal to zero. The expansion is reversible. Hence:

$$w = -nRT \ln\left(\frac{V_2}{V_1}\right) = -(2.00 \text{ mol})(8.315 \text{ J K}^{-1}\text{mol}^{-1})(298 \text{ K}) \ln\left(\frac{36.00}{9.00}\right)$$

The work comes out to equal -6.87 kJ. The first law provides that $\Delta E = q + w$. Therefore q of the gas is $+6.87$ kJ.

8-53 During any adiabatic process $q = 0$. During this *reversible* adiabatic expansion of an ideal gas:

$$T_1 V_1^{\gamma-1} = T_2 V_2^{\gamma-1}$$

where γ is c_p/c_v and the subscripts refer the initial and final states of the gas. In this problem, V_1 is 20.0 L, V_2 is 60.0 L, γ is 5/3, and T_1 is 300 K. Solving for T_2 and substituting gives:

$$T_2 = T_1 \left(\frac{V_1}{V_2}\right)^{\gamma-1} = (300 \text{ K}) \left(\frac{20.0 \text{ L}}{60.0 \text{ L}}\right)^{2/3} = 144.22 \text{ K}$$

To the correct number of significant digits this is 144 K.
Meanwhile, the ΔE of the ideal gas depends solely upon its change in temperature:

$$\Delta E = nc_{\text{v}}\Delta T = (2.00\ \text{mol})\left(\frac{3}{2}\ 8.315\ \text{J K}^{-1}\text{mol}^{-1}\right)(-155.78\ \text{K}) = -3.89\ \text{kJ}$$

This number is also w, the work done on the gas, because $\Delta E = q + w$ and q is zero in this process. Finally, ΔH of an ideal gas also depends entirely on the change in temperature:

$$\Delta H = nc_{\text{p}}\Delta T = (2.00\ \text{mol})\left(\frac{5}{2}\ 8.315\ \text{J K}^{-1}\text{mol}^{-1}\right)(-155.78\ \text{K}) = -6.48\ \text{kJ}$$

Notice that $\Delta H = \gamma\Delta E$ for this reversible adiabatic process.

8-55 a) The maximum theoretical efficiency of an engine operating between two temperatures is attained when the engine operates reversibly. This maximum efficiency is, according to text equation 8-16 (page 304):

$$\epsilon = 1 - \frac{T_\ell}{T_\text{h}}$$

In this problem, T_ℓ is 300 K and T_h is 450 K so ϵ is 0.333.
b) The efficiency of the engine is the ratio of the net work it *performs* to the heat that it *absorbs:*

$$\epsilon = \frac{-w_{\text{net}}}{q}$$

The minus sign is necessary because of the convention that $+w$ is work absorbed. If 1500 J of heat is absorbed per cycle from the 400 K reservoir and ϵ is 0.333, then w_{net} is -500 J in each turn of the cycle. It follows from the first law that the engine discards 1000 J of heat ($q = -1000$ J) in the low-temperature reservoir during each cycle.
c) The engine has absorbed 1500 J of heat during one portion of the cycle of operation. It must lose this number of joules of energy by the time it completes the cycle (for which ΔE is zero). Of the 1500 J, 1000 J goes to the 300 K reservoir as heat. Accordingly, 500 J appears as work done by the engine. That is, $w = -500$ J.

8-57 A refrigerator is a machine that consumes work to create a temperature difference. This is the reverse of a heat engine, which take advantage of a temperature difference to create work. The maximum heat q absorbed from the interior of a refrigerator is:

$$q = w_{\text{net}}\left(\frac{T_\ell}{T_\text{h} - T_\ell}\right)$$

where T_ℓ and T_h are the low and high temperatures. Solve this equation for w_{net} and substitute a T_ℓ of 263.15 K (−10°C), a T_h of 293.15 K (20°C), and a q of 800 J:

$$w_{net} = q\left(\frac{T_h - T_\ell}{T_\ell}\right) = 800 \text{ J}\left(\frac{293.15 - 263.15}{263.15}\right) = 91 \text{ J}$$

Naturally, more work than this would be required to extract the 800 J if the refrigerator operated at less than 100 percent thermodynamic efficiency, that is, irreversibly. The q is 800 J per cycle, and the w is 91 J per cycle. The total heat discharged per cycle into the room (at 293.15 K) is 800 + 91 = 891 J

8-59 The law of Dulong and Petit states that all metals have a molar heat capacity of approximately 25 J K^{-1}mol^{-1}. The molar heat capacity equals the specific heat of a substance multiplied by its molar mass. Hence:

$$c = c_s\mathcal{M} \approx 25 \text{ J K}^{-1}\text{mol}^{-1}$$

The experimental specific heat of indium of 0.233 J g^{-1}K^{-1}. A molar mass of 76 g mol^{-1} combined with this number gives a molar heat capacity for indium of only 17.7 J K^{-1}mol^{-1}. This violates the law of Dulong and Petit badly. The modern value of $\mathcal{M}$ = 114.8 g mol^{-1} for indium works much better in the law of Dulong and Petite.[1]

8-61 a) The work done *on* the system is $-P\Delta V$. Since the gas is ideal and P is constant, $P\Delta V = nR\Delta T$ for the system consisting of the argon gas. This is convenient because ΔT is known. It is −100 K. Then:

$$w = -nR\Delta T = -(2.00 \text{ mol})(8.315 \text{ J K}^{-1}\text{mol}^{-1})(-100 \text{ K}) = +1.66 \times 10^3 \text{ J}$$

b) The process goes on at constant pressure so the heat absorbed is q_p:

$$q_p = nc_p\Delta T = (2.00 \text{ mol})\left(\frac{5}{2}8.315 \text{ J K}^{-1}\text{mol}^{-1}\right)(-100 \text{ K}) = -4.16 \times 10^3 \text{ J}$$

c) The energy change of the system is the sum of the work done on the system and the heat it absorbs: $\Delta E = q + w = -4157 + 1663 = -2494$ J. This rounds off to -2.49×10^3 kJ. Note the use of un-rounded answers from parts a) and b) in the addition.

d) The ΔH of the system is q_p, which was computed in a previous part. Hence, ΔH is −4.16 kJ.

---

[1]For solids and liquids the distinction between c_p and c_v is unimportant, especially in an approximate relationship. For this reason there is no subscript on c in this problem.

8-63 The amount of work done by the gas on the paddle mechanism is the negative of the work absorbed by the gas:

$$w = -(-P\Delta V) = +(1.00 \text{ atm})(13.00 - 5.00 \text{ L}) = 8.00 \text{ L atm}$$

This work amounts to 811 J because 1 L atm is 101.325 J. All of this work is converted to heat in the 1.00 L of water. Hence the heat absorbed by the water is +811 J. At the given density, the 1.00 L of water weighs 1.00×10^3 g. Therefore:

$$\Delta T = \frac{q}{c_s M} = \frac{811 \text{ J}}{(4.18 \text{ J g}^{-1}\text{K}^{-1})(1.00 \times 10^3 \text{ g})} = 0.194 \text{ K}$$

8-65 The quantity of glucose $C_6H_{12}O_6$ must be converted to a chemical amount. Then the ΔH of the oxidation of the candy bar inside the person is:

$$14.3 \text{ g } C_6H_{12}O_6 \times \left(\frac{1 \text{ mol } C_6H_{12}O_6}{180.16 \text{ g } C_6H_{12}O_6}\right) \times \left(\frac{-2820 \text{ kJ}}{1 \text{ mol } C_6H_{12}O_6}\right) = -223.8 \text{ kJ}$$

The amount of heat absorbed by the surroundings of the reaction (which are the person's body) is therefore +228 kJ. When this amount of heat is absorbed by 50 kg of water:

$$\Delta T = \frac{q}{c_s M} = \frac{228 \times 10^3 \text{ J}}{(4.18 \text{ J g}^{-1}\text{K}^{-1})(50 \times 10^3 \text{ g})} = 1.1 \text{ K}$$

8-67 Near the boiling points, one can determine which liquid is a better coolant by merely inspecting the specific heats. He(l) absorbs 4.25 J of heat per gram as it heats up 1 K in temperature. N_2(l) absorbs only 1.95 J of heat per gram as it warms by the same amount. Therefore, near their boiling points liquid helium is a better coolant. *At* their boiling points, the two liquefied gases cool by vaporization. Liquid N_2 is better because it absorbs much more heat per gram in vaporization than He(l).

8-69 The chemical amount of the silane under the conditions of temperature and pressure stated in the problem is:

$$n = \frac{PV}{RT} = \frac{(0.658 \text{ atm})(0.250 \text{ L})}{(0.08206 \text{ L atm mol}^{-1}\text{K}^{-1})(298 \text{ K})} = 6.727 \times 10^{-3} \text{ mol}$$

The combustion of so much silane in a constant-volume process *absorbs* −9.757 kJ of heat at 25°C. This value is accordingly $\Delta E°$ of the combustion reaction (because $\Delta E = q_v$). On a molar basis:

$$\Delta E° = \frac{-9.757 \text{ kJ}}{6.727 \times 10^{-3} \text{ mol}} = -1450 \text{ kJ mol}^{-1}$$

Now, compute $\Delta H°$ of the combustion reaction:

$$\Delta H° = \Delta E° + RT\Delta n_g$$
$$\Delta H° = -1450 \text{ kJ mol}^{-1} + (8.315 \text{ J K}^{-1}\text{mol}^{-1})(298.15 \text{ K})(-3 \text{ mol}) = -1458 \text{ kJ mol}^{-1}$$

The enthalpy of the combustion reaction is the sum of the enthalpies of formation of the products minus the sum of the enthalpies of formation of the reactants. Taking values from Appendix D:

$$-1458 \text{ kJ mol}^{-1} = \underbrace{(-910.94)}_{SiO_2\ quartz} + 2\underbrace{(-285.83)}_{H_2O(l)} - 1\underbrace{(\Delta H_f°)}_{SiH_4(g)}$$

Solving gives $\Delta H_f° = -25 \text{ kJ mol}^{-1}$ for gaseous silane.

8-71 a) The combustion of the liquid fuel isooctane is represented by the equation: $C_8H_{18}(l) + 25/2\, O_2(g) \rightarrow 8\, CO_2(g) + 9\, H_2O(l)$.

b) The combustion of 0.542 g of isooctane is obviously exothermic because isooctane is a fuel. The heat released by the combustion goes to raise the temperature of the calorimeter vessel and of the water inside it:

$$q_{\text{tot}} = q_{\text{(absorbed by water)}} + q_{\text{(absorbed by calorimeter)}}$$

The calorimeter has a ΔT of 28.670 − 20.450 = 8.220°C. A Celsius degree is the same size as a kelvin so this ΔT is also 8.220 K. The water in the calorimeter has the same ΔT as the structure of the calorimeter itself. The heat absorbed by a system in a change at constant pressure equals its heat capacity at constant pressure times its change in temperature. Apply this fact to both the calorimeter and the water it contains. Use the given heat capacity of the calorimeter (48 J K^{-1}) and the known specific heat of water (4.184 J g^{-1} K^{-1}). Then, the heat absorbed by that system that is the surroundings of the reaction equals:

$$q = \underbrace{(8.22 \text{ K})(48 \text{ J K}^{-1})}_{\text{calorimeter}} + \underbrace{(4.184 \text{ J g}^{-1}\text{K}^{-1})(750 \text{ g})(8.22 \text{ K})}_{\text{water}} = 2.62 \times 10^4 \text{ J}$$

The heat absorbed by the reaction is the negative of this, or -2.62×10^4 J. No work is accomplished either by or upon the reaction because it goes on at constant volume. Hence:

$$\Delta E = q + w = q_v + 0 = -2.62 \times 10^4 \text{ J}$$

c) The molar mass of C_8H_{18} is 114.23 g mol^{-1} making the 0.542 g of isooctane that is actually burned far less than one mole. An entire mole of isooctane would *absorb* proportionately more heat during this combustion:

$$\frac{-2.62 \times 10^4 \text{ J}}{0.542 \text{ g}} \times 114.23 \text{ g mol}^{-1} = -5.52 \times 10^6 \text{ J mol}^{-1} = -5520 \text{ kJ mol}^{-1}$$

d) The ΔH of any process is $\Delta H = \Delta E + \Delta(PV)$. Assuming the gases are ideal and the liquids have nearly zero volume, $\Delta(PV) = \Delta n_g RT$, where Δn_g is the change in the number of moles of gas. In this reaction, $\Delta n_g = 8 - 12.5 = -4.5$ mol. Therefore:

$$\Delta(PV) = (-4.5 \text{ mol})(8.315 \text{ J K}^{-1}\text{mol}^{-1})(298.15 \text{ K}) = -11.16 \text{ kJ}$$

The temperature starts at 20.450 and ends up at 28.670°. Take it to be constant during the reaction and equal to 298 K. Any error that this approximation introduces is probably much less than the uncertainty in the calorimetric measurements. Therefore:

$$\Delta H = \Delta E + \Delta(PV) = -5520 + (-11.16) = -5530 \text{ kJ} = \Delta H^\circ$$

e) The standard enthalpy change of the reaction as written above is:

$$\Delta H^\circ = 8\,\Delta H_f^\circ(\text{CO}_2(g)) + 9\,\Delta H_f^\circ(\text{H}_2\text{O}(l)) - \Delta H_f^\circ(\text{isooctane})$$

This follows from Hess's law. The ΔH_f°'s of the products are -393.51 kJ mol^{-1} and -285.83 kJ mol^{-1} respectively. The ΔH° of the reaction is -5530 kJ so:

$$-5531 \text{ kJ} = 8(-393.51 \text{ kJ}) + 9(-285.83 \text{ kJ}) - \Delta H_f^\circ(\text{isooctane})$$

Solving gives ΔH_f°(isooctane) equal to -191 kJ mol^{-1}.

8-73 a) For $C_2H_2(g) + 5/2\,O_2(g) \rightarrow 2\,CO_2(g) + H_2O(g)$, we combine ΔH_f°'s:

$$\Delta H^\circ = 2\,\underbrace{(-393.51)}_{\text{CO}_2(g)} + 1\,\underbrace{(-241.82)}_{\text{H}_2\text{O}(g)} - 1\,\underbrace{(226.73)}_{\text{C}_2\text{H}_2(g)} - 5/2\,\underbrace{(0.00)}_{\text{O}_2(g)} = -1255.57 \text{ kJ}$$

b) The total heat capacity of the mixture of the two gases equals the molar heat capacity of the first multiplied by the number of moles of the first plus the molar heat capacity of the second multiplied by the number of moles of the second:

$$nc_p = (2.00 \text{ mol})\,\underbrace{(37 \text{ J mol}^{-1}\text{K}^{-1})}_{\text{CO}_2} + (1.00 \text{ mol})\,\underbrace{(36 \text{ J mol}^{-1}\text{K}^{-1})}_{\text{H}_2\text{O}} = 110 \text{ J K}^{-1}$$

c) Assume for convenience that one mole of $C_2H_2(g)$ is burned. Then the product gases, which are 2 mol of $CO_2(g)$ and 1 mol of $H_2O(g)$, absorb 1255.57 kJ of heat in the flame. For these product gases (which comprise the flame):

$$\Delta T = \frac{q}{c_p} = \frac{1.25557 \times 10^3 \text{ J}}{110 \text{ J K}^{-1}} = 1.14 \times 10^4 \text{ K} = 11400 \text{ K}$$

A change of one degree Celsius is the same as a change of one kelvin. Therefore, $\Delta T = 11400$°C. If the temperature before combustion is near room temperature (25°C), the final temperature is 11400°C.

8-75 Define the system, which is a closed system, as the contents of the engine cylinder. Before the explosive combustion of the n-octane, the temperature is 600 K, the volume is 0.150 L, and the pressure is 12.0 atm. Apply the ideal-gas equation to the mixed contents of the cylinder before the combustion:

$$n_{\text{octane}} + n_{\text{air}} = \frac{PV}{RT} = \frac{(12.0\text{ atm})(0.150\text{ L})}{(0.08206\text{ L atm mol}^{-1}\text{K}^{-1})(600\text{ K})} = 0.03656\text{ mol}$$

Also, the cylinder holds n-octane and air in a 1 to 80 molar ratio, i.e.:

$$80n_{\text{octane}} = n_{\text{air}}$$

Solving the two simultaneous equations gives:

$$n_{\text{octane}} = 4.514 \times 10^{-4}\text{ mol} \quad \text{and} \quad n_{\text{air}} = 0.03611\text{ mol}$$

According to the problem, the system does not change its volume during the actual combustion of the fuel, so w is zero. Furthermore, q is zero (the combustion happens so fast that there is no time for heat to be lost or gained). Since w and q both equal zero the first law assures that ΔE of the system equals zero during the process. Imagine the combustion to occur in two stages: step a: the reaction goes at a constant temperature of 600 K; step b: the product gases heat up at constant volume. The sum of these two changes is the overall change that takes place within the cylinder. Therefore:

$$\Delta E_{\text{sys}} = 0 = \Delta E_a + \Delta E_b \quad \text{which means} \quad \Delta E_a = -\Delta E_b$$

The problem offers data pertaining to enthalpy changes, not energy changes, in the two steps. Deal with this by substituting for the ΔE_a and ΔE_b in terms of ΔH's:

$$\Delta H_a - \Delta(PV)_a = -\big(\Delta H_b - \Delta(PV)_b\big)$$

Step a involves ideal gases, takes place at a constant temperature, and changes the chemical amount of gas. Therefore $\Delta(PV)_a$ equals $\Delta n_g RT$, where Δn_g is the change in the chemical amount of gases during the reaction. Step b is heating of the ideal gases inside the cylinder. The term $\Delta(PV)_b$ therefore equals $nR\Delta T$ where n is the chemical amount of gases present *after* the reaction. Also, for the change in temperature that comprises step b, ΔH_b is equal to $nc_p\Delta T$, as long as the heat capacity c_p is independent of temperature. Substitution of these relations gives:

$$\Delta H_a - \Delta n_g RT = -(nc_p\Delta T - nR\Delta T)$$

In this equation T is 600 K, and ΔT is the temperature change during the heating. The approach will be to compute all the other quantities and then use this equation to get ΔT and thence the final temperature.

Air is 80 percent N_2 and 20 percent O_2 on a molar basis. From this fact and the fact that the fuel burns according to the balanced equation:

$$C_8H_{18}(g) + 12\frac{1}{2}\, O_2(g) \rightarrow 8\, CO_2(g) + 9\ H_2O(g)$$

the chemical amounts of the different gases within the cylinder both before and after the reaction can be calculated. Before the reaction:

$$\begin{aligned} n_{N_2} &= 0.80 \times (0.036109) = 0.02889 \text{ mol} \\ n_{O_2} &= 0.20 \times (0.036109) = 0.007222 \text{ mol} \\ n_{\text{octane}} &= 4.514 \times 10^{-4} \text{ mol} \end{aligned}$$

After the reaction:

$$\begin{aligned} n_{CO_2} &= 8 \times (4.514 \times 10^{-4}) = 0.003611 \text{ mol} \\ n_{H_2O} &= 9 \times (4.514 \times 10^{-4}) = 0.004063 \text{ mol} \\ n_{N_2} &= 0.02889 \text{ mol} \\ n_{O_2} &= (7.222 \times 10^{-3}) - (12.5 \times 4.514 \times 10^{-4}) = 0.001579 \text{ mol} \\ n_{\text{octane}} &= 0 \text{ mol} \end{aligned}$$

Notice that the n-octane is the limiting reactant and that the nitrogen does not react. The sum of the second five quantities is n, the total chemical amount of gases in the cylinder after the reaction. It is 0.03814 mol. The original chemical amount of gases in the system equals 0.03656 mol. Hence:

$$\Delta n_g = 0.03814 - 0.03656 = +0.001583 \text{ mol}$$

The enthalpy change of combustion of one mole of gaseous n-octane can be computed using the ΔH_f°'s of the products and reactants:

$$\Delta H = 9 \underbrace{(-241.8)}_{H_2O(g)} + 8 \underbrace{(-393.5)}_{CO_2(g)} - 1 \underbrace{(-57.4)}_{n\text{-octane}(g)} - 12.5 \underbrace{(0)}_{O_2(g)} = -5266.8 \text{ kJ}$$

This is *not* ΔH_a, the enthalpy change of the combustion inside the cylinder. Because only 4.514×10^{-4} mol of n-octane is in the cylinder:

$$\Delta H_a = -5266.8 \text{ kJ mol}^{-1} \times (4.514 \times 10^{-4} \text{ mol}) = -2.377 \text{ kJ}$$

The overall heat capacity at constant pressure in the cylinder after the reaction is the sum of the nc_p values for the four product gases, as in the solution to problem 8-73b:

$$nc_p = (0.001579) \underbrace{(35.2)}_{O_2} + (0.02889) \underbrace{(29.8)}_{N_2} + (0.004063) \underbrace{(38.9)}_{H_2O}$$

$$+(0.003611)\underbrace{(45.5)}_{CO_2} = 1.24 \text{ J K}^{-1}$$

Each term consists of the chemical amount of the gas (in moles) multiplied by its molar heat capacity (in J $mol^{-1}K^{-1}$).
Now, there is enough information to calculate ΔT. Solve the equation derived previously and make the various substitutions:

$$\Delta T = \frac{\Delta H_a - \Delta n_g RT}{nR - nc_p}$$

$$\Delta T = \frac{-2377 \text{ J} - (0.001583 \text{ mol})(8.315 \text{ J K}^{-1}\text{mol}^{-1})(600 \text{ K})}{(0.03814 \text{ mol})(8.315 \text{ J K}^{-1}\text{mol}^{-1}) - 1.24 \text{ J K}^{-1}} = 2580 \text{ K}$$

The maximum temperature inside the cylinder is therefore $600 + 2580 = 3180$ K. This is 2910°C.

8-77 a) The gases trapped inside the cylinder of the motorcycle's engine have volume V_1 when the piston is fully withdrawn and a smaller volume V_2 when the piston is thrust home. The compression ratio is 8:1 so $V_1 = 8V_2$. The area of the base of the engine's cylinder is πr^2, where r is the radius of the base. The volume of a cylinder is the area of its base times its height, h:

$$V_1 = Ah \quad \text{and} \quad V_2 = A(h - 12.00 \text{ cm})$$

which employs the fact that full compression shortens h by 12.00 cm. Because r is 5.00 cm, A is 78.54 cm^2. Substituting for V_1 and V_2 in terms of A and h gives:

$$Ah = 8A(h - 12.00 \text{ cm})$$

The A's cancel, making solution for h easy. The result is 13.714 cm. With h known it is easy to compute $V_1 = 1.077$ L, and $V_2 = 0.1347$ L. The temperature and pressure of the fuel mixture are 353 K (80°C) and 1.00 atm when the mixture enters the cylinder with fully withdrawn piston (V_1). Assuming the fuel mixture is an ideal gas:

$$n_{\text{mixture}} = \frac{(1.00 \text{ atm})(1.077 \text{ L})}{(0.08206 \text{ L atm mol}^{-1}\text{K}^{-1})(353 \text{ K})} = 0.0372 \text{ mol}$$

The molar ratio of air to fuel (C_8H_{18}) is 62.5 to 1. If n_{fuel} is the chemical amount of the fuel and n_{air} is the chemical amount of air, then:

$$n_{\text{fuel}} + n_{\text{air}} = 0.0372 \text{ mol} \quad \text{and} \quad n_{\text{air}} = 62.5 n_{\text{fuel}}$$

Solving these two simultaneous equations establishes that at the start the cylinder contains 0.0366 mol of air and 5.86×10^{-4} mol of n-octane fuel.

During the compression stroke this system undergoes an irreversible adiabatic compression to one-eighth of its initial volume. None of the relationships that govern *reversible* adiabatic processes strictly applies here. Assume however that the compression is near to reversible. If it is, then:

$$T_1 V_1^{\gamma-1} \approx T_2 V_2^{\gamma-1} \quad \text{where } \gamma = \frac{35 \text{ J K}^{-1}\text{mol}^{-1}}{26.7 \text{ J K}^{-1}\text{mol}^{-1}} = 1.31$$

Assuming reversibility, the temperature after the compression stroke is:

$$T_2 = T_1 \left(\frac{V_1}{V_2}\right)^{\gamma-1} = (353 \text{ K}) \left(\frac{1.077 \text{ L}}{0.1347 \text{ L}}\right)^{0.31} = (353 \text{ K})(8)^{0.31} = 673 \text{ K}$$

b) The compressed gases occupy a volume of 0.135 L just before they are ignited, as calculated above.

c) The pressure of the compressed fuel mixture just before ignition is P_2. It can be computed by applying the ideal-gas equation to the system with $T_2 = 673$ K, $V_2 = 0.1347$ L, and $n = 0.0372$ mol. It equals 15.3 atm. Alternatively, one can compute P_2 using the formula for a reversible adiabatic change:

$$P_1 V_1^{\gamma} = P_2 V_2^{\gamma}$$

Use of the formula emphasizes that the value 15.3 atm depends on the assumption of a *reversible* adiabatic compression stroke.

d) The ΔH for the combustion of gaseous n-octane by the reaction:

$$C_8H_{18}(g) + 25/2\, O_2(g) \rightarrow 8\, CO_2(g) + 9\, H_2O(g)$$

is -5266.8 kJ mol^{-1} (see problem 8-75). The combustion mixture inside the cylinder contains 5.86×10^{-4} mol of n-octane. Consequently, the ΔH of combustion in this system is:

$$\Delta H = \left(\frac{-5266.8 \text{ kJ}}{1 \text{ mol}}\right) \times (5.86 \times 10^{-4} \text{ mol}) = -3.09 \text{ kJ}$$

After the combustion, the cylinder contains CO_2, H_2O and unreacted O_2 and N_2. Referring to the balanced chemical equation, it is clear that the combustion consumes 5.86×10^{-4} mol of octane and $12.5 \times (5.86 \times 10^{-4})$ mol of O_2 to produce $8 \times (5.86 \times 10^{-4})$ mol of CO_2 and $9 \times (5.86 \times 10^{-4})$ mol of H_2O. The effect of the reaction is to increase the chemical amount of gases in the cylinder by $3.5 \times (5.86 \times 10^{-4})$ mol. This is Δn_g for the reaction. The original quantity of gases is 0.0372 mol. After the combustion there is 0.0393 mol of gases.

The *energy* (not enthalpy) released from the reaction all goes to heat up the gaseous contents of the cylinder as long as no heat escapes to the cylinder walls and no work is done until the power stroke starts. Therefore:

$$\Delta T = \frac{\Delta H_{\text{react}} - \Delta n_g RT}{nR - nc_p}$$

In this equation, which is derived in more detail in problem 8-75, every quantity but ΔT is known:

$$\Delta T = \frac{-3090 \text{ J} - (0.002051 \text{ mol})(8.315 \text{ J K}^{-1}\text{mol}^{-1})(673 \text{ K})}{(0.0393 \text{ mol})(8.315 \text{ J K}^{-1}\text{mol}^{-1}) - (0.0393 \text{ mol})35 \text{ J K}^{-1}\text{mol}^{-1}} = 2960 \text{ K}$$

The temperature inside the cylinder rises by 2960 K to a maximum of 3630 K (3.6×10^3 K).

e) Assume that the expansion stroke is not only adiabatic but reversible. Then the formula:

$$T_2 = T_1 \left(\frac{V_1}{V_2}\right)^{\gamma - 1}$$

applies. In this case, T_1 is 3630 K. The ratio $V_1 \,/\, V_2$ is 1 to 8 because now the initial state is the *small* volume state just before the expansion stroke of the piston. The exponent $\gamma - 1$ is still 0.31, as established in a preceding part.

$$T_2 = (3630 \text{ K}) \left(\frac{1}{8}\right)^{0.31} = 1900 \text{ K}$$

This is the temperature of the exhaust gases.

8-79 A refrigerator is placed in a closed room, plugged in and left with its door open. Imagine the system to be the closed room. If it is closed, then the room is probably fairly well insulated. Electrical energy is flowing into the room yet q and w across the wall of the room are close to zero. Therefore the internal energy E of the room increases and the temperature of the room also must increase.

It is true that a region of the room just at the open refrigerator door will get cooler. But the refrigerator is pouring heat into the room from the coils at its rear. Once the room comes to thermal equilibrium its temperature will be higher than before the refrigerator was plugged in.

Chapter 9

Spontaneous Change and Equilibrium

A spontaneous change is one that occurs by itself, given enough time, without outside intervention. Spontaneous change is directional. If a process is spontaneous in one direction, then the reverse process is never spontaneous. At equilibrium, all potential for spontaneous change has been exhausted.

Entropy, Spontaneous Change, and the Second Law

The **entropy** S of a system is a state function of that system, just as are the energy, pressure, volume, and temperature. It is related to the *disorder* or randomness of the system (see text page 327). The directionality of physical and chemical changes is understood in terms of the changes in entropy that accompany them.

A **spontaneous** process occurs without outside intervention. The driving force for a spontaneous process is an increase in the entropy of the universe. In any process, the entropy change of the universe is the sum of the entropy changes of the system under consideration and its surroundings:

$$\Delta S_{\text{univ}} = \Delta S_{\text{sys}} + \Delta S_{\text{surr}}$$

Either of the terms on the right may be negative in a spontaneous process. The *sum* of the terms, the total entropy change of the universe is always positive in a spontaneous change. **Problem 9-7** illustrates this point, as do **problems 9-9** and **9-59,** which concern the changes in entropy when a piece of hot iron is plunged into cold water. The iron of course spontaneously cools. Its entropy *decreases.* The increase in the entropy of the surroundings of the iron (the water) more than compensates.

To compute the difference in entropy ΔS_{sys} between two states of a system, it is only necessary to imagine some one **reversible** path connecting them. The path can be as outlandishly unrealizable as necessary (or desired). It is used only for computational purposes. As the change proceeds along the imagined reversible path, the system absorbs heat from its surroundings; ΔS is the sum of each little quantity of heat absorbed divided by the temperature at which it is absorbed:

$$\Delta S = \int \frac{dq_{\text{rev}}}{T}$$

The q in this equation may be positive or negative (recall that a negative q corresponds to the evolution of heat by the system.) On the other hand, the T in the equation is the absolute temperature, which is *never* negative. Therefore, for a system gaining heat on a reversible path, the entropy change is positive. For a system losing heat on a reversible path, the entropy change is negative.

The change in the entropy of a system during any process may be positive, negative, or, in the case that q is zero, equal to zero. This last case is important enough to merit a special name: an **adiabatic** change.

The same net change between two states may be achieved in a great number of ways, or paths. Every different path may have a different total q. The above method of calculation gives correct values for ΔS only if reversible paths are chosen.

Mapping out a reversible path may seem difficult, especially if the final temperature differs from the original. The best tactic is to break down the total change into a combination of steps that occur at constant pressure **(isobarically)**, at constant volume **(isochorically)**, and at constant temperature **(isothermally)**. Such an approach is used in **problem 9-57b.**

Calculation of Entropy Changes

The tactic of imagining the change to proceed by a series of steps succeeds because simple formulas allow one to calculate entropy changes along isochoric, isobaric and isothermal path segments. Here is a summary of formulas for computing ΔS under various circumstances:

Reversible gain (loss) of heat at constant T. Direct application of the formula defining ΔS is easy because T is a constant:

$$\Delta S = \frac{1}{T} q_{\text{rev}}$$

This situation arises when a system is exchanging work with its surroundings as heat flows in or out. Also, when substances freeze, melt, boil, or condense they stay at a constant temperature while absorbing or emitting heat. For example, as water boils it absorbs heat at a temperature equal to its boiling point.

This heat equals its ΔH of vaporization. The entropy change of vaporization therefore is:

$$\Delta S_{\text{vap}} = \frac{\Delta H_{\text{vap}}}{T_{\text{b}}}$$

This fact is used in **problem 9-27.** Most liquids have nearly the same molar entropy of vaporization. **Trouton's rule** states: $\Delta S_{\text{vap}} = 88 \pm 5 \text{ J mol}^{-1} \text{K}^{-1}$. This rule in useful in estimates of boiling point and enthalpy of vaporization (see **problems 9-3** and **9-53**).

The entropy change of fusion (melting) is:

$$\Delta S_{\text{fus}} = \frac{\Delta H_{\text{fus}}}{T_{\text{f}}}$$

which has the same form as for vaporization. There is however is no rule like Trouton's for fusion.

Reversible heating (cooling) at constant *P*. The change in entropy is:

$$\Delta S = nc_{\text{p}} \ln\left(\frac{T_2}{T_1}\right)$$

Whenever a system cools, its entropy diminishes along with its temperature. The entropy change of the system is negative if its temperature drops and positive if it rises. This formula is used in **problem 9-57.** To use it, one must either assume that c_{p} is not a function of temperature or get information on how c_{p} varies with temperature.

Reversible heating (cooling) at constant *V*. The formula is very similar to the previous one. Note that at constant volume c_{v} appears instead of c_{p}:

$$\Delta S = nc_{\text{v}} \left(\ln \frac{T_2}{T_1}\right)$$

Reversible expansion (compression) at constant *T*. The entropy of a system always increases with its size as long as the temperature does not change. The entropy change upon expansion *of an ideal gas* is:

$$\Delta S = nR \ln\left(\frac{V_2}{V_1}\right) \qquad \text{ideal gas}$$

The formula:

$$\Delta S = nR \ln\left(\frac{P_1}{P_2}\right) \qquad \text{ideal gas}$$

is an immediate consequence (by combination with the ideal-gas law). If a system that is partly solid or liquid and partly gas is expanded or compressed, the volume change can almost certainly be assigned entirely to the gas. These formulas can then be used, as long as the gas can be approximated as ideal.

Even the most complex process involving simultaneous changes in T, V, and P can be imagined to proceed along a path involving a change first with T, then with P, and then with V constant. Coming up with such paths is a common theme in problems involving the entropy function. See **problem 9-55.** When a path is finally sketched, the above formulas allow the computation of the change in entropy.

Example: 10.00 L of an ideal monatomic gas at 10.00 atm and 273.15 K is expanded and cooled to a final volume of 63.93 L and a final temperature of 174.8 K. Compute the entropy change of the gas. **Solution:** During the process, T, P, and V all change. Applying the ideal-gas equation to the system in its original state shows that there is 4.457 mol of gas. None of the gas escapes during the change, so there is 4.457 mol in the final state. The ideal gas equation applies to the final state as well as the initial. Using it, the final pressure is 1.000 atm. In summary: $P_1 = 10.0$ atm, $P_2 = 1.00$ atm; $V_1 = 10.00$ L, $V_2 = 69.93$ L; $T_1 = 273.15$ K, $T_2 = 174.8$ K; $n =$ 4.457 mol. A reversible path connecting these two states is: a) cool the gas reversibly at constant T from T_1 to T_2; b) expand the gas reversibly at constant T from its intermediate volume (the volume attained after the cooling step) to V_2. During the isobaric cooling (step a), the volume of the gas falls to 6.399 L. The entropy change is:

$$\Delta S_a = nc_p \ln \frac{T_2}{T_1} = (4.457 \text{ mol}) \left(\frac{5}{2} 8.315 \text{ J K}^{-1}\text{mol}^{-1} \right) \ln \frac{174.8}{273.15} = -41.35 \text{ J K}^{-1}$$

In step b of the reversible path, the gas expands isothermally (at $T = 174.8$ K) from 6.3993 L to 63.99 L, and the entropy change is:

$$\Delta S_b = nR \ln \left(\frac{V_2}{V_1} \right) (4.457 \text{ mol})(8.315 \text{ J K}^{-1}\text{mol}^{-1}) \ln \left(\frac{63.99}{6.399} \right) = 85.33 \text{ J K}^{-1}$$

The overall entropy change for the process is the sum $85.33 + (-41.35) = 43.98$ J K^{-1}.

Many additional reversible paths connect the final and initial states. The problem could also have been solved by imagining first an isothermal expansion from 10.0 to 63.99 L followed by an isochoric cooling from 273.15 to 174.8 K.

Adiabatic Does Not Necessarily Mean Isentropic

For changes that are *both* adiabatic *and* reversible $\Delta S = 0$. Such changes are **isentropic.** It is a common error to suppose that all adiabatic processes are isentropic.

Suppose that a spontaneous adiabatic process occurs. Like all spontaneous changes, it is irreversible. After sizing up the final and initial states, it will turn out to be impossible to imagine a reversible path achieving the same change unless the path has a non-adiabatic segment.

Irreversible Processes

Reversibility is an idealization. All real processes are more or less **irreversible.** If a comparison is made between an idealized, reversible path connecting two states and a real, irreversible path then:

$$q_{\text{irrev}} < q_{\text{rev}} \quad \text{and} \quad -w_{\text{irrev}} < -w_{\text{rev}}$$

Suppose that as a practical matter one wishes to get work out of a system by some kind of process. The maximum work is extracted (largest $-w$) only if the process proceeds reversibly. The *actual* work performed by any real system during a real process is always less than this maximum: all real changes are irreversible. Similarly, the system will absorb the maximum heat only if the process goes on reversibly. The heat absorbed in a real (irreversible) process will be less than this maximum. **Problem 9-59** shows how an irreversible process becomes more nearly reversible as the method by which it is performed is changed. The process in that problem approaches reversibility as a limit. Reversibility is always an unattainable ideal.

The Second Law

In many applications a paramount question is whether a proposed change can occur. The entropy function is pivotal in answering this question for chemical (and all other) processes. The **second law of thermodynamics** states:

1. **In a reversible process the total entropy of a system plus its surroundings is unchanged.**
2. **In an irreversible process the total entropy of a system plus its surroundings increases.**
3. **In all human experience, processes for which ΔS_{univ} is negative are impossible.**

This all means that in real processes:

$$\Delta S_{\text{univ}} = \Delta S_{\text{sys}} + \Delta S_{\text{surr}} > 0$$

To see if a proposed process can occur, calculate the entropy change of the system. Then calculate the entropy change of the surroundings (regarding them momentarily as a system). If the sum of these two answers exceeds zero, then the proposed change can occur. This is actually done in **problem 9-9.**

Entropy and Irreversibility: A Statistical Interpretation

The entropy is a measure of the randomness of a system. Spontaneous change occurs from states of low probability (ordered states) to states of high probability (disordered states). **Statistical thermodynamics** identifies and compares such states on a molecular basis. Consider the expansion of an ideal gas into a vacuum. Originally, the gas is confined behind a closed stopcock in the left-hand bulb (of volume V) of a two-bulb container. The right-hand bulb (of identical volume V) is empty. The stopcock is opened. Very soon, the gas has the volume $2V$. Everybody knows that the gas expands spontaneously. But why? The energy of the ideal gas does not change during the expansion. The gas absorbs neither work nor heat. As far as the *first* law of thermodynamics is concerned the ideal gas could just as well stay in the left-hand bulb.

To appreciate why the gas expands, suppose for simplicity that there are only six molecules of the gas. How can the six molecules be distributed between the two sides?

Number of Molecules on Left	Number of Molecules on Right	Number of Ways Attainable
6	0	1
5	1	6
4	2	15
3	3	20
2	4	15
1	5	6
0	6	1

There are 64 ways to distribute the six molecules between the two sides. This is the sum of the numbers in the last column of the table. These give only 7 distinguishable states of the gas, corresponding to the 7 lines in the table. **The state that is most likely to occur is the one that can be achieved in the greatest number of ways.**

The chance of finding all 6 molecules on the left is only 1/64. The chance of finding 3 on the left and 3 on the right is 20/64. The latter state is 20 times more probable than the former. The gas is 20 times more likely when left to itself to fill the entire volume evenly than to stay on the left. For a large number of molecules there is a huge number of ways to achieve an even distribution and far fewer ways to achieve a state with all the molecules on the left or right. **See problem 9-15** for a related calculation.

Entropy

Nature proceeds spontaneously toward states that have the highest probability of occurring. These states have higher entropies. Entropy is closely associated with probability. The entropy of a system depends on the number of arrangements available to the system as it exists in a given state. Such arrangements are the **microstates** of the system. A microstate is a particular distribution of molecules among the positions and momenta accessible to them. Every microstate is equally likely to be occupied by the system. The number of microstates of a system is symbolized Ω.

The entropy of a system depends on the number of its microstates according to the equation:

$$S = k_{\text{B}} \ln \Omega$$

where k_{B} is Boltzmann's constant. In problems, actual numerical values for Ω are rarely available. Fortunately, ΔS values can be calculated even if values for S are not available. The *change* in the entropy of a system between two states depends on the *ratio* of the number of microstates:

$$\Delta S = k_{\text{B}} \ln \left(\frac{\Omega_2}{\Omega_1}\right)$$

If state 2 of the six-molecule gas is the state with 3 molecules on the left and three on the right and state 1 is the state with all 6 molecules on the left, then Ω_2/Ω_1 is 20/1 and the difference in entropy between the states is:

$$\Delta S = k_{\text{B}} \ln \left(\frac{20}{1}\right) = (1.381 \times 10^{-23}\ \text{J K}^{-1})(2.996) = 4.137 \times 10^{-23}\ \text{J K}^{-1}$$

This approach is used in solving **problems 9-15** and **9-61a.**

Entropy and Disorder

The greater the disorder of a system, then the greater its entropy. Disorder should not be conceived of in terms of a motionless array of atoms or molecules, but in terms of all the possible motions of the particles and the ways in which their arrangements change with time. These are the microstates of the system. Their number depends on factors like the number of atoms per molecule and the strength of the bonds between atoms. As a result two different substances in general have different entropies even if both are solids (or liquids or gases) at the same temperature. Some helpful generalizations:

- The entropy of substances increases when they melt. In solids every atom or molecule is at or near a prescribed position. In liquids the particles may move around more.

- Substances in their gaseous state have more entropy than in their liquid state at the same temperature. There are more microstates for a gas than a liquid. Many more arrangements of atoms or molecules correspond to the same observed state.

- Gases at low pressure have greater entropy than at high pressure.

- A dilute solution of a given quantity of a substance has greater entropy than a concentrated solution.

- When one substance dissolves in another, ΔS of the system usually is positive.

- The entropy of a system always increases with increasing temperature.

Absolute Entropies and the Third Law

Unlike energy and enthalpy, for which there is no natural zero point, the entropy function *does* have an obvious zero of reference. The **third law of thermodynamics** concerns the setting of this zero for the measurement of entropies. The third law states: **in any thermodynamic process involving only pure phases in their equilibrium states, the entropy change ΔS approaches zero as T approaches 0 K.**

This experimental law means that the entropy of any pure substance in its equilibrium state approaches zero as T approaches 0 K. It provides a natural reference for the tabulation of entropies. See Appendix D.

Standard-State Entropies

Appendix D gives the absolute entropies at 1 atm pressure and 298.15 K for a variety of substances. The S° values (note the absence of a Δ) of the pure solids, liquids and gases are all positive because they are all referred to a zero entropy at 0 K. The values are experimental. They were obtained by measuring c_P as a function of temperature for each substance. Then the increase from 0 K (T_1) to 298.15 K (T_2) of the entropy of each substance was computed using the formula:

$$\Delta S = n \int_{T_1}^{T_2} \frac{c_P}{T} dT$$

Problem 9-63 displays a plot of c_P/T versus T for two substances. The area under such a curve is proportional to ΔS. The solution-phase entropies in Appendix D are not absolute entropies. They are instead measured relative to an arbitrary standard. Some of them are negative.

In a chemical reaction the entropy change is the sum of the standard entropies of the products minus the sum of the standard entropies of the reactants:

$$\Delta S^\circ = \sum S^\circ(\text{products}) - \sum S^\circ(\text{reactants})$$

The use of this equation to get ΔS° values is like the use of ΔH°_f's (tabulated in Appendix D) to get ΔH°'s of reaction. Remember these points:

- The tabulated S° values are *molar* entropies. If a reaction forms, say, 2 mol of $NH_3(g)$ (for which S° is 192.3 J $K^{-1}mol^{-1}$), the contribution of the ammonia to the entropy of the products is 384.6 J K^{-1}. If it forms 3 mol of $NH_3(g)$ then ammonia's contribution is 576.9 J K^{-1}. **See problems 9-19** and **9-21.**

- The S° of an element in its standard state is *not* equal to zero. See **problem 9-21** and **9-29**. The mistake of taking elemental S°'s to equal zero occurs because elemental ΔH°_f's *are* zero.

- Standard entropies are tabulated in *joules* per Kelvin per mole (J $K^{-1}mol^{-1}$). In contrast, ΔH°_f's are tabulated in kilojoules per mole (kJ mol^{-1}). These are entirely different units. The similarity in the magnitudes of the values in the S° and ΔH°_f columns in Appendix D is accidental.

- The ΔS° of a reaction may be positive, negative or even zero. A negative ΔS° does *not* mean that a reaction is forbidden to occur by the second law of thermodynamics. For example, in **problem 9-29a**, the ΔS° for the rusting of iron at room conditions comes out to be negative. Iron still rusts.

The Gibbs Free Energy

So far, predicting whether a given process can occur requires the calculation of ΔS_{univ}, the sum of ΔS of the surroundings and ΔS of the system. The universe is big and complex, and getting its ΔS for every proposed process is hard. It is possible, at a price, to avoid figuring ΔS_{univ} when evaluating a proposed process. The method requires the definition of a new function, the **Gibbs free energy,** G**:**

$$G = H - TS$$

The Gibbs free energy is a state function. Hence, for any change in a system:

$$\Delta G_{sys} = \Delta H_{sys} - \Delta(TS)_{sys}$$

When T does not change it may be placed in front of the Δ:

$$\Delta G_{sys} = \Delta H\text{sys} - T\Delta S_{sys} \quad (T\text{ constant})$$

Under an even more restricted set of conditions, *constant temperature and pressure,* a simple criterion for spontaneity emerges:

If ΔG_{sys} for a proposed process at constant temperature and pressure is negative, then the process is spontaneous.

What if ΔG_{sys} is positive or zero?

- If ΔG_{sys} is positive, then the proposed process at constant temperature and pressure is *not* spontaneous, but the reverse process is.

- If ΔG is 0, then the system is at equilibrium at constant temperature and pressure, and no change occurs.

The price paid for the convenience and simplicity of this criterion is the restriction to constant temperature and pressure. Processes at constant temperature and pressure are common in chemistry, so the price is not heavy.

A ΔG is a difference between an enthalpy and an entropy term. Its value at constant T and P represents a compromise between the tendency toward minimum energy in the system and the tendency toward maximum entropy. The following table shows this compromise works for the four possible combinations of sign of ΔH and ΔS.

ΔH	ΔS	Outcome
Positive	Negative	Always nonspontaneous
Negative	Positive	Always spontaneous
Positive	Positive	Spontaneous at high T
Negative	Negative	Spontaneous at low T

This point is discussed further in the solution to **problem 9-33.**

Free Energy and Chemical Equilibrium

Standard-State Free Energies

Just as with other thermodynamic quantities, a *standard* free-energy change is defined. It is the free-energy change measured between reactants and products in their standard states at 298.15 K and 1 atm. A superscript zero to the right of the symbol ("ΔG°") designates that it is a standard change. The standard free energy change of a reaction is computed in two ways. The first way is from ΔH° and ΔS° values:

$$\Delta G^\circ = \Delta H^\circ - T\Delta S^\circ$$

The solution to **problem 9-29** consists of such calculations. The other way is from tabulated values of standard free energies of formation:

$$\Delta G^\circ = \sum \Delta G_f^\circ \text{ (products)} - \sum \Delta G_f^\circ \text{ (reactants)}$$

Problem 9-37 includes typical calculations of this second type. The procedure in this second type of calculation is identical in every respect to the procedure for getting $\Delta H°$'s of reaction from ΔH_f°'s. Text Appendix D contains the necessary free energy of formation data. **Warning:** the most common error in calculations of $\Delta G°$ is to subtract a $T\Delta S$ term in joules from a ΔH term that is in kilojoules.

Problem 9-69 emphasizes an important point of working with $\Delta H°$, $\Delta S°$ and ΔG. The free-energy change of a process is very strongly dependent on the temperature. The enthalpy change and entropy change are only weakly dependent on temperature. This means that in problem-solving, ΔH can often be approximated by $\Delta H°$ and ΔS can often be approximated by $\Delta S°$, but ΔG at temperatures other than 298 K can never be approximated by $\Delta G°$. Assuming temperature independent for $\Delta S°$ and $\Delta H°$ is common, particularly when the temperature does not change very much. See **problems 9-29** and **9-31.**

Free Energy and Phase Transitions

The equation $Li(s) \rightarrow Li(g)$ represents the sublimation of lithium, a phase transition. At 298.15 K, ΔG for this process is positive. It turns out (using data from Appendix D) that subliming one mole of metallic lithium requires 126.71 kJ of free energy at room conditions. Obviously, as the temperature increases, the quantity $T\Delta S$ for this process becomes larger and larger. Since $T\Delta S$ is subtracted from ΔH to give ΔG, it follows that ΔG slides down toward zero as the temperature increases. When ΔG for a process is equal to zero, then there is equilibrium between the two sides. For the sublimation of lithium, ΔG becomes equal to zero when the temperature reaches 1504 K, assuming that ΔS and ΔH are not themselves affected by the change in temperature and remain equal to $\Delta S°$ and $\Delta H°$ respectively.

The Equilibrium Constant

For a chemical reaction involving gases such as $aA(g) + bB(g) \rightleftharpoons cC(g) + dD(g)$, the associated equilibrium-constant expression is:

$$\frac{(P_C/P_{ref})^c(P_D/P_{ref})^d}{P_A/P_{ref})^a(P_B/P_{ref})^b} = K$$

The text derives the fundamentally important equation:

$$\boldsymbol{\Delta G° = -RT \ln K} \qquad \textbf{Memorize this relationship}$$

Although not derived in the text for reactions involving dissolved species, the relationship is valid in such cases. Furthermore, it applies to reactions involving *both* gases and dissolved species. The equation is so very important because it allows calculation of the equilibrium position of a chemical reaction from tabulated calorimetric data alone. Such data are in Appendix D.

Problems fall into these types:

Calculate K for a reaction at 25°C. After balancing the chemical equation, obtain the ΔG_f°'s of all of the reactants and products from Appendix D. Combination of these values gives ΔG° for the reaction. This intermediate answer (usually in kJ) the free-energy change observed every time the chemical reaction takes place as expressed in the balanced equation. Regard it as having the units kJ per mole of the chemical reaction as written. Multiplication by 1000 J kJ^{-1} converts it to J mol^{-1}. Subsequent division by 8.315 J $mol^{-1}K^{-1}$ and then by 298.15 K gives $-\ln K$. See **problem 9-37.**

Calculate ΔG°, given K. This is just the reverse use of the fundamental equation. Sometimes either ΔH° or ΔS° is given and the problem extends to the calculation of whichever one is missing. See **problems 9-45a** and **9-49**.

Calculate K at T not equal to 25o. Obtain ΔH° and ΔS° for the reaction by combination of the data in Appendix D. Do not bother to calculate ΔG° from the data in Appendix D because all ΔG's are strongly dependent on temperature. The value of ΔG_{298}° has nothing to do with K at temperatures other than 298 K (25°C). In contrast, ΔH's and ΔS's are usually only weakly dependent on temperature. The equilibrium constant is calculated using:

$$-RT\ln K_T = \Delta G_T^\circ = \Delta H^\circ - T\Delta S^\circ \quad (\Delta H^\circ \text{ and } \Delta S^\circ \text{ independent of } T)$$

See **problem 9-47.**

Given K's at two T's, calculate ΔH° and ΔS°. The K's are related to the ΔG° values:

$$-RT_1\ln K_1 = \Delta G_{T_1}^\circ = \Delta H^\circ - T_1\Delta S^\circ$$
$$-RT_2\ln K_2 = \Delta G_{T_2}^\circ = \Delta H^\circ - T_2\Delta S^\circ$$

Substitution of the two T's and two K's allows calculation of the two unknowns, ΔH° and ΔS°. If the two equations are combined by eliminating ΔS°:

$$\ln\frac{K_2}{K_1} = -\frac{\Delta H^\circ}{R}\left(\frac{1}{T_2} - \frac{1}{T_1}\right)$$

This is the **van't Hoff equation.** It is valid as long as ΔH° and ΔS° equal ΔH and ΔS in the range between T_1 and T_2. In fact, ΔH and ΔS do change with T (see **problem 9-69**). The van't Hoff equation is a route from equilibrium data to both ΔH° and ΔS° **See problem 9-45b.**

When $\Delta H°$ is positive, a reaction is endothermic, and its equilibrium constant increases with increasing temperature. When $\Delta H°$ is negative, a reaction is exothermic, and its equilibrium constant decreases with increasing temperature.

Problems may combine the thermodynamic avenue to K's with concentration and partial pressure data. Thus, an elaborate problem might give enough concentration data to calculate K at two different temperatures. Then it would ask for $\Delta H°$ and $\Delta S°$ for the reaction.

In all problems, use the activities of gases and solutes in solution. In practice, this means to take pressures of gases in atmospheres and concentrations of dissolved species in mol L^{-1}. This is correct done because the standard state (reference state) for all of the thermodynamic values in Appendix D is either a pressure of 1 atm or a concentration of 1 M.

Detailed Solutions to Odd-Numbered Problems

9-1 At the melting point of tungsten W(s) and W(l) coexist in equilibrium. This means that at 3410°C (3683 K), ΔG equals zero for the process of fusion W(s) $\rightarrow$ W(l). Fusion occurs at constant temperature, so:

$$\Delta G_{\text{fus}} = 0 = \Delta H_{\text{fus}} - T_{\text{fus}}\Delta S_{\text{fus}}$$

Solving for ΔS and substituting gives:

$$\Delta S_{\text{fus}} = \frac{\Delta H_{\text{fus}}}{T_{\text{fus}}} = \frac{35.4 \times 10^3 \text{ J mol}^{-1}}{3683 \text{ K}} = 9.61 \text{ J K}^{-1}\text{mol}^{-1}$$

Take care with the units in such calculations. Make sure that T is in kelvins, and that ΔH and ΔS both include either kJ or J.

9-3 Trouton's rule states that most liquids have approximately the same molar entropy of vaporization, which equals $\Delta S_{\text{vap}} = 88 \pm 5 \text{ J mol}^{-1}\text{K}^{-1}$. When a substance is at its boiling point, the gaseous phase at a pressure of 1 atm and the liquid phase are in equilibrium: $\Delta G_{\text{vap}} = 0$. Thus, if the normal boiling point is known, Trouton's rule can be used to estimate the molar enthalpy of vaporization:

$$\Delta H_{\text{vap}} = T_{\text{b}}\Delta S_{\text{vap}}$$

In the case of acetone:

$$\Delta H_{\text{vap}} \approx (329.35 \text{ K})(88 \text{ J mol}^{-1}\text{K}^{-1}) = 29 \times 10^3 \text{ J mol}^{-1}$$

9-5 The internal energy of an ideal gas depends solely on its absolute temperature T. In an isothermal process, T does not change. Assume that the system of 4.00 mol of hydrogen behaves ideally during its isothermal expansion. Then, ΔE equals zero. Consequently, ΔH also equals zero. To appreciate why, write the definition of ΔH:

$$\Delta H = \Delta E + \Delta(PV)$$

and observe that for a given portion of an ideal gas $\Delta(PV) = nR\Delta T$. But ΔT is zero. This means that $\Delta(PV) = 0$.
The work done *on* the gas during the reversible isothermal expansion from 12.0 L to 30.0 L is:

$$w = -nRT\ln\left(\frac{V_2}{V_1}\right) = -4.00\text{ mol}(8.315\text{ J K}^{-1}\text{mol}^{-1})(400\text{ K})\ln\left(\frac{30.0}{12.0}\right)$$

Completion of the arithmetic gives w equal to -12.2 kJ. From the first law, if $\Delta E = 0$ then $q = -w$. This means the gas absorbs 12.2 kJ of heat during its expansion, just enough to balance off the 12.2 kJ of work that it performs: $q = +12.2$ kJ. Finally, $\Delta S = q_{rev}/T$ for an isothermal process, and q_{rev} is the q just computed. Hence:

$$\Delta S = \frac{+12.2 \times 10^3\text{ J}}{400\text{ K}} = +30.5\text{ J K}^{-1}$$

9-7 Break down the process into the obvious three steps and calculate ΔS_{sys} for each. Then add up the three contributions. The steps are: I. warming of ice; II. melting of ice; III. warming of melted ice. As the text shows, ΔS for any temperature change at constant pressure is given by the equation:

$$\Delta S = nc_p\ln\left(\frac{T_2}{T_1}\right)$$

This allows the calculation of ΔS of the system for the first and third steps:

$$\Delta S_{\text{I}} = (1.00\text{ mol}(38\text{ J K}^{-1}\text{mol}^{-1})\ln(273.15/253.15) = 2.9\text{ J K}^{-1}$$
$$\Delta S_{\text{III}} = (1.00\text{ mol}(75\text{ J K}^{-1}\text{mol}^{-1})\ln(293.15/273.15) = 5.3\text{ J K}^{-1}$$

In the second step of the process, T is constant (273.15 K), and ΔS of the system is the quantity of heat absorbed reversibly (q_{rev}) divided by this temperature:

$$\Delta S_{\text{II}} = \frac{6007\text{ J}}{273.15\text{ K}} = 21.99\text{ J K}^{-1}$$

The *total* ΔS of the system is: $\Delta S_{sys} = \Delta S_{\text{I}} + \Delta S_{\text{II}} + \Delta S_{\text{III}} = +30.2\text{ J K}^{-1}$. The entire process is reversible so the entropy of the universe remains constant: $\Delta S_{univ} = \Delta S_{sys} + \Delta S_{surr} = 0$ which means $\Delta S_{surr} = -30.2\text{ J K}^{-1}$.

9-9 Hot iron is plunged into cool water. The final temperature is 28.9°C. This process is far from reversible. Nevertheless, the ΔS of the iron and the ΔS of the water may be computed using the same equation that was used for the reversible heat flow in problem 9-7. This is legitimate because entropy is a state function. Its change depends only on the original and final states of the system, not on how the path along which the change occurs. As computed in Example 8-3 (text page 281-2), the iron cools from 373.15 K to 289.65 K. The 72.4 g of iron ($\mathcal{M}$ = 55.847 g mol^{-1}) is 1.296 mol. Hence:

$$\Delta S_{\text{Fe}} = nc_{\text{p}} \ln\left(\frac{T_2}{T_1}\right) = (1.296 \text{ mol})(25.1 \text{ J K}^{-1}\text{mol}^{-1}) \ln\left(\frac{289.65 \text{ K}}{373.15 \text{ K}}\right) = -8.24 \text{ J K}^{-1}$$

There is 100.0 g of water, which is 5.55 mol. The c_{p} of water is 75.3 J K^{-1}mol^{-1}, and the water is warmed from 283.15 K to 289.65 K. Hence, substituting as before:

$$\Delta S_{\text{H}_2\text{O}} = 5.55 \text{ mol}(75.3 \text{ J K}^{-1}\text{mol}^{-1}) \ln\left(\frac{289.65}{283.15}\right) = +9.49 \text{ J K}^{-1}$$

The overall ΔS of the system is +1.25 J K^{-1}, which is the sum of the ΔS's for the water and the iron.

9-11 a) The number of available microstates is the number of possible ways for a number to come up on one die times the number of possible ways for a number to come up on the other die. Each die has six faces and therefore 6 available microstates. The total number of available microstates is 36.

b) The probability that one die will show a six is 1/6. The same is true for the other die. Thus, the probability that two sixes will be show up at the same time is (1/6)(1/6) = 1/36.

9-13 The driving force for the reaction $H_2O(l) + D_2O(l) \rightarrow 2\,HOD(l)$ is an increase in entropy. Two moles of HOD have a larger entropy than a mixture of one mol of H_2O and one mol of D_2O because there is a much larger number of ways for the available H's, D's and O's to be assembled into a collection of HOD molecules than into a collection of H_2O's and D_2O's. Note well that a mixture of HOD, H_2O, and D_2O will have an even larger entropy than either pure products or pure reactants. The reaction comes to equilibrium in an intermediate state corresponding closely to just such a mixture.

9-15 Before the stopcock is opened, the number of microstates available to a single H_2 (or He) is proportional to the volume of the glass bulb: $\Omega = cV$ where c is a constant. There are N_0 molecules of H_2 and N_0 atoms of He. The number of possible microstates for each gas is:

$$\Omega_{\text{H}_2} = (cV)^{N_0} \quad \text{and} \quad \Omega_{\text{He}} = (cV)^{N_0}$$

The number of microstates of the entire system, still before the valve is opened, is the *product* of the Ω's:

$$\Omega_{\text{sys}} = \Omega_{\text{H}_2}\Omega_{\text{He}} = (cV)^{2N_0}$$

This is the number of microstates for which all of the H_2 occupies the first bulb and all of the He occupies the second bulb. By symmetry it is also the number of microstates in which all of the H_2 is in the *second* bulb and all of the He is in the first. *After* the stopcock is opened, $2N_0$ molecules occupy a volume of $2V$ and:

$$\Omega_{\text{sys}} = (c\,2V)^{2N_0}$$

The probability p of the "cross-diffused" result, the state in which the H_2 and He trade places, is the number of ways in which it can be constituted divided by the number of ways in which the mixed system can be constituted:

$$p = \frac{(cV)^{2N_0}}{(c2V)^{2N_0}} = 2^{-2N_0}$$

Take the logarithm of both sides of this equation:

$$\log p = -2N_0 \log 2 = -2N_0(0.301) = -3.62 \times 10^{23}; \text{ hence, } p = 10^{-3.62\times 10^{23}}$$

9-17 If the amount of disorder in a system increases when a process occurs, the change in entropy ΔS of that system is positive.

a) When NaCl melts it goes from a highly ordered solid to a relatively disordered liquid state. This means an increase in disorder and therefore $\Delta S > 0$.

b) When a building is demolished it goes from a ordered state to a highly disordered state; thus $\Delta S > 0$.

c) In this case a highly disordered system (air) has order imposed on it by the separation into three distinct subsystems. An increase in order means $\Delta S < 0$.

9-19 a) The ΔS° of the reaction as written equals the standard molar entropies (the S°'s) of the products each multiplied by its chemical amount in the balanced equation minus the S°'s of the reactants each multiplied by its chemical amount in the balanced chemical equation:

$$\Delta S^\circ = 2\underbrace{(239.95)}_{\text{NO}_2(g)} + 2\underbrace{(69.91)}_{\text{H}_2\text{O}(l)} - 1\underbrace{(121.21)}_{\text{N}_2\text{H}_4(l)} - 3\underbrace{(205.03)}_{\text{O}_2(g)} = -116.58 \text{ J K}^{-1}$$

where the numbers in parenthesis are standard molar entropies (with units of J $K^{-1}mol^{-1}$) and the other numbers are the numbers of moles from the balanced equation.

b) In the process $N_2H_4(l) \rightarrow N_2H_4(g)$, the disorder in the N_2H_4 increases, and therefore $\Delta S > 0$. In other words S° for $N_2H_4(g)$ is more positive than S° for $N_2H_4(l)$. This will cause ΔS° of the reaction with oxygen to be algebraically smaller than when liquid N_2H_4 reacts.

9-21 The computations follows the method used in problem 9-19.

$$\text{For LiCl}: \quad \Delta S^\circ = 2\underbrace{(59.33)}_{\text{LiCl}(s)} - 2\underbrace{(29.12)}_{\text{Li}(s)} - 1\underbrace{(222.96)}_{\text{Cl}_2(g)} = -162.54 \text{ J K}^{-1}$$

$$\text{For NaCl}: \quad \Delta S^\circ = 2\underbrace{(72.13)}_{\text{NaCl}(s)} - 2\underbrace{(51.21)}_{\text{Na}(s)} - 1\underbrace{(222.96)}_{\text{Cl}_2(g)} = -181.12 \text{ J K}^{-1}$$

$$\text{For KCl}: \quad \Delta S^\circ = 2\underbrace{(82.59)}_{\text{KCl}(s)} - 2\underbrace{(64.18)}_{\text{K}(s)} - 1\underbrace{(222.96)}_{\text{Cl}_2(g)} = -186.14 \text{ J K}^{-1}$$

$$\text{For RbCl}: \quad \Delta S^\circ = 2\underbrace{(95.90)}_{\text{RbCl}(s)} - 2\underbrace{(76.78)}_{\text{Rb}(s)} - 1\underbrace{(222.96)}_{\text{Cl}_2(g)} = -184.72 \text{ J K}^{-1}$$

$$\text{For CsCl}: \quad \Delta S^\circ = 2\underbrace{(101.17)}_{\text{CsCl}(s)} - 2\underbrace{(85.23)}_{\text{CsCl}(s)} - 1\underbrace{(222.96)}_{\text{Cl}_2(g)} = -191.08 \text{ J K}^{-1}$$

The ΔS°'s become increasingly negative moving down the group although there is an exception at RbCl.

9-23 By the second law, $\Delta S_{\text{univ}} = \Delta S_{\text{sys}} + \Delta S_{\text{surr}} > 0$. In this example, $\Delta S_{\text{sys}} = -44.7$ J K^{-1}. Thus, ΔS_{surr} must be greater than $+44.7$ J K^{-1}.

9-25 a) The melting of the ammonia goes on at a constant temperature. Hence $\Delta G = \Delta H - T\Delta S$. Compute the required ΔG on a molar basis:

$$\Delta G = 5.65 \text{ kJ mol}^{-1} - (170 \text{ K})(0.0289 \text{ kJ mol}^{-1}\text{K}^{-1}) = 0.74 \text{ kJ mol}^{-1}$$

For 1.00 mol of NH_3, ΔG is 0.74 kJ.

b) The value of ΔG is now 3.60 times larger than in part a), because the amount of ammonia is 3.60 times more. The answer is 2.65 kJ.

c) At 170 K, $\Delta G > 0$. This means that the melting of ammonia is not spontaneous at 170 K.

d) If solid and liquid ammonia are in equilibrium, then ΔG is zero for the solid$\rightleftharpoons$liquid process. The values of the molar enthalpy and the molar entropy are given, so the temperature can be calculated:

$$\Delta G = \Delta H - T\Delta S = 0$$

$$T = \frac{\Delta H}{\Delta S} = \frac{5.65 \times 10^3 \text{ J mol}^{-1}}{28.9 \text{ J K}^{-1}\text{mol}^{-1}} = 196 \text{ K}$$

9-27 When one mole of ethanol is vaporized at its normal boiling point, ΔH is 38.7 kJ. The vaporization goes on at constant pressure so that $q_p = \Delta H$ and q is +38.7 kJ. Moreover, the vaporization occurs isothermally and reversibly, so q is also q_{rev}, and:

$$\Delta S = \frac{q_{rev}}{T} = \frac{38.7 \text{ kJ}}{351.1 \text{ K}} = +0.110 \text{ kJ K}^{-1} = 110 \text{ J K}^{-1}$$

Next is the calculation of ΔE. From the definition of enthalpy:

$$\Delta E = \Delta H - \Delta(PV)$$

At constant pressure $\Delta(PV) = P\Delta V = P(V_2 - V_1)$. V_2 is the volume of one mole of ethanol vapor at 351.1 K and V_1 is the volume of one mole of liquid ethanol at the same temperature. The vapor behaves ideally:

$$V_2 = \frac{nRT}{P} = \frac{1.00 \text{ mol}(0.08206 \text{ L atm mol}^{-1}\text{K}^{-1})(351.15 \text{ K})}{1.00 \text{ atm}} = 28.8 \text{ L}$$

The volume of the liquid (V_1) is less than 0.1 L, which makes it negligibly small compared to 28.8 L. Therefore:

$$P\Delta V = 1 \text{ atm}(28.8 - 0) = 28.8 \text{ L atm} \times \left(\frac{0.101325 \text{ kJ}}{1.00 \text{ L atm}}\right) = 2.92 \text{ kJ}$$

Substitute these values into the expression for ΔE:

$$\Delta E = \Delta H - \Delta(PV) = 38.7 \text{ kJ} - 2.92 \text{ kJ} = +35.8 \text{ kJ}$$

By expanding against a constant pressure the system performs +2.92 kJ of pressure-volume work on its surroundings. This is the only kind of work possible. The total work done on the system is −2.92 kJ.

For all reversible processes at constant T and P, $\Delta G = 0$. This can be verified in this case:

$$\Delta G = \Delta H - T\Delta S = 38.7 - (351.15 \times 0.110) = 0 \text{ kJ}$$

9-29 In each case calculate ΔH° and ΔS° by using data from Appendix D. Then use the formula $\Delta G = \Delta H - T\Delta S$ to calculate the temperature at which $\Delta G = 0$. The assumption is that ΔH equals ΔH° and ΔS equals ΔS°.

a) $$\Delta H^\circ = 2\underbrace{(-824.2)}_{Fe_2O_3(s)} - 4\underbrace{(0)}_{Fe(s)} - 3\underbrace{(0.00)}_{O_2(g)} = -1648.4 \text{ kJ}$$

$$\Delta S^\circ = 2\underbrace{(87.40)}_{Fe_2O_3(s)} - 4\underbrace{(27.28)}_{Fe(s)} - 3\underbrace{(205.03)}_{O_2(g)} = -549.44 \text{ J mol}^{-1}$$

The relationship $\Delta G = \Delta H - T\Delta S$ can now be used to compute ΔG at any temperature if it is assumed that ΔH and ΔS at that temperature are stay equal to $\Delta H°$ and $\Delta S°$. Remember to convert $\Delta S°$ to kJ K^{-1} or $\Delta H°$ to J when such a calculation is carried out. The problem asks for the temperature ranges for which the reaction is spontaneous. The changeover from spontaneity to non-spontaneity occurs at $\Delta G = 0$. Simply calculate the temperature that makes this happen:

$$T = \frac{\Delta H° - 0}{\Delta S°} = \frac{\Delta H°}{\Delta S°} = \frac{-1648.1 \text{ kJ}}{-0.54941 \text{ kJ K}^{-1}} = 3000 \text{ K}$$

Because ΔH and ΔS are both negative, the reaction is spontaneous *below* 3000 K. Above 3000 K the increased magnitude of the $-T\Delta S$ term makes ΔG positive.
b) Perform the calculations for b) as in a): $\Delta H = -98.89$ kJ. Since ΔH and ΔS are both negative, the reaction will be spontaneous between the temperatures of 0 K and 1052 K.
c) $\Delta H = -401.59 \text{ kJ} - (-365.56 \text{ kJ}) = -36.03$ kJ
$\Delta S = 597.18 \text{ J K}^{-1} - 151.08 \text{ J K}^{-1} = 446.10 \text{ J K}^{-1}$. In this case, if ΔH is divided by ΔS the answer is negative. A negative absolute temperature is meaningless because 0 K is the lowest temperature which can be reached. This reaction is in fact spontaneous at all temperatures (see the discussion on text page 332).

9-31 The reduction reaction is: $WO_3(s) + 3\,H_2(g) \rightarrow W(s) + 3\,H_2O(g)$. Calculate $\Delta H°$ and $\Delta S°$ for this process using the data in Appendix D. The procedures are the same as in previous problems.

$$\Delta H° = 1\underbrace{(0.00)}_{W(s)} + 3\underbrace{(-241.82)}_{H_2O(g)} - 1\underbrace{(-842.87)}_{WO_3(s)} - 3\underbrace{(0.00)}_{H_2(g)} = +117.41 \text{ kJ}$$

$$\Delta S° = 1\underbrace{(32.64)}_{W(s)} + 3\underbrace{(188.72)}_{H_2O(g)} - 1\underbrace{(75.90)}_{WO_3(s)} - 3\underbrace{(130.57)}_{H_2(g)} = +131.19 \text{ J K}^{-1}$$

Because $\Delta H°$ and $\Delta S°$ are both positive, the reaction becomes spontaneous above a temperature computed as follows:

$$T = \frac{\Delta H°}{\Delta S°} = \frac{+117.41 \times 10^3 \text{ J}}{+131.19 \text{ J K}^{-1}} = 895 \text{ K}$$

High temperature favors the reduction of tungsten(VI) oxide by gaseous hydrogen. The answer is only approximate because it depends on the assumption that $\Delta H°$ and $\Delta S°$ are independent of temperature between 298 K and 895 K. In fact, both are slightly dependent on temperature.
As shown on text page 336:

$$\ln K = \frac{\Delta S°}{R} - \frac{\Delta H°}{RT}$$

When K equals one, $\ln K$ equals zero. Solving for T shows that the required temperature is:

$$T = \frac{\Delta H^\circ}{\Delta S^\circ}$$

This is the temperature just computed. Thus K becomes equal to one at 895 K. This reaction *can* proceed to a measurable extent at temperatures below the 895 K required to make $K = 1$. However, when K is less 1, the reactants are preponderant at equilibrium.

9-33 A process can be quite exothermic, but still not be spontaneous. A process at constant T and P is spontaneous if $\Delta G < 0$. Consider that $\Delta G = \Delta H - T\Delta S$. This means spontaneity is a compromise between two terms. If ΔS is negative, the $-T\Delta S$ term in this equation always eventually wins out, as T gets larger, and forces ΔG to go positive. The more negative the ΔS, then the lower the required temperature. Helium is only slightly soluble in $H_2O(l)$ because $-T\Delta S$ for the process of solution has quite a large positive value in the range of temperatures in which H_2O is a liquid.

9-35 Insert the data in Appendix D into the usual form:

$$\Delta G^\circ = 2\underbrace{(51.29)}_{NO_2(g)} + 3\underbrace{(-228.59)}_{H_2O(g)} - 2\underbrace{(-16.48)}_{NH_3(g)} - 7/2\underbrace{(0)}_{O_2(g)} = -550.23 \text{ kJ}$$

Substitute this answer in the equation:

$$\ln K = \frac{-\Delta G^\circ}{RT} = \frac{-(-550.23 \times 10^3 \text{ J mol}^{-1})}{8.315 \text{ J K}^{-1}\text{mol}^{-1}(298.15 \text{ K})} = 221.95$$

Hence, $K = e^{221.95} = 2.5 \times 10^{96}$. The ΔG° of the reaction is properly reported in kJ, but the value used in the calculation of K is in kJ mol^{-1}. The extra per mole refers to "per mole of the reaction as it is written." If the equation were rewritten with all the coefficients doubled, then ΔG° would be twice as many kJ's and the equilibrium constant K would be squared.

9-37 The data in Appendix D allow calculation of ΔG° for each reaction. The corresponding K is then computed using $\Delta G^\circ = -RT\ln K$. The form of the equilibrium-constant expression comes from the balanced equation.

a)

$$\Delta G^\circ = 1\underbrace{(-371.08)}_{SO_3(g)} - 1\underbrace{(-300.19)}_{SO_2(g)} - 1\underbrace{(0)}_{O_2(g)} = -70.89 \text{ kJ}$$

$$\ln K = \frac{-70.89 \times 10^3 \text{ kJ mol}^{-1}}{-8.315 \text{ J K}^{-1}\text{mol}^{-1}(298.15 \text{ K})} = 28.6$$

$$K = e^{28.6} = 2.6 \times 10^{12} = \frac{P_{SO_3}}{P_{SO_2}P_{O_2}^{1/2}}$$

b)

$$\Delta G^\circ = 2\underbrace{(-1015.5)}_{Fe_3O_4(s)} + 1/2\underbrace{(0)}_{O_2(g)} - 3\underbrace{(-742.2)}_{Fe_2O_3(s)} = +195.6 \text{ kJ}$$

$$\ln K = \frac{195.6 \times 10^3 \text{ kJ mol}^{-1}}{-8.315 \text{ J K}^{-1}\text{mol}^{-1}(298.15 \text{ K})} = -78.90$$

$$K = e^{-78.90} = 5.4 \times 10^{-35} = P_{O_2}^{1/2}$$

c)

$$\Delta G^\circ = 1\underbrace{(65.49)}_{Cu^{2+}(aq)} + 2\underbrace{(-131.23)}_{Cl^-(aq)} - 1\underbrace{(-175.7)}_{CuCl_2(s)} = -21.27 \text{ kJ}$$

$$\ln K = \frac{-21.27 \times 10^3 \text{ kJ mol}^{-1}}{-8.315 \text{ J K}^{-1}\text{mol}^{-1}(298.15 \text{ K})} = 8.58$$

$$K = e^{8.58} = 5.3 \times 10^3 = [Cu^{2+}][Cl^-]^2$$

9-39 The process is the dilution of aqueous ammonia from 0.100 M to 0.020 M at 298.15 K. The free energy change for n moles of ammonia going from c_1 to c_2 is:

$$\Delta G = nRT \ln \frac{c_2}{c_1}$$

In this case:

$$\frac{\Delta G}{n} = (8.315 \text{ J K}^{-1}\text{mol}^{-1})(298.15 \text{ K}) \ln \left(\frac{0.20 \text{ M}}{1.00 \text{ M}}\right) = -4.0 \times 10^3 \text{ J mol}^{-1}$$

9-41 The ΔH of the reaction is clearly less than zero. The hydration of ethylene to give ethanol is thus exothermic. Think of the heat as a reaction product:

$$C_2H_4(g) + H_2O(g) \rightarrow C_2H_5OH(g) + \text{ Heat}$$

By LeChatelier's principle, the equilibrium production of ethanol is maximized by running the reaction at low temperature (which allows the product heat to escape better). There are two moles of gas on the reactant side and only one mol of gas of the product side. High pressures will force the equilibrium from left to right, thus maximizing the yield of ethanol.

9-43 Use the van't Hoff equation to calculate ΔH for the reaction from the two K's and their corresponding temperatures:

$$\ln \frac{K_2}{K_1} = -\frac{\Delta H^\circ}{R}\left(\frac{1}{T_2} - \frac{1}{T_1}\right)$$

$$\ln\left(\frac{0.00121}{6.8}\right) = -8.634 = -\frac{\Delta H^\circ}{8.315 \text{ J K}^{-1}\text{mol}^{-1}}\left(\frac{1}{473.15 \text{ K}} - \frac{1}{298.15 \text{ K}}\right)$$

$$\Delta H^\circ = -5.8 \times 10^4 \text{ J mol}^{-1}$$

9-45 a) The change in Gibbs free energy at 25°C ($\Delta G°$) is related to the equilibrium constant K by $\Delta G° = -RT \ln K$. Upon substitution:

$$\Delta G°_{298} = -RT \ln K = -\frac{8.315 \text{ J}}{\text{mol K}}(298 \text{ K}) \ln(9.3 \times 10^9) = -56.9 \times 10^3 \frac{\text{J}}{\text{mol}}$$

b) Use the van't Hoff equation and the two values of K to obtain $\Delta H°$. Then $\Delta S°$ can be calculated from the above $\Delta G°$ and the equation: $\Delta G° = \Delta H° - T\Delta S°$. The van't Hoff equation becomes:

$$\ln\left(\frac{3.3 \times 10^7}{9.3 \times 10^9}\right) = -\frac{\Delta H°}{8.315 \text{ J K}^{-1}\text{mol}^{-1}}\left(\frac{1}{398 \text{ K}} - \frac{1}{298 \text{ K}}\right)$$

Completion of the arithmetic gives $\Delta H°$ equal to -55.6 kJ mol^{-1}. This means that $\Delta H°$ during the reaction of 1/2 mol of Cl_2 and 1/2 mol of F_2 to give 1 mol of ClF is -55.6 kJ. Finally:

$$\Delta S° = \frac{\Delta H° - \Delta G°}{T} = \frac{-55.63 \times 10^3 \text{ J} - (-56.88 \times 10^3 \text{ J})}{298 \text{ K}} = 4.2 \text{ J K}^{-1}$$

9-47 Use the van't Hoff equation to calculate the equilibrium constant (call it K_2) at 600 K from the constant K_1 at 298 K and the standard enthalpy change $\Delta H°$, both of which are given:

$$\ln\left(\frac{K_2}{K_1}\right) = -\frac{\Delta H°}{R}\left(\frac{1}{T_2} - \frac{1}{T_1}\right)$$

$$\ln\left(\frac{K_2}{5.9 \times 10^5}\right) = -\frac{-92.2 \times 10^3 \text{ J mol}^{-1}}{8.315 \text{ J K}^{-1}\text{mol}^{-1}}\left(\frac{1}{600 \text{ K}} - \frac{1}{298 \text{ K}}\right)$$

Solution gives $K_2 = 4.3 \times 10^{-3}$

9-49 a) Use the van't Hoff equation to calculate ΔH_{vap}. In this case, the equilibrium constants at the two temperatures are just the vapor pressures of the liquid:

$$\ln\left(\frac{K_2}{K_1}\right) = \ln\left(\frac{P_2}{P_1}\right) = -\frac{\Delta H_{vap}}{R}\left(\frac{1}{T_2} - \frac{1}{T_1}\right)$$

Substitution gives:

$$\ln\left(\frac{4.2380 \text{ atm}}{0.4034 \text{ atm}}\right) = \frac{-\Delta H_{vap}}{8.315 \text{ J K}^{-1}\text{mol}^{-1}}\left(\frac{1}{273.15 \text{ K}} - \frac{1}{223.15 \text{ K}}\right)$$

Solution gives $\Delta H_{vap} = 23.8$ kJ mol^{-1}.

b) The normal boiling point T_b of a liquid is defined as the temperature at which the vapor pressure of the liquid is 1.000 atm. Therefore, set P_1 equal to 1.000 atm and T_1

equal to T_b in the previous (see text page 340). Using the vapor pressure at 273.15 K for P_2 and T_2 gives:

$$\ln\left(\frac{4.2380\text{ atm}}{1.000\text{ atm}}\right) = -\frac{23{,}800\text{ J}}{8.315\text{ J K}^{-1}\text{mol}^{-1}}\left(\frac{1}{273.15} - \frac{1}{T_b}\right); T_b = 240\text{ K}$$

Inserting 0.4034 atm for P_2 and 223.15 K for T_2 gives the same answer.

9-51 The text (page 342) develops the relationship:

$$\Delta T_b = \frac{RT_b^2}{\Delta H_{\text{vap}}} X_2$$

where X_2 is the mole fraction of the *solute*, T_b and ΔH_{vap} are the boiling temperature and enthalpy change of vaporization of the *solvent* and ΔT_b is the boiling point elevation. In this problem, 2.00 g of the solute $C_{14}H_{10}$ ($\mathcal{M}$ = 178.23 g mol^{-1}) is dissolved in 100 g of C_6H_6 (benzene, $\mathcal{M}$ = 78.115 g mol^{-1}). The solute contributes 0.01122 mol to a total of 1.29142 mol in the solution. Hence X_2 is 0.008688. Substitute this and the other given values into the above equation, remembering that T_b is 353.35 K and that ΔH_{vap} and RT_b both must both be expressed in the same units. Then $\Delta T_b = 0.230$ K. The boiling point of the solution is $353.35 + 0.230 = 353.58$ K, which is 80.4°C.

9-53 The liquid and gaseous forms of a substance are in equilibrium at its normal boiling point if the pressure of the gaseous form is 1.000 atm. Hence, $\Delta G = 0$ for the process $C_2H_5OH(l) \rightleftharpoons C_2H_5OH(g, 1\text{ atm})$. If:

$$\Delta G = \Delta H_{\text{vap}} - T\Delta S_{\text{vap}} = 0 \quad \text{then} \quad \Delta S_{\text{vap}} = \frac{\Delta H_{\text{vap}}}{T}$$

Substitution of the values from the problem gives:

$$\Delta S_{\text{vap}} = \frac{38.4 \times 10^3\text{ J}}{351.4\text{ K}} = 109\text{ J K}^{-1}\text{mol}^{-1}$$

Trouton's rule states that ΔS_{vap} is close to 88 J K^{-1}mol^{-1} for all liquids. The ΔS_{vap} for ethanol is 25% higher than predicted by Trouton's rule.

9-55 a) The compression of the oxygen is reversible and adiabatic. Therefore, ΔS_{sys} equals zero.

b) If 2.60 mol of oxygen is compressed reversibly and adiabatically from a state (P_1, V_1) to a state (P_2, V_2) then:

$$P_1V_1^\gamma = P_2V_2^\gamma \quad \text{or} \quad \frac{P_1}{P_2} = \left(\frac{V_2}{V_1}\right)^\gamma$$

where γ is the ratio of c_p to c_v of the system. The exponent γ in this case is 29.4 J mol^{-1}K^{-1} divided by 21.09 J mol^{-1}K^{-1} or 1.394. The original volume (V_1) of the oxygen is 64.0 L, computed from the ideal gas equation with $T_1 = 300$ K. With P_2 equal to 8.00 atm, substitution gives:

$$\frac{P_1}{P_2} = \left(\frac{V_2}{V_1}\right)^{\gamma} = \frac{1.00\text{ atm}}{8.00\text{ atm}} = \left(\frac{V_2}{64.0\text{ L}}\right)^{1.394} \qquad \text{so that } V_2 = 14.4\text{ atm}$$

Knowing P_2 and V_2 (with ideality assumed) gives T_2. In summary:

$$P_1 = 1.00\text{ atm} \quad V_1 = 64.0\text{ L} \quad T_1 = 300\text{ K}$$
$$P_2 = 8.00\text{ atm} \quad V_2 = 14.4\text{ L} \quad T_2 = 540\text{ K}$$

The problem traces an alternative path between state 1 and state 2. The oxygen is first heated to T_2 at constant pressure and then compressed reversibly and isothermally to P_2. Compute all of the state variables in the *intermediate* state (subscripted i), after the isochoric heating but before the isothermal compression:

$$P_i = 1.00\text{ atm} \quad T_i = 540\text{ K} \quad V_i = 115.2\text{ L}$$

The volume comes from the ideal-gas law, with $n = 2.60$ mol. The entropy change during the constant pressure heating is:

$$\Delta S_{1\to i} = nc_p \ln\left(\frac{T_i}{T_1}\right) = (2.60\text{ mol})(29.4\text{ J K}^{-1}\text{mol}^{-1})\ln\left(\frac{540}{300}\right) = 44.9\text{ J K}^{-1}$$

The entropy change during the isothermal compression is:

$$\Delta S_{i\to 2} = nR\ln\left(\frac{V_2}{V_i}\right) = (2.60\text{ mol})(8.315\text{ J K}^{-1}\text{mol}^{-1})\ln\left(\frac{14.40}{115.2}\right) = -45.0\text{ J K}^{-1}$$

The ΔS for the overall process is the sum of these values. it is zero, allowing for round-off errors.

9-57 a) If the motion of air masses through the atmosphere is adiabatic and reversible, then $q_{rev} = 0$ and ΔS equals zero.

b) During the upward displacement of an air mass one can expect its temperature and pressure to drop simultaneously. Break down this overall process into two parts: a temperature change at constant pressure (step I) and a pressure change at constant temperature (step II). The original values of temperature and pressure are T_0 and P_0 and the final values are T and P. For the two steps:

$$\Delta S_{\text{I}} = nc_p \ln\frac{T}{T_0} \qquad\qquad \Delta S_{\text{II}} = nR\ln\frac{P_0}{P}$$

In the first step, ΔS is *less* than zero. because cooling a system reduces its entropy. In the second step, ΔS is *greater* than zero. This step is the expansion of the air mass. The sum of the ΔS's must be zero because the overall process, the sum of the two steps, is isentropic:

$$\Delta S = c_p \ln\left(\frac{T}{T_0}\right) + R\ln\left(\frac{P_0}{P}\right) = 0$$

c) According to the problem, $\ln(P/P_0)$ is approximately equal to $-\mathcal{M}gh/RT$. It follows that:

$$-\ln\left(\frac{P}{P_0}\right) = \ln\left(\frac{P_0}{P}\right) \approx \frac{+\mathcal{M}gh}{RT}$$

Substitute this (approximate) equality into the final expression in part b) and rearrange to give:

$$T\ln\left(\frac{T}{T_0}\right) \approx \frac{-\mathcal{M}gh}{c_p}$$

All of the quantities on the right-hand side of this equation are given in the problem:

$$T\ln\left(\frac{T}{T_0}\right) \approx \frac{(0.029 \text{ kg mol}^{-1})(9.8 \text{ m s}^{-2})(5.9\times 10^3 \text{ m})}{29 \text{ J K}^{-1}\text{mol}^{-1}} = -57.8 \text{ K}$$

This means that T, the temperature on top of the mountain, fulfills the equation:

$$T\ln\frac{T}{T_0} \approx -57.8 \text{ K}$$

where T_0 is the sea-level temperature (311 K). It is easiest to determine T by guessing a few trial values and using a calculator. It is quickly seen that $T = 246$ K ($-27°$C) satisfies the approximate equation.

9-59 In problem 9-10, a 1.000 mol piece of iron at 100°C is plunged into a large reservoir of water at 0°C. It loses 2510 J to the water as its temperature falls from 373 K to 273 K. Its entropy decreases. The change is:

$$\Delta S_{\text{Fe}} = nc_p \ln\frac{T_2}{T_1} = (1.00 \text{ mol})(25.1 \text{ J K}^{-1}\text{mol}^{-1})\ln\left(\frac{373.15}{273.15}\right) = -7.83 \text{ J K}^{-1}$$

a) The piece of iron is first cooled from 100 to 50°C and then from 50 to 0°C using two water reservoirs. It loses 1255 J of heat to the first reservoir and 1255 J of heat to the second. The entropy change of the first reservoir, which *absorbs* 1255 J of heat and which is so big it stays at 323.15 K, is:

$$\Delta S_{\text{I}} = \frac{q}{T} = \frac{1255 \text{ J}}{323.15 \text{ K}} = 3.88 \text{ J K}^{-1}$$

The entropy change of the second reservoir, which also absorbs 1255 J of heat but at 273.15 K, is larger:

$$\Delta S_{\text{II}} = \frac{q}{T} = \frac{1255\ \text{J}}{273.15\ \text{K}} = 4.60\ \text{J K}^{-1}$$

These two reservoirs are the surroundings of the iron:

$$\Delta S_{\text{surr}} = 3.88 + 4.60 = 8.48\ \text{J K}^{-1}$$

The ΔS of the iron is still -7.83 J K^{-1} because only the path by which it cooled has changed. It still ends up in the same final state.

$$\Delta S_{\text{univ}} = \Delta S_{\text{H}_2\text{O}} + \Delta S_{\text{Fe}} = 8.48\ \text{J K}^{-1} - 7.83\ \text{J K}^{-1} = 0.65\ \text{J K}^{-1}$$

b) Each of the four reservoirs absorbs 627.5 J, one-fourth of the total heat the iron gives up. The entropy changes of the four reservoirs are:

$$\Delta S_{\text{I}} = \frac{627.5\ \text{J}}{348.15\ \text{K}} = 1.80\ \text{J K}^{-1} \quad \Delta S_{\text{II}} = \frac{627.5\ \text{J}}{323.15\ \text{K}} = 1.94\ \text{J K}^{-1}$$

$$\Delta S_{\text{III}} = \frac{627.5\ \text{J}}{298.15\ \text{K}} = 2.11\ \text{J K}^{-1} \quad \Delta S_{\text{IV}} = \frac{627.5\ \text{J}}{273.15\ \text{K}} = 2.30\ \text{J K}^{-1}$$

ΔS_{surr} is the sum of the ΔS's of the four reservoirs. It is 8.15 J K^{-1}. The ΔS of the iron is still -7.83 J K^{-1}.

$$\Delta S_{\text{univ}} = \Delta S_{\text{H}_2\text{O}} + \Delta S_{\text{Fe}} = 8.15 - 7.83\ \text{J K}^{-1} = 0.32\ \text{J K}^{-1}$$

c) Using four reservoirs makes the process more nearly reversible as evidenced by the smaller ΔS_{univ}. Making the process exactly reversible would require an infinite series of reservoirs each one absorbing an infinitesimal quantity of heat from the iron at a temperature infinitesimally less than the previous reservoir. In such a case, the ΔS's of all the reservoirs would add up to only $+7.83$ J K^{-1}, and ΔS_{univ} for the process would be zero.

9-61 a) Several different ideal gases, each occupying its own original volume V_i and all at the same temperature and pressure, are allowed to mix. To get an expression for ΔS for this process, compute ΔS for each of the gases *separately* and then add up all of these contributions. Work with the i-th gas. This gas starts at V_1 and expands to V_2. In both of these states its entropy depends on the number of microstates Ω:

$$S_1 = k_{\text{B}} \ln \Omega_1 \qquad S_2 = k_{\text{B}} \ln \Omega_2$$

The *change* in entropy of the i-th gas is:

$$\Delta S_i = S_2 - S_1 = k_{\text{B}} \ln \left(\frac{\Omega_2}{\Omega_1}\right)$$

Both before and after the expansion the number of microstates available to one molecule of the gas is proportional to the volume. The number of microstates available to *all* the molecules is proportional to the volume raised to a power $n_i N_0$, the total number of molecules. where N_0 is Avogadro's number, and n_i is the number of moles of the i-th gas. The change in entropy for the i-th gas now is:

$$\Delta S_i = k_B \ln(cV_2)^{n_i N_0} - k_B \ln(cV_1)^{n_i N_0} = n_i N_0 k_B \ln\left(\frac{V_2}{V_1}\right)$$

Now, focus on the term (V_2/V_1). By Boyle's law, it is equal to (P_1/P_2), the ratio of the original pressure of the i-th gas to the final partial pressure of the i-th gas in the mixture. This latter pressure is, by Dalton's law:

$$P_2 = X_i P_{\text{tot}}$$

where X_i is the mole fraction of the i-th gas. But P_1, the original pressure of the i-th gas, *equals* P_{tot}. All of the gases started at the same pressure. Therefore:

$$\frac{V_2}{V_1} = \frac{P_1}{P_2} = \frac{P_{\text{tot}}}{X_i P_{\text{tot}}} = \frac{1}{X_i}$$

Substituting this result into the expression for ΔS_i gives:

$$\Delta S_i = n_i N_0 k_B \ln\left(\frac{1}{X_i}\right) = -n_i N_0 k_B \ln X_i$$

Next, $N_0 k_B$ is equal to R and n_i is equal to $X_i n$, where n is the total number of moles of gas. Inserting these values gives:

$$\Delta S_i = -X_i n\, N_0 k_B \;\ln X_i = -nRX_i \ln X_i$$

Finally, add up the contributions of all of the gases to get the overall ΔS:

$$\Delta S = \sum_i \Delta S_i = -nR \sum_i X_i \ln X_i$$

b) This part requires first the calculation of the total chemical amount of mixed gas and the mole fractions of O_2, N_2 and Ar in the mixture. Beyond that, there is nothing more than careful substitution in the formula derived in part a). Dividing 50 g by the molar masses of the three gases gives the chemical amount of each. The mole fraction of each is its chemical amount divided by the total chemical amount in the mixture. The results of these calculations are:

Gas	Chemical Amount	Mole Fraction	$X \ln X$
O_2	1.563 mol	0.3399	−0.3668
N_2	1.784	0.3879	−0.3673
Ar	1.252	0.2722	−0.3542

The total chemical amount of gas is, of course, the sum of the numbers in the second column of the above table. It is 4.6 mol. The sum of the entries in the *last* column of the table is the summation term in the equation:

$$\Delta S = -nR\sum_i X_i \ln X_i$$

Using $R = 8.315$ J $K^{-1}mol^{-1}$ gives a ΔS of 42 J K^{-1}. This is the entropy change of mixing at any temperature and pressure as long as the assumption of ideal gas behavior holds. It is not necessary to know that the mixing occurs as STP.
c) Separating the components of air is the reverse of mixing them. The entropy change of separation is therefore the negative of the entropy change of mixing. To solve the problem, compute ΔS_{sys} for the process of mixing and then change its sign. Table 3-1 (text page 88) gives the volume percentages of the various gases in the air. Assuming that the ideal gas equation holds, the mole fractions (X's) of the gases are just these numbers divided by 100. In the following table the mole fraction is written and the quantity $X \ln X$ is calculated for each gas:

Gas	Mole Fraction (X)	$X \ln X$
N_2	0.78110	−0.19297
O_2	0.20953	−0.32747
Ar	0.00934	−0.04365
Ne	0.00001818	−0.0001976

Continuing the table to include atmospheric gases of smaller concentration than that of Ne will not provide $X \ln X$ values significantly different from zero. The sum of the values in the last column is −0.56429. The entropy change of mixing is $-nR$ times this number. With a volume of 100 L at 298.15 K and 1 atm, the chemical amount n of air is 4.09 mol, according to the ideal-gas equation. The entropy change of mixing therefore is +19.2 J K^{-1}. The entropy change of separation of the components *of the system* is −19.2 J K^{-1}. The problem asks simply for the "entropy change." The total entropy change of the universe cannot be calculated because there is no information about the surroundings of the system. It is of course certain that ΔS of the universe exceeds zero.

9-63 The absolute entropy is proportional to the area under such a curve. Gold has the higher absolute entropy at 200 K.

9-65 a) Higher temperature makes the conversion of rhombic to monoclinic sulfur a spontaneous process. According to the discussion on text page 332, both $\Delta H°$ and $\Delta S°$ must then be positive: if the two had unlike signs, then there would be no "cross-over temperature"; if the two were both negative, then the conversion would be favored by lower temperature.

b) Equilibrium between rhombic and monoclinic sulfur at constant temperature and pressure means $\Delta G = \Delta H - T\Delta S = 0$. Rearranging and substituting the values from the problem gives:

$$\Delta S = \frac{\Delta H}{T} = \frac{400 \text{ J}}{368.5 \text{ K}} \quad \text{so} \quad \Delta S = 1.09 \text{ J K}^{-1}$$

9-67 The reaction is $3\,CO_2(g) + Si_3N_4(s) \rightarrow 3\,SiO_2(s) + 2\,N_2(g) + 3\,C(s)$. The problem and Appendix D supply the necessary ΔG_f° data for the computation:

$$\Delta G^\circ = 3\underbrace{(-856.67)}_{SiO_2(s)} + 2\underbrace{(0)}_{N_2(g)} + 3\underbrace{(0)}_{C(s)} - 3\underbrace{(-394.36)}_{CO_2(g)} - 1\underbrace{(-642.6)}_{Si_3N_4(s)} = -744.3 \text{ kJ}$$

9-69 a) The reaction of interest is $2CuCl_2(s) \rightarrow 2\,CuCl(s) + Cl_2(g)$. Appendix D supplies ΔH_f° and S° values for the computation of its ΔH° and ΔS°

$$\Delta H^\circ = 2\underbrace{(-137.2)}_{CuCl(s)} + 1\underbrace{(0)}_{Cl_2(g)} - 2\underbrace{(-220.1)}_{CuCl_2(s)} = 165.8 \text{ kJ}$$

$$\Delta S^\circ = 2\underbrace{(86.2)}_{CuCl(s)} + 1\underbrace{(222.96)}_{Cl_2(g)} - 2\underbrace{(108.07)}_{CuCl_2(s)} = 179.2 \text{ J K}^{-1}$$

b) $\Delta G_{590} = \Delta H^\circ - T\Delta S^\circ = 165.8 \text{ kJ} - (590 \text{ K})(0.1792 \text{ kJ K}^{-1}) = 60.1 \text{ kJ}$
c) Use the experimental values at 590 K instead of assuming that the values at 298 K still are exactly correct:

$$\Delta G_{590} = \Delta H_{590} - T\Delta S_{590} = 158.36 \text{ kJ} - (590 \text{ K})(0.17774 \text{ kJ K}^{-1}) = 53.5 \text{ kJ}$$

The answer using ΔH° and ΔS° is about 12% larger than the actual ΔG.

9-71 The plan is to use the ΔG_f°'s from Appendix D to calculate ΔG° for the reaction, then use ΔG° to calculate the equilibrium constant for the reaction. The equilibrium constant is related to the partial pressure of oxygen through the equilibrium constant expression. First the ΔG°:

$$\Delta G^\circ = 1\underbrace{(0)}_{O_2(g)} + 1/2\underbrace{(0)}_{O_2(g)} - 1\underbrace{(-211.7)}_{NiO(s)} = 211.7 \text{ kJ}$$

THen the K:

$$\ln K = \frac{-\Delta G^\circ}{RT} = \frac{-(211.7 \times 10^3 \text{ J mol}^{-1})}{(8.315 \text{ J K}^{-1}\text{mol}^{-1})(298.15 \text{ K})} = -85.44; \quad K = 7.8 \times 10^{-38}$$

Because two of the three substances involved in the reaction are pure solids, the equilibrium constant expression is very simple: $\sqrt{P_{O_2}} = K$. It follows that the equilibrium pressure of oxygen in this system is the square of the equilibrium constant. It equals 6.1×10^{-75} atm. The decomposition of NiO(s) to its elements occurs to a *very* slight extent at room temperature!

9-73 a) The proper numbers to plot are the second-to-last and last lines of the following table. The $\ln K$ goes on the vertical axis and $1/T$ goes on the horizontal axis.

T (K)	276.8	288.3	298	308	323.2
K	1160	841	689	533	409
$1/T$ (K^{-1})	0.00361	0.00347	0.00336	0.00325	0.00309
$\ln K$	7.06	6.73	6.54	6.28	6.01

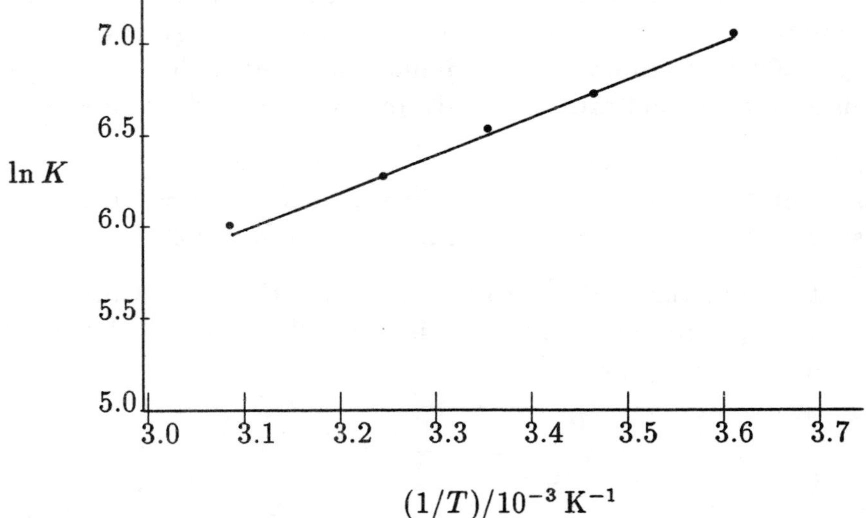

The van't Hoff equation is:

$$\ln K = \frac{-\Delta H^\circ}{R}\left(\frac{1}{T}\right) + \frac{\Delta S^\circ}{R}$$

Comparing this equation to the general equation for a straight line ($y = mx + b$) reveals that the slope m of the straight line that results from plotting $\ln K$ versus $1/T$ is $-\Delta H^\circ/R$. The slope m of the best straight line in the graph is 2020.3 K. Therefore:

$$\Delta H^\circ = -mR = -(2020.3 \text{ K})(8.315 \text{ J K}^{-1}\text{mol}^{-1}) = -16.8 \times 10^3 \text{ J mol}^{-1}$$

9-75 Let the original pressure of $N_2(g)$ before any of its molecules break down equal P_0. Let the fraction that has broken down at equilibrium equal α. The value of α is 0.0065 at 5000 K but rises to 0.116 at 6000 K, according to the problem. Represent the approach to equilibrium in the usual way:

	$N_2(g) \rightleftharpoons$	$2\,N(g))$
Init. Pressure (atm)	P_0	0
Change in Pressure (atm)	$-\alpha P_0$	$+2\alpha P_0$
Equil. Pressure (atm)	$P_0 - \alpha P_0$	$2\alpha P_0$

The *total* pressure of the equilibrium mixture is 1.000 atm at both 5000 K and at 6000 K:

$$P_{N_2} + P_N = (P_0 - \alpha P_0) + 2\alpha P_0 = P_0(1+\alpha) = 1.000 \text{ atm}$$

Since α is given for both temperatures, the original pressure of N_2 at both temperatures is readily computed from the preceding equation: P_0 was 0.9935 atm at 5000 K and 0.8961 atm at 6000 K. Combine these original pressures with α as indicated above to get the equilibrium partial pressures of N and N_2 at the two temperatures:

Temperature	5000 K	6000 K
Equil. Pressure of N_2 (atm)	$P_0(1-\alpha) = 0.9871$	$P_0(1-\alpha) = 0.7921$
Equil. Pressure of N (atm)	$2\alpha P_0 = 0.0129$	$2\alpha P_0 = 0.2079$

It is easy to verify that the pressures add up to 1.00 atm at both temperatures. These equilibrium partial pressures are important because they allow the calculation of K of the reaction at the two temperatures:

$$\frac{P_N^2}{P_{N_2}} = K$$

Substitution gives K_{5000} equal to 1.69×10^{-4} and K_{6000} equal to 5.46×10^{-2}. Next, use the van't Hoff equation to estimate ΔH for the reaction from the two K's and their temperatures:

$$\ln \frac{K_2}{K_1} = -\frac{\Delta H}{R}\left(\frac{1}{T_2} - \frac{1}{T_1}\right)$$

A ΔH appears in the preceding (instead of $\Delta H°$) because one need only assume that ΔH for the dissociation is constant between 5000 and 6000 K and not all the way from 298 to 6000 K. Substitute:

$$\ln\left(\frac{5.46 \times 10^{-2}}{1.69 \times 10^{-4}}\right) = 5.78 = -\frac{\Delta H}{8.315 \text{ J K}^{-1}\text{mol}^{-1}}\left(\frac{1}{6000 \text{ K}} - \frac{1}{5000 \text{ K}}\right)$$

The computed ΔH is 1.44×10^3 kJ mol^{-1} only an estimate assuming that ΔH and ΔS are constant over the big span between 5000 K and 6000 K is likely to be wrong.

9-77 As shown on text page 342, the following relationship holds for the boiling-point elevation of dilute solutions:

$$\Delta T_b = \frac{RT_b^2}{\Delta H_{vap}} X_2$$

where T_b is the boiling point of the pure solvent, ΔH_{vap} is the enthalpy of vaporization of the pure solvent, and X_2 is the mole fraction of the solute. Dividing both sides of this equation by T_b and recognizing that $T_b/\Delta H_{vap}$ is equal to $1/\Delta S_{vap}$ gives:

$$\frac{\Delta T_b}{T_b} = R\left(\frac{T_b}{\Delta H_{vap}}\right) X_2 = R\left(\frac{1}{\Delta S_{vap}}\right) X_2 = \left(\frac{R}{\Delta S_{vap}}\right) X_2$$

Comparison of this result with the equation given in the problem (which was $\Delta T_b/T_b = CX_2$) establishes C:

$$C = \frac{R}{\Delta S_{vap}} = \frac{8.315 \text{ J K}^{-1}\text{mol}^{-1}}{88 \text{ J K}^{-1}\text{mol}^{-1}} = 0.094$$

As a combination of other constants, C is unrelated to the properties of either solute or solvent.

Chapter 10

Electrochemistry

Balancing Oxidation-Reduction Equations

The criteria for balancing such equations are the same as for other chemical equations: the number of each kind of atom shown may not change from left to right nor may the net electrical charge change from left to right. Two situations are common:

- All of the reactants and products are specified. The whole task is then finding a set of coefficients. This is entirely a mechanical exercise. The only chemical knowledge required is familiarity with chemical symbolism.

- The reaction takes place in aqueous solution where $H_2O(l)$, $H_3O^+(aq)$, and $OH^-(aq)$ may enter the equation as either reactants or products. Such cases, which require the completion of an equation in addition to balancing, require additional chemical knowledge.

When adding reactants or products to equations to complete them, bear certain ideas in mind:

- The formula $H^+(aq)$ is equivalent to $H_3O^+(aq)$. The two are both valid representations of the hydrogen ion in aqueous solution. The second is the first with an H_2O added in. The textbook uses mostly H_3O^+, but either is acceptable, based on convenience.

- In acid solution the concentration of $OH^-(aq)$ is low. It is therefore not realistic to use it as a reactant in a balanced equation. Instead use H_2O as a source of O in the -2 oxidation state. If $OH^-(aq)$ is formed in an oxidation-reduction reaction in acidic solution, then it will react with the plentiful $H_3O^+(aq)$ ion to give water. Recognize this fact in balanced equations.

- In basic solution, $H_3O^+(aq)$ is scarce and should not be used as a source of hydrogen in the +1 oxidation state. Instead use H_2O. If $H_3O^+(aq)$ forms as a product, then it will react with the plentiful $OH^-(aq)$ to give water. Recognize this in balanced equations.

The text gives a set of steps to balance the oxidation-reduction equations in either case. These steps are built around the concept of **half-reactions**. A half-reaction is represented by a half-equation in which electrons appear explicitly as reactants (reduction half-equations) or as products (oxidation half-equations). To arrive at an overall equation, write a balanced oxidation half-equation that releases the same number of electrons that is absorbed by a balanced reduction half-equation. Then to add the two. The electrons cancel out. See **problem 10-1**. Note certain common errors in balancing redox equations:

- Inserting H_2, O_2, H_2O_2 and other species in which H and O have oxidation numbers other than +1 and −2 respectively in attempting to complete half-equations. This always causes failure.

- Failing to recognize that two or three or even more elements in a compound can change oxidation number in the same reaction.

- Not understanding that a single substance can *disproportionate,* that is, be both oxidized and reduced in the same reaction.

Disproportionation

Some species can reduce and oxidize themselves. This means that some of the molecules (or atoms or ions) of the species lose electrons to give one set of products and other identical molecules (or atoms or ions) gain electrons to give a different set of products. The phenomenon is called **disproportionation**. For example, the $Cu^+(aq)$ ion disproportionates to give $Cu(s)$, and $Cu^{2+}(aq)$ ion. Once it is understood that the same species can "go both ways," dealing with equations if which there is disproportionation presents no special problems. The oxidation half-equation and the reduction half-equation show the same species on the left. The coefficients of these species are added together when the half-equations are combined to make a full equation.

Electrochemical Cells

Oxidation-reduction reactions can occur without direct contact between the reactants. In an **electrochemical cell**, oxidation and reduction go on at **electrodes** that are well separated in space. One electrode collects the electrons from the species being oxidized. The electrons pass through a wire and are delivered to the species being reduced at a second electrode. The oxidation and reduction half-reactions are said to occur in two compartments of the cell. The reduction half-reaction occurs at the **cathode** and the oxidation half-reaction occurs at the **anode.** The electrodes are in contact with an **electrolyte,** often an aqueous solution.

The transfer of electrons in an electrochemical cell quickly stops unless some means of maintaining electrical neutrality within the electrolyte is provided. The design of electrochemical cells therefore includes a **salt bridge** to allow the transfer of counter-ions from the vicinity of one electrode to the vicinity of the other. The diagram shows the essential parts of an electrochemical cell:

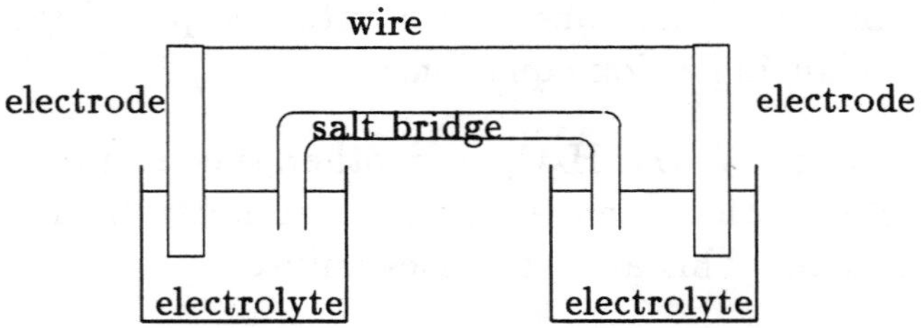

Suppose the electrode on the left is a piece of metallic magnesium dipping into a solution of $Mg(NO_3)_2$ and that the electrode on the right is a piece of metallic silver dipping into a solution of $AgNO_3$. At the moment the connection is completed by attaching the wire, magnesium begins to be oxidized. Atoms of magnesium into solution in the anode compartment (on the left) as Mg^{2+} ions. Simultaneously, NO_3^- ions flow from right to left in the salt bridge to maintain electrical neutrality in the anode compartment (see **problem 10-9).** Silver ions are reduced in the cathode compartment; metallic silver plates out on the cathode.

Suppose that the two solutions are both 1.00 M. This quickly changes as electrons flow. The concentration of Mg^{2+} ion in the left compartment increases from 1.00 M; the concentration of Ag^+ ion in the right compartment decreases from 1.00 M. If an **ammeter** is inserted into the line connecting the two electrodes, the direction and magnitude of the flow of electrons through the wire can be measured.

Galvanic and Electrolytic Cells

If a **voltmeter** is inserted into the line connecting the two electrodes, it is possible to measure the **difference in electrical potential** $\Delta\mathcal{E}$ between the two sides of the electrochemical cell. Differences in electrical potential are measured in **volts** and are often called **voltages.** They are easily measured and are related to the tendency of the reaction spontaneously to proceed. This is quite different from the size of the electrical current through the outside wire, which is a measure of the rate of the reaction. A positive difference in potential means there is an intrinsic "push" for the transfer of electrons through the wire from one side to the other: oxidation in one compartment and reduction in the other compartment tend to occur. An electrochemical cell operating spontaneously is a **galvanic cell.** A negative potential difference under these circumstances would mean that the reaction was trying to push electrons from through the wire in the reverse direction.

It is possible to force a spontaneous reaction to run in a reverse direction by applying a sufficiently large outside voltage. An electrochemical cell in which a non-spontaneous reaction is being forced to occur by the application of a potential difference from outside the cell is an **electrolytic cell.** Such a cell uses electrical energy to carry out chemical reactions that would not otherwise occur.

Both galvanic and electrolytic cells employ oxidation-reduction reactions. For example, the overall reaction:

$$\mathrm{Cu}^{2+}(aq) + Zn(s) \rightarrow \mathrm{Cu(s)} + \mathrm{Zn}^{2+}(aq)$$

is the sum of the half-reactions:

$$\begin{array}{ll} \mathrm{Cu}^{2+}(aq) + 2e^- \rightarrow \mathrm{Cu}(s) & \text{reduction} \\ \mathrm{Zn}(s) \rightarrow \mathrm{Zn}^{2+}(aq) + 2e^- & \text{oxidation} \end{array}$$

Breaking a balanced equation down into balanced half-equations is an important skill:

1. Determine the oxidation numbers of all elements in all compounds to learn which elements are oxidized and which reduced.

2. Write down an oxidation half-equation and a reduction half-equation separately.

3. Insert electrons so that the two sides of the resultant half-equations have the same net charge.

Also, writing half-equations helps in balancing oxidation-reduction reactions.

The following are useful generalizations about half-equations:

- The oxidation half-reaction always involves the *loss* of electrons. It always takes place at the *anode* of an electrochemical cell.

- The reduction half-reaction always involves the *gain* of electrons. It always takes place at the *cathode* of an electrochemical cell.

- Electrons liberated at the anode pass through the outside circuit and are consumed by the reduction half-reaction at the cathode. The number of electrons liberated by the oxidation must exactly equal the number absorbed by the reduction.

In solving problems it is vital to know the definitions of cathode and anode. Thus, in **problem 10-1,** the only way to know which half-reaction occurs at which electrode is to recognize which is an oxidation and to know that oxidation always occurs at the anode:

$$\textbf{Oxidation} \Longleftrightarrow \textbf{Anode} \qquad \textbf{Reduction} \Longleftrightarrow \textbf{Cathode}$$

Do not bother to learn the sign of the electrical polarity of cathode and anode in electrochemical cells. It depends on whether the cell is galvanic or electrolytic. It is a mistake to memorize that "the cathode is negative and the anode is positive." The statement is true enough for electrolytic cells, but exactly wrong for galvanic cells.

Faraday's Laws

When channeled through an outside circuit, electrons are easily regulated. Their rate of flow (the **electric current**) is readily measured with an ammeter and easily varied by changing the electrical resistance of the outside circuit. Electric current is measured in amperes (A). A current of one ampere is one coulomb flowing through an electrical circuit every second:

$$I = \frac{Q}{t}$$

It is easy to measure electrical current (with an ammeter). Consequently, this definition crops up extensively in problems. (see **problem 10-17b.)**

The electron is in effect just another chemical reactant. **Faraday's laws** apply the rules of stoichiometry to the electron as chemical reactant (or product). The laws are:

1. In any cell the mass of a given substance produced or consumed at an electrode is proportional to the quantity of electrical charge Q passed through the cell.
2. Equivalent masses of different substances are produced or consumed at an electrode by the passage of a given quantity of electrical charge through the cell.

Once the electron is accepted as a chemical reactant a mole of electrons assumes the same role in chemical calculations as a mole of sodium or $ZnCl_2$ or other substance. **See problem 5b.** Although electrons cannot be weighed out, they are easily tracked because they are charged. **A mole of electrons is 96485.31 coulomb.** This quantity of charge is the **Faraday constant $\mathcal{F}$:**

$$\mathcal{F} = 96485.31 \text{ C mol}^{-1}$$

Numerous problems use Faraday's laws. Observe the following key points in applying the laws:

- The quantity of charge passing through a cell is the average current times the time it flows. If t is in seconds (s) and I in amperes (A), then their product is the quantity of electricity (total electrical charge) passing through the cell in coulombs (C). See **problem 10-13b.**

- To convert a quantity of electricity from electrical units (coulombs) to chemical units (moles) use the faraday constant: memorize the fact that 1 mole is 96 485 C.

- Balanced half-equations show the chemical changes at the electrodes of a cell. Treat the electron in a half-equation just like any other reactant or product. In solving stoichiometry problems, use half-equations on the same basis as full equations; treat electrons just like any other chemical substance. See **problems 10-11, 10-13c, 10-19,** and **10-53.**

- An **equivalent mass** of a substance is the amount produced or consumed in an electrochemical cell by the passage of 1 mol of electrons (96 485 C). The following table shows the relationship of molar mass and equivalent mass:

Reaction	Molar Mass	Equivalent Mass
$Ag^+(aq) + e^- \rightarrow Ag(s)$	107.9 g mol^{-1}	107.9 g per equiv
$Cu^{2+}(aq) + 2e^- \rightarrow Cu(s)$	63.54	31.8
$Al^{3+}(aq) + 3e^- \rightarrow Al(s)$	26.98	9.0
$2\,Cl^-(aq) \rightarrow Cl_2(g) + 2e^-$	70.91	35.5

Problem 10-15 involves equivalent masses.

For examples of the use of Faraday's laws see **problems 10-17, 10-69** and **10-71.**

Gibbs Free Energy and Cell Voltage

Usually, the only kind of work that chemical systems can do on their surroundings is pressure-volume work (text page 276-7). Electrochemical cells are the major exception. They are able to produce or consume **electrical work,** in addition to pressure-volume work. A galvanic cell is a chemical system that is not at equilibrium. It develops a difference in electrical potential between its two electrodes. The potential difference ($\Delta\mathcal{E}$) is measured in volts (V). It represents the intrinsic tendency of the cell to come to equilibrium. Completing an electrical circuit between the two electrodes allows the spontaneous flow of electricity from one to the other. If suitably harnessed, a flow of electricity can perform work. The amount of electrical work possible when an amount of charge $\mathcal{Q}$ moves through a potential difference $\Delta\mathcal{E}$ is:

$$w_{\text{elec}} = -\mathcal{Q}\Delta\mathcal{E}$$

This equation defines the joule in terms of electrical quantities. The work w is in joules when Q is in coulombs (C) and $\Delta\mathcal{E}$ is in volts (V): **a joule (J) is the amount of energy required to push one coulomb of charge through a potential difference of one volt**

$$1 \text{ joule} = 1 \text{ volt}\cdot\text{coulomb}$$

Combining the equation for w_{elec} with the definition of electrical current I gives:

$$w_{\text{elec}} = -It\Delta\mathcal{E}$$

From this equation, a joule is an ampere·volt·second.

Redox reactions are either spontaneous or non-spontaneous. Only a spontaneous oxidation-reduction reaction can generate a measurable potential difference (voltage) between the two electrodes of electrochemical cell. This $\Delta\mathcal{E}$ is related to the free-energy change of the reaction in the cell:

$$\boldsymbol{\Delta G = -n\mathcal{F}\Delta\mathcal{E}}$$

Again, $\mathcal{F}$ is the faraday constant, and n is the chemical amount of electrons transferred. The ΔG comes out in joules if $\Delta\mathcal{E}$ is in volts, $\mathcal{F}$ $\mathcal{F}$ is in C mol^{-1}, and n is in moles. However, ΔG comes out in J mol^{-1} if n is regarded as the number of moles of electrons transferred per mole of chemical reaction. The two results echo the distinction between an extensive and an intensive property. Running an redox reaction in an industrial-scale cell transfers many thousands of moles of electrons. Doing the same reaction in a single small cell transfers only a few tenths of a mole of electrons even though the balanced reaction is the same in both cases. The *extensive* property ΔG (measured in J) is large in the first case but small in the second. If n is viewed as the number of moles of electrons transferred *per mole of reaction,* then its units are "mol mol^{-1}", making it unit-less and causing ΔG to come out in J mol^{-1}. A ΔG in J mol^{-1} refers to a chemical reaction rather than to events in an actual physical cell. It is an *intensive* property.

In an electrolytic cell the chemical reaction is non-spontaneous. This corresponds to a positive ΔG and therefore to a negative $\Delta\mathcal{E}$. A negative $\Delta\mathcal{E}$ means a positive voltage must be impressed from outside to force the reaction to run. The reverse reaction has a positive $\Delta\mathcal{E}$ and runs spontaneously.

It is not necessary to allow any current to flow in a galvanic cell to measure its voltage. If no current flows, then no chemical reaction occurs. The potential difference between the electrodes then tells the *tendency* or *drive* for the chemical reaction to occur. It is thus quite well-named.

If current is allowed to flow, a chemical change take place. Only then does a galvanic cell have even a chance of doing any electrical work on its surroundings. If the current flows infinitely slowly, the chemical change goes reversibly and can produce the maximum electrical work. Continue to focus on the work produced. Recall that the work produced by a system is $-w$ because $+w$ is the work absorbed. Then:

$$-w_{\text{elec, max}} = -\Delta G = n\mathcal{F}\Delta\mathcal{E}$$

This equation is used to solve **problem 10-19.** The operation of a practical cell is always irreversible and produces *less* than $-w_{\text{max}}$:

$$-w_{\text{elec, irrev}} < -\Delta G$$

If some galvanic cell has a ΔG of -25 kJ, then, by the above equations, it can produce a maximum of $+25$ kJ of electrical work. The catch is that extracting this maximum work would require reversible operation, which is an unattainable ideal. In actual operation, *less* than 25 kJ of electrical work would appear in the surroundings. Depending on how the output of the cell was used, as little as 0 J of electrical work could appear. All of free energy change might be diverted to simple resistive heating, for example.

Standard States and Cell Voltages

The above equation applies to cell reactions going on at any combination of T and P as long as T and P are constant. It certainly also applies when the products and reactants in the equation are in their standard states at a temperature of 298.15 K and a pressure of 1 atm. Under these restricted circumstances, the superscript zero (naught) is added to the symbols:

$$\Delta G^\circ = -n\mathcal{F}\Delta\mathcal{E}^\circ$$

Solutes are in their standard states when their concentration is 1 M under the standard conditions. Gases are in their standard states when their partial pressure is 1.00 atm.

The above equation means that measurement of $\Delta\mathcal{E}^\circ$ values gives equilibrium constants:

$$\boldsymbol{\Delta G^\circ = -RT\ln K = -n\mathcal{F}\Delta\mathcal{E}^\circ} \text{ from which } \boldsymbol{\ln K = \frac{n\mathcal{F}\Delta\mathcal{E}^\circ}{RT}}$$

This relationship is used in **problem 10-43.** In practice, $\Delta\mathcal{E}^\circ$ is measured in volts. Multiplication by $\mathcal{F}$ (which is 96485 C mol^{-1}) and n, the number of moles of electrons transferred per mole of reaction as written, gives ΔG° in J mol^{-1}. Division of this value by $-R$ (which is 8.315 J K^{-1}mol^{-1}) and T on the Kelvin scale gives $\ln K$.

Standard Reduction Potentials

Appendix E (text) tabulates **standard reduction potentials $\mathcal{E}^\circ$**. Each standard potential is associated with a half-equation. The potentials are thus for *half-cells.* A half-cell can be visualized as an electrode compartment and its contents in an electrochemical cell.

The data in Appendix E are needed in many problems, so the structure of the table must be understood:

1. The $\mathcal{E}^\circ$'s are standard *reduction* potentials. They refer to reduction half-reactions (those having half-equations with electrons shown on the *left*). **The larger the reduction potential of a species, the greater is its relative tendency to be reduced.**

2. A species on the *left* of a given half-cell reaction and in its standard state will spontaneously oxidize a standard-state species on the *right* of any half-cell reaction located *below* it in the table.

3. A reduction potential becomes an oxidation potential when the direction of a half-cell reaction is reversed. Simply change the sign of the associated potential. All half-cell reactions may be written in the reverse direction. The larger the oxidation potential of a species then the greater is its relative tendency to be oxidized. This point is used in **problem 10-77.**

4. The standard reduction potentials are *intensive* properties. They are not affected when the coefficients in the half-reaction are changed. **Example:** The half-reaction $Cu^{2+}(aq) + 2e^- \rightarrow Cu(s)$ has the same $\mathcal{E}^\circ$ as $2\,Cu^{2+}(aq) + 4e^- \rightarrow 2\,Cu(s)$

5. A standard half-cell reduction potential, $\mathcal{E}^\circ$, cannot be directly measured. Listings of such potentials are derived by measuring $\Delta\mathcal{E}^\circ$ values for different pairings of half-reactions and arbitrarily assigning a $\mathcal{E}^\circ$ of exactly 0 V to the *reference* half-reaction: $H_3O^+(aq) + e^- \rightarrow 1/2\,H_2(g) + H_2O(l)$.

The standard potential difference developed by a galvanic cell is the *difference* between the standard reduction potential of the half-reaction taking place at the cathode and that of the half-reaction taking place at the anode:

$$\boldsymbol{\Delta\mathcal{E}^\circ = \mathcal{E}^\circ(\text{cathode}) - \mathcal{E}^\circ(\text{anode})}$$
$$\boldsymbol{\Delta\mathcal{E}^\circ = \mathcal{E}^\circ(\text{reduction}) - \mathcal{E}^\circ(\text{oxidation})}$$

The procedure for getting $\Delta\mathcal{E}^\circ$ in applications is:

1. Balance the overall oxidation-reduction equation.

2. Break down the equation into a balanced reduction half-equation and a balanced oxidation half-equation.

3. Find these half-equations in Appendix E and note their standard reduction potentials. The oxidation half-reaction will be the *reverse* of one of the tabulated half-reactions. **Caution:** Find the correct half-equations, including the physical state of all reactants and products. Some half-equations resemble each other fairly closely, but have quite different standard reduction potentials.

4. Subtract the tabulated reduction potential, $\mathcal{E}^\circ$, for the half-reaction taking place at the anode (the oxidation) from the tabulated reduction potential for the half-reaction taking place at the cathode (the reduction).

This procedure is followed in **problems 10-21, 10-25b, 10-35, 10-45,** and **10-49.**

Addition and Subtraction of Half-Cell Reactions

As just pointed out, when two half-equations are combined to give a whole equation, in which no electrons appear explicitly, then the $\mathcal{E}^\circ$ values are combined by simple subtraction:

$$\Delta\mathcal{E}^\circ = \mathcal{E}^\circ(\text{cathode}) - \mathcal{E}^\circ(\text{anode})$$

If $\Delta\mathcal{E}^\circ$ comes out negative, then the reaction is **non-spontaneous** as written and the reverse reaction is spontaneous.

When two half-equations are combined to make a new *half-equation* the calculation is not so simple. Then:

$$\mathcal{E}_3^\circ = \frac{n_1\mathcal{E}_1^\circ - n_2\mathcal{E}_2^\circ}{n_3}$$

where the subscripts 1 and 2 refer to the reduction potentials and quantities of electrons transferred in the two half-reactions being combined, and where subscript 3 refers to the resultant half-reaction. See **problem 10-25a.** Note that this equation, if used when half-reactions combine to make a whole reaction, has $n_1 = n_2 = n_3$, and the simple subtractive relationship already cited is the result. See **problem 10-27.**

Oxidizing and Reducing Agents

This section introduces a useful tool in studying the descriptive chemistry of the elements. It is the **reduction potential diagram.** Such diagrams consist of a list of species containing a particular element either by itself or in combination with hydrogen, oxygen or both. The diagrams always concern reactions taking place in aqueous solution. The species are written in order of decreasing oxidation state of the subject element from left to right across the page. Lines connecting pairs of species stand for the properly balanced half-equation for the reduction of the form on the left to the one on the right. The corresponding standard reduction potentials appear above the lines.

The reduction potentials are for half-reactions going on in water. The pH of the solution sometimes affects reduction potentials sharply. Reduction potential diagrams always include a statement of whether the solution is acidic (pH 0) or basic (pH 14). The balanced half-reactions must take the relative availability of $H_3O^+(aq)$ or $OH^-(aq)$ into account, too.

It is essential to be able to determine the oxidation numbers of all the elements appearing in a reduction potential diagram. This skill is a preliminary to computing even *more* reduction potentials. Thus, the compactly displayed information in a reduction potential diagram can be expanded into many balanced chemical half-equations and equations. **Example:** The reduction potential diagram for the element oxygen in acid is given on text page 369. The reduction potential diagram in *base* is:

$$O_3 \xrightarrow{1.24\text{ V}} O_2 \xrightarrow{-0.086\text{ V}} HO_2^- \xrightarrow{0.87\text{ V}} OH^- \qquad \text{(base pH 14)}$$

Write out and balance half-equations for *all* of the half-reactions represented either explicitly or implicitly in the diagram and state their standard reduction potentials. **Solution:** Four different species appear in this reduction potential diagram. Therefore, there are six possible reduction half-equations. In base, $H_3O^+(aq)$ is scarce. Only $H_2O(l)$ and $OH^-(aq)$ should appear explicitly in the balanced half-equations. Balance the three half-reactions for which the reduction potential appears in the diagram:

$$\begin{array}{ll} O_3(g) + H_2O(l) + 2\,e^- \rightarrow O_2(g) + 2\,OH^-(aq) & \mathcal{E}^\circ = 1.24\text{ V} \\ O_2(g) + H_2O(l) + 2\,e^- \rightarrow HO_2^-(aq) + OH^-(aq) & \mathcal{E}^\circ = -0.086\text{ V} \\ HO_2^-(aq) + H_2O(l) + 2\,e^- \rightarrow 3\,OH^-(aq) & \mathcal{E}^\circ = 0.87\text{ V} \end{array}$$

The first reduction consists of the oxygen atoms in $O_3(g)$ (ozone) gaining two electrons and simultaneously abstracting a H^+ from water to give the hydroxide ion.

Three more half-equations come from the pair-wise combinations of the first three:

$$\begin{array}{ll} O_3(g) + 3\,H_2O(l) + 6\,e^- \rightarrow 6\,OH^-(aq) & \mathcal{E}^\circ = 0.68\text{ V} \\ O_2(g) + 2\,H_2O(l) + 4\,e^- \rightarrow 4\,OH^-(aq) & \mathcal{E}^\circ = 0.40\text{ V} \\ O_3(g) + 2\,H_2O(l) + 4\,e^- \rightarrow HO_2{}^-(aq) + 3\,OH^-(aq) & \mathcal{E}^\circ = 0.58\text{ V} \end{array}$$

The reduction potentials were computed using the rules of Section 10-3. For instance, the reduction potential in the first half-reaction is:

$$\mathcal{E}^\circ = \frac{2(1.24) + 2(-0.086) + 2(0.87)}{6} = 0.68\text{ V}$$

This half-reaction involves the gaining of 6 electrons and is the sum of the three half-reactions given in the diagram, each of which involves a two-electron gain. Compare to **problem 10-27.**

Members of the second group of three half-equations and reduction potentials could be represented explicitly in the reduction potential diagram by lines directly connecting the species which react. This is seen in the diagram on text page 369.

The half-reactions represented in a reduction potential diagram can also be combined to give *whole* reactions.

The $\Delta\mathcal{E}°$ for a whole reaction is:

$$\Delta\mathcal{E}° = \mathcal{E}°(\text{cathode}) - \mathcal{E}°(\text{anode})$$

Thus, in base, $O_3(g)$ reacts with $HO_2^-(aq)$ to give $O_2(g)$ with a standard cell potential of $1.24-(-0.086) = 1.34$ V. Moreover, $OH_2^-(aq)$ reacts with itself with a cell potential $\Delta\mathcal{E}°$ of $0.87 - (-0.086) = 0.96$ V. The products are $OH^-(aq)$ and $O_2(g)$. This is a disproportionation.

• A species in a reduction potential diagram disproportionates spontaneously whenever a reduction potential on a line leading from it to the right exceeds a reduction potential on a line leading from it to the left.

The reduction potential diagram for oxygen in base (page 278 of this Guide) lists different voltages than the diagram for oxygen in acid (text page 369). The differences are not random. For example, the reduction potential for the reduction of ozone in *acid:*

$$O_3(g) + 2\,H_3O^+(aq) + 2\,e^- \longrightarrow O_2(g) + 2\,H_2O(1)$$

is 2.07 V, substantially higher than the 1.24 V reduction potential for the same reduction taking place in base:

$$O_3(g) + H_2O(l) + 2\,e^- \longrightarrow O_2(g) + 2\,OH^-(aq$$

Why is it much easier to reduce $O_3(g)$ in acid than in base? In base, the half-reaction has to generate OH^- in a solution with an OH^- concentration of 1 M. Making the solution acidic reduces the OH^- concentration and so favors the products (LeChatelier's principle). The larger voltage in acid is thus to be expected.

Eventually, as a basic solution is acidified, the OH^- disappears from among the products in the equation, and the best representation becomes the acid half-reaction, which has H_3O^+ on the left-hand side.

The voltage for the $O_3(g)$ to $O_2(g)$ reduction at pH 14 is computed by putting $[H_3O^+] = 10^{-14}$ M in the Nernst equation (see next section) for the half-reaction in acid. The voltage at pH 0 is computed by putting $[OH^-] = 10^{-14}$ M in the Nernst equation for the half-reaction in base.

In reduction potential diagrams, care is taken to represent the correct predominant species according to the pH. Thus, in standard acid (at pH 0), H_2O_2 is the major species containing oxygen in the -1 oxidation state. In standard base (pH 14) HO_2^-, the conjugate base of H_2O_2, is the predominant species.

Reduction potential diagrams are a compact means of communicating great quantities of chemical information. They can replace paragraphs of descriptive text. Often, reduction potential diagrams of members of a group in the periodic table are displayed one after another to highlight trends.

Concentration Effects and the Nernst Equation

Standard differences in potential calculated from Appendix E are exactly correct only for cells in which all of the reactants and products are present in their standard states (activity = 1). This is because potential difference depend on the *activity* of the reactants and products in the cell reaction as well as their identity. To compute $\Delta\mathcal{E}$ (without a superscript) requires a knowledge of the concentrations (in mol L^{-1}) of all the solutes in the chemical reaction taking place in the cell and the partial pressures (in atm) of all the gases. Accept these numbers as equal to the activities of the solutes and gases. Assume that the activities of pure solids and liquids are 1. Then apply the **Nernst equation:**

$$\Delta\mathcal{E} = \Delta\mathcal{E}^\circ - \frac{RT}{n\mathcal{F}} \ln Q$$

where Q is the reaction quotient of the chemical equation. The second term on the right side of the Nernst equation is a correction factor for the non-standard activities of the reactants and products in the reaction. Suppose a galvanic cell employs the reaction:

$$\mathrm{Mg}(s) + 2\,\mathrm{Ag}^+(aq) \rightarrow \mathrm{Mg}^{2+}(aq) + 2\,\mathrm{Ag}(s)$$

For standard potential difference $\Delta\mathcal{E}^\circ$ is $\mathcal{E}^\circ$ (cathode) minus $\mathcal{E}^\circ$ (anode). The expression for Q for this reaction is:

$$Q = \frac{[\mathrm{Mg}^{2+}]}{[\mathrm{Ag}^+]^2}$$

and the Nernst equation is:

$$\Delta\mathcal{E} = 3.175\ \mathrm{V} - \left(\frac{RT}{2\mathcal{F}}\right) \ln \frac{[\mathrm{Mg}^{2+}]}{[\mathrm{Ag}^+]^2}$$

Note that n is 2 because 2 mol of electrons is transferred per mole of the balanced oxidation-reduction reaction. If $[\mathrm{Mg}^{2+}] = [\mathrm{Ag}^+] = 1$ M, Q is equal to 1 and $\ln Q$ is zero. The $\Delta\mathcal{E}$ as it would be read from a voltmeter, then equals $\Delta\mathcal{E}^\circ$, which is 3.175

V. If the anode and cathode are connected by a wire, a current spontaneously flows through the wire, and the reaction proceeds. The concentration of Mg^{2+} increases and the concentration of Ag^+ decreases. As these changes take place, Q increases and $\Delta\mathcal{E}$ becomes less than 3.175 V. As current continues to flow, the $\Delta\mathcal{E}$ ultimately trickles all the way down to 0.00 V since in the Nernst equation an ever-larger correction factor is subtracted from 3.175 V.

At $\Delta\mathcal{E} = 0$ the cell is *dead* as far as further release of chemical free energy by this reaction is concerned. The ΔG of the system is zero because $\Delta\mathcal{E}$ is zero. The free energy of the cell is at a minimum. A dead galvanic cell is at equilibrium. Meanwhile, neither $\Delta\mathcal{E}^\circ$ nor ΔG° has changed.

At equilibrium the reaction quotient Q equals K, the equilibrium constant. Measurement of $\Delta\mathcal{E}^\circ$'s of cells gives K's because:

$$0.00 = \Delta\mathcal{E}^\circ - \frac{RT}{n\mathcal{F}} \ln K \quad \text{from which} \quad \ln K = \frac{n\mathcal{F}}{RT}\Delta\mathcal{E}^\circ$$

The calculation of a K is illustrated in **problem 10-43.**

In problem solving, a convenient form of the Nernst equation is:

$$\Delta\mathcal{E} = \Delta\mathcal{E}^\circ - \left(\frac{0.0592 \text{ V}}{n}\right) \log_{10} Q$$

This equation combines numerical values of R, T and $\mathcal{F}$ and converts from natural logarithms to base-10 logarithms. This specialized form should *not* be used unless the temperature is 298.15 K. Like ΔG, $\Delta\mathcal{E}$ is strongly dependent on temperature. Taking the magic number approach (shoe-horning "0.0592 V" into every problem) often results in ignoring this dependence. The Nernst equation occurs constantly in problems in electrochemistry.

The Nernst equation applies to half-reactions and their standard half-cell potentials just as well as it does to whole reactions. The half-reaction: $H_3O^+(aq) + e^- \rightarrow 1/2\,H_2(g) + H_2O(l)$ has the Nernst equation:

$$\mathcal{E} = \mathcal{E}^\circ - \frac{RT}{\mathcal{F}} \ln\left(\frac{P_{H_2}^{1/2}}{[H_3O^+]}\right)$$

Note the use of $\mathcal{E}^\circ$ instead of $\Delta\mathcal{E}^\circ$. This exact expression is used in **problem 10-63.**

The operation of the **pH meter** is an important application of the dependence of cell potential on concentration. A **reference** electrode and an H_3O^+(aq)-sensitive electrode are placed in the aqueous solution under investigation. The observed cell potential depends linearly on $[H_3O^+]$ in the solution, according to the Nernst equation. At 25°C:

$$\Delta\mathcal{E} = \Delta\mathcal{E}^\circ\,(\text{ref}) - 0.0592 \log[H_3O^+]$$

where $\Delta\mathcal{E}^\circ$ (ref) is a constant that depends on which half-reaction is selected for use in the reference electrode **Problem 10-47.**

Batteries and Fuel Cells

Galvanic cells **(batteries)** have many important applications. **Primary cells** generate electrical work but cannot be recharged. **Secondary cells,** or accumulators, are cells that generate electrical work and can also be recharged by using electrical energy from an external source to reverse the oxidation-reduction reaction that discharged them.

"Re-charging" does not involve putting electrons back into a cell. A cell does not in fact lose any electrons even if completely discharged. Instead it means forcing the electrons in the cell back to their former positions of higher chemical potential energy.

Important non-rechargeable galvanic cells are the **Leclanché cell** (see **problem 10-85**), the **alkaline dry cell,** and the **zinc-mercuric oxide cell.**

Rechargeable Batteries

The **lead storage battery** is a rechargeable battery. As it is discharged the following takes place:

$$\mathrm{Pb}(s) + \mathrm{SO_4^{2-}}(aq) \rightarrow \mathrm{PbSO_4}(s) + 2\,e^- \qquad \text{anode}$$
$$\mathrm{PbO_2}(s) + \mathrm{SO_4^{2-}}(aq) + 4\,\mathrm{H_3O^+}(aq) + 2\,e^- \rightarrow \mathrm{PbSO_4}(s) + 6\,\mathrm{H_2O} \qquad \text{cathode}$$
$$\mathrm{Pb}(s) + \mathrm{PbO_2}(s) + 2\,\mathrm{SO_4^{2-}}(aq) + 4\,\mathrm{H_3O^+}(aq) \rightarrow 2\,\mathrm{PbSO_4}(s) + 6\,\mathrm{H_2O}(l)$$

Other rechargeable batteries are the **nickel-cadmium cell**, which uses a basic electrolyte, and the **sodium-sulfur cell.**

Fuel Cells

Unlike a battery, a closed system to which additional quantities of reactants cannot be added, a **fuel cell** continuously "burns" fuel electrochemically. Fuel cells offer a theoretical advantage over traditional means of using fuels. When a fuel is burned in the usual way, the conversion of the heat released, $-\Delta H$, to work encounters thermodynamic limitations on its efficiency (text page 382-3). In a fuel cell the free energy of the chemical reaction is converted directly to electrical energy, and this limitation is evaded.

Corrosion and Its Prevention

Corrosion in metals can very often be understood as the operation of short-circuited electrochemical cells, in which stress or exposure to water and air creates cathodic and anodic regions in a single material and permits the oxidation and dissolution of the metal. **Passivation** protects metals against corrosion. Often, a thin coating of a metal oxide forms on the surface of the metal and slows further electrochemical reactions. Another way of preventing corrosion is to use a **sacrificial anode** made out of a metal that is oxidized preferentially to the metal being protected. See **problem 10-61.**

Electrolysis of Water and Aqueous Solutions

In pure water, $H_3O^+(aq)$ is not in its standard state (1 M) but instead has a concentration of 10^{-7} M. This lowers $\mathcal{E}$ for the half-reaction:

$$2\,H_3O^+(aq) + 2\,e^- \rightarrow H_2(g)\,1\text{ atm} + 2\,H_2O$$

from 0.00 V, the standard half-cell potential, to −0.414 V. Similarly, it lowers $\mathcal{E}$ for the reduction of oxygen to water:

$$1/2\,O_2(g)\,(1\text{ atm}) + 2\,H_3O^+(aq) + 2\,e^- \rightarrow 3\,H_2O(l)$$

from 1.229 V to 0.815 V. The two lowered reduction potentials were calculated using the Nernst equation. Both results make sense from the point of view of LeChatelier's principle. In the two reactions, reactant $H_3O^+(aq)$ is present in less than 1 M (standard) concentration so there is a decreased drive to the right.

The potential difference for the process composed by subtracting the second of these half-equations from the first equals the second voltage subtracted from the first:

$$H_2O(l) \rightarrow H_2(g)\,(1\text{ atm}) + 1/2\,O_2(g)\,(1\text{ atm}) \quad \Delta\mathcal{E} = -0.414 - (0.815) = -1.229\text{ V}$$

The negative $\Delta\mathcal{E}^\circ$ confirms that water does not spontaneously decompose to $H_2(g)$ and $O_2(g)$ at ordinary conditions. The superscript appears on $\Delta\mathcal{E}$ because all reactants and products are in their standard states. The decomposition of water requires an outside voltage of 1.229 V in an electrolytic cell to force it to occur. When such a cell is constructed and run, $H_2(g)$ forms at the cathode and $O_2(g)$ at the anode.

These facts have important applications in the electrolysis of aqueous solutions:

- A species in neutral aqueous solution can be reduced only if its reduction potential exceeds −0.414 V. (Note: The number −0.50 for example is *less* than −0.414 because of the minus sign). If this condition is not met, then $H_3O^+(aq)$ is reduced instead of the species in question.

- A species in neutral aqueous solution can be oxidized only if its reduction potential is *less* than 0.815 V. Otherwise $H_2O(l)$ is oxidized to $O_2(g)$ instead.

These criteria are applied in **problem 10-63.**

Detailed Solutions to Odd-Numbered Problems

10-1 a) Identify the half-reactions taking place:

$$e^- + VO_2^+(aq) \rightarrow VO^{2+}(aq) \quad \text{and} \quad SO_2(g) \rightarrow SO_4^{2-}(aq) + 2\,e^-$$

Vanadium is reduced from the +5 to the +4 state, and sulfur is oxidized from the +4 to the +6 state so there is 1 e^- in the reduction half-equation and 2 e^- in the oxidation half-equation. The two half-equations are balanced as to V and S respectively, but neither is balanced as to charge nor the element oxygen. Balance as to charge by adding H_3O^+ (the solution is acidic):

$$e^- + 2\,H_3O^+(aq) + VO_2^+(aq) \rightarrow VO^{2+}(aq)$$
$$SO_2(g) \rightarrow SO_4^{2-}(aq) + 4\,H_3O^+(aq) + 2\,e^-$$

Balance each half-equation as to oxygen by adding water to one side or the other:

$$e^- + 2\,H_3O^+(aq) + VO_2^+(aq) \rightarrow VO^{2+}(aq) + 3\,H_2O(l)$$
$$6\,H_2O(l) + SO_2(g) \rightarrow SO_4^{2-}(aq) + 4\,H_3O^+(aq) + 2\,e^-$$

Confirm that the two statements are indeed balanced as to the element hydrogen. Multiply each half-equation by a factor to make the electron loss equal to the electron gain. Then add the equations:

$$2\,e^- + 4\,H_3O^+(aq) + 2\,VO_2^+(aq) \rightarrow 2\,VO_2^+(aq) + 6\,H_2O(l)$$
$$6\,H_2O(1) + SO_2(g) \rightarrow SO_4^{2-}(aq) + 4\,H_3O^+(aq) + 2\,e^-$$
$$\text{Answer}: \; 2\,VO_2^+(aq) + SO_2(g) \rightarrow 2\,VO_2^+(aq) + SO_4^{2-}(aq)$$

Use the same approach in the other parts of the problem:

b) $Br_2(aq) + SO_2(g) + 6\,H_2O(l) \rightarrow 2\,Br^-(aq) + SO_4^{2-}(aq) + 4\,H_3O^+(aq)$

c) $Cr_2O_7^{2-}(aq) + 3\,Np^{4+}(aq) + 2\,H_3O^+ \rightarrow 2\,Cr^{3+}(aq) + 3\,NpO_2^{2+}(aq) + 3\,H_2O(aq)$

d) $5\,HCOOH(aq) + 2\,MnO_4^-(aq) + 6\,H_3O^+(aq) \rightarrow 5\,CO_2(g) + 2\,Mn^{2+}(aq) + 14\,H_2O(l)$

e) $3\,Hg_2HPO_4(s) + 2\,Au(s) + 8\,Cl^-(aq) + 3\,H_3O^+(aq) \rightarrow 6\,Hg(l) + 3\,H_2PO_4^-(aq) + 2\,AuCl_4^-(aq) + 3\,H_2O(l)$

10-3 The steps are the same as in the balancing in problem 10-1 except that now OH^- is used to balance charge (see page 353 of the text).

a) $10\,OH^-(aq) + 2\,Cr(OH)_3(s) + 3\,Br_2(aq) \rightarrow 2\,CrO_4^{2-}(aq) + 6\,Br^-(aq) + 8\,H_2O(l)$

b) $ZrO(OH)_2(s) + 2\,SO_3^{2-}(aq) \rightarrow Zr(s) + 2\,SO_4^{2-}(aq) + H_2O(l)$

c) $7\,HPbO_2^-(aq) + 2\,Re(s) \rightarrow 7\,Pb(s) + 2\,ReO_4^-(aq) + H_2O(l) + 5\,OH^-(aq)$

d) $4\,HXeO_4^-(aq) + 8\,OH^-(aq) \rightarrow 3\,XeO_6^{4-}(aq) + Xe(g) + 6\,H_2O(l)$

e) $N_2H_4(aq) + 2\,CO_3^{2-}(aq) \rightarrow N_2(g) + 2\,CO(g) + 4\,OH^-(aq)$

10-5 a) $Fe^{2+}(aq) \rightarrow Fe^{3+}(aq) + e^-$ (oxidation)
$H_2O_2(aq) + 2\,H_3O^+(aq) + 2e^- \rightarrow 4\,H_2O$ (reduction)
b) $5\,H_2O(l) + SO_2(g) \rightarrow HSO_4^-(aq) + 3H_3O^+(aq) + 2\,e^-$ (oxidation)
$MnO_4^-(aq) + 8\,H_3O^+(aq) + 5\,e^- \rightarrow Mn^{2+}(aq) + 4\,H_2O(l)$ (reduction)
c) $ClO_2^-(aq) \rightarrow ClO_2(g) + e^-$ (oxidation)
$ClO_2^-(aq) + 4\,H_3O^+(aq) + 4\,e^- \rightarrow Cl^-(aq) + 6\,H_2O(l)$ (reduction)

10-7 The oxidation state of nitrogen both increases and decreases in this reaction, which is a disproportionation. The two half-reactions are:
$HNO_2(aq) + 4\,H_2O(l) \rightarrow NO_3^-(aq) + 3\,H_3O^+(aq) + 3\,e^-$ oxidation
$HNO_2(aq) + H_3O^+(aq) \rightarrow NO(g) + 2\,H_2O(l)$ reduction
Combining the two half-equations gives:
$3\,HNO_2(aq) \rightarrow NO_3^-(aq) + 2\,NO(g) + H_3O^+(aq)$.

10-9 Electrons flow from the left electrode to the right as Cr(II) is oxidized to Cr(III). In the salt bridge negative ions flow from right to left and positive ions from left to right.

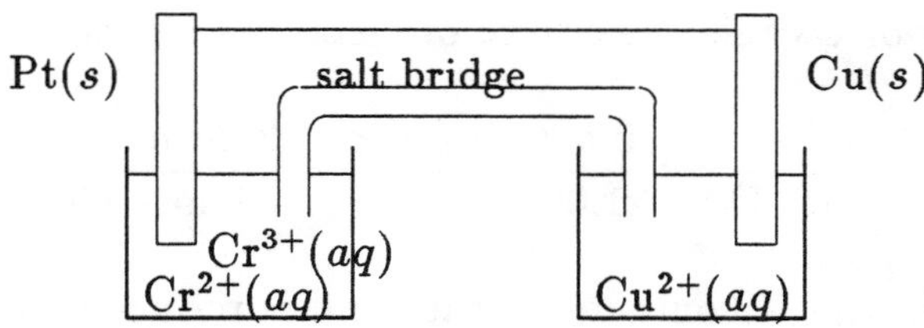

10-11 Formation of 1 mol of Sn(*s*) from $Sn^{4+}(aq)$ ions requires 4 mol of electrons: $Sn^{4+}(aq) + 4\,e^- \rightarrow Sn(s)$. Hence:

$$6.95 \times 10^4\ \text{C} \times \left(\frac{1\ \text{mol}\ e^-}{96\,485.3\ \text{C}}\right) \times \left(\frac{1\ \text{mol Sn}(s)}{4\ \text{mol}\ e^-}\right) = 0.180\ \text{mol Sn}$$

10-13 At the anode, Zn(*s*) is being oxidized to $Zn^{2+}(aq)$; at the cathode, $Cl_2(g)$ is being reduced to $Cl^-(aq)$.
a) $Zn(s) + Cl_2(g) \rightarrow Zn^{2+}(aq) + 2\,Cl^-(aq)$.
b) An ampere is a coulomb per second. Hence:

$$25.0\ \text{min} \times \left(\frac{60\ \text{s}}{1\ \text{min}}\right) \times \left(\frac{0.800\ \text{C}}{1\ \text{s}}\right) = 1.20 \times 10^3\ \text{C}$$

In moles:

$$25.0\ \text{min} \times \left(\frac{60\ \text{s}}{1\ \text{min}}\right) \times \left(\frac{0.800\ \text{C}}{1\ \text{s}}\right) \times \left(\frac{1\ \text{mol}\ e^-}{96\,485.3\ \text{C}}\right) = 0.0124\ \text{mol}\ e^-$$

c) Two moles of electrons are produced by oxidation of one mole of Zn(s). Passage of 0.0124 mol of electrons therefore means 0.00622 mol of Zn(s) is oxidized. This is a loss of 0.407 g (using a molar mass of 65.38 g mol^{-1} for zinc) from the zinc anode.
d) Two moles of electrons is equivalent in this reaction to one mole of $Cl_2(g)$. Passage of 0.0124 mol of electrons accordingly requires reduction of 0.00622 mol of $Cl_2(g)$. Use the ideal-gas law to calculate the volume of this chemical amount of chlorine at 25°C (298 K) and a pressure of 1 atm:

$$V = \frac{nRT}{P} = \frac{(0.00622 \text{ mol})(0.08206 \text{ L atm mol}^{-1}\text{K}^{-1})(298 \text{ K})}{1 \text{ atm}} = 0.152 \text{ L} .$$

10-15 Calculate the ratio of the chemical amounts of oxygen and copper generated by the operation of the cell. The molar mass of oxygen is 32.0 g mol^{-1}, and the molar mass of copper is 63.54 g mol^{-1}, The cell therefore forms 0.500 mol of O_2 as it forms 1.00 mol of Cu. A balanced half-equation for the oxidation of water to gaseous

oxygen is $3\,H_2O(l) \rightarrow 1/2\,O_2(g) + 2\,H_3O^+(aq) + 2\,e^-$. This equation states that the production of 1/2 mol of O_2 releases 2 mol of electrons. Hence:

$$\frac{2 \text{ mol } e^-}{1/2 \text{ mol O}_2} \times \frac{0.500 \text{ mol O}_2}{1.00 \text{ mol Cu}} = \frac{2.00 \text{ mol } e^-}{1 \text{ mol Cu}}$$

The copper starts off in the +2 oxidation state and is reduced as follows: $Cu^{2+}(aq) + 2e^- \rightarrow Cu(s)$.

10-17 a) In the electrolysis of molten KCl the half-cell reactions are:

$$\begin{array}{ll} K^+(l) + e^- \rightarrow K(l) & \text{reduction; cathode} \\ Cl^-(l) \rightarrow 1/2\,Cl_2(g) + e^- & \text{oxidation; anode} \end{array}$$

The sum of these two half-equations represents the overall cell reaction:

$$K^+ + Cl^- \rightarrow K(l) + 1/2\,Cl_2(g)$$

b) A current of 2.00 A is a current of 2.00 C s^{-1}; 500 hours is 1.80×10^4 s. The amount of electricity passing through the cell is the product of the average current and the length of time it flows or 3.60×10^4 C. Since 9.6485×10^4 C is a mole of electrons, 0.373 mol of electrons passes through the cell. As one mole of electrons passes through the cell, one mole of K forms, and one half mole of Cl_2 forms. Therefore 0.0187 mol of Cl_2 and 0.373 mol of K are generated during the 5.00 hour electrolysis run. These amounts are 13.2 g of Cl_2 ($\mathcal{M}$ = 70.91 g mol^{-1}) and 14.6 g of K ($\mathcal{M}$ = 39.102 g mol^{-1}). A tacit assumption is that the cell has at least 27.8 g of KCl in it.

10-19 The balanced half-equation $Ag^+ + e^- \longrightarrow Ag(s)$ relates the number of moles of electrons that is transferred to the number of moles (and thus the number of grams) of silver that plates out. Use this stoichiometry to calculate the number of moles of electrons transferred:

$$n_{e^-} = 1.00 \text{ g Ag} \times \left(\frac{1 \text{ mol Ag}}{107.9 \text{ g Ag}}\right) \times \left(\frac{1 \text{ mol } e^-}{1 \text{ mol Ag}}\right) = 0.00927 \text{ mol } e^-$$

Maximum electrical work is absorbed by the cell only if it is operated reversibly (see text page 363):

$$w_{\text{elec,max}} = w_{\text{elec,rev}} = \Delta G = -n\mathcal{F}\Delta\mathcal{E}$$

In this case, all concentrations remain at their standard-state values so the equation can be modified:

$$w_{\text{elec,max}} = \Delta G^\circ = -n\mathcal{F}\Delta\mathcal{E}^\circ$$

Substitution gives:

$$w_{\text{elec,max}} = -(0.00927 \text{ mol})(96\,485 \text{ C mol}^{-1})(1.03 \text{ V}) = -921 \text{ J}$$

The negative sign in this answer means that the surroundings do a maximum of −921 J of work on the cell. This is equivalent to saying the cell performs a maximum of +921 J of work on its surroundings.

10-21 a) The balanced equations are:

anode	$Co(s) \longrightarrow Co^{2+}(aq) + 2\,e^-$
cathode	$Br_2(l) + 2\,e^- \longrightarrow 2\,Br^-(aq)$
overall reaction	$Co(s) + Br_2(l) \longrightarrow Co^{2+}(aq) + 2\,Br^-(aq)$

b) The Co/Co^{2+} reaction is at the anode; the Br_2/Br^- reaction is at the cathode. The $\Delta\mathcal{E}^\circ$ of the cell equals the standard reduction potential at the cathode minus the standard reduction potential at the anode:

$$\Delta\mathcal{E}^\circ = \mathcal{E}^\circ(\text{cathode}) - \mathcal{E}^\circ(\text{anode}) = 1.065 - (-0.28) = 1.34 \text{ V}$$

10-23 a) The In^{3+}/In half-reaction must be the reduction half-reaction because metallic indium plates out when the cell runs. The Zn^{2+}/Zn half reaction must proceed as an oxidation:

$$\text{anode}: Zn(s) \longrightarrow Zn^{2+}(aq) + 2\,e^- \quad \text{cathode}: In^{3+}(aq) + 3\,e^- \longrightarrow In(s)$$

b) The standard potential difference of the cell (0.425 V) equals the standard reduction potential at the anode subtracted from the standard reduction potential at the cathode. According to Appendix E, the reduction standard reduction at the anode (the zinc) is −0.763 V. Hence:

$$\Delta\mathcal{E}^\circ = 0.425\ \text{V} = \mathcal{E}^\circ(\text{cathode}) - \mathcal{E}^\circ(\text{anode}) = \mathcal{E}^\circ(\text{cathode}) - (-0.763\ \text{V})$$

Solving gives $\mathcal{E}^\circ$(cathode) equal to −0.338 V. The standard potential difference $\Delta\mathcal{E}^\circ$ of a reaction is a measure of the driving force for that reaction. It is independent of the quantities (but *not* the concentrations) of reactants and products present, as long as some of each is present. This means that multiplying the coefficients in the balanced equation for the reaction in an electrochemical cell by a constant does not change $\Delta\mathcal{E}^\circ$.

10-25 a) The standard potential for the half-reaction: $Mn^{3+}(aq) + 3e^- \longrightarrow Mn(s)$ is *not* the simple sum of the half-cell potentials for the half-reactions:

$$\begin{array}{ll} Mn^{2+}(aq) + 2e^- \longrightarrow Mn(s) & \mathcal{E}^\circ = -1.029\ \text{V} \\ Mn^{3+}(aq) + e^- \longrightarrow Mn^{2+}(aq) & \mathcal{E}^\circ = 1.51\ \text{V} \end{array}$$

even though the target half-reaction *is* the sum of these two half-reactions. Instead the potential is a *weighted average:*

$$\mathcal{E}_3^\circ = \frac{n_1\mathcal{E}_1^\circ + n_2\mathcal{E}_2^\circ}{n_3}$$

where the three subscripted n's are the number of electrons transferred in the two half-reactions being combined and in the half-reaction that results from the combination. Substitution gives:

$$\mathcal{E}^\circ = \frac{2(-1.029\ \text{V}) + 1(1.51\ \text{V})}{3} = -0.183\ \text{V}$$

b) The disproportionation: $3\,Mn^{2+}(aq) \longrightarrow Mn(s) + 2\,Mn^{3+}(aq)$ combines the reduction of $Mn^{2+}(aq)$ to $Mn(s)$ and the oxidation of $Mn^{2+}(aq)$ to $Mn^{3+}(aq)$. It is represented by the second of the following half-equations subtracted from the first:

$$\begin{array}{ll} 2\,e^- + Mn^{2+}(aq) \longrightarrow Mn(s) & \mathcal{E}^\circ = -1.029\ \text{V} \\ 2\,Mn^{3+}(aq) + 2\,e^- \longrightarrow 2\,Mn^{2+}(aq) & \mathcal{E}^\circ = 1.51\ \text{V} \end{array}$$

Note that the coefficients in the second half-equation are all twice the coefficients appearing in Appendix E. Because the final disproportionation reaction is a whole reaction, not a half-reaction, the doubling has no effect on the computation of standard potential difference of the disproportionation reaction, which proceeds:

$$\Delta\mathcal{E}^\circ = \mathcal{E}^\circ(\text{reduction}) - \mathcal{E}^\circ(\text{oxidation}) = -1.029 - 1.51 = -2.539\ \text{V}$$

To see why this is so, write a weighted-average formula like the one in the preceding part and use it to compute $\Delta\mathcal{E}^\circ$. Note carefully what happens when the values of n_1, n_2 and n_3 are substituted:

$$\Delta\mathcal{E}^\circ = \frac{n_1\mathcal{E}_1^\circ - n_2\mathcal{E}_2^\circ}{n_3} = \frac{2(-1.029) - 2(1.51)}{2} = -1.029 - 1.51 = -2.539 \text{ V}$$

In combining half-reactions to give a whole reaction n_1, n_2, and n_3 are *always* equal to each other. This is *never* the case in combining half-reactions to give another half-reaction.

Solutions of $Mn^{2+}(aq)$ do not spontaneously disproportionate to $Mn(s)$ and $Mn^{3+}(aq)$; the standard voltage for the process is negative and large.

10-27 Look in Appendix E to find two half-reactions that involve Fe^{2+}, Fe^{3+}, and Fe and that can be combined to give a reaction having a positive $\Delta\mathcal{E}^\circ$ (a spontaneous reaction). The only two candidates are:

$$Fe^{3+}(aq) + e^- \longrightarrow Fe^{2+}(aq) \quad \mathcal{E}^\circ = +0.770 \text{ V}$$
$$Fe^{2+}(aq) + 2\,e^- \longrightarrow Fe(s) \quad \mathcal{E}^\circ = -0.409 \text{ V}$$

It is clear that if the second half-reaction is at the anode, then the standard potential difference is positive:

$$\Delta\mathcal{E}^\circ = \mathcal{E}^\circ(\text{cathode}) - \mathcal{E}^\circ(\text{anode}) = 0.770 - (-0.409) = 1.179 \text{ V}$$

The corresponding reaction is $2\,Fe^{3+}(aq) + Fe(s) \longrightarrow 3\,Fe^{2+}(aq)$.

10-29 Powdered metallic aluminum should act as a reducing agent. A reducing agent is itself oxidized as it acts. The +3 oxidation state of Al is well-known; it is the obvious product when aluminum gives up electrons. There are no common negative oxidation states of Al; such states would have to result if Al served as an oxidizing agent. Finally, according to Appendix E the reduction of Al^{3+} to $Al(s)$ has a large negative $\mathcal{E}^\circ$. If Al^{3+} is hard to reduce, then $Al(s)$ is easy to oxidize. Not only is powered aluminum a reducing agent, it is a strong one.

10-31 The stronger an oxidizing agent is, the easier it is to reduce (see problem 10-29). The more powerful oxidizing agent will have the algebraically larger reduction potential. For these two elements the standard reduction potentials are:

$$Cl_2(g) + 2\,e^- \longrightarrow 2\,Cl^-(aq) \quad \mathcal{E}^\circ = +1.3583 \text{ V}$$
$$Br_2(l) + 2\,e^- \longrightarrow 2\,Br^-(aq) \quad \mathcal{E}^\circ = +1.065 \text{ V}$$

Since the Cl_2/Cl^- couple has a larger $\mathcal{E}^\circ$ than the Br_2/Br^- couple, $Cl_2(g)$ is the stronger oxidizing agent and the better disinfectant.

10-33 a) A species tends spontaneously to disproportionate if and only if the reduction potential for the process connecting it to the lower oxidation state exceeds the reduction potential for the process connecting it to the higher oxidation state. This means that the reduction potential that lies immediately to its right in a reduction potential diagram (see text page 369) must be larger than one that appears immediately to its left. The data in the problem allows construction of the diagram:

$$H_2PO_2^- \xrightarrow{-2.05} P_4 \xrightarrow{-0.89} PH_3 \quad \text{basic solution}$$

Thus, P_4 will spontaneously disproportionate at a pH of 14.

b) A reducing agent is itself oxidized as it acts; the stronger a reducing agent is, the easier it is to oxidize. The half-reaction involving P_4 has the more negative standard reduction potential so P_4 is easier to oxidize than PH_3; P_4 is the stronger reducing agent. Another way to make the comparison is to reverse both half-reactions so that they represent oxidations. The signs of the $\mathcal{E}°$'s are thereby changed. The potential for the oxidation of P_4 then exceeds that for the oxidation of PH_3 showing that P_4 is the better reducing agent.

10-35 The overall reaction in this cell is:

$$2\,Cr^{2+}(aq) + Pb^{2+}(aq) \rightarrow Pb(s) + 2\,Cr^{3+}(aq)$$

and the potential difference is:

$$\Delta\mathcal{E}° = \mathcal{E}°(\text{reduction}) - \mathcal{E}°(\text{oxidation}) = \mathcal{E}°(Pb^{2+}/Pb) - \mathcal{E}°(Cr^{3+}/Cr^{2+})$$
$$\Delta\mathcal{E}° = -0.1263 - (-0.41) = 0.28 \text{ V}$$

The reduction potentials used here are strictly valid only if the reactants and products are in their standard states. Typically, however, electrochemistry is not performed under these conditions. In this cell none of the solute concentrations equals 1 M. The Nernst equation allows for the correction of standard cell potentials ($\Delta\mathcal{E}°$'s) to potentials at non-standard conditions ($\Delta\mathcal{E}$'s). The equation is

$$\Delta\mathcal{E} = \Delta\mathcal{E}° - \frac{RT}{n\mathcal{F}} \ln Q = \Delta\mathcal{E}° - \frac{0.0592 \text{ V}}{n} \log Q \quad (\text{at } 25°)$$

where Q is the reaction quotient. For this particular reaction and initial set of conditions:

$$Q = \frac{[Cr^{3+}]^2}{[Cr^{2+}]^2[Pb^{2+}]} = \frac{(0.0030)^2}{(0.15)(0.20)^2}$$

and n equals 2. Substitution of these values gives:

$$\Delta\mathcal{E} = 0.2837 \text{ V} - \frac{0.0529 \text{ V}}{2} \log\left(\frac{(0.0030)^2}{(0.15)(0.20)^2}\right) = 0.37 \text{ V}$$

10-37 Platinum is included in the cell notation because it is being used as an electrode. The Nernst equation can be applied to half-reactions as well as to cell reactions. The half-reaction indicated by the cell notation is:

$$Cr^{3+}(aq) + e^- \rightarrow Cr^{2+}(aq) \qquad \mathcal{E}^\circ = -0.41 \text{ V}$$

for which the Nernst equation is:

$$\mathcal{E} = \mathcal{E}^\circ - \frac{RT}{n\mathcal{F}} \ln Q = -0.41 \text{ V} - \left(\frac{0.0529 \text{ V}}{1}\right) \log\left(\frac{0.0019}{0.15}\right) = -0.31 \text{ V}$$

By using "0.0529 V", one automatically assumes the temperature is 25°C.

10-39 The I_2/I^- half-reaction is at the cathode, the site of reduction. Hence the H_3O^+/H_2 half-reaction is an oxidation and takes place at the anode. The overall cell reaction is:

$$2\ H_2O(l) + I_2(s) + H_2(g) \rightarrow 2\,H_3O^+(aq) + 2\,I^-(\text{aq})$$

for which $\Delta\mathcal{E}^\circ$ is 0.535 V. This is calculated by combination of the reduction potentials from Appendix E:

$$\Delta\mathcal{E}^\circ = \mathcal{E}^\circ(\text{cathode}) - \mathcal{E}^\circ(\text{anode}) = 0.535(I_2/I^-) - (0.000)(H_3O^+/H_2) = 0.535 \text{ V}$$

The measured cell voltage, 0.841 V, is related to the H_3O^+ concentration, to the other concentrations, and to the partial pressures in the cell by the Nernst equation:

$$0.841 \text{ V} = 0.535 \text{ V} - \left(\frac{0.0592 \text{ V}}{2}\right) \log\left(\frac{[H_3O^+]^2[I^-]^2}{P_{H_2}}\right)$$

The concentration of iodide ion is 1.00 M; the pressure of the $H_2(g)$ is 1 atm. Substitution gives:

$$0.841 - 0.535 = 0.306 \text{ V} = -\frac{0.0592 \text{ V}}{2} \log\left(\frac{[H_3O^+]^2(1.00)^2}{1.00}\right)$$

Solving for $\log H_3O^+$ gives -5.17. The pH is $-\log[H_3O^+]$, so it is 5.17.

10-41 a) Calculate $\Delta\mathcal{E}^\circ$ by breaking the given equation into half-equations, looking up the standard reduction potentials of the half-equations and subtracting the reduction potential at the anode from the reduction potential at the cathode. In the equation, Cr^{3+} is oxidized to $Cr_2O_7^{2-}$ under acidic conditions and $HClO_2$ is reduced to HClO under acidic conditions. Appendix E gives the half-equations:

$$Cr_2O_7^{2-}(aq) + 14\,H_3O^+(aq) + 6\,e^- \rightarrow 2\,Cr^{3+}(aq) + 21\,H_2O(l) \quad \Delta\mathcal{E}^\circ = 1.33 \text{ V}$$
$$3\,HClO_2(aq) + 6\,H_3O^+(aq) + 6\,e^- \rightarrow 3\,HClO(aq) + 9\,H_2O(l) \quad \Delta\mathcal{E}^\circ = 1.64 \text{ V}$$

If the first reaction occurs at the anode and the second at the cathode, then the standard potential difference is positive:

$$\Delta\mathcal{E}^\circ = \mathcal{E}^\circ(\text{cathode}) - \mathcal{E}^\circ(\text{anode}) = 1.64 - 1.33 = 0.31\text{ V}$$

b) The concentration of Cr^{3+} ion is related to the measured cell potential by the Nernst equation. For the above reaction, the Nernst equation at 25° C becomes:

$$0.15\text{ V} = 0.31\text{ V} - \frac{0.0592\text{ V}}{6}\log\left(\frac{[\text{HClO}]^3[\text{Cr}_2\text{O}_7^{2-}][\text{H}_3\text{O}^+]^8}{[\text{HClO}_2]^3[\text{Cr}^{3+}]^2}\right)$$

Insert the various concentrations, which are given in the problem. Recall that at a pH of 0 the concentration of H_3O^+ ion is 1.00 M. Then:

$$0.15\text{ V} = 0.31\text{ V} - \frac{0.0529\text{ V}}{6}\log\left(\frac{(0.20)^3(0.80)(1.00)^8}{(0.15)^3[\text{Cr}^{3+}]^2}\right)$$

Solving for the concentration of Cr^{3+} is now straightforward:

$$\frac{6(0.15-0.31)}{0.0592} = -\log\left(\frac{1.896}{[\text{Cr}^{3+}]^2}\right) \quad \text{hence} \quad [\text{Cr}^{3+}] = 1\times10^{-8}\text{ M}$$

10-43 We need the K for the reaction in problem 10-41:

$$3\,\text{HClO}_2(aq) + 2\,\text{Cr}^{3+}(aq) + 12\,\text{H}_2\text{O}(l) \rightarrow 3\,\text{HClO}(aq) + \text{Cr}_2\text{O}_7^{2-}(aq) + 8\,\text{H}_3\text{O}^+(aq)$$

and we know the $\Delta\mathcal{E}^\circ$ of the reaction to be 0.31 V. The two are related:

$$\ln K = \frac{n\mathcal{F}\Delta\mathcal{E}^\circ}{RT} = \frac{6(9.6485\times10^4\text{ C mol}^{-1})(0.31\text{ V})}{(8.315\text{ J K}^{-1}\text{mol}^{-1})(298.15\text{ K})} = 72.4 \qquad K = 3\times10^{31}$$

This equilibrium constant is so very large that at equilibrium the reaction mixture consists essentially completely of products. In other words, $HClO_2$ and Cr^{3+} tend to react until one of the other is all used up. The two chromium-containing ions are the only colored species in the reaction. To determine the color of the solution, assume complete reaction and calculate whether Cr^{3+} ion is consumed or is in excess. There are 2.00 mol of $HClO_2$ and 1.0 mol of Cr^{3+} ion, and the two react in a 3-to-2 ratio. The green Cr^{3+} ion is the limiting reactant. Because all of the green ion is used up, the solution will be orange.

10-45 A disproportionation reaction is one in which one species is both oxidized and reduced. In this instance In^+ is oxidized to In^{3+} and also reduced to In. Proceed by calculating the $\Delta\mathcal{E}^\circ$ that would be developed if the overall reaction were caused to take place in an electrochemical cell. Then get K from $\Delta\mathcal{E}^\circ$. The equations are:

cathode	$2\,In^+(aq) + 2\,e^- \rightarrow 2\,In(s)$	$\mathcal{E}^\circ = -0.21$ V
anode	$In^{3+}(aq) + 2\,e^- \rightarrow In^+(aq)$	$\mathcal{E}^\circ = -0.40$
cell reaction	$3\,In^+(aq) \rightarrow In^{3+}(aq) + 2\,In(s)$	$\Delta\mathcal{E}^\circ - 0.21 - (-0.40) = 0.19$ V

As always, the $\Delta\mathcal{E}^\circ$ of the net reaction is the cathode $\mathcal{E}^\circ$ minus the anode $\mathcal{E}^\circ$. The calculation continues as in problem 10-43:

$$\ln K = \frac{n\mathcal{F}\Delta\mathcal{E}^\circ}{RT} = \frac{2(96\,485\text{ C mol}^{-1})(0.19\text{ V})}{(8.315\text{ J K}^{-1}\text{mol}^{-1})(298.15\text{ K})} = 15 \quad K = e^{15} = 3\times10^6$$

10-47 At the anode H_2 is oxidized to H_3O^+; at the cathode H_2 is reduced to H_3O^+. The sum of this oxidation and reduction is the reaction:

$$H_2(g,\text{anode}) + 2\,H_3O^+(aq,\text{cathode}) \rightarrow H_2(g,\text{cathode}) + 2\,H_3O^+(aq,\text{anode})$$

This reaction transfers 2 mol of electrons for every 1 mol of H_2. The standard potential difference for the galvanic cell is zero! The non-zero potential difference that is observed is caused by the unequal concentrations of H_3O^+ on the two sides. The Nernst equation (at 25°C) becomes:

$$0.150\text{ V} = 0.00\text{ V} - \frac{0.0592\text{ V}}{2}\log\left(\frac{[H_2]_{\text{cathode}}[H_3O^+]^2_{\text{anode}}}{[H_2]_{\text{anode}}[H_3O^+]^2_{\text{cathode}}}\right)$$

$$0.150 = -\frac{0.0592\text{ V}}{2}\log\left(\frac{(1.00)[H_3O^+]^2_{\text{anode}}}{(1.00)(1.00)^2_{\text{cathode}}}\right)$$

Solving for the H_3O^+ concentration at the anode gives $[H_3O^+] = 0.00293$ M. The pH at the anode is therefore 2.53. Buffer action at the anode is founded in the reaction:

$$HA(aq) + H_2O(l) \rightleftharpoons H_3O^+(aq) + A^-(aq) \quad K_a = \frac{[H_3O^+][A^-]}{[HA]}$$

According to the problem, the concentrations of A^- and HA in the buffer solution are both 0.10 M at the same time the pH is 2.53. Substitution in the K_a expression gives:

$$K_a = \frac{[H_3O^+][A^-]}{[HA]} = \frac{(0.00293)(0.10)}{(0.10)} = 0.0029$$

10-49 **a)** The cell reaction can be broken down into the half-reactions:

$$\begin{array}{lll} \text{cathode} & Br_2(l) + 2\,e^- \rightarrow 2\,Br^-(aq) & \mathcal{E}^\circ = 1.065 \text{ V} \\ \text{anode} & 2\,H_3O^+(aq) + 2\,e^- \rightarrow H_2(g) + 2\,H_2O(l) & \mathcal{E}^\circ = 0.000 \text{ V} \end{array}$$

The standard potential difference is then:

$$\Delta\mathcal{E}^\circ = \mathcal{E}^\circ(\text{cathode}) - \mathcal{E}^\circ(\text{anode}) = 1.065 - 0.000 = 1.065 \text{ V}$$

b) The concentration of Br^- in the cell is related to the measured cell potential and other concentrations (and partial pressures) through the Nernst equation, which for this cell (at 25° C) becomes:

$$1.710 \text{ V} = 1.065 \text{ V} - \left(\frac{0.0592}{2}\right) \log\left(\frac{[Br^{2-}]^2[H_3O^+]^2}{P_{H_2}}\right)$$

At pH 0 the concentration of H_3O^+ is 1.00 M. The partial pressure of H_2 is 1.0 atm. Substitution of these values gives:

$$1.710 \text{ V} = 1.065 \text{ V} - \left(\frac{0.0592 \text{ V}}{2}\right) \log\left(\frac{[Br^-]^2(1.00)^2}{(1.0)}\right)$$

Solving for the concentration of bromide ion gives 1.27×10^{-11} M which is, to two significant figures, 1.3×10^{-11} M.
c) The dissolution of AgBr(*s*) is governed by a solubility-product constant expression:

$$AgBr(s) \rightleftharpoons Ag^+(aq) + Br^-(aq) \quad K_{sp} = [Ag^+][Br^-]$$

The Br^- ion in the cell is at equilibrium with AgBr(*s*) and $Ag^+(aq)$ ion. The concentrations of both ions are known: the concentration of Br^- was computed in the previous part, and the concentration of Ag^+ is 0.060 M. Hence: $K_{sp} = (0.060)(1.27 \times 10^{-11}) = 7.6 \times 10^{-13}$.

10-51 The half-reactions in a lead-acid cell are:

$$\begin{array}{ll} \text{cathode} & PbO_2(s) + SO_4^{2-}(aq) + 4\,H_3O^+(aq) + 2\,e^- \rightarrow PbSO_4(s) + 6\,H_2O(l) \\ \text{anode} & Pb(s) + SO_4^{2-}(aq) \rightarrow PbSO_4(s) + 2\,e^- \end{array}$$

The standard reduction potentials are 1.685 V (cathode) and −0.356 V (anode). The potential difference is 1.685 − (−0.356) = 2.041 V. When electrochemical cells are connected in series, their voltages add. The voltage generated by six such cells in series would be 6(2.04 V) = 12.2 V.

10-53 The "quantity of charge furnished" should be read to mean the quantity of charge the storage battery is capable of forcing through an outside resistance before its voltage falls to zero.

a) Oxidation of 1 mol of spongy lead requires that 2 mol of electrons pass through the circuit: $Pb(s) + SO_4^{2-}(aq) \rightarrow PbSO_4(s) + 2\,e^-$. Hence:

$$10 \times 10^3 \text{ g Pb} \times \left(\frac{1 \text{ mol Pb}}{207 \text{ g Pb}}\right) \times \left(\frac{2 \text{ mol } e^-}{1 \text{ mol Pb}}\right) \times \left(\frac{96.5 \times 10^4 \text{ C}}{1 \text{ mol } e^-}\right) = 9.3 \times 10^6 \text{ C}$$

b) The maximum amount of electrical work that a battery can perform is related to the maximum amount of charge that it pushes through the outside circuit and the voltage difference between its two electrodes. The maximum work on the battery by the surroundings is:

$$w_{\text{elec,max}} = -Q\Delta\mathcal{E}$$

This maximum work is obtained only if the discharge of the battery proceeds reversibly. Assume that the voltage of this lead-acid battery does not change as it is discharged. Writing the Nernst equation for the reaction (which is the sum of the half-reaction in problem 10-51) shows that this is tantamount to assuming that the concentration of sulfuric acid in the battery stays constant. Then:

$$w_{\text{elec,max}} = -(9.3 \times 10^6 \text{ C})(12 \text{ V}) = -1.1 \times 10^8 \text{ V C} = -1.1 \times 10^8 \text{ J}$$

The work done by the battery on the surroundings is therefore $+1.1 \times 10^8$ J.

10-55 The amounts of Pb and PbO_2 in the battery diminish during discharge. Without reactants, the battery cannot function. Therefore, simply replacing the dilute H_2SO_4 with concentrated H_2SO_4 is not enough to recharge the battery. The $PbSO_4$ that accumulates must be removed and fresh PbO and PbO_2 added.

10-57 The maximum amount of electrical work done in an electrochemical reaction at constant T and P equals the change in free energy for that reaction. This quantity is in turn related to the potential difference:

$$w_{\text{elec,max}} = \Delta G = -Q\Delta\mathcal{E}$$

If all of the gases in this fuel cell are kept at a pressure of 1 atm and the temperature is 298.15 K, then all reactants and products of the cell process are in their standard states. The previous equation becomes:

$$w_{\text{elec,max}} = \Delta G^\circ = -Q\Delta\mathcal{E}^\circ$$

The chemical equation for the reaction in the fuel cell can be broken down into half-equations:

anode	$H_2(g) + 2\,OH^-(aq) \rightarrow 2\,H_2O(l) + 2e^-$
cathode	$1/2\,O_2(g) + H_2O(l) + 2\,e^- \rightarrow 2\,OH^-$

The difference between the standard reduction potentials (Appendix E) is the standard potential difference:

$$\mathcal{E}^\circ(\text{cathode}) - \mathcal{E}^\circ(\text{anode}) = 0.401(\text{O}_2/\text{OH}^-) - (-0.828)(\text{H}_2\text{O}/\text{H}_2) = 1.229 \text{ V}$$

Save this potential difference to replace $\Delta\mathcal{E}^\circ$ in the equation for the maximum work. The Q in the equation for the maximum work is the charge transferred as the reaction occurs. To get it, compute the chemical amount of electrons transferred per gram of water and convert to coulombs:

$$1 \text{ g H}_2\text{O} \times \left(\frac{1 \text{ mol H}_2\text{O}}{18.015 \text{ g H}_2\text{O}}\right) \times \left(\frac{2 \text{ mol } e^-}{1 \text{ mol H}_2\text{O}}\right) \times \left(\frac{96\,485 \text{ C}}{1 \text{ mol } e^-}\right) = 1.0711 \times 10^4 \text{ C}$$

The maximum electrical work per gram of water produced in this reaction is therefore:

$$w_{\text{elec,max}} = -Q\mathcal{E}^\circ = -(1.0711 \times 10^4 \text{ C})(1.23 \text{ V}) = -1.317 \times 10^4 \text{ J}$$

At 60% efficiency only six-tenths of this maximum work flows. This is -7.9×10^3 J per gram of water formed. The work maximum work generated in the surroundings is the negative of this or $+7900 \text{ J g}^{-1}$.

10-59 Iron is oxidized and water is reduced. The overall reaction is:

$$\text{Fe}(s) + 2\,\text{H}_2\text{O}(l) \rightarrow \text{Fe}^{2+}(aq) + 2\,\text{OH}^-(aq) + \text{H}_2(g)$$

The standard potential difference for this reaction is:

$$\Delta\mathcal{E}^\circ = \mathcal{E}^\circ(\text{reduction}) - \mathcal{E}^\circ(\text{oxidation}) = 0.409 - 0.828 = -0.419 \text{ V}$$

where the $\mathcal{E}^\circ$'s come from Appendix E. The negative $\Delta\mathcal{E}^\circ$ means the reaction is not spontaneous as written. If the pH becomes low enough, the reduction in the concentration of OH^- will make the reaction spontaneous. A low concentration of Fe^{2+} ion (relative to 1.00 M) and a low partial pressure (relative to 1.00 atm) of gaseous H_2 would also tend to make the reaction spontaneous. Since all three of the products would have far lower than standard concentrations (or pressures) under practical circumstances, corrosion might well proceed by this reaction.

10-61 In a remote theoretical sense, metallic sodium could be used as a sacrificial anode—it is far more easily oxidized than iron according to the standard reduction potentials (see Appendix E). In practice, however, the water of the sea would very rapidly oxidize it. Metallic sodium reacts violently with H_2O to form $H_2(g)$. Sodium would be a bad material as a sacrificial anode to protect the hull of a ship.

10-63 a) The product at the cathode could be either gaseous hydrogen from the reduction of 1.0×10^{-5} M $H_3O^+(aq)$ or metallic nickel from the reduction of 1.00 M $Ni^{2+}(aq)$. Direct observation of an operating cell would of course settle the issue at a glance. Note that the candidates for reduction at the *cathode* are both *cations.* In electrolytic cells, positively charged cations migrate toward the negatively charged cathode. The reduction potential for 1 M $Ni^{2+}(aq)$ to $Ni(s)$ is −0.23 V, as tabulated in Appendix E. The reduction potential of the H_3O^+ must be adjusted from the tabulated value of 0.00 V because the H_3O^+ is not in its standard state (1 M) but instead is 1.0×10^{-5} M. The Nernst equation (at 25°C) for this half-reaction is:

$$\mathcal{E} = \mathcal{E}^\circ - \frac{0.0592 \text{ V}}{1} \log\left(\frac{P_{H_2}^{1/2}}{[H_3O^+]}\right) = 0.0 - (0.0592 \text{ V}) \log\left(\frac{1}{10^{-5}}\right) = -0.296 \text{ V}$$

This reduction potential is algebraically less than the −0.23 V for the reduction of $Ni^{2+}(aq)$. Therefore the $Ni^{2+}(aq)$ is reduced first. The fact that the $H_3O^+(aq)$ is dilute rather than in its standard state lowers its reduction potential (makes it harder to reduce).

b) A current of 2.00 amperes for 10 hours is equivalent to 2.00 C s^{-1} for 36000 s. Therefore 7.20×10^4 C passes through the cell. Then:

$$7.20 \times 10^4 \text{ C} \times \left(\frac{1 \text{ mol } e^-}{96\,485 \text{ C}}\right) \times \left(\frac{1 \text{ mol Ni}(s)}{2 \text{ mol } e^-}\right) \times \left(\frac{58.71 \text{ g Ni}}{1 \text{ mol Ni}}\right) = 21.9 \text{ g Ni}$$

The volume of the electrolyte has to be large enough that removal of 21.9 g of nickel does not lower the concentration of $Ni^{2+}(aq)$ to the point that H_3O^+ starts to be reduced.

c) If the pH is 1.0, then $[H_3O^+]$ is 0.10 M. The Nernst equation for the reduction of H_3O^+ to $H_2(g)$ at 1 atm and 25°C becomes:

$$\mathcal{E} = 0.00 - 0.0592 \text{ V} \log\left(\frac{1.00}{1 \times 10^{-1}}\right) = -0.0592 \text{ V}$$

At this pH, $H_2(g)$ rather than $Ni(s)$ tends to form at the cathode. The increased concentration of $H_3O^+(aq)$ makes the reduction potential for H_3O^+ algebraically greater than −0.23 V, which is the value for reduction of 1.00 M $Ni^{2+}(aq)$.

10-65 $2\,Al(s) + 6\,H_2O(l) + 2\,OH^-(aq) \rightarrow 2\,Al(OH)_4^-(aq) + 3\,H_2(g)$

10-67 The desired reaction is the reverse of the non-spontaneous reaction that is forced to occur by the application of the outside voltage. It is:

$$NaClO_3(aq) + 3\,H_2(g) \rightarrow NaCl(aq) + 3\,H_2O(l)$$

10-69 **a)** The spontaneous chemical reaction in the cell combines the oxidation of zinc and the reduction of Ni^{2+}: $Zn(s) + Ni^{2+}(aq) \rightarrow Zn^{2+}(aq) + Ni(s)$.
Compute the chemical amount of zinc and the chemical amount of nickel(II) ion available for reaction. Whichever is less is the limiting reactant:

$$32.68 \text{ g Zn} \times \left(\frac{1 \text{ mol Zn}}{65.39 \text{ g Zn}}\right) = 0.4998 \text{ mol Zn}$$

$$0.575 \text{ L } Ni^{2+} \text{ solution} \times \left(\frac{1.00 \text{ mol } Ni^{2+}}{1 \text{ L solution}}\right) = 0.575 \text{ mol } Ni^{2+}$$

Zinc is the limiting reactant. See problem 1-65 to review how this fact is determined.
b) The cell is discharged when the zinc is all gone:

$$0.4998 \text{ mol Zn} \times \left(\frac{2 \text{ mol } e^-}{1 \text{ mol Zn}}\right) \times \left(\frac{9.6485 \times 10^4 \text{ C}}{1 \text{ mol } e^-}\right) \times \left(\frac{1 \text{ s}}{0.0715 \text{ C}}\right) = 1.35 \times 10^6 \text{ s}$$

c) The cell produces as many moles of $Ni(s)$ as it consumes moles of $Zn(s)$:

$$0.4998 \text{ mol Ni} \times \left(\frac{58.70 \text{ g Ni}}{1 \text{ mol Ni}}\right) = 29.34 \text{ g Ni}$$

d) The reaction has reduced 0.4998 mol of Ni^{2+} ion when it comes to a stop for want of zinc. There remains $0.575 - 0.4998 = 0.075$ mol of $Ni^{2+}(aq)$. This remaining nickel ion is still dissolved in the original 575 mL of solution. Its concentration is $0.075 \text{ mol}/0.575 \text{ L} = 0.13 \text{ mol L}^{-1}$.

10-71

$$1.83 \text{ g Zn} \times \left(\frac{1 \text{ mol Zn}}{65.38 \text{ g Zn}}\right) \times \left(\frac{2 \text{ mol } e^-}{1 \text{ mol Zn}}\right) \times \left(\frac{9.6485 \times 10^4 \text{ C}}{1 \text{ mol } e^-}\right)$$

$$\times \left(\frac{100 \text{ C total}}{0.25 \text{ C}}\right) = 2.16 \times 10^6 \text{ C total}$$

10-73 In the conversion $Al_2O_3 \rightarrow Al$, aluminum passes from the +3 oxidation state to the 0 oxidation state; 3 mol of electrons is transferred for every 1 mol of aluminum formed. The total charge Q transferred for the year's supply of Al is:

$$1 \times 10^{10} \text{ kg} \times \left(\frac{1 \text{ mol Al}}{0.02697 \text{ kg}}\right) \times \left(\frac{3 \text{ mol } e^-}{1 \text{ mol Al}}\right) \times \left(\frac{9.65 \times 10^4 \text{ C}}{1 \text{ mol } e^-}\right) = 1.07 \times 10^{17} \text{ C}$$

The energy required to transfer charge through an electrolysis cell is the amount of charge times the voltage at which the cell runs:

$$E = Q\Delta\mathcal{E} = (1.07 \times 10^{17} \text{ C})(5.0 \text{ V}) = 5.4 \times 10^{17} \text{ J}$$

The cost of the energy is:

$$5.4 \times 10^{17}\ \text{J} \times \left(\frac{1\ \text{kWh}}{3.6 \times 10^{6}\ \text{J}}\right) \times \left(\frac{\$0.10}{1\ \text{kWh}}\right) = \$1 \times 10^{10} = \$10\ \text{billion}$$

10-75 a) The problem describes in detail the identity and state of the reacting species in two half-cells connected by a salt bridge and a wire to make a galvanic cell. Translate the descriptions into balanced half-equations:

$$O_2(g) + 2\,H_3O^+(aq) + 2e^- \rightarrow H_2O_2(aq) + 2\,H_2O(l) \qquad \mathcal{E}^\circ = 0.682\ \text{V}$$
$$MnO_2(s) + 4\,H_3O^+(aq) + 2e^- \rightarrow Mn^{2+}(aq) + 6\,H_2O(l) \qquad \mathcal{E}^\circ = 1.208\ \text{V}$$

The problem makes it clear that all reactants and products in the cell are in their standard states. The above standard reduction potentials apply without correction to this cell. In a galvanic cell the two half-cells must be combined in such a way that the potential difference exceeds 0.00 V. This means that reduction occurs in the half-cell with the algebraically larger $\mathcal{E}^\circ$. The first half-reaction is the oxidation and the second is the reduction; the overall reaction is the second half-reaction minus the first:

$$H_2O_2(aq) + MnO_2(s) + 2\,H_3O^+(aq) \rightarrow O_2(g) + Mn^{2+}(aq) + 4\,H_2O(l)$$

b) Since all participating species are in their standard states:

$$\text{cell voltage} = \Delta\mathcal{E}^\circ = \mathcal{E}^\circ(\text{cathode}) - \mathcal{E}^\circ(\text{anode}) = 1.208 - 0.682 = +0.526\ \text{V}$$

10-77 A reducing agent is itself oxidized as it acts. The greater the strength of a reducing agent, then the easier it is to oxidize. A useful tactic in comparing ease of oxidation is to rewrite reduction half-equations as oxidations. When this is done the sign of the half-cell potential is reversed. Contrast the equations in problems 10-33 and 10-34 with the following:

$$4\,PH_3(g) + 12\,OH^-(aq) \rightarrow P_4(s) + 12\,H_2O(l) + 12\,e^- \qquad \mathcal{E}^\circ = +0.89\ \text{V}$$
$$HS^-(aq) + OH^-(aq) \rightarrow S(s) + H_2O(l) + 2\,e^- \qquad \mathcal{E}^\circ = +0.51\ \text{V}$$

The $\mathcal{E}^\circ$ for oxidation is more positive for the first half-reaction. Therefore $PH_3(g)$ is a stronger reducing agent than $HS^-(aq)$.

10-79 The cell reaction is $2\,Ag^+(aq) + Zn(s) \rightarrow 2\,Ag(s) + Zn^{2+}(aq)$. The standard voltage for the reaction is:

$$\Delta\mathcal{E}^\circ = \underbrace{(0.7996)}_{Ag^+/Ag} - \underbrace{(-0.7628)}_{Zn^{2+}/Zn} = 1.5624\ \text{V}$$

since clearly Ag^+ is reduced (cathode) and Zn is oxidized (anode). The required voltage of 1.50 V is somewhat less than the standard voltage of the cell. Even if 1 M solutions of $Ag^+(aq)$ and $Zn^{2+}(aq)$ were available, using them would not give the required voltage. The concentrations of one or both of the available solutions must be adjusted (by dilution) so that the cell generates 1.50 V. The Nernst equation relates the actual potential difference and standard potential difference: At 25°C the Nernst equation for this cell is:

$$1.50 = 1.5624 - \frac{0.0592}{2}\log\left(\frac{[Zn^{2+}]}{[Ag^+]^2}\right) \quad \text{which gives} \quad \frac{[Zn^{2+}]}{[Ag^+]^2} = 10^{2.108} = 128$$

This equation must be satisfied to generate the required voltage. If $[Ag^+]$ equals 0.010 M, then $[Zn^{2+}]$ must equal 0.0128 M. There are many other combinations of concentrations that give the required potential difference, but this one is probably the simplest to prepare.

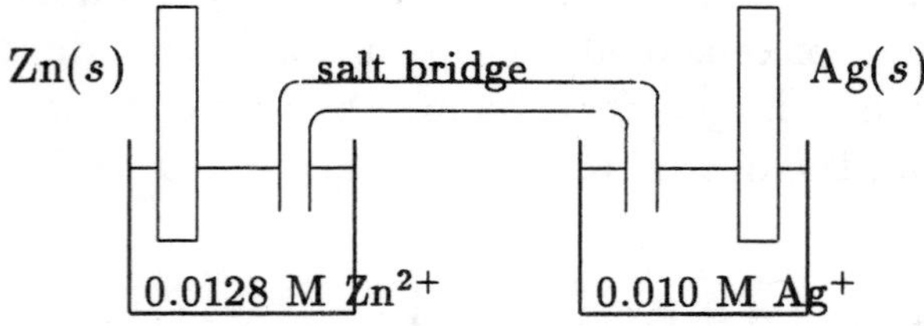

10-81 a) The standard potential difference $\Delta\mathcal{E}°$ for this cell equals zero because the reactions at the anode and cathode are each other's reverses. An actual potential difference exists only because the concentration of the $Pb^{2+}(aq)$ is different in the two half-cells. The spontaneous "reaction" in this cell is entirely a consequence of the natural tendency of solutions at different concentrations to mix to form a solution of intermediate concentration. The change is represented by the sum of the two half-equations:

$$Pb^{2+}(aq,\ 0.10\,M) + 2\,e^- \rightarrow Pb(s)$$
$$Pb(s) \rightarrow Pb^{2+}(aq,\ 0.010\,M) + 2\,e^-$$

The initial voltage generated by this cell can be calculated using the Nernst equation:

$$\Delta\mathcal{E} = \Delta\mathcal{E}° - \frac{0.0592\text{ V}}{n}\log Q = 0.00\text{ V} - \frac{0.0592\text{ V}}{2}\log\left(\frac{[Pb^{2+}]_{\text{less conc}}}{[Pb^{2+}]_{\text{more conc}}}\right)$$

$$\Delta\mathcal{E} = 0.00 - \frac{0.0592}{2}\log\left(\frac{0.010}{0.10}\right) = +0.030\text{ V}$$

b) Addition of the sulfate ion drastically lowers the concentration of lead ion in the first half-cell by means of:

$$Pb^{2+}(aq) + SO_4^{2-}(aq) \rightarrow PbSO_4(s) \qquad K_{sp} = 1.1\times10^{-8}$$

At equilibrium in this half-cell:

$$1.1 \times 10^{-8} = [Pb^{2+}][SO_4^{2+}] = [Pb^{2+}](0.10)$$

from which the equilibrium concentration of lead ion is 1.1×10^{-7} M. Now, the Nernst equation becomes:

$$\Delta\mathcal{E} = 0.00 \text{ V} - \frac{0.0592 \text{ V}}{2} \log \left(\frac{[Pb^{2+}]_{\text{less conc}}}{[Pb^{2+}]_{\text{more conc}}} \right)$$

$$\Delta\mathcal{E}^\circ = 0.00 \text{ V} - \frac{0.0592 \text{ V}}{2} \log \left(\frac{1.1 \times 10^{-7}}{0.010} \right) = 0.15 \text{ V}$$

10-83 A gas confined at high pressure expands spontaneously to a low pressure if a path is available. In a pressure cell, the free-energy change of this spontaneous expansion appears as electrical work. In this case, the gas is Cl_2. At the cathode, $Cl_2(g)$ is reduced to 1 M $Cl^-(aq)$. At the anode, 1 M $Cl^-(aq)$ is oxidized to $Cl_2(g)$:

$$1/2\, Cl_2(g) + e^- \rightarrow Cl^-(aq) \quad \text{cathode}$$
$$Cl^-(aq) \rightarrow 1/2\, Cl_2(g) + e^- \quad \text{anode}$$

The cathode reaction *removes* $Cl_2(g)$ so the reduction must occur in the half-cell held at the *higher* pressure of $Cl_2(g)$, the 0.50 atm half-cell. The overall reaction is:

$$Cl_2 \text{ (0.50 atm)} \rightarrow Cl_2 \text{ (0.01 atm)}$$

The $\Delta\mathcal{E}^\circ$, the standard potential difference for this reaction, is 0.00 V. The Nernst equation at 25°C gives the actual potential difference:

$$\Delta\mathcal{E} = \Delta\mathcal{E}^\circ - \frac{0.0592 \text{ V}}{2} \log \left(\frac{0.010}{0.50} \right) = 0.0000 - (-0.050) = 0.050 \text{ V}$$

10-85 The reactions in the cell are:

anode $\quad Zn(s) \rightarrow Zn^{2+}(aq) + 2\,e^-$
cathode $\quad 2\,MnO_2(s) + 2\,NH_4^+(aq) + 2\,e^- \rightarrow Mn_2O_3(s) + 2\,NH_3(aq) + H_2O(l)$

Electrons are released at the anode and taken up at the cathode. Thus, electrons flow from the Zn electrode to the graphite electrode, at which they reduce MnO_2. The stud on the top of a common flashlight battery is positive, and the bottom is negative.

10-87 a) Oxygen is reduced to water: $O_2(g) + 4\,H_3O^+ + 4\,e^- \rightarrow 6\,H_2O(l)$. This takes place at the nickel cathode, but the nickel does not react. This is implied by the presence of H_3O^+ and O_2 in the shorthand representation of the cell.
b) The overall reaction in the cell consumes ethanol and oxygen to give CO_2 and water: $C_2H_5OH(l) + 3\,O_2(g) \rightarrow 2\,CO_2(g) + 3\,H_2O(l)$. The plan now is to compute ΔG° for this reaction and use it to get $\Delta\mathcal{E}^\circ$:

$$\Delta G^\circ = 2\underbrace{(-394.36)}_{CO_2(g)} + 3\underbrace{(-237.18)}_{H_2O(l)} - 1\underbrace{(-174.89)}_{C_2H_5OH(l)} = -1325.37 \text{ kJ}$$

$$\Delta\mathcal{E}^\circ = \frac{-\Delta G^\circ}{n\mathcal{F}} = \frac{-(-1325.37 \times 10^3 \text{ J})}{(12 \text{ mol})(96\,485 \text{ C mol}^{-1})} = 1.1447 \text{ J C}^{-1} = 1.1447 \text{ V}$$

c) The standard potential difference is the standard reduction potential at the cathode minus the standard reduction potential at the anode. The $\mathcal{E}^\circ$ at the cathode is for the reduction of gaseous oxygen to water in standard acid. For the half-reaction, the Nernst equation is:

$$\Delta\mathcal{E}^\circ = \mathcal{E}^\circ(H_3O^+, O_2/H_2O) - \mathcal{E}^\circ(CO_2, H_3O^+/C_2H_5OH)$$

$$1.1447 \text{ V} = 1.229 \text{ V} - \mathcal{E}^\circ(CO_2, H_3O^+/C_2H_5OH)$$

Hence $\mathcal{E}^\circ$ is 0.084 V for the ethanol half-cell.

10-89 a) In the first cell, water is oxidized: $6\,H_2O(l) \rightarrow O_2(g) + 4\,H_3O^+ + 4\,e^-$. In the second cell, metallic nickel is oxidized: $Ni(s) \rightarrow Ni^{2+}(aq) + 2\,e^-$.
b) The charge $\mathcal{Q}$ passing through the cells is the current times the elapsed time. Compute this in coulombs and use Faraday's constant to convert to moles:

$$0.10 \text{ C s}^{-1} \times 10 \text{ hr} \times \left(\frac{3600 \text{ s}}{1 \text{ hr}}\right) \times \left(\frac{1 \text{ mol } e^-}{96\,485 \text{ C}}\right) = 0.0373 \text{ mol } e^-$$

Next, set up calculations using ratios from the half-equations and treating the electron like any other product or reactant. In the first cell:

$$0.0373 \text{ mol } e^- \times \left(\frac{1\ O_2}{4 \text{ mol } e^-}\right) \times \left(\frac{32.00 \text{ g } O_2}{1 \text{ mol } O_2}\right) = 0.30 \text{ g } O_2$$

$$0.0373 \text{ mol } e^- \times \left(\frac{4\ H_3O^+}{4 \text{ mol } e^-}\right) \times \left(\frac{1.008 \text{ g } H_3O^+}{1 \text{ mol } H_3O^+}\right) = 0.038 \text{ g } H_3O^+$$

In the second cell:

$$0.0373 \text{ mol } e^- \times \left(\frac{1\ Ni}{2 \text{ mol } e^-}\right) \times \left(\frac{58.69 \text{ g Ni}}{1 \text{ mol Ni}}\right) = 1.1 \text{ g Ni}$$

Chapter 11

Chemical Kinetics

Chemical kinetics is the study of the **rates** and **mechanisms** of chemical reactions. Most chemical reactions do not occur as represented in chemical equations but instead by a series of much simpler steps that when added together give the net equation. A **mechanism** is a sequence of such **elementary steps.**

Rates of Chemical Reactions

As a chemical reaction progresses, the concentrations or partial pressures of the reactants fall, and the concentrations or partial pressures of the products increase. The **average reaction rate** is the change in the concentration of a reactant or product divided by the time interval Δt over which it occurs. The **instantaneous rate** is the rate obtained in the limit as smaller and smaller values of Δt are considered.

The instantaneous rate of the general chemical reaction:

$$a\mathrm{A} = b\mathrm{B} \rightarrow c\mathrm{C} + d\mathrm{D}$$

is determined by monitoring the disappearance of a reactant or the appearance of a product. It is defined as:

$$\text{rate} = -\frac{1}{a}\frac{d[\mathrm{A}]}{dt} = -\frac{1}{b}\frac{d[\mathrm{B}]}{dt} = +\frac{1}{c}\frac{d[\mathrm{C}]}{dt} = +\frac{1}{d}\frac{d[\mathrm{D}]}{dt}$$

In this equation, the first d in the last term stands for the coefficient of the product D in the chemical equation, and the others are part of the derivative. Because the concentration of A falls as the reaction proceeds, the time rate of change of this concentration, $d[\mathrm{A}]/dt$, is negative. The same holds for all reactants. The minus signs are put in to keep the rate positive.

In general, the instantaneous rate of a reaction changes from moment to moment. The **initial rate** of a reaction is the rate at the moment that the reactants are mixed. It is the rate at $t = 0$.

The experimental determination of reaction rates requires a thermostat, because rates are quite sensitive to temperature, a clock to measure the time, and a means to monitor the concentration of at least one product or reactant. In some cases it is possible to *quench* (stop) a reaction by suddenly cooling it and then to analyze for the concentration of a reactant or product. An alternative is to monitor some physical property as a continuous function of time. The total pressure is often used. See **problems 11-1** and **11-48.** The color of the reaction mixture is another property frequently monitored.

Rate Laws

Rates of reactions almost always change with time. Most reactions slow down, but reactions that start slow and pick up speed are also known. One reason for a reaction to slow down is that any observed rate of a reaction is really a **net rate,** the difference between the rate of the forward reaction and of the reverse reaction. If pure reactants are mixed, there is at first no reverse reaction because there are no products available to react. Back-reaction takes hold only later.

Reaction Order

The forward rate of a reaction may change with time even if back-reaction is prevented by quick removal of the products. In general the forward rate depends on the concentrations of the reactants. The dependence is expressed in the **rate law** of the chemical reaction. For the general chemical equation written above, the common form of the rate law is:

$$\textbf{rate} = -\frac{1}{a}\frac{d[\mathbf{A}]}{dt} = k[\mathbf{A}]^m[\mathbf{B}]^n$$

The exponents m and n may equal positive or negative whole numbers or fractions or zero. A species whose concentration has an exponent of zero has no effect on the rate of the reaction (although some of it does have to be present if it is a reactant). A species with a negative exponent in a rate law makes the reaction go more slowly as its concentration is raised. See **problem 11-8** for an illustration of both of these points.

The constant of proportionality k is the **rate constant.** Rate constants are strongly affected by the temperature (see Section 11-5, text page 422). The exponents m and n determine the **order** of the reaction: m is the order with respect to reactant A, and n is the order with respect to reactant B. The **overall order** of the reaction is the sum of all the exponents in the rate law.

The Units of k

The overall order of a reaction determines the units of its rate constant. A rate (not rate constant) is measured in units of moles per liter per second (mol $L^{-1}s^{-1}$). In the rate law for a first-order reaction, k is multiplied by a concentration (mol L^{-1}). The units of k must be s^{-1} to make the rate come out as required. In a second-order reaction k is multiplied by *two* concentrations (with overall units mol^2 L^{-2}. The units of k now must be L $mol^{-1}s^{-1}$ to make the units of the rate come out right. See **problem 11-9b.**

Determining a Rate Law

Rate laws are experimental results. In one kind of experiment used to get rate laws, the concentrations of all but one reactant are held constant in a series of runs, while the initial rate is measured with different concentrations of the one reactant under study. The initial rate is the rate observed before possible back-reaction among the products has a chance to confuse the issue. If doubling the concentration of the reactant under test doubles the initial rate, then the reaction is first-order in the test reactant. If the same doubling quadruples the rate, then the reaction is second order in that one reactant. See **problems 11-7 and 11-9c.**

Integrated Rate Laws

The instantaneous rate of consumption (or production) of a reactant (or product) in a chemical reaction changes as the reaction proceeds. An **integrated rate law** takes account of this and still expresses the concentration of the product or reactant directly as a function of time. Although there are mathematical methods by which to integrate nearly all of the many experimental rate laws that occur, the necessary calculus can be quite complex. Fortunately, first-order and second-order reactions are the most common and have simple integrated forms. It is best simply to learn them.

First-order Integrated Rate Law. Consider the reaction: A $\longrightarrow$ products. Assume taht it is first order. Let $\boldsymbol{c}$ equal the concentration of A at any time, $\boldsymbol{t}$, after the reaction starts. Let $\boldsymbol{c}_0$ equal the original concentration of A (the concentration at time 0). Then:

$$\boldsymbol{c = c_0 e^{-kt}}$$

Other forms of this integrated rate law are:

$$\ln c - \ln c_0 = -kt \quad \text{and} \quad \ln\left(\frac{c}{c_0}\right) = -kt$$

According to the equation, the change in $\ln c$ over equal intervals of time is always the same in a first-order reaction. In other words, the equation predicts that for a first-order reaction a plot of $\ln c$ against time will be a straight line.

The **half-life** $t_{1/2}$ of a first-order reaction is the time required for the concentration of a reactant to decrease to half of its initial value. Big rate constants mean fast reactions which require but little time to consume half the initial concentration of a reactant. Thus the half-life is inversely proportional to the rate constant:

$$\text{half life} = t_{1/2} = \frac{\ln 2}{k}$$

See problems 11-13 and 11-15.

Second-order Integrated Rate Law. For the second-order reaction 2 A $\rightarrow$ products, the rate law is:

$$\text{rate} = -\frac{1}{2}\frac{d[\text{A}]}{dt} = k[\text{A}]^2$$

and the integrated rate law is:

$$\frac{1}{c} = \frac{1}{c_0} + 2kt$$

where c is the concentration of A at any time and c_0 is its original concentration. This integrated rate equation predicts that a plot of $1/c$ versus time will be a straight line. Knowing the rate constant and the integrated rate law for a reaction it is possible to predict the concentrations of reactants at any time (see **problems 11-13 and 11-17**). It is also possible to figure out the time required for some specified change in concentration **(problems 11-11 and 11-19).**

Integrated rate laws for reactions of the same order may differ somewhat if the stoichiometries of the reactions differ. The integrated rate law for the second order reaction A + B $\rightarrow$ products is:

$$\frac{1}{c} = \frac{1}{c_0} + kt$$

where c and c_0 refer to the concentration of A. The extra factor of 2 in equation 11-4 (developed on page 410 of the text) reflects the fact that the balanced equation has the form 2 A $\rightarrow$ products. This point is important in **problem 11-19.**

Reaction Mechanisms

A reaction mechanism consists of a sequence of **elementary steps** that tell exactly what molecules must collide and in what sequence to convert the reactants into products. A mechanism also includes an indication of the rates of all the steps. This indication consists of a numerical rate constant for each step or a word or two (fast, slow, very fast, and so forth) giving a rough idea of the relative rates of the steps. Only the actual rate constants themselves are a complete statement of the rates and relative rates of the steps.

A mechanism may include **intermediates.** Intermediates are species produced in a early step and later consumed. It may also include **catalysts,** which are species that come from outside, join in an early step in the mechanism and are regenerated in their original form by a later step. The sum of all of the steps of a reaction mechanism equals the balanced chemical reaction. See **problems 11-23, 11-55,** and **11-57.**

Elementary Reactions

An elementary step is also called an elementary reaction. Elementary reactions involve the collision *in a single event* of the reactant particles to give the product particles. Such reactions have **molecularity.** Collision of two molecules to give products is a **bimolecular** elementary reaction. The simultaneous collision of three particles is a **termolecular** elementary reaction. An **unimolecular** reaction involves only a single reactant molecule. Only elementary reactions have molecularity. The order of elementary reactions equals their molecularity. The vast majority of chemical reactions proceed by way of mechanisms that are the sums of several elementary steps. The concept of molecularity does not apply to such non-elementary reactions.

In any mechanism, the product of the rate constants for the forward steps divided by the product of the rate constants for the reverse step is equal to the equilibrium constant of the overall reaction. See **problem 11-49.**

Rate Determining Steps

If one of the elementary steps in a mechanism is much slower than the rest, it acts as a bottle-neck and determines the overall rate of reaction. With this in mind the procedure for writing the rate law predicted by a mechanism is:

1. Write the rate law for the slow step. This requires nothing more than taking the coefficients of the species in the slow step as the exponents in the rate law.

2. Use algebra to eliminate the concentrations of any intermediates that may occur in the slow-step rate law. Intermediates often appear in mechanisms but may never appear in experimental rate laws. Deducing a rate law from a mechanism is a waste of effort unless the result is in a form directly comparable with experiment.

Many mechanisms have one or more fast equilibria preceding the slow step. **Problems 11-27** shows how to get rid of concentrations of intermediates in writing rate laws from mechanisms.

The Steady-State Approximation

If all the steps in a mechanism have about the same rate, then no single one of them is rate-determining. The **steady-state approximation** is useful in such cases. In this approximation, it is assumed that the concentrations of intermediates remain constant over most of the course of the reaction. Suppose an intermediate is produced by the first step of a mechanism and consumed by the second. The net rate of change of its concentration is its rate of production minus its rate of consumption. In the steady-state approximation assumes that this net rate of change equals zero. See **problem 11-55.** The net rate of change may involve several different terms **(see problem 11-55).** In every case, the steady state approximation furnishes an equation that gives the concentrations of intermediates in terms of the concentrations of non-intermediates and rate constants. In this way it allows the elimination of the concentrations of intermediates as a rate law is deduced from a mechanism. **See problem 11-65.**

Chain Reactions

A chain reaction proceeds through a series of elementary steps, some of which are repeated many times. The repetition arises when a step in a mechanism consumes a reactive intermediate but simultaneously generates another intermediate. Reaction of this intermediate then generates some more of the first intermediate. Such reactions have three stages: **initiation,** the production of the first intermediate; **propagation,** the use-one, make-one weaving of the chain, and **termination,** in which two reactive intermediates collide and thereby take each other of the chain-generating scheme. A reaction in which exactly new intermediate is generated for every one used proceeds at a steady rate (see **problem 11-57).** A **branching chain reaction** features an increase in the number of intermediates and accelerates as it proceeds. In the branching chain reaction diagrammed in Figure 12-12 (text page 467), which is a nuclear reaction, the intermediate is the neutron.

Effect of Temperature on Reaction Rates

Molecules must collide in order to react, but not all collisions lead to reactions. Most molecular collisions are not energetic enough. Only those collisions for which the collision energy exceeds some minimum can result in reaction. The minimum is the **activation energy, E_a,** of the reaction.

The **Arrhenius equation** gives the temperature dependence of rate constants (and thus of reaction rates themselves):

$$k = Ae^{-E_a/RT}$$

The activation energy has the same units as RT; kJ mol^{-1} are common. The pre-exponential factor A is a constant having the same units as k. See **problem 11-39a.** If E_a were zero, then the exponent in the Arrhenius equation would equal zero and, because any number raised to the zero power is 1, k would equal A.

A useful form of the Arrhenius equation is:

$$\ln\left(\frac{k_1}{k_2}\right) = +\frac{E_a}{R}\left(\frac{1}{T_2} - \frac{1}{T_1}\right) = -\frac{E_a}{R}\left(\frac{1}{T_1} - \frac{1}{T_2}\right)$$

where k_1 is the rate constant at T_1 and k_2 is the rate constant at T_2. Note that if k_1 and k_2 are exchanged on the left in the above equation, then the signs in front of (E_a/R) both reverse.

Clearly, measuring the rate constant at just two temperatures could give the activation energy. In practice, rate constants are measured at several different temperatures and the activation energy is determined from the slope of the line in a plot of $\ln k$ versus $1/T$ (see**problem 11-35**).

Reactions have activation energies because formation of a product requires more than a mere encounter between the reactant molecules. As reactant molecules approach each other along a **reaction path,** they must possess sufficient energy to scale an energy barrier or hill associated with an unstable **activated complex** or **transition state**. Only then do products form. The activation energy for a forward reaction is in general different from the activation energy for the reverse reaction. The difference between $E_{a,f}$ and $E_{a,r}$ is the thermodynamic ΔE of the reaction.

The activation energy of an elementary step is always positive. For overall reactions consisting of two or more elementary steps, E_a can be negative as well as positive. A negative E_a means that the observed rate constant diminishes with increasing temperature. This occurs when a fast exothermic equilibrium (with equilibrium constant K) precedes the rate-determining step (with rate constant k) in the reaction's mechanism. The observed rate constant is such a case equals Kk. See **problem 11-27.** Although k rises with temperature, K for an exothermic reaction gets smaller and can outweigh the increase in k in determining the behavior of Kk.

Collision Theory of Gas Phase Reactions

Results (Chapter 3) on the rate of collisions among like molecules in the gaseous phase provide insight into the rates of gas-phase reactions. Rate constants can be predicted knowing the diameter d and molar mass $\mathcal{M}$ of the molecules that are colliding and the activation energy of a collision:

$$k = 2Pd^2N_0\sqrt{\frac{\pi RT}{\mathcal{M}}}e^{-E_a/RT}$$

The factor P is a steric factor to account for the fact that some of the collisions that are energetic enough to cause reaction are ineffective because the molecules collide in the wrong orientation. The units of this expression require some care (see **problem 10-43**).

Kinetics of Catalysis

A catalyst is a substance that enters into a chemical reaction and speeds it up, but is not itself consumed in the reaction. Catalysts do not appear in the balanced overall equation representing a reaction, but do appear in the rate law. Catalysts accelerate chemical reactions by providing new reaction pathways that have lower activation energies.

An **inhibitor** is the opposite of a catalyst. It **slows** the rates of chemical reactions. See **problem 10-65.** In a sense a catalyst and intermediate are opposites, too. An intermediate is generated within a reaction's mechanism and consumed there; it may *not* appear in the rate law for the reaction. A catalyst joins a mechanism from without and is regenerated before the mechanism is over; it *may* appear in the rate law for the reaction. See **problems 11-23** and **11-63.**

In **homogeneous catalysis** the catalyst is present in the same phase as the reactants. In **heterogeneous catalysis** the reactants are in one phase and the catalyst is in another phase.

Enzyme Catalysis

Enzymes catalyze many chemical reactions that take place in living organisms. They are high-molecular-mass proteins that bind at their **active sites** to reactant molecules (called **substrates** of the enzyme) and promote specific transformations of the reactant molecules. The kinetics of enzyme-catalyzed reactions involve no kinetic concepts that do not also appear in non-enzyme catalysis. **Problem 11-65** shows how the rate of an enzyme-catalyzed reaction is lowered by the presence of an inhibitor.

Detailed Solutions to Odd-Numbered Problems

11-1 If the CO_2 behaves ideally, its concentration in the vessel at the start is 0.02125 mol L^{-1} (using $n/V = P/RT$). Then, 1800 s (30:00 minutes) later, its concentration is 0.02636 mol L^{-1}. The average change in concentration is the final concentration minus the initial concentration divided by the elapsed time. It is 2.84×10^{-6} mol $L^{-1}s^{-1}$.

11-3 Figure 11-3 on text page 401 presents the concentration of NO in a system as a function of time. The instantaneous rate of production of NO in the system equals the slope of the tangent to this line. The slope of the tangent changes constantly.

Sketch in the tangent line at t equal to 200 s. Reading from the graph (as shown in the Figure for a t of 150 s) gives 5.3×10^{-5} mol $L^{-1}s^{-1}$ as the instantaneous rate.

11-5 The rate of a reaction is expressed in terms of the rate of disappearance of a reactant, or formation of a product:

$$\text{rate} = -\frac{1}{1}\frac{d[N_2]}{dt} = -\frac{1}{3}\frac{d[H_2]}{dt} = +\frac{1}{2}\frac{d[NH_3]}{dt}$$

11-7 a) The way in which the rate of the reaction responds to the changes in concentration of the two reactants gives the order of the reaction. The reaction is first order in H_2 because the rate varies as to the first power of the concentration of H_2 and second order in NO because the rate varies as to the second power of the concentration of NO. The rate expression is

$$\text{rate} = k[H_2][NO]^2$$

The reaction has an overall order of 3. The units of the rate constant, k, are $\text{mol}^{-2}\,L^2\,s^{-1}$ (see text page 406 for details). This can also be written as $M^{-2}s^{-1}$.
b) Multiplying $[H_2]$ by 2 would double the rate, and multiplying [NO] by 3 would increase the rate by a factor of 9. The combined effect would be to increase the rate by a factor of 18.

11-9 a) Consider the second and third data lines in the table. When $[C_5H_5N]$ is held constant and $[CH_3I]$ is doubled, the rate doubles. Consider the first and second lines in the table. When both concentrations are doubled, the rate is increased by a factor of 4. We therefore conclude:

$$\text{rate} = k[C_5H_5N][CH_3I]$$

b) Calculate k by substituting the data from any one of the three data points into the rate equation. Using the first line of the table:

$$7.5 \times 10^{-7}\text{ mol L}^{-1}\text{s}^{-1} = k(1.00 \times 10^{-4}\text{ mol L}^{-1})(1.00 \times 10^{-4}\text{ mol L}^{-1})$$

Hence k is 75 L $\text{mol}^{-1}\text{s}^{-1}$.
c) Substitute the known k and the two given concentrations into the rate equation. Remember that "M" stands for mol L^{-1}.

$$\text{rate} = (75\text{ mol}^{-1}\text{L s}^{-1)})(5.0 \times 10^{-5}\text{ mol L}^{-1})(2.0 \times 10^{-5}\text{ mol L}^{-1})$$

$$= 7.5 \times 10^{-8}\text{ mol L}^{-1}\text{s}^{-1}$$

11-11 The problem asks for the time it will take for the pressure to fall to half of its initial value. This is exactly the half-life. In a first-order process the half-life depends solely upon the rate constant. In this case:

$$t_{1/2} = \frac{\ln 2}{k} = \frac{0.69315}{2.2 \times 10^{-5}\ \text{s}^{-1}} = 3.2 \times 10^{4}\ \text{s}$$

11-13 The dehydration is first order in the chemical amount of the $LaPO_4{\cdot}\frac{1}{2}H_2O(s)$. This means that the chemical amount n of this solid reactant at any time t is related to the original amount n_0 by the equation:

$$n = n_0 e^{-kt}$$

which is an adaptation of the first-order integrated rate law given on text page 407. The mass of the reactant is directly proportional to its chemical amount, so the following equation can also be written:

$$m = m_0 e^{-kt}$$

The time t in this case is 45 minutes, which is 2.7×10^3 s, and the original mass m_0 of the reactant is 5.00 g. Substitution gives

$$m = m_0 e^{-kt} = (5.00\ \text{g}) \exp\bigl(-2.3 \times 10^{-4}\ \text{s}^{-1}\bigr)(2.7 \times 10^{3}\ \text{s}) = 2.7\ \text{g}$$

11-15 This first-order reaction follows the integrated rate law:

$$[C_2H_5Cl] = [C_2H_5Cl]_0 e^{-kt}$$

Dividing both sides of this equation by the original concentration of the C_2H_5Cl and taking the natural logarithm of both sides gives:

$$\ln \frac{[C_2H_5Cl]}{[C_2H_5Cl]_0} = -kt$$

Substitution of two concentrations and the time gives:

$$\ln \frac{0.0016}{0.0098} = -k(340\ \text{s})$$

which is easily solved to give k equal to 5.3×10^{-3} s.

11-17 For this second-order process, the integrated rate law is:

$$\frac{1}{[\text{I}]} = \frac{1}{[\text{I}]_0} + 2kt \text{ from which: } \frac{1}{[\text{I}]} = \frac{1}{1.00 \times 10^{-4}\ \text{M}} + 2(8.2 \times 10^{9}\ \text{M}^{-1}\text{s}^{-1})(2.0 \times 10^{-6}\ \text{s})$$

Solving gives $[\text{I}] = 2.34 \times 10^{-5}$ M.

11-19 The reaction is the neutralization of $OH^-(aq)$ with $NH_4^+(aq)$. Aqueous acid-base reactions are generally fast. This reaction is no exception because the room-temperature rate constant is huge: $k = 3.4 \times 10^{10}\ M^{-1}s^{-1}$. The answer will be a very short time. If 1.00 L of 0.0010 M NaOH and 1.00 L of 0.0010 M NH_4Cl are mixed, then *after* the mixing, but *before* the reaction can start each reactant is 5.0×10^{-4} M. The kinetics are second-order overall:

$$\text{rate} = \frac{-d[OH^-]}{dt} = k[OH^-][NH_4^+]$$

Throughout the reaction $[OH^-] = [NH_4^+]$. Let this concentration be represented by c. Then:

$$\frac{-dc}{dt} = kc^2$$

Integrating this equation and inserting the initial condition gives:

$$\frac{1}{c} = \frac{1}{c_0} + kt$$

This equation lacks the factor of 2 that appears in text equation 11-4 because the stoichiometry of the reaction in this problem does not have a factor of 2. For c_0 equal to 1.0×10^{-5} M and k equal to $3.4 \times 10^{10}\ M^{-1}s^{-1}$, the equation becomes:

$$\frac{1}{1.0 \times 10^{-5}\ M} = \frac{1}{5.0 \times 10^{-4}\ M} + (3.4 \times 10^{10}\ M^{-1}s^{-1})t$$

$$9.8 \times 10^4\ M^{-1} = (3.4 \times 10^{10}\ M^{-1}s^{-1})t$$

$$t = 2.9 \times 10^{-6}\ s$$

11-21 a) Two particles collide in an elementary reaction. The reaction is therefore bimolecular. Its rate law is: rate $= k[HCO][O_2]$.
b) Three particles collide in an elementary reaction. The reaction is therefore termolecular: rate $= k[CH_3][O_2][N_2]$.
c) A single particle decomposes spontaneously in an elementary reaction. The reaction is unimolecular: rate $= k[HO_2NO_2]$.

11-23 a) The first step is unimolecular, and the three subsequent steps are bimolecular. This is determined simply by counting the number of interacting particles on the left sides of the four equations. Molecularity has meaning only in reference to elementary reactions.
b) The overall reaction is the sum of the steps: $H_2O_2 + O_3 \rightarrow H_2O + 2\,O_2$.
c) The intermediates are O, ClO, CF_2Cl, and Cl. These species are produced in the course of the reaction and later consumed. The CF_2Cl_2 is a catalyst. It enters the

reaction mechanism from outside and is regenerated in the course of the reaction. The overall reaction could not proceed according to this mechanism without the presence and participation of CF_2Cl_2, but the compound is, on a net basis, neither consumed nor generated. Incidentally, the CF_2Cl_2 is a chlorofluorocarbon (see text page 792). It and other chlorofluorocarbons catalyze the destruction of ozone in the stratosphere by the mechanism given in this problem and by related mechanisms.

11-25 The equilibrium constant of this elementary reaction equals the ratio of the rate constant of the forward reaction to the rate constant of the reverse reaction:

$$K = \frac{k_f}{k_r} = 5.0 \times 10^{10} = \frac{1.3 \times 10^{10} \text{ L mol}^{-1}\text{s}^{-1}}{k_r}$$

Solving for k_r gives 0.26 L $\text{mol}^{-1}\text{s}^{-1}$. The reaction in question is in fact just the reverse of the original elementary reaction, so the answer is 0.26 L $\text{mol}^{-1}\text{s}^{-1}$.

11-27 a) The rate-limiting elementary step in a mechanism determines the overall reaction rate. In this case the slow step is: $C + E \rightarrow F$ so a preliminary version of the rate law is:

$$\text{rate} = k_2[C][E]$$

Unfortunately, the expression involves the concentration of C, an intermediate. This is unacceptable. To eliminate [C] in the rate law, consider how C is formed. It arises in the first step of the mechanism, a fast equilibrium. For that first step:

$$k_1[A][B] = k_{-1}[C][D]$$

Solve this equation for the concentration of C and substitute into the preliminary rate law:

$$\text{rate} = \frac{k_1 k_2}{k_{-1}} \frac{[A][B][E]}{[D]}$$

This rate law, which is the answer, predicts that accumulation of product D in the system slows down the reaction. The overall reaction is the sum of the two steps: $A + B + E \rightarrow D + F$.

b) The overall reaction in this case is: $A + D \rightarrow B + F$. For the two fast equilibria:

$$k_1[A] = k_{-1}[B][C] \quad \text{and} \quad k_2[C][D] = k_{-2}[E]$$

The last, slow step is the rate-determining step:

$$\text{rate} = k_3[E]$$

But E is an intermediate and its concentration may not appear in the final rate expression. To eliminate [E], solve the second of the preceding pair of equations of [E] and substitute:

$$\text{rate} = \frac{k_2 k_3}{k_{-2}}[C][D]$$

This expression *still* contains the concentration of an intermediate (C now). Eliminate [C] by solving the first of the pair for [C] and substituting:

$$\text{rate} = \frac{k_1 k_2 k_3}{k_{-1} k_{-2}} \frac{[\text{A}][\text{D}]}{[\text{B}]} = k_{\text{expt}} \frac{[\text{A}][\text{D}]}{[\text{B}]}$$

where the experimental k is the algebraic composite of the several step-wise rate constants. The reaction is first-order in both A and D, is -1 order in B and first-order overall.

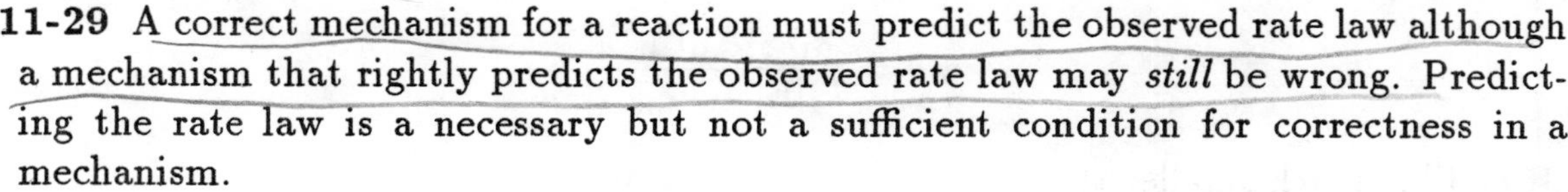

11-29 A correct mechanism for a reaction must predict the observed rate law although a mechanism that rightly predicts the observed rate law may *still* be wrong. Predicting the rate law is a necessary but not a sufficient condition for correctness in a mechanism.

The reaction of HCl with propane is first-order in propane and third order in HCl. This is an experimental fact. Mechanism (a) proposes the rapid formation of the intermediate H from 2 HCl's followed by slow combination of the H with CH_3CHCH_2. This mechanism thus predicts only second-order kinetics in HCl and must be wrong. Mechanism (b) proposes *two* fast equilibria. The first involves HCl only and the second involves HCl and propane. The slow step is the combination of the two intermediates that these equilibria produce. The rate is accordingly proportional to all the concentrations of all the reactants in the two fast equilibria. HCl occurs three times among these reactants and propane occurs once so this mechanism *is* consistent with the observed rate law.

Mechanism (c) involves HCl in the fast production of two different intermediates which then slowly combine to give the product (and to regenerate some HCl). It predicts second-order kinetics in HCl (2 HCl's consumed to furnish reactants for the slow step). It is therefore not consistent with observation.

11-31 The three mechanisms predict the three rate laws:

$$\text{rate} = k_1[\text{NO}_2\text{Cl}] \quad \text{rate} = \frac{k_1 k_2}{k_{-1}} \frac{[\text{NO}_2\text{Cl}]^2}{[\text{Cl}_2]} \quad \text{rate} = \frac{k_1 k_2 k_3}{k_{-1} k_{-2}} \frac{[\text{NO}_2\text{Cl}]^2}{[\text{NO}_2]}$$

The cluster of constants in each rate law can be regarded as a single new rate constant "k_{expt}." Only mechanism (a) is consistent with experiment.

11-33 The overall reaction is $\text{A} + \text{B} + \text{E} \rightarrow \text{D} + \text{F}$. It proceeds first by an equilibrium between A and B giving D and the intermediate C, and then by the consumption of C in reaction with E to give F. The rate of appearance of C is $k_1[\text{A}][\text{B}]$, and the rate of *disappearance* of C is $k_{-1}[\text{C}][\text{D}] + \text{k}_2[\text{C}][\text{E}]$. This latter is a sum because C disappears both by its *back* reaction (with D) and by *further* reaction (with E). As the concentration of C becomes constant (the steady state approximation) the rate of disappearance of C must become equal its rate of appearance:

$$k_1[\mathrm{A}][\mathrm{B}] = k_{-1}[\mathrm{C}][\mathrm{D}] + k_2[\mathrm{C}][\mathrm{E}]$$

The rate of the reaction can be expressed in terms of the rate of appearance of a product, for example, F:

$$\text{rate} = \frac{d[\mathrm{F}]}{dt} = k_2[\mathrm{E}][\mathrm{C}]$$

Solve the steady-state equation for [C] and substitute the answer in the preceding:

$$\text{rate} = \frac{k_1 k_2[\mathrm{A}][\mathrm{B}][\mathrm{E}]}{k_2[\mathrm{E}] + k_{-1}[\mathrm{D}]}$$

If $k_2[\mathrm{E}]$ is much smaller than $k_{-1}[\mathrm{D}]$ then this expression becomes:

$$\text{rate} \approx \frac{k_1 k_2}{k_{-1}} \frac{[\mathrm{A}][\mathrm{B}][\mathrm{E}]}{[\mathrm{D}]}$$

The second-step rate constant k_2 is much smaller than k_{-1} when the first step of the reaction is a fast equilibrium.

11-35 a) According to the Arrhenius equation, the rate constant of an elementary reaction depends on the absolute temperature and activation energy E_a:

$$k = Ae^{-E_a/RT}$$

where A is a constant. Taking the natural logarithm of both sides:

$$\ln k = \ln A - \frac{E_a}{RT}$$

This means that a plot of $\ln k$ versus the reciprocal of T should be a straight line with a slope of $-E_a/R$ and an intercept (when $1/T = 0$) of $\ln A$. Two points determine a line. A quick way to estimate E_a is to select any two of the four data points in the problem (such as the first two), insert the values in the above equation:

$$\ln(5.49 \times 10^6) = \ln A - \frac{E_a}{(5000\ K)R} \quad \text{and} \quad \ln(9.86 \times 10^6) = \ln A - \frac{E_a}{(1000\ K)R}$$

and then solve for E_a (by eliminating $\ln A$ between the equations). This procedure gives $E_a = 432$ kJ mol^{-1}. Note that E_a has the same units as RT. Now, selecting *another* pair of points (for example the second two) and doing the same thing gives a slightly different answer: $E_a = 392$ kJ mol^{-1}. The discrepancy means that the experimental data do not fall exactly on a straight line. The best way to use all data is to perform a *least-squares fit,* mathematically determining the slope of the

straight line that comes closest to all four data points. Many electronic calculators are equipped to complete the necessary calculations almost without effort. Based on the minimization of the sum of the squares of the deviations, E_a is 425 kJ mol^{-1}.
b) As $1/T$ goes to zero, $\ln k$ approaches $\ln A$. Recall that using just the first two data points gave $E_a = 432$ kJ mol^{-1}. Substituting this value and the k and T values of the first data point into the Arrhenius equation gives $\ln A$ equal to 25.9. Using an E_a of 392 kJ mol^{-1} and the third or fourth (k, T) pair makes $\ln A$ equal 25.5. The least-squares fitting gives $\ln A = 25.76$ corresponding to $A = 1.54 \times 10^{11}$ M^{-1}s^{-1}. This is the best answer. The units of A are always the same as the units of k.

11-37 (a) Start by calculating $\ln A$ for this reaction using the Arrhenius equation and the values of k and E_a given in the problem. The T is equal to 303.2 K (30.0°C). The (rearranged) equation is:

$$\ln A = \ln k + \frac{E_a}{RT} = \ln(1.94 \times 10^{-4}) + \left(\frac{1.61 \times 10^5 \text{ J mol}^{-1}}{(8.315 \text{ J K}^{-1}\text{mol}^{-1})(303.2 \text{ K})}\right) = 55.31$$

The value of A itself might be calculated at this point, but it is not needed. Simply put $\ln A$ back into the Arrhenius equation with T equal to 313.2 (40.0°C):

$$\ln k = \ln A - \frac{E_a}{RT} = 55.31 - \left(\frac{1.61 \times 10^5 \text{ J mol}^{-1}}{(8.315 \text{ J K}^{-1}\text{mol}^{-1})(313.2 \text{ K})}\right) = -6.51$$

Taking the antilogarithm of -6.51 gives k equal to 1.49×10^{-3} M^{-1}s^{-1}. A small increase in temperature (a mere 10 K) increases the rate constant of the reaction by a factor of almost 8.
b) This reaction is second order, but the following reasoning applies to reactions of any order. The larger the rate constant of a reaction, the more rapidly it goes. Faster reactions require less time to reach any designated point in their progress. Increasing the temperature of this reaction from 30.0 to 40.0°C increases k from 1.94×10^{-4} M^{-1}s^{-1} to 14.9×10^{-4} M^{-1}s^{-1}, which is a factor of 7.68. The time to reach the half-way mark in the reaction is therefore reduced by a factor of 7.68. The 50 percent conversion requires only 1300 s at 40.0°C instead of the 10,000 s it requires at 30.0°C.

11-39 a) Solve the Arrhenius equation for A and substitute the quantities given in the problem:

$$A = \frac{k}{e^{-E_a/RT}} = \frac{0.41 \text{ s}^{-1}}{\exp(-1.61 \times 10^5 \text{ J mol}^{-1}/(8.315 \text{ J K}^{-1}\text{mol}^{-1})(600 \text{ K}))}$$

$$= 4.3 \times 10^{13} \text{ s}^{-1}$$

b) Assume that neither the activation energy E_a nor the Arrhenius A changes with temperature. Substitute these values into the Arrhenius equation with T equal to

1000 K:

$$k = 4.25 \times 10^{13}\ \text{s}^{-1} \exp\left(\frac{-1.61 \times 10^5\ \text{J mol}^{-1}}{(8.315\ \text{J K}^{-1}\text{mol}^{-1})(1000\ \text{K})}\right) = 1.7 \times 10^5\ \text{s}^{-1}$$

11-41 The activation energy is the difference in energy between the initial state and the activated complex. The activated complex is 3.5 kJ mol^{-1} higher in energy than the reactants, which are HO(g) plus HCl(g). The products, H_2O plus Cl(g), are themselves 66.8 kJ mol^{-1} lower in energy than the reactants. It follows that the activated complex is 70.3 kJ mol^{-1} higher in energy than the products. To pass from the products to the activated complex requires an activation energy of 70.3 kJ mol^{-1}.

11-43 Equation 11-7 (text page 427), relates the rate constant of a second-order reaction between like molecules to the molecular diameter and molar mass. Rewrite this equation inserting the steric factor P, which accounts for the fact that some collisions with sufficient energy may have the wrong orientation for successful reaction:

$$k = P2d^2 N_0 \sqrt{\frac{\pi RT}{\mathcal{M}}}\, e^{-E_a/RT}$$

This equation has the form of the Arrhenius equation. Term-by-term comparison of it and the Arrhenius equation identifies the pre-exponential part with A in the Arrhenius equation:

$$A = P2d^2 N_0 \sqrt{\frac{\pi RT}{\mathcal{M}}}$$

All of the quantities on the right side of this equation are available. Express them in SI base units, if they are not already in such units, and insert them into the equation. This includes in particular the molar mass of the NOCl, which is 0.06546 kg mol^{-1}. The computation then proceeds:

$$A = (0.16)(2)(3.0 \times 10^{-10})^2(6.022 \times 10^{23})\sqrt{\frac{(3.1416)(8.315)(298)}{0.06546}} = 5.98 \times 10^6$$

The units in this computation were omitted. It is wise to repeat the algebra on the units alone:

$$\text{units of } A = \text{m}^2(\text{mol}^{-1})\sqrt{\frac{(\text{J K}^{-1}\text{mol}^{-1})(\text{K})}{\text{kg mol}^{-1}}} = (\text{m})^2(\text{mol}^{-1})\sqrt{\frac{(\text{kg m}^2\text{s}^{-2})}{\text{kg}}}$$

$$= \text{m}^2\ \text{mol}^{-1}\,(\text{m s}^{-1}) = \text{m}^3\text{mol}^{-1}\text{s}^{-1}$$

The answer is therefore 6.0×10^6 m^3 mol^{-1}s^{-1}. The cubic meter is too large for use as a unit of volume in laboratory-scale chemistry. It equals 1000 L. Converting from cubic meters to liters gives a final answer of 6.0×10^9 L mol^{-1}s^{-1}.

11-45 The rate expression for the process is: rate = [hemoglobin][O_2]. Straightforward substitution of the given data yields the answer:

$$\text{rate} = (4 \times 10^7\ \text{M}^{-1}\text{s}^{-1})(2 \times 10^{-9}\ \text{M})(5 \times 10^{-5}\ \text{M}) = 4 \times 10^{-6}\ \text{M s}^{-1}$$

11-47 a) For this first-order process, $\ln[\text{In}^+] = -kt + \ln[\text{In}^+]_0$. Prepare a table of the natural logarithm of the concentration of In^+ versus time:

t (s)	0	240	480	720	1000	1200	10000
$\ln[\text{In}^+]$	−4.80	−5.05	−5.30	−5.55	−5.80	−5.80	−5.80

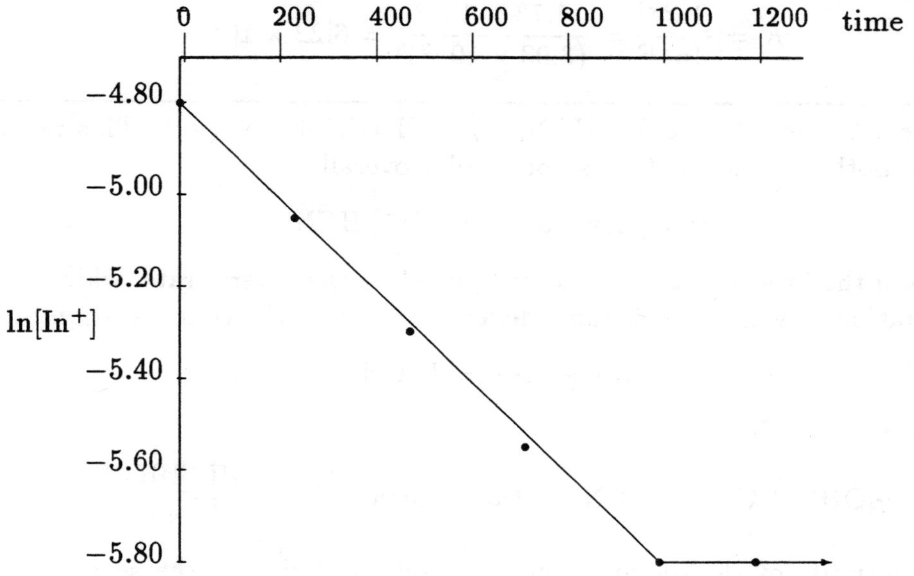

0 200 400 600 800 1000 1200 time

−4.80

−5.00

−5.20

ln[In+]

−5.40

−5.60

−5.80

A common error is to make the values on the y-axis go up the axis as they become more negative. They should go down the axis as they become more negative. Then the plot consists of a straight line that slopes from northwest to southeast and levels off above 1000 s. The initial slope of the line is $-k$ and the intercept is $\ln[\text{ I}^+]_0$. The slope can be read off the graph, but a least-squares analysis gives a somewhat better value. From such an analysis of the data, the slope is -1.01×10^{-3} s^{-1}. The rate constant is therefore 1.01×10^{-3} s^{-1}.

b) The half-life can be determined from the rate constant,

$$t_{1/2} = \frac{\ln 2}{k} = \frac{0.6931}{1.01 \times 10^{-3}\ \text{s}^{-1}} = 686\ \text{s}$$

c) At 10000 seconds, the concentration of In^+ has persisted unchanged for 9000 seconds. The reaction has clearly reached equilibrium. The concentration of In^+ is 3.03×10^{-3} M, and the amount of In^+ in the 1.00 L solution is accordingly 3.03×10^{-3} mol. The equilibrium concentration of In^{3+} is calculated by subtracting this amount from the original amount of In^+, and using the stoichiometric relation between the In^+ and In^{3+} as follows:

$$n_{In^{3+}} = \frac{8.23 \times 10^{-3} - 3.03 \times 10^{-3}}{3} = 1.73 \times 10^{-3} \text{ mol}$$

The equilibrium concentration of In^{3+} in the 1.00 L system is therefore 1.73×10^{-3} M. Substitute in the usual equilibrium expression:

$$K = \frac{[In^{3+}]}{[In^+]^3} = \frac{1.73 \times 10^{-3}}{(3.03 \times 10^{-3})^3} = 6.22 \times 10^4$$

11-49 The reaction is: $OH^-(aq) + HCN(aq) \rightarrow H_2O(l) + CN^-(aq)$. This reaction is first-order in both OH^- and HCN, second-order overall:

$$\text{rate (forward)} = k_f[OH^-][HCN]$$

At equilibrium the forward rate is exactly equaled by the reverse rate, which, because the concentration of water is constant, depends solely on the concentration of CN^-:

$$\text{rate (reverse)} = k_r[CN^-]$$

It follows that:

$$k_f[OH^-][HCN] = k_r[CN^-] \quad \text{from which:} \quad \frac{k_r}{k_f} = \frac{[OH^-][HCN]}{[CN^-]}$$

Note that the units of the quantity on the right-hand side of the second equation are mol L^{-1}. Numerically, the right-hand side of the preceding equation equals the equilibrium constant K_b of the reaction:

$$CN^-(aq) + H_2O(l) \rightleftharpoons HCN(aq) + OH^-(aq)$$

The K_b of CN^- ion is related to K_a, the acid ionization constant of its conjugate acid HCN:

$$\frac{1.0 \times 10^{-14}}{4.93 \times 10^{-10}} = 2.03 \times 10^{-5}$$

(where the K_a comes from text Table 6-2). Therefore:

$$\frac{k_r}{k_f} = 2.03 \times 10^{-5} \text{ mol L}^{-1}$$

Substitution of 3.7×10^{-9} L $mol^{-1}s^{-1}$ for k_f gives $k_r = 7.5 \times 10^4$ s^{-1}.

11-51 Both reactions are third-order, but [M] can be treated as a constant since it is much larger than the original concentrations of the I and Br and is not changed by the progress of the reaction. The integrated rate laws for the two reactions are then:

$$\frac{1}{[\mathrm{I}]} - \frac{1}{[\mathrm{I}]_0} = 2k_{\mathrm{I}}[\mathrm{M}]t \quad \text{and} \quad \frac{1}{[\mathrm{Br}]} - \frac{1}{[\mathrm{Br}]_0} = 2k_{\mathrm{Br}}[\mathrm{M}]t$$

After one half-life, $[\mathrm{I}] = 1/2\ [\mathrm{I}]_0$ and $[\mathrm{Br}] = 1/2\ [\mathrm{Br}]_0$ so:

$$\frac{1}{[\mathrm{I}]_0} = 2k_{\mathrm{I}}[\mathrm{M}]t_{1/2,\mathrm{I}} \quad \text{and} \quad \frac{1}{[\mathrm{Br}]_0} = 2k_{\mathrm{Br}}[\mathrm{M}]t_{1/2,\mathrm{Br}}$$

Dividing the first equation by the second gives:

$$\frac{[\mathrm{Br}]_0}{[\mathrm{I}]_0} = \left(\frac{k_{\mathrm{I}}}{k_{\mathrm{Br}}}\right)\left(\frac{t_{1/2,\mathrm{I}}}{t_{1/2,\mathrm{Br}}}\right)$$

Substitute $[\mathrm{I}]_0 = 2[\mathrm{Br}]_0$ and $k_{\mathrm{I}} = 3.0k_{\mathrm{Br}}$:

$$\frac{1}{2} = 3.0\left(\frac{t_{1/2,\mathrm{I}}}{t_{1/2,\mathrm{Br}}}\right) \text{ from which } \left(\frac{t_{1/2,\mathrm{I}}}{t_{1/2,\mathrm{Br}}}\right) = \frac{1}{6.0} = 0.17$$

11-53 The two reactions are both reactions of hydrogen with a halogen to form a hydrohalic acid. However, the reactions must proceed by different mechanisms since the rate laws are different:

$$\text{rate (bromine)} = k[\mathrm{H_2}][\mathrm{Br_2}]^{1/2} \qquad \text{rate (iodine)} = k[\mathrm{H_2}][\mathrm{I_2}]$$

The currently accepted mechanisms for both reactions involve a fast equilibrium to split the halogen molecule into its two atoms followed by reaction of one such atom with H_2. In the case of Br, this elementary process is rate-limiting, but in the case of I, it is fast. The third steps in the mechanisms differ.

$\mathrm{H} + \mathrm{Br_2} \rightarrow \mathrm{HBr} + \mathrm{Br}$ fast for bromine reaction

$\mathrm{H_2I} + \mathrm{I} \rightarrow 2\,\mathrm{HI}$ rate – limiting for iodine reaction

Also, in the iodine reaction, the intermediate H_2I is thought to exist, whereas the analogous H_2Br does not appear in the proposed mechanism for the bromine reaction.

11-55 The reaction is the decomposition of ozone by light: $2\,\mathrm{O_3}$ + light $\rightarrow 3\,\mathrm{O_2}$. The mechanism involves the production of the intermediate O atom from O_3 (in the first step), and its consumption either to regenerate O_3 (the second step) or to make $2\,\mathrm{O_2}$

in an encounter with an O_3 molecule (the third step). The change in the concentration of O with time is:

$$\frac{d[\mathrm{O}]}{dt} = k_1[\mathrm{O_3}] - k_2[\mathrm{O}][\mathrm{O_2}][\mathrm{M}] - k_3[\mathrm{O}][\mathrm{O_3}]$$

This equation states that the rate of change of the concentration of O equals its rate of production minus its rate of consumption. The steady-state approximation is that [O] comes to a *steady* value. If [O] is steady it is unchanging and $d[\mathrm{O}]/dt = 0$. Then:

$$k_1[\mathrm{O_3}] - k_2[\mathrm{O}][\mathrm{O_2}][\mathrm{M}] - k_3[\mathrm{O}][\mathrm{O_3}] = 0$$

so that

$$[\mathrm{O}] = \frac{k_1[\mathrm{O_3}]}{k_2[\mathrm{O_2}][\mathrm{M}] + k_3[\mathrm{O_3}]}$$

All of this concerns the intermediate. The rate of the overall reaction is:

$$\text{rate} = \frac{1}{3}\frac{d[\mathrm{O_2}]}{dt} = k_3[\mathrm{O}][\mathrm{O_3}]$$

Substituting the expression for the concentration of the intermediate O into this rate law gives:

$$\text{rate} = \frac{k_3 k_1[\mathrm{O_3}]^2}{k_2[\mathrm{O_2}][\mathrm{M}] + k_3[\mathrm{O_3}]}$$

Divide the top and bottom of the fraction by k_3:

$$\text{rate} = \frac{k_1[\mathrm{O_3}]^2}{(k_2/k_3)\,[\mathrm{O_2}][\mathrm{M}] + [\mathrm{O_3}]}$$

Clearly, only the ratio k_2/k_3 affects the rate, not the individual values. This ratio tells how much of the intermediate O cycles back to O_3 relative to how much goes on to give the product.

11-57 The initiation step is $CH_3CHO \rightarrow CH_3 + CHO$. The propagation steps involve the intermediates CH_3 (the methyl radical) and CH_2CHO (the acetyl radical). The propagation step is the combination of the second and third elementary reactions in the problem. The CH_3 is consumed in the second reaction and regenerated in the third, which forms the product CO. This chain tends to continue indefinitely to consume all available CH_3CHO except when cut by the termination reaction, which forms the by-product CH_3CH_3 when two methyl radicals encounter each other and react.

11-59 a) Assume that the rate constants of the reactions occurring in the cooking depend on the temperature according to the Arrhenius equation. Since the food cooks two times faster at 112°C (385 K) than at 100°C (373 K), the rate constant must be twice as large at 385 K: $k_{385} = 2\,k_{373}$. An Arrhenius equation can be written for each of these k's. Hence:

$$Ae^{-E_a/385R} = 2\,Ae^{-E_a/373R}$$

Cancelling out the A's, taking the natural logarithm of both sides and rearranging gives:

$$\frac{-E_a}{385\text{ K}} - \frac{-E_a}{373\text{ K}} = R\ln 2 = (8.315\text{ J K}^{-1}\text{mol}^{-1})(0.6931)$$

Solving for E_a gives an activation energy equal to 69.0 kJ mol^{-1}.

b) The rates of the cooking reactions at 94.4°C (367.6 K) depend on the rate constant at that temperature. The following equation involving the rate constant for cooking at 367.6 K comes from dividing the Arrhenius equation at 367.6 K by the Arrhenius equation at 373 K:

$$\frac{k_{367.6}}{k_{373}} = \frac{e^{-E_a/367.6R}}{e^{-E_a/373R}}$$

Taking the natural logarithm of both sides gives

$$\ln\left(\frac{k_{367.6}}{k_{373}}\right) = \frac{-E_a}{367.6R} - \frac{-E_a}{373R} = \frac{-69.0\times10^3\text{ J mol}^{-1}}{8.315\text{ J K}^{-1}\text{mol}^{-1}}\left(\frac{1}{367.6} - \frac{1}{373}\right) = -0.339$$

where the value of E_a has been taken from part a). Take the antilogarithm of both sides:

$$\frac{k_{367.6}}{k_{373}} = e^{-0.339} = 0.712$$

Since the rate constant at 94.4°C is smaller by the factor 0.712 times than the rate constant at 100°C, the food takes 1/0.712 = 1.404 times longer to cook. Instead of 10 minutes, the cooking requires 14 minutes.

11-61 The rate constants of the forward and reverse reactions in the equilibrium between gaseous hydrogen and iodine on the one side and gaseous hydrogen iodide gas on the other are related to the equilibrium constant by the equation:

$$K = \frac{k_f}{k_r}$$

Write Arrhenius equations for both the forward reaction and the reverse reaction:

$$k_f = Ae^{-E_{a,f}/RT} \qquad \text{and} \qquad k_r = Ae^{-E_{a,r}/RT}$$

Divide the first equation by the second and take the logarithm of both sides:

$$\ln\left(\frac{k_f}{k_r}\right) = \frac{-E_{a,f}}{RT} - \frac{-E_{a,r}}{RT} = \frac{1}{RT}(E_{a,r} - E_{a,f})$$

The left side of this equation equals $\ln K$, the logarithm of the equilibrium constant. But $\ln K$ of the reaction at the temperature T is related to the standard free-energy change at that temperature by the thermodynamic equation $\Delta G^\circ_T = -RT\ln K$. Substitution gives

$$-\frac{\Delta G^\circ_T}{RT} = \frac{1}{RT}(E_{a,r} - E_{a,f})$$

Multiplication of both sides by $-RT$ gives (at the constant temperature T):

$$\Delta G^\circ_T = (E_{a,f} - E_{a,r})$$

Assuming that ΔH° and ΔS° do not change going from 298 to 1000 K, the standard free-energy change of this reaction at 1000 K is:

$$\Delta G^\circ_{1000} = \Delta H^\circ - 1000\,\Delta S^\circ$$

Use the standard enthalpies of formation of the reactants and products to compute ΔH°:

$$\Delta H^\circ = 2\underbrace{(26.48)}_{\text{HI}(g)} - 1\underbrace{(62.44)}_{\text{I}_2(g)} - 1\underbrace{(0)}_{\text{H}_2(g)} = -9.48 \text{ kJ}$$

Do the same with absolute entropies to get ΔS°:

$$\Delta S^\circ = 2\underbrace{(206.48)}_{\text{HI}(g)} - 1\underbrace{(260.58)}_{\text{I}_2(g)} - 1\underbrace{(130.57)}_{\text{H}_2(g)} = 21.81 \text{ J K}^{-1}$$

Combining these values gives ΔG° at 1000 K:

$$\Delta G^\circ_{1000} = -9.48 \times 10^3 \text{ J} - (1000 \text{ K})(21.81\text{J K}^{-1}) = -31.29 \times 10^3 \text{ J}$$

Now, two of the three quantities in the relationship $\Delta G^\circ_{1000} = (E_{a,f} - E_{a,r})$ are known. Substitute them:

$$-31.29 \text{ kJ mol}^{-1} = (165 \text{ kJ mol}^{-1} - E_{a,r})$$

Solving gives 196 kJ mol^{-1} as the activation energy of the reverse reaction.
An easy way to obtain k_r is to calculate the equilibrium constant K_{1000} and use the equation: $K = \frac{k_f}{k_r}$ because k_f is known. Thus:

$$\ln K = -\frac{-31.29 \times 10^3 \text{ J mol}^{-1}}{8.315 \text{ J K}^{-1}\text{mol}^{-1}(1000 \text{ K})} = 3.763 \quad \text{and} \quad K = 43.08$$

It follows that:

$$k_r = \frac{k_f}{K} = \frac{240 \text{ L mol}^{-1}\text{s}^{-1}}{43.08} = 5.6 \text{ L mol}^{-1}\text{s}^{-1}$$

11-63 The CF_2Cl_2 enters into the chemical reaction and presumably speeds it up, but is not itself consumed in the reaction. It does not appear in the balanced overall equation representing the reaction because it is consumed in one step of the mechanism but regenerated later. It is a catalyst.

11-65 The mechanism of enzyme catalysis, as modified to allow for the action of an inhibitor, is:

$E + S \rightleftharpoons ES$	fast equilibrium	k_1 and k_{-1}
$ES \rightarrow E + P$	slow	k_2
$E + I \rightleftharpoons EI$	fast equilibrium	k_3 and k_{-3}

This is a case of competitive inhibition. The inhibitor (symbolized I) competes with the substrate S in binding to the enzyme E. The generation of product P is thereby slowed because P can form from S only through the complex ES. Follow the general pattern of the derivation in Section 11-7, but allow for the complication of the inhibitor. Label the total concentration of enzyme $[E]_0$. The enzyme is present in one of three states: free, bound to the inhibitor, or bound to the substrate:

$$[E]_0 = [E] + [EI] + [ES]$$

Set up the equilibrium expression for the third step of the mechanism, letting k_3/k_{-3} equal K_3:

$$K_3 = \frac{[EI]}{[E][I]}$$

Solve this expression for [EI], substitute into the first equation, and solve for [E] as follows:

$$[E]_0 = [E] + K_3[E][I] + [ES]$$

$$[E]_0 = [E]\big(1 + K_3[I]\big) + [ES]$$

$$[E] = \frac{[E]_0 - [ES]}{1 + K_3[I]}$$

Now, make the steady-state approximation for ES, the intermediate. This approximation is that the ES is generated as fast as it is consumed—its concentration does not change with time:

$$0 = \frac{d[ES]}{dt} = k_1[E][S] - k_{-1}[ES] - k_2[ES]$$

Substitute the previous expression for [E] into this equation. Then solve for [ES]:

$$[ES] = \frac{k_1[E]_0[S]}{k_1[S] + \big(1 + K_3[I]\big)(k_{-1} + k_2)}$$

The rate of the reaction is the rate of the slow step, which equals $k_2[\mathrm{ES}]$. Therefore:

$$\text{rate} = k_2[\mathrm{ES}] = \frac{k_2k_1[\mathrm{E}]_0[\mathrm{S}]}{k_1[\mathrm{S}] + (1 + K_3[\mathrm{I}])(k_{-1} + k_2)} = \frac{k_2[\mathrm{E}]_0[\mathrm{S}]}{[\mathrm{S}] + K_m(1 + K_3[\mathrm{I}])}$$

where K_m is defined as $(k_{-1}+k_2)/k_1$. This K_m is called the Michaelis-Menten constant and is introduced (although not named) on text page 433. Any concentration of the inhibitor increases the denominator of this expression and therefore lowers the initial rate of the reaction.

Chapter 12

Fundamental Particles and Nuclear Chemistry

Building Blocks of the Atom

Several important experiments reveal the constituents of the atom and give numerical values for the parameters (like charge and mass) that characterize them. These experiments include:

The Thomson Experiment. This experiment determined the charge-to-mass ratio of the electron. J. J. Thomson and his predecessors knew that an electrical current flows across the gap between two electrodes inserted in an evacuated tube and held at a large potential difference. The current flows although little or no air or other gas remains in the tube to conduct the current. The current is carried by **cathode rays,** so called because they emanate from the negatively charged cathode. Cathode rays are streams of **electrons,** the fundamental particles of electricity.

An electric field deflects a beam of cathode rays that passes through it. Similarly, a magnetic field established in the region of a beam of cathode rays deflects the trajectory of the beam. In the Thomson experiment a beam of cathode rays in an evacuated tube is passed through an electric and magnetic field simultaneously. By adjusting the strengths and directions of the two fields (E and H), Thomson attained a *balance* between their deflecting effects and caused the beam to pass undeviated through the tube. Knowing the strengths of the two fields allowed calculation of the velocity of the cathode rays ($v = E/H$). Turning off the magnetic field allowed the electric field alone to act on the cathode rays and deflect them. Measuring their displacement s as they left the field and the length ℓ of the region of the field allowed calculation of the ratio of charge to mass of the electrons (cathode rays):

$$\frac{e}{m_e} = \frac{2sE}{\ell^2 H^2} = 1.7588196 \times 10^{11} \text{ C kg}^{-1}$$

In the equation, the SI units of E are N C^{-1}, the SI units of H are N s $m^{-1}C^{-1}$ (where N stands for the newton, the unit of force and C stands for the coulomb, the unit of charge). Naturally, s and ℓ are in meters. Refresh skills in the cancellation of units (as in **problem 3-27a**) by verifying that the units of e/m_e are indeed C kg^{-1}.

The Millikan Oil-Drop Experiment. This experiment determined the charge of the electron. Judging from the large size of the ratio e/m, electrons are either very highly charged or very light. Determination of either e or m_e tells which, and the oil-drop experiment determines e.

The experiment starts with a fine spray of oil droplets. Electrons are lost or gained by some droplets, either by friction or other means, giving them an electrostatic charge. The droplets fall under the influence of gravity but are caught and stopped by a properly arranged electrical field of variable strength.

If a droplet is held motionless, the electrical and gravitational forces acting on it are in balance. The gravitational force depends on the mass of the droplet, which can be determined from its terminal velocity as it falls through the air in the absence of the electric field, and g, the known acceleration of gravity. The electrical force depends on the known strength of the electric field and the charge on the oil droplet. The only unknown, the charge on the oil droplet, is thus determined.

Millikan observed (as in **problem 12-41**) that the magnitude of the charge on many different droplets was always an integral multiple of the same basic value. He suggested that different droplets carry different integral numbers of electrons, all with the same fundamental charge. The modern value is $1.6021773 \times 10^{-19}$ C. Using e/m_e in a quick computation gives m_e) of the electron. It is 9.109390×10^{-31} kg.

The Rutherford Experiment The experiment led to the nuclear model of the structure of the atom. Rutherford and his co-workers investigated the scattering of α-particles (helium atoms with both 2 electrons removed) as beams of them impinged on thin foils of gold and other metallic elements. Most of the α-particles passed through the foils as if they were not there or suffered only slight deflection. Significantly, however, *the foils deflected some α-particles through large angles.*

This unexpected result implied that the mass of the foils was concentrated in small, dense, positively-charged **nuclei** that were totally missed by most of the α-particles but closely approached by those few that ended up scattered through large angles. A detailed analysis based on this model predicted the number of scattered α-particles as a function of angle. The prediction matched with the experimental results.

The Rutherford model pictures the nucleus with charge $+Ze$ and radius on the order of 10^{-15} m surrounded by Z extra-nuclear electrons distributed at distances on the order of 10^{-10} m.

Protons and Neutrons

An atomic nucleus consists of Z protons which account for the nuclear charge by contributing one unit of positive charge ($+1.602 \times 10^{-19}$ C) each. The lightest atom ($Z = 1$) is hydrogen. Avogadro's number (1 mole) of hydrogen atoms weighs about 1.00 g so the mass of a proton is about

$$\frac{1.00 \times 10^{-3} \text{ kg}}{6.022 \times 10^{23}} = 1.66 \times 10^{-27} \text{ kg}$$

The precise value of the proton mass is 1.672623×10^{-27} kg.

For elements other than hydrogen the atomic mass always *exceeds* the mass contributed by Z protons. Indeed, an early conclusion was: "the number of elementary charges composing the center of the atoms is equal to half the atomic mass." Rutherford accounted for the discrepancy in mass by assuming the existence of additional, electrically neutral particles, **neutrons,** in the nuclei of non-hydrogen atoms. Later experiments by James Chadwick detected the neutron. The neutron mass is 1.674929×10^{-27} kg, indeed close to the proton mass. The neutron charge is zero.

Chemistry concerns itself with the electrons of an atom and the way they are affected in interactions with other atoms. The positive charge on the atomic nucleus dictates the number of the negatively-charged electrons surrounding it and so dictates its chemistry. Thus Z, the atomic number, determines the chemical identity of a nucleus. The 100-plus different chemical symbols are each synonymous with a different Z. The **mass number,** A of a nuclide is the sum of Z and the number of neutrons in the nucleus. It is the number of **nucleons** (protons and neutrons) comprising the nucleus. The mass number has but little influence on the chemistry of the atom.

Isotopes

As just stated, a nuclear species is characterized by its atomic number Z and its mass number A. Nuclei having the same Z but different A's are **isotopes.** Such nuclei have the same number of protons but different numbers of neutrons. They differ only slightly in their chemical behavior.

A full symbolization of an atom or ion is comprised of the chemical symbol augmented by a left superscript, a left subscript and a right superscript. The left superscript is A, the mass number, the left subscript is Z, the atomic number, and the right superscript is the electrical charge on the particle, if any. For example, two isotopes of lithium are ${}^{6}_{3}\mathrm{Li}$ and ${}^{7}_{3}\mathrm{Li}$. If an atom of the first lithium isotope loses an electron the result is the ${}^{6}_{3}\mathrm{Li}^{+}$ ion. Chemists often omit the left subscript since the chemical symbol conveys the same information to a user of the periodic table of the elements.

Atomic Mass Units (Amu)

The masses of protons and neutrons are so small (about 10^{-27} kg) that it is wise to define a new unit of mass that is better scaled for use in talking about them.

• The **atomic mass unit** is defined as 1/12 of the mass of a single neutral ${}^{12}\mathrm{C}$ atom.

One mole of ${}^{12}\mathrm{C}$ weighs exactly 12 g. Therefore:

$$\mathbf{1\ amu} = \frac{\mathbf{12}}{\boldsymbol{N_o}}\ \mathbf{g} = \mathbf{1.660540 \times 10^{-24}\ g}$$

The following table summarizes the symbols and masses of some important particles:

Particle	Symbols	Mass (amu)	Charge (e)
Proton	${}^{1}_{1}p$, ${}^{1}_{1}\mathrm{H}^{+}$	1.00727647	+1
Neutron	${}^{1}_{0}n$	1.00866490	0
Electron	e^{-}, ${}^{0}_{-1}\beta$	0.00054857990	-1
Positron	e^{+}, ${}^{0}_{1}\beta$	0.00054857990	+1
Alpha	${}^{4}_{2}\mathrm{He}$, ${}^{4}_{2}\alpha$	4.001506	+2

Nuclear Stability

The mass of an atom is not equal to the sum of the masses of the protons, neutrons and electrons that make it up. Instead, for all elements, the mass of the atom is *less* than the mass of the component electrons, protons and neutrons. A change in mass is related to a change in energy by the Einstein equation:

$$\boldsymbol{\Delta E = c^2 \Delta m}$$

In problem-solving in nuclear chemistry, Δm values come out in amu. The conversion of a Δm (in amu) to a ΔE is so frequent that it is worthwhile to memorize the conversion factor. An amu is equivalent to 1.4924×10^{-10} J, but, more importantly:

1 amu is equivalent to 931.494 $\times$ 10^6 eV or 931.494 MeV.

An eV is an **electron-volt,** the energy required to accelerate an electron through a potential difference of one volt. Energies of nuclear reactions are usually expressed in MeV, the mega-electron-volt, equal to a million electron-volts. See **problems 12-9,** and 12-53.

Since the formation of an atom from its component parts has a negative Δm, a nuclear reaction such as:

$$20\,{}^1_1\mathrm{H} + 20\,{}^1_0n \rightarrow {}^{40}_{20}\mathrm{Ca}$$

(see **problem 12-5**) releases energy. The quantity of energy released tells how tightly the **nucleons** (protons and neutrons) in the product nucleus are bound.

The **binding energy** E_{B} of a nucleus is the negative of the energy change of its formation from its component nucleons. The **binding energy per nucleon** is the binding energy of a nucleus divided by the sum of the number of neutrons and protons in the nucleus. See **problem 12-5**. The binding energy per nucleon is greatest for ^{56}Fe.

The preceding equation is a **nuclear equation.** Nuclear equations resemble ordinary chemical equations in requiring balancing. The criteria for balance however differ:

1. In a balanced nuclear equation the sum of the subscripts (atomic numbers) must be same on the two sides of the equation.

2. The sum of the superscripts (mass numbers) must be the same on the two sides of the equation. See **problems 12-33** and **12-63.**

3. The electrical charge represented on the two sides of the equation must be the same.

Nuclear Decay Processes

Spontaneous nuclear decay occurs only if the energy of the products is less than the energy of the reactants. Given the Einstein equation this implies that the total mass of the products must be less than the mass of the reactants:

$$\Delta E < 0 \quad \text{and} \quad \Delta m < 0$$

This criterion for spontaneity is equivalent to the thermodynamic criterion (at constant temperature and pressure) $\Delta G < 0$. In nuclear decay, a parent nucleus gives rise to a daughter nucleus, and energy leaves the system as kinetic energy of emitted particles and as electromagnetic radiation.

Although radioactive decay involves the conversion of one kind of nucleus into another, it is more convenient in problems to use the masses of whole atoms (nucleus plus electrons) in calculations. It is whole-atom masses that are given in tables (Table 12-1, text page 452, for example). The following are important processes of nuclear decay:

$^{0}_{-1}\beta$-Decay. This is the decay mode of **proton-deficient** nuclei because it increases the number of protons in the nucleus. A neutron in the unstable nucleus converts into a proton, emitting an electron (beta particle). In this decay mode a massless **antineutrino** ($\bar{\nu}$) is also emitted. The mass number A stays the same and the atomic number Z increases by one. The criterion for $^{0}_{-1}\beta$-decay is:

$$m\,(\text{daughter atom}) - m\,(\text{parent atom}) = \Delta m < 0$$

The masses in this equation are whole-atom masses (Table 12-1), not the masses of bare nuclei. The mass of the β^{-}-particle (electron) is *not* explicitly added in with the mass of the products because it is already counted in the mass of the neutral daughter atom. (see **problem 12-35).** If Δm is negative, then ΔE of the nuclear reaction is also negative, and the reaction is spontaneous. The antineutrino and the β^{-} share in carrying off a total of $-\Delta E$ in energy. If the antineutrino should get zero kinetic energy, then the β^{-} will have its maximum kinetic energy, $-\Delta E$.

β^{+} Decay. If a nucleus is neutron deficient it may emit a positron (β^{+}) and convert a proton into a neutron. The mass number A remains unchanged, but the atomic mass Z decreases by one. The process also emits a neutrino. Thus:

$$^{40}_{21}\text{Sc} \rightarrow {}^{40}_{20}\text{Ca} + {}^{0}_{1}\beta + \nu$$

The criterion for β^{+} decay is:

$$m\,[\text{daughter atom}] - m\,[\text{parent atom}] + 2\,m_e = \Delta m < 0$$

where m_e is 0.00054858 amu, the mass of the electron (and positron). The $2\,m_e$ term appears because the use of neutral atom masses leaves one electron and one positron unaccounted for. See **problem 12-9.** The energy equivalent of 2×0.00054858 amu, using the conversion factor 931.494 MeV amu^{-1}, equals 1.0220 MeV. This is a lot of energy and neglecting it gives seriously wrong answers. See **problem 12-53b.**

Electron Capture. A neutron-deficient nucleus may capture one of its extra-nuclear electrons, converting a proton into a neutron. For example:

$$^{7}_{4}\text{Be} + e^{-} \rightarrow {}^{7}_{3}\text{Li} + \nu$$

This process achieves the same change as β^+ decay, that is, Z decreases by one while A stays the same. The energy (mass) criterion for electron capture is less stringent than for β^+-decay:

$$m\,[\text{daughter atom}] - m\,[\text{parent atom}] = \Delta m < 0$$

The additional $2\,m_e$ term that appears in the case of β^+ decay is now absent. The mass of the electron on the left-hand side of the nuclear equation does not appear explicitly because it is included in the mass of the neutral parent atom, and nothing but a neutrino is emitted.

α-Decay. A nucleus may emit an alpha particle, ${}^4_2\text{He}$, decreasing Z by 2 and A by 4. Thus:

$${}^{226}_{88}\text{Ra} \rightarrow {}^{222}_{86}\text{Rn} + {}^4_2\text{He}$$

The mass (energy) criterion for α decay is:

$$m\,[\text{daughter atom}] + m[\,{}^4_2\text{He atom}] - m\,[\text{parent atom}] = \Delta m < 0$$

Many problems use the above relationships. **Problem 12-49** varies the usual pattern by giving ΔE for two processes and the mass of the parent atom and asking for the mass of the daughter atom. **Problem 12-13** uses none of the formal criteria but requires the ability to apply the fundamental criterion for spontaneous decay, a decrease in mass.

In nuclear decay problems use the conversion factor 931.494 MeV amu^{-1} to switch between mass units and energy units at will. As a practical matter, it may be easier to do arithmetic on the masses in amu, just because the numbers are smaller, and convert to MeV as a last step. In computing energy and mass differences, do not round off prematurely. Carry extra digits at the intermediate stages, and round off at the last step.

Kinetics of Radioactive Decay

All unstable nuclei decay by a first-order process. The kinetics of radioactive decay therefore follow the equation:

$$N = N_0 e^{-kt}$$

where N_0 is the number of nuclei originally present, N is the number present at time t and k is the rate constant. The half-life of a reaction is:

$$t_{1/2} = \frac{\ln 2}{k}$$

Radioactivity is measured by detecting the high energy particles that the decay process produces. The decay rate at any moment is the **activity** $\boldsymbol{A}$of the sample, in terms of the number of disintegrations per second. The activity depends both on the number of atoms of the radioactive nucleus (i.e. the mass of the sample) and the rate constant of the decay:

$$A = kN$$

These definitions are used in **problems 12-19, 12-21, 12-25,** and **12-35. Specific activity** is the activity per specific quantity of sample (gram or mole). See **problem 12-59.** The unit of activity is the **curie.** One curie (Ci) is 3.7×10^{10} radioactive disintegrations per second. The activity and specific activity themselves decay with time:

$$A = A_0 e^{-kt}$$

Once A and k are known the number of nuclei at any time can be calculated. See **problem 12-21.**

In solving problems the following points are helpful:

- Having the half-life is the same as having the rate constant for the decay (and vice-versa) because the product of the two is $\ln 2$ (0.69315). The units of k and $t_{1/2}$ are each other's reciprocals.
- Avoid confusing the units of time. Convert all half-lives (rate constants) in the same problem to the same unit of time (reciprocal time), preferably seconds (reciprocal seconds). See **problem 12-35.**
- All of the concepts of chemical kinetics (Chapter 11) apply to nuclear decay kinetics. In particular, the steady-state approximation (see text page 419) is prominent when a nuclide decays in a series of steps. See **problem 12-61b.**
- Many problems involve radioactive dating. Dating using ^{14}C is a major example. These problems offer only one additional facet that first-order chemical kinetics do not: *activities* (the actual current rate of decay) rather than concentrations of reacting species are usually given.

Radiation in Biology and Medicine

The curie measures the radioactivity of a species in terms of the number of nuclear decay events per second. Different kinds of events emit different kinds of particles. These particles differ in energy and penetrating power. In biology and medicine the important factor is not the number of disintegrations per second in a radioactive source but the quantity of energy deposited in living tissue. The **rad** is therefore defined as the amount of radiation that deposits 10^{-2} J in a 1 kg mass.

The **rem** is the unit of radioactive **dosage.** It is the number of rads absorbed by tissue multiplied by a "fix-up factor," the **relative biological effectiveness,** that accounts for variables such as the type of tissue and type of radiation and the dose rate.

Nuclear Fission

Nuclear fission is the splitting apart of a nucleus into less massive nuclei. Fission is exothermic. The binding energy per nucleon in products exceeds binding energy per nucleon in reactants The fission of several elements, in particular uranium, is induced by the absorption of neutrons. When fission generates two or more new neutrons, the nuclear reaction is a branching chain reaction. It can lead to explosion. In nuclear power reactor, neutron generation moderated by presence (in control rods) of ^{112}Cd or ^{10}B, which capture neutrons well. When the rate of neutron production equals the rate of consumption in a fission reactor, energy is generated at a controlled rate. The products of fission of ^{235}U include 34 different elements. Some are in radioactive isotopes that present disposal problems.

Nuclear Fusion and Nucleosynthesis

In nuclear fusion, light nuclides merge to form heavier ones. The fusion of light atoms is thermodynamically spontaneous, but has a very large activation energy. Fusion reactions are called thermonuclear because colliding atoms must have high kinetic energy (high T). Fusion reactions in stars lead to nucleosynthesis of elements. Hydrogen burning in stars proceeds by the reaction series:

$$2\,{}^{1}_{1}\mathrm{H} \rightarrow {}^{2}_{1}\mathrm{H} + {}^{0}_{1}\beta + \nu$$
$${}^{2}_{1}\mathrm{H} + {}^{1}_{1}\mathrm{H} \rightarrow {}^{3}_{2}\mathrm{He} + \gamma$$
$$2\,{}^{3}_{2}\mathrm{He} \rightarrow {}^{4}_{2}\mathrm{He} + 2\,{}^{1}_{1}\mathrm{H}$$

Detailed Solutions to Odd-Numbered Problems

12-1 a) The atomic number Z of Pu is 94. Hence, an atom of Pu has 94 protons in its nucleus. An atom of ^{239}Pu has a total of 239 nucleons in its nucleus. Those of the 239 that are not protons are neutrons, so there are 145 neutrons. The requested ratio is 145/94, which is 1.54.
b) The term ion implies an electrically charged species. The nucleus of a Pu atom has a charge of +94. The extranuclear electrons of the Pu^- ion must therefore contribute a charge of −95. The charge on each electrons is −1, so there must be 95 electrons. The unit of charge in this discussion is the magnitude of the charge on the electron, which is exactly equal to the magnitude of the charge on the proton.

12-3 The atomic number of americium is 95; americium has 95 protons in its nucleus. In the neutral atom there are also exactly 95 electrons because the negative charge of the electrons balances the positive charge of the protons. Of the 241 nucleons, those that are not protons are neutrons. There are accordingly 146 neutrons.

12-5 a) The total binding energy of the calcium isotope equals the negative of the energy equivalent of the mass change in the following reaction, which represents the formation of the isotope from its component particles:

$$20\,{}^{1}_{1}\mathrm{H} + 20\,{}^{1}_{0}n \longrightarrow {}^{40}_{20}\mathrm{Ca}$$

$$\Delta m = m[{}^{40}_{20}\mathrm{Ca}] - 20\,m[{}^{1}_{1}\mathrm{H}] - 20\,m[{}^{1}_{0}n]$$

$$\Delta m = 39.962589 - 20(1.00782504) - 20(1.00866497) = -0.367209 \text{ amu}$$

As shown on text page 452, the energy equivalent of 1 amu is 931.494 MeV. This Δm therefore corresponds to:

$$\Delta E \times \frac{931.494 \text{ MeV}}{1 \text{ amu}} = -342.053 \text{ MeV}$$

The binding energy E_{B} is the negative of this, +342.053 MeV. To compute the ΔE in joules, convert Δm to kilograms and substitute it into the Einstein equation $\Delta E = c^2\Delta m$ along with the speed of light in meters per second:

$$\Delta E(-0.367209 \text{ amu})\left(\frac{1.660540 \times 10^{-27} \text{ kg}}{1 \text{ amu}}\right)(2.9979246 \times 10^{8} \text{ m s}^{-1})^2$$

The answer is -5.48030×10^{-11} J so the binding energy is $+5.48030 \times 10^{-11}$ J. The binding energy of a mole of atoms is this answer, which is for one atom, multiplied by Avogadro's number. The result is 3.30031×10^{13} J mol^{-1}, which is 3.30031×10^{10} kJ mol^{-1}. Since there are 40 nucleons, the E_{B} per nucleon is $342.053/40 = 8.55132$ MeV.

To save time in subsequent calculations observe that according to the Einstein equation:

$$\Delta m = \frac{1 \text{ amu}}{\text{atom}} \quad \text{is equivalent to } \Delta E = \frac{8.9875519 \times 10^{10} \text{ kJ}}{\text{mol}}$$

b) Compute the difference in mass between the product ^{87}Rb and the reactants $50\,{}^{1}_{0}n$ and $37\,{}^{1}_{1}$H. It is −0.8135878 amu. By computations like those in part a) then, the E_{B} for one atom of ${}^{87}_{37}$Rb is 757.852 MeV, the binding energy per mole of these atoms is 7.3122×10^{10} kJ mol^{-1}, and the binding energy per nucleon is E_{B} divided by 87 or 8.7109 MeV.

c) In ^{238}U there are 92 protons and 146 neutrons; Δm in the assembly of the atom is −1.934194 amu. Accordingly, E_{B} is 1801.69 MeV, which amounts to 17.384×10^{10} kJ mol^{-1}. The binding energy per nucleon is 7.57013 MeV.

12-7 The mass of a single ^{8}Be atom is 8.0053052 amu whereas the mass of two ^{4}He atoms is twice 4.0026033 amu, which is 8.0052066 amu. Because the system has a larger mass when organized as a ^{8}Be atom, the ^{8}Be atom is less stable than the pair of ^{4}He atoms. In other terms, the Δm for the nuclear reaction: ${}^8_4\text{Be} \rightarrow 2\,{}^4_2\text{He}$ is negative. The Δm is -9.86×10^{-5} amu.

12-9 Represent the decay as: ${}^8_5\text{B} \rightarrow {}^8_4\text{Be} + {}^0_1\beta + \nu$. The difference in mass between the two sides of the reaction *appears* to be:

$$\Delta m = \underbrace{8.0053052}_{^8\text{Be}} + \underbrace{0.00054858}_{^0_1\beta} - \underbrace{8.024612}_{^8\text{B}} = -0.0187582 \text{ amu}$$

using the numbers from Table 12-1 (text page 452) for the boron atom, beryllium atom, and positron. At this point a trap looms. The tabulated masses include the electrons that surround the nuclei of the atoms. Beryllium has only four electrons, but boron has five. The computation at this stage has wrongly caused one electron to vanish. This must be corrected by including another electron mass to the mass of the products. The true change in mass is:

$$\Delta m = 8.0053052 + 2(0.00054858) - 8.024612 = -0.01820964 \text{ amu}$$

This mass change has an energy equivalent of -16.962 MeV. Therefore 16.962 MeV is released by the reaction of 1 atom of ^{8}B.

12-11 Positron emission (loss of ${}^0_1\beta$) and electron capture by a nucleus always lower the atomic number by one; electron emission (loss of ${}^{\;0}_{-1}\beta$) raises the atomic number by one.

a) ${}^{39}_{17}\text{Cl} \rightarrow {}^{39}_{18}\text{Ar} + {}^{\;0}_{-1}\beta + \bar{\nu}$ **b)** ${}^{22}_{11}\text{Na} \rightarrow {}^{22}_{10}\text{Ne} + {}^0_1\beta + \nu$

c) ${}^{224}_{88}\text{Ra} \rightarrow {}^{220}_{86}\text{Rn} + {}^4_2\text{He}$ **d)** ${}^{82}_{38}\text{Sr} + {}^{\;0}_{-1}e \rightarrow {}^{82}_{37}\text{Rb} + \nu$

12-13 The equation representing the decay of the neutron is:

$${}^1_0n \rightarrow {}^1_1p + {}^{\;0}_{-1}\beta$$

The other particle is a ${}^{\;0}_{-1}\beta$ (an electron) because only a ${}^{\;0}_{-1}\beta$ balances the nuclear equation. Using the data in Table 12-1 (text page 452), the mass of the products is 1.00782504 amu and the mass of the reactant is 1.00866497 amu. Hence, Δm is -0.00083993 amu. The energy equivalent of 1 amu is 931.494 MeV so $\Delta E = -0.7824$ MeV. As the neutron decays it emits an electron with a maximum kinetic energy of 0.7824 MeV.

12-15 The answer requires the translation of the written description of the process into nuclear equations: ${}^{30}_{14}\text{Si} + {}^1_0n \rightarrow {}^{31}_{14}\text{Si} \rightarrow {}^{31}_{15}\text{P} + {}^{\;0}_{-1}\beta + \bar{\nu}$.

12-17 ${}^{210}_{84}\text{Po} \rightarrow {}^{206}_{82}\text{C} + {}^4_2\text{He}$ $\quad alphap + {}^9_4\text{Be} \rightarrow {}^{12}_6\text{C} + {}^1_0n$.

12-19 Compute how many atoms are present in 0.0010 g of ^{209}Po:

$$0.0010 \text{ g} \times \left(\frac{1 \text{ mol}}{209 \text{ g}}\right) \times \left(\frac{6.02214 \times 10^{23} \text{ atom}}{1 \text{ mol}}\right) = 2.88 \times 10^{18} \text{ atom } ^{209}\text{Po}$$

The activity A, which is the instantaneous rate of disintegration in the polonium, depends on the number of atoms of polonium present:

$$A = -\frac{dN}{dt} = kN$$

The constant in this equation is a first-order rate constant. It equals $\ln 2/t_{1/2}$ where $t_{1/2}$ is the half-life. Substituting:

$$A = \left(\frac{\ln 2}{t_{1/2}}\right) N = \left(\frac{0.6931}{103 \text{ yr}}\right) 2.88 \times 10^{18} = 1.94 \times 10^{16} \text{yr}^{-1}$$

The activity is put on a per minute basis as follows:

$$A = 1.94 \times 10^{16} \text{ yr}^{-1} \times \left(\frac{1 \text{ yr}}{365.2 \text{ day}}\right) \times \left(\frac{1 \text{ day}}{1440 \text{ min}}\right) = 3.7 \times 10^{10} \text{ min}^{-1}$$

That is, 3.7×10^{10} atoms of polonium decay per minute.

12-21 a) The observed activity (rate of decay) at the moment of preparation of the sample of radioactive oxygen is directly proportional to N_0, the number of atoms of oxygen present in the freshly prepared sample:

$$A_0 = kN_0$$

The rate constant k is ($\ln 2/29$ s), which equals 0.0239 s^{-1}. Then:

$$N_0 = \frac{A_0}{k} = \frac{2.5 \times 10^4 \text{ s}^{-1}}{0.0239 \text{ s}^{-1}} = 1.046 \times 10^6 \text{ atom} = 1.0 \times 10^6 \text{ atom } ^{19}\text{O}$$

b) The number of ^{19}O atoms that remains at any time t is:

$$N = N_0 e^{-kt}$$

In this case t is 2.00 min or 120 s. Substitution of k and N_0 gives:

$$N = (1.046 \times 10^6) \exp\big(-(0.0239 \text{ s}^{-1})(120 \text{ s})\big) = 5.9 \times 10^4 \text{ atom}$$

12-23 First calculate N, the number of atoms in 44 mg of ^{219}At:

$$N = 44 \times 10^{-3}\ \text{g} \times \left(\frac{1\ \text{mol}}{219\ \text{g}}\right) \times \left(\frac{6.022 \times 10^{23}\ \text{atoms}}{1\ \text{mol}}\right) = 1.21 \times 10^{20}\ \text{atoms}$$

Then compute the activity of this amount of astatine using the half-life as given:

$$A = kN = \left(\frac{\ln 2}{t_{1/2}}\right) N = \left(\frac{0.6931}{54\ \text{s}}\right)(1.21 \times 10^{20}) = 1.6 \times 10^{18}\ \text{s}^{-1}$$

12-25 The activity of ^{14}C decays according to the same law as the number of atoms of ^{14}C:

$$A = A_0 e^{-kt} \quad \text{which gives:} \quad -kt = \ln\left(\frac{A}{A_0}\right)$$

For the papyrus, A is 9.2 disintegrations min^{-1}g^{-1} and A_0 is 15.3 disintegrations min^{-1}g^{-1}. (It is assumed that the activity of ^{14}C in the biosphere has not changed since Egyptian times.) Also, k is $\ln 2/t_{1/2}$ or 1.21×10^{-4} yr^{-1}. Substituting gives:

$$-(1.21 \times 10^{-4}\ \text{yr}^{-1})\,t = \ln\left(\frac{9.2\ \text{min}^{-1}\text{g}^{-1}}{15.3\ \text{min}^{-1}\text{g}^{-1}}\right) \quad \text{so that:} \quad t = 4.2 \times 10^3\ \text{yr}$$

12-27 First-order kinetics govern the radioactive decay of both ^{235}U and ^{238}U:

$$N(^{235}\text{U}) = N_0(^{235}\text{U})e^{-kt} \quad \text{and} \quad N(^{238}\text{U}) = N_0(^{238}\text{U})e^{-kt}$$

If in the beginning the two isotopes were equally abundant, the current difference in abundance is caused entirely by the faster decay of ^{235}U. (The half-life of ^{235}U is briefer by a factor of about 6.) The current ratio of abundances is 137.7 to 1. In equation form:

$$N_0\,(^{238}\text{U}) = N_0\,(^{235}\text{U}) \quad \text{and} \quad N\,(^{238}\text{U}) = 137.7N\,(^{235}\text{U})$$

This means that:

$$\frac{137.7}{1} = \frac{N_0\,(^{238}\text{U})\,e^{-k_{238}t}}{N_0\,(^{235}\text{U})\,e^{-k_{235}t}} = \frac{e^{-k_{238}t}}{e^{-k_{235}t}}$$

where k_{238} is the rate constant for the decay of ^{238}U and k_{235} is the rate constant for the decay of ^{235}U. Take the natural logarithm of both sides of the equation:

$$\ln 137.7 = (-k_{238}t) - (-k_{235}t) = t(k_{235} - k_{238})$$

For each isotope $k = \ln 2/t_{1/2}$ so:

$$\ln 137.7 = t\left(\frac{\ln 2}{t_{1/2,235}} - \frac{\ln 2}{t_{1/2,238}}\right)$$

The half-lives of the isotopes are 7.13×10^8 years and 4.51×10^9 years respectively. Substitution in the equation gives t equal to 6.0×10^9 years. The supposed supernova occurred about 6 billion years ago. This is about 1.5 billion years before the estimated time of the formation of the solar system.

12-29 Positron emission is accompanied by emission of a neutrino:
$^{11}_{6}\text{C} \rightarrow {}^{11}_{5}\text{B} + {}^{0}_{1}\beta + \nu \qquad {}^{15}_{8}\text{O} \rightarrow {}^{15}_{7}\text{N} + {}^{0}_{1}\beta + \nu$

12-31 Assume that all of the ^{11}C and ^{15}O decay before any is excreted or else that equal chemical amounts of the two radioisotopes are excreted. The ^{15}O deposits 1.74 times more energy per kilogram of body mass than the ^{11}C because its positrons, which are emitted in equal number, are on the average more energetic by the factor $1.72/0.99 = 1.74$.

12-33 The ^{238}U captures a neutron to form ^{239}U, which then loses two ${}^{0}_{-1}\beta$'s to form $^{239}_{94}$Pu. The overall nuclear reaction is:
$^{238}_{92}\text{U} + {}^{1}_{0}n \rightarrow {}^{239}_{92}\text{U} \rightarrow {}^{239}_{94}\text{Pu} + 2\,{}^{0}_{-1}\beta + 2\,\bar{\nu}$.

12-35 **a)** The balanced nuclear reaction is $^{90}_{38}\text{Sr} \rightarrow {}^{90}_{40}\text{Zr} + 2\,{}^{0}_{-1}\beta + 2\,\bar{\nu}$.
b) The overall nuclear reaction is two consecutive beta decays. The change in mass in beta decay is:

$$\Delta m = m\,(\text{daughter atom}) - m\,(\text{parent atom})$$

as explained on text page 454-5. Therefore, to get the Δm of the overall reaction, simply subtract the mass of an atom of ^{90}Sr from the mass of an atom of ^{90}Zr. The masses are the beta particles are automatically accounted for when this is done. The required isotopic masses are listed in the problem. The result is a Δm of -0.0030 amu. The corresponding energy at 1 amu equivalent to 931.494 MeV is 2.8 MeV.
c) This part is just like problem 12-19. Compute N, the number of atoms in 1.00 g of ^{90}Sr:

$$1.00\text{ g} \times \left(\frac{1\text{ mol}}{89.9073\text{ g}}\right) \times \left(\frac{6.02214 \times 10^{23}\text{ atom}}{1\text{ mol}}\right) = 6.698 \times 10^{21}\text{ atom }^{90}\text{Sr}$$

The activity A, which is the instantaneous rate of disintegration of the strontium, depends on the number of strontium atoms present:

$$A = -\frac{dN}{dt} = kN$$

The k in this equation equals $\ln 2/t_{1/2}$ where $t_{1/2}$ is the half-life. Substituting:

$$A = \left(\frac{\ln 2}{t_{1/2}}\right) N = \left(\frac{0.6931}{28.1\text{ yr}}\right)(6.698 \times 10^{21}) = 1.65 \times 10^{20}\text{ yr}^{-1}$$

This is the number of disintegrations per year at the moment that the ^{90}Sr is released. The problem asks for the activity on a per second basis:

$$A = 1.65 \times 10^{20}\ \text{yr}^{-1} \times \left(\frac{1\ \text{yr}}{365.2\ \text{day}}\right) \times \left(\frac{1\ \text{day}}{86400\ \text{s}}\right) = 5.23 \times 10^{12}\ \text{s}^{-1}$$

d) The activity of the ^{90}Sr falls off with time as the number of strontium atoms persisting in the 1.00 g sample diminishes. The relationship is:

$$A = A_0 e^{-kt} = A_0 \exp\left(-\frac{\ln 2\, t}{t_{1/2}}\right)$$

Substituting the initial activity from part d), the specified time of 100 yr and the half-life of 28.1 yr gives

$$A = (5.23 \times 10^{12}\ \text{s}^{-1}) \exp\left(\frac{-0.6931(100\ \text{yr})}{28.1\ \text{yr}}\right) = 4.44 \times 10^{11}\ \text{s}^{-1}$$

The activity of the isotope falls off to about 8.5% of its original value in 100 years.

12-37 The lighter isotopes of uranium happen to decay faster than the heavier isotopes. The quicker breakdown of the light isotopes leaves heavy isotopes behind, causing the average atomic mass of the uranium to increase with time. This assumes that none of the lighter isotopes are the products of decay of the heavier isotopes.

12-39 The change in mass Δm when one atom of ^{235}U gains a neutron and then undergoes fission as specified in the problem is the mass of the products minus the mass of the reactants:

$$\Delta m = 1\underbrace{(93.919)}_{^{94}\text{Kr}} + 1\underbrace{(138.909)}_{^{139}\text{Ba}} + 3\underbrace{(1.0086649)}_{^{1}_{0}n} - 1\underbrace{(235.043925)}_{^{235}\text{U}} - 1\underbrace{(1.0086649)}_{^{1}_{0}n}$$

The difference in mass comes out to 0.1986 amu. When converted to kJ mol^{-1} using the equivalence established in problem 12-5a, the result is -1.785×10^{10} kJ mol^{-1}. The problem asks for the energy change per gram of ^{235}U:

$$-1.785 \times 10^{10}\ \text{kJ mol}^{-1} \times \left(\frac{1\ \text{mol}\ ^{235}\text{U}}{235.04\ \text{g}\ ^{235}\text{U}}\right) = -7.59 \times 10^{7}\ \text{kJ g}^{-1}$$

The energy released in the surroundings is the negative of the energy change of the system. It is $+7.59 \times 10^{7}$ kJ g^{-1}.

12-41 The data consist of nine charges arranged in ascending order. These are the charges found on nine different oil drops in a series of runs of the Millikan experiment. Clearly, the data are only the magnitudes of the charges on the oil drops. If electrons are in excess on the drops, the sign of the charge is in each case negative.
a) As the hint suggests, start by computing the differences between the successive charges in the table of data. The eight differences fall into two categories: those close to 1.6×10^{-19} C and those close to 3.2×10^{-19} C. If the smaller difference is assumed to equal the quantum of charge, which is the charge on a single electron, then the nine oil drops have respectively 4, 5, 7, 8, 10, 11, 12, 14, and 16 electrons in excess of the number required for electrical neutrality.
b) Dividing the charges on the oil drops by the appropriate integers from the list gives nine charges that are close to each other:

1.641×10^{-19} C	1.641×10^{-19} C	1.643×10^{-19} C
1.642×10^{-19}	1.648×10^{-19}	1.635×10^{-19}
1.643×10^{-19}	1.644×10^{-19}	1.636×10^{-19}

The mean of these nine charges is 1.641×10^{-19} C.
c) It is conceivable based on these data alone that the actual charge on the electron is half (or a third or a fourth or other submultiple) of the above charge. There is for example a real chance in this small set of data that only oil drops with an even number of excess electrons were studied. The best way to check against this possibility is to repeat the experiment for a large number of oil drops.

12-43 Moseley established an empirical relationship between the characteristic frequency ν of the x-rays emitted by different elements when bombarded by electrons and the atomic number Z of the elements. This relationship is:

$$\sqrt{\nu} = a(Z - b)$$

A plot of $\sqrt{\nu}$ versus Z for the known elements left neat gaps on the straight line that just accommodated the missing elements predicted by Mendeleev (such as eka-manganese at $Z = 43$). The pattern held at high Z as well as low Z. (Incidentally, Moseley's constant a was 4.980×10^{7} s$^{-1/2}$ and his constant b was nearly exactly 1.)

12-45 The mass of the electron and positron are both equal to 0.00054858 amu. The total mass converted to energy when a positron encounters an electron is therefore twice this or 0.00109716 amu. The equivalent energy of this mass is:

$$0.00109716 \text{ amu} \times \left(\frac{931.494 \text{ MeV}}{1 \text{ amu}}\right) = 1.02200 \text{ MeV}$$

The energy of each γ-particle is half this, or 0.51100 MeV.

12-47 **a)** The $\Delta H°$ $N_2H_4(l) + O_2(g) \rightarrow N_2(g) + 2\,H_2O(g)$ is the sum of the ΔH_f°'s of the products less the sum of the ΔH_f°'s of the reactants, according to Hess's law:

$$\Delta H° = 2\,\underbrace{(-241.82)}_{H_2O(g)} - 1\,\underbrace{(0.00)}_{O_2(g)} - 1\,\underbrace{(-50.63)}_{N_2H_4(l)} = -534.27 \text{ kJ}$$

where the numbers come from Appendix D (recall that elements in their standard states have ΔH_f°'s of zero).

b) The standard energy change of the system during the constant-pressure reaction is:

$$\Delta E° = \Delta H° - P\Delta V$$

The $P\Delta V$ term equals $\Delta n_g RT$, if it is assumed that the gases in the reaction are ideal and that the volume of the liquid hydrazine is negligible (see text page 291). As a mole of $N_2H_4(l)$ burns in gaseous oxygen at 298.15 K, Δn_g of the system is +2 mol, making $\Delta n_g RT$ equal to 4.96 kJ. Hence, for the combustion of one mole of $N_2H_4(l)$:

$$\Delta E° = \Delta H° - P\Delta V = -534.27 - 4.96 = -539.23 \text{ kJ}$$

c) The ΔEo of the reaction is −539.23 kJ per mole of hydrazine that is burned. From the Einstein relationship:

$$\Delta m = \frac{\Delta E}{c^2} = \frac{-539.23 \times 10^3 \text{ J}}{(2.9979 \times 10^8 \text{ m s}^{-1})^2} = -5.9998 \times 10^{-12} \text{ kg}$$

The chemical reaction of 32 g of hydrazine and 32 g of oxygen occasions a mass loss by the system of about 6 nanograms. This is too small to detect.

12-49 **a)** The nuclear reaction that produces ^{64}Ni from ^{64}Cu is represented:

$$^{64}_{29}\text{Cu} \rightarrow {}^{64}_{28}\text{Ni} + {}^{0}_{1}\beta + \nu$$

The ΔE for this reaction is:

$$\Delta E = \left(m(^{64}_{28}\text{Ni}) - m(^{64}_{29}\text{Cu})\right)c^2 + 2\left(m(^{0}_{1}\beta)\right)c^2$$

The last term must be included because a positron is lost from the daughter-parent atom pair and because the neutral daughter atom has one fewer electron than the parent atom. The masses of a positron and electron are the same. The value of ΔE is given in the problem as −0.65 MeV. Convert this to amu using the equivalency "931.494 MeV ≡ 1 amu" (text page 452). Then rewrite the previous equation in terms of masses:

$$-0.65 \text{ Mev} \times \left(\frac{1 \text{ amu}}{931.494 \text{ MeV}}\right) = \left(m(^{64}_{28}\text{Ni}) - m(^{64}_{29}\text{Cu})\right) + 2\,(0.00054858 \text{ amu})$$

Solving for the difference in mass between daughter and parent gives -0.00179 amu. The parent ^{64}Cu weighs 63.92976 amu, so the daughter ^{64}Ni weighs 63.92797 amu.
b) The beta decay of ^{64}Cu is: $^{64}_{29}\text{Cu} \rightarrow {}^{64}_{30}\text{Zn} + {}^{0}_{-1}\beta + \bar{\nu}$. For this process:

$$\Delta E = [m(^{64}_{30}\text{ Zn}) - m(^{64}_{29}\text{Cu})]c^2$$

The ΔE for the process is given as -0.58 MeV. Convert this to amu and use it in the previous equation rewritten in terms of masses:

$$-0.58 \text{ Mev} \times \left(\frac{1 \text{ amu}}{931.494 \text{ MeV}}\right) = \Delta m = \left(m(^{64}_{30}\text{Zn}) - m(^{64}_{29}\text{Cu})\right)$$

Solving for the difference in mass between daughter and parent gives -0.000623 amu. The parent ^{64}Cu weighs 63.92976 amu, so the daughter ^{64}Zn weighs 63.92914 amu.

12-51 Only an element having Z larger by 2 can decay directly to Ac by alpha emission. Since Ac is element 89, this would be element 91, protactinium. Note that element 91 is in fact named as the parent of actinium ("proto-actinium"). Only an element with Z less by 1 can decay directly to Ac by beta emission. This would be element 88, radium. The fact that compounds of radium contain no actinium tends to rule out beta emission by radium as a significant source of actinium.

12-53 a) The formation of a ^{30}P atom is represented $15\,^{1}_{1}\text{H} + 15\,^{1}_{0}n \rightarrow {}^{30}_{15}\text{P}$. Note that the mass of the electrons is included in the mass of the $^{1}_{1}$H atoms. The mass of the product is 29.97832 amu, and the mass of the reactants is 15(1.00782504 + 1.00866497) or 30.24735 amu. The difference between these masses is -0.26903 amu, or, in the equivalent energy units, -250.60 MeV. The binding energy is the negative of this figure; the binding energy per nucleon is then $+250.60/30 = 8.353$ MeV per nucleon.
b) The equation for position emission by ^{30}P is: $^{30}_{15}\text{P} \rightarrow {}^{30}_{14}\text{Si} + {}^{0}_{1}\beta + \nu$. The change in mass in this process is:

$$\Delta m = [m(^{30}_{14}\text{Si}) - m(^{30}_{15}\text{P})] + 2\,(0.00054858 \text{ amu})$$

$$\Delta m = 29.97376 - 29.97832 + 0.0010972 = -0.003463 \text{ amu}$$

The negative sign means that the process is spontaneous. The change in energy of the system is also negative:

$$\Delta E = (-0.003463 \text{ amu}) \times \frac{931.494 \text{ MeV}}{1 \text{ amu}} = -3.226 \text{ MeV}$$

The kinetic energy of the products equals $-\Delta E$ and is distributed among them. The positron has its maximum kinetic energy when the other products get none. This maximum is $+3.226$ MeV. Note that the energy equivalent of the mass of the

positron and the extra electron in the mass-change equation is 1.022 MeV. Omitting this term introduces an unacceptable error (about 32%).
c) The quick way to get the fraction of ^{30}P atoms left after 450 s is to recognize that 450 s is exactly three 150 s half-lives. The fraction is then obviously $(1/2)^3$ or 1/8. The rate constant is $\ln 2/t_{1/2}$; it is $4.62 \times 10^{-3}\ \mathrm{s}^{-1}$.

12-55 The fraction of the isotope that remains at any time equals the number of atoms that remains divided by the number of atoms originally present. This fraction can be formed by dividing both sides of the equation for first-order decay by N_0:

$$\frac{N}{N_0} = e^{-kt} \quad \text{or} \quad \ln\left(\frac{N}{N_0}\right) = -kt$$

a) If 1% of the radium has decayed, then 0.99 is the fraction still present. Hence $N/N_0 = 0.99$ and:

$$\ln 0.99 = -kt = -\left(\frac{\ln 2}{t_{1/2}}\right)t = -\left(\frac{\ln 2}{1622\ \text{yr}}\right)t \ \text{ from which } \ t = 24\ \text{yr}$$

b) The procedure is identical using 0.10 as the fraction of the radium remaining. The answer is 5400 yr.

12-57 For every atom of ^{238}U that decays, eight atoms of He are produced. The number of moles of He can be computed using the ideal-gas law:

$$n = \frac{PV}{RT} = \frac{(1\ \text{atm})(9.0 \times 10^{-8}\ \text{L})}{(0.08206\ \text{L atm mol}^{-1}\text{K}^{-1})(273.15\ \text{K})} = 4.015 \times 10^{-9}\ \text{mol}$$

The number of atoms of ^{238}U that has decayed is therefore:

$$4.015 \times 10^{-9}\ \text{mol He} \times \left(\frac{6.022 \times 10^{23}\ \text{atom He}}{1\ \text{mol He}}\right) \times \left(\frac{1\ \text{atom}\ ^{238}\text{U}}{8\ \text{atom He}}\right) = 3.02 \times 10^{14}\ \text{atom}$$

The number of atoms of ^{238}U still present per gram of rock is also readily computed:

$$2.0 \times 10^{-7}\ \text{g}\ ^{238}\text{U} \times \left(\frac{1\ \text{mol}}{238.0\ \text{g}}\right) \times \left(\frac{6.022 \times 10^{23}\ \text{atom}}{1\ \text{mol}}\right) = 5.06 \times 10^{14}\ \text{atom}$$

The initial number of atoms of ^{238}U is simply the sum of these two numbers. It is 8.08×10^{14} atoms. The decay process is described by the equation:

$$\ln\left(\frac{N}{N_0}\right) = -kt = -\left(\frac{\ln 2}{t_{1/2}}\right)t$$

Substitution gives:

$$\ln\left(\frac{5.06 \times 10^{14}}{8.08 \times 10^{14}}\right) = -\left(\frac{\ln 2}{4.47 \times 10^{9}\ \text{yr}}\right) t$$

Solving for t gives the approximate age of the rock as 3.0 billion years.

12-59 The activity of ^{14}C equals the rate constant for its decay multiplied by the number of atoms present. Consider 1.00 g of carbon from the biosphere. The total number of carbon atoms is:

$$N_{\text{tot}} = 1.00\ \text{g} \times \left(\frac{1\ \text{mol}}{12.01115\ \text{g}}\right) \times \left(\frac{6.022 \times 10^{23}\ \text{atom}}{1\ \text{mol}}\right) = 5.014 \times 10^{22}\ \text{atom}$$

The number of atoms of ^{14}C per gram of carbon is given by the equation:

$$N = \frac{A}{k} = A\left(\frac{t_{1/2}}{\ln 2}\right)$$

where $t_{1/2}$ is the half-life of the ^{14}C and A is its specific activity. Expressing $t_{1/2}$ in minutes so that the units of A cancel gives:

$$N = \left(\frac{15.3\ \text{min}^{-1}}{1\ \text{g}}\right) \times \left(\frac{3.014 \times 10^{9}\ \text{min}}{\ln 2}\right) = 6.65 \times 10^{10}\ \text{g}^{-1}$$

b) The preceding result means that 1.00 g of ordinary carbon contains 6.65×10^{10} atoms of ^{14}C. There is a vastly larger number of atoms of C of all isotopes in 1.00 gram of C. It is 5.014×10^{22} atoms. The required fraction is the first number divided by the second. The answer is 1.32×10^{-12}.

12-61 a) To produce ^{228}Ra, the ^{232}Th must emit an α-particle in the first step of its decay:

$$^{232}_{90}\text{Th} \longrightarrow {}^{228}_{88}\text{Ra} + {}^{4}_{2}\text{He}$$

In the next step, the ^{228}Ra emits a beta particle (an electron) to give ^{228}Ac:

$$^{228}_{88}\text{Ra} \longrightarrow {}^{228}_{89}\text{Ac} + {}^{0}_{-1}\beta + \bar{\nu}$$

The decay products over the two steps are an α-particle, an ^{228}Ac atom, an electron and an anti-neutrino. These products have the same total mass as an ^{228}Ac (228.03117 amu) plus a 4_2He atom (4.0026033 amu). The sum is 232.03377 amu. The mass of the beta particle (electron) is included in the mass tabulated for a neutral actinium atom, and the anti-neutrino is massless. The mass of the reactant is 232.038054 amu so Δm for the process is -0.00428 amu. This mass has an energy equivalent of -3.98

MeV. Hence, 3.98 MeV is the energy lost by the system consisting of one Th atom and taken away as kinetic energy by the product particles.
b) The thorium decays to radium in a first-order process and the radium goes on to decay to actinium in another first-order process. Let k_1 be the rate constant for the first step and k_2 the rate constant for the second. When the radium is present in a steady-state amount then:

$$\frac{dN_{\rm Ra}}{dt} = 0 = k_1 N_{\rm Th} - k_2 N_{\rm Ra} \qquad \text{from which} \quad N_{\rm Ra} = \frac{k_1}{k_2} N_{\rm Th}$$

For each first-order process, the rate constant is $\ln 2$ divided by the half-file $t_{1/2}$. Applying this fact to the previous equation:

$$N_{\rm Ra} = \left(\frac{t_{1/2,2}}{t_{1/2,1}}\right) N_{\rm Th} = \left(\frac{6.7 \text{ yr}}{1.39 \times 10^{10} \text{ yr}}\right) N_{\rm Th} = 4.8 \times 10^{-10} N_{\rm Th}$$

The number of radium atoms is 4.8×10^{-10} times the number of thorium atoms.

12-63 The incomplete nuclear equation:

$${}^{10}_{5}\text{B} + {}^{1}_{0}n \rightarrow \,?\, + {}^{4}_{2}\text{He}$$

must be balanced by the insertion of a single symbol for the question mark. The mass number on this symbol must be 7 and the atomic number 3. The element of atomic number 3 is lithium. Hence the other atom that is formed is a ${}^{7}_{3}\text{Li}$ atom.

12-65 The earth orbits the sun at a radius R of 1.50×10^8 km. Imagine a sphere of this radius surrounding the sun. The surface area of this immense sphere is $4\pi R^2$. Radiation from the sun streams out in all directions, cutting through this sphere. The earth intercepts a small proportion of this radiation. Call the 6371 km radius of the earth r. From the point of view of the sun the earth appears as a tiny disk of area πr^2. This disk, minuscule in comparison to the area of the big sphere, intercepts a fraction f of the total radiation in proportion to the area of the big sphere that it covers:

$$f = \frac{\pi r^2}{4\pi R^2} = \frac{1}{4}\left(\frac{r}{R}\right)^2 = 4.5 \times 10^{-10}$$

The surface area of the hemisphere of the earth that is exposed to the sun's rays at any time is $2\pi r^2$. Assume that the radiant flux of 0.135 J s^{-1}cm^{-2} is the *average* value over the earth's exposed hemisphere. Then the earth receives the radiant power:

$$P_{\rm (earth)} = (0.135 \text{ J s}^{-1}\text{cm}^{-2}) \times 2\pi(6.371 \times 10^8 \text{ cm})^2 = 3.44 \times 10^{17} \text{ J s}^{-1}$$

The total power output of the sun is $P(\text{earth})$ divided by the fraction of the sun's radiant power that hits the earth:

$$P(\text{sun}) = \frac{P(\text{earth})}{f} = \frac{3.44 \times 10^{17} \text{ J s}^{-1}}{4.5 \times 10^{-10}} = 7.6 \times 10^{26} \text{ J s}^{-1}$$

The mass equivalent of energy is given by the equation $\Delta E = c^2 \Delta m$. The sun loses 7.6×10^{26} J each second, so its ΔE is negative. The value of the equivalent change in mass is:

$$\Delta m = \frac{\Delta E}{c^2} = \frac{-7.6 \times 10^{26}\ \text{J}}{(3.0 \times 10^8\ \text{m s}^{-1})^2} = -8.5 \times 10^9\ \text{kg}$$

The sun burns (in the thermonuclear sense) 8.5 million metric tons of matter per second.

Chapter 13

Quantum Mechanics and the Hydrogen Atom

Waves and Light

A beam of light is electromagnetic radiation. It consists of electric and magnetic fields (symbolized by E and H respectively) oscillating perpendicular to the direction in which the beam propagates and perpendicular to each other. Several parameters characterize such radiation:

The wavelength λ. The distance between successive peaks in the intensity of the oscillating electric field (or magnetic field) at any instant is λ, the wavelength of the electromagnetic radiation.

The frequency λ. Electromagnetic radiation is a periodic or cyclic disturbance. As radiation passes a given point, a maximum in the electric field is followed by a minimum, then another maximum, etc. The number of complete cycles occurring each second is the frequency of the radiation, represented by the symbol ν , the Greek nu. The units of ν are reciprocal seconds, s^{-1}. A reciprocal second is also called a *Hertz*, Hz. Sometimes the unit "cps" for "cycles per second" is encountered. One cps is one reciprocal second.

The *reciprocal* of the frequency of a wave is the **period** of the wave. The period is the time required for the wave to complete one cycle. Naturally, it has units of time. See **problem 13-1.**

The amplitude. The amplitude of any wave disturbance is the maximum size of the excursion the wave makes during its oscillation. In classical wave theory, the **intensity** of a wave is proportional to the square of its amplitude. For electromagnetic waves:

$$\text{Intensity} \propto (E^2_{\mathbf{max}} + H^2_{\mathbf{max}})$$

The speed of propagation. A traveling wave completes ν oscillations per second and with each such cycle advances by a distance λ , the distance between its successive crests. It follows that:

$$\text{speed of propagation} = \lambda\nu$$

This equation applies to all kinds of traveling waves. It is used in **problems 13-1, 13-7,** and **13-41,** among others. The speed of propagation of electromagnetic waves (more simply, **the speed of light**) in a vacuum is a universal constant c. It is 2.9979×10^8 m s^{-1}. For light:

$$\boldsymbol{c = \lambda\nu} \qquad \text{for light in a vacuum}$$

This equation is the basis for **problem 13-3,** and is used in **problem 13-15** as well as many others. It is best to memorize it. There is a simple inverse relationship between λ and ν: the higher the frequency of light, the shorter the wavelength. Visible light has wavelengths ranging from approximately 4×10^{-7} to 7×10^{-7} m (equivalent to 400 to 700 nm). See Figure 13-4, text page 483. The frequency of visible light is on the order of 10^{14} s^{-1}.

Paradoxes in Classical Physics

Classical physics (19th century physics) treated electricity, magnetism and electromagnetic radiation separately from **mechanics** which dealt with the motions of particles and their interactions. Key concepts of mechanics include the **kinetic energy,** the energy associated with the motion of particles, and the **potential energy** the energy associated with their positions. For a single particle:

$$KE = \frac{p^2}{2m} = 1/2\, mv^2$$

where p is the **momentum** of the particle, m is its mass and v is its velocity. The potential energy depends on the location of the particle. The total energy of the particle is the sum of its potential and kinetic energies:

$$\text{total energy} = KE + PE$$

Classical physics is highly successful in describing many phenomena. Nevertheless it makes wrong predictions in some important experiments:

Blackbody radiation. A heated object such as an iron bar or tungsten light-bulb filament emits radiation. A **blackbody** is an idealized version of such an emitter. Classical physics predicts that the intensity of the light emitted by a blackbody depends directly on the absolute temperature and inversely on the fourth power of the light's wavelength.

$$\text{intensity} \propto \frac{T}{\lambda^4}$$

At long wavelengths this formula agrees with observation. At short wavelengths it gives values that are far too large. As λ gets smaller and smaller the predicted intensity goes to infinity. This result is called the **ultraviolet catastrophe.**

The photoelectric effect. When a beam of light falls onto a metal or other material, **photoelectrons** are often ejected from the surface. Classical physics predicts the maximum kinetic energy of such light-ejected electrons to depend on the intensity of the beam of light. Instead, the maximum kinetic energy depends on the frequency of the light.

$$KE = 1/2m_e v^2 = h(\nu - \nu_0)$$

where h is a constant and ν_0 is a threshold frequency. Light of frequency less than ν_0 does not cause the emission of photoelectrons, no matter how great its intensity.

Stability of the atom. In the Rutherford model of the atom, a massive and small nucleus is surrounded by electrons, which are charged particles. If the charges move in orbits about the nucleus, then classical physics predicts that they must emit electromagnetic radiation. If they actually did this, they would lose energy and quickly spiral into the nucleus. But, if the charges are not in motion, there is nothing to keep them from being pulled into the nucleus. Thus, according to classical physics, the Rutherford model of the atom cannot be correct. Yet, Rutherford's experimental results are conclusive.

Quantum Explanation of Paradoxes

The various paradoxes of classical physics are all resolved by the introduction of the central idea of quantum mechanics:

- **Energy is not continuous, but instead is quantized in discrete packets.**

Radiation transmits energy. The quantum idea, when applied to radiation (light) means that the energy of light is absorbed, emitted or converted in individual packets, or quanta. The quantum of radiation is the **photon.** It is the particle of light. Its energy is directly related to its frequency:

$$\boldsymbol{E = h\nu = \frac{hc}{\lambda}}$$

where h is Planck's constant (6.626×10^{-34} J s). This is the Planck equation. It occurs again and again in problems and important applications. See **problems 13-13,** and **13-31.**

Planck's constant is extremely small. Consequently, the grain-size of particles of light is small. For many purposes light is as good as continuous.

Photons have zero rest mass but *do* have momentum p. Their energy is:

$$E = pc$$

The preceding two equations give the energy of light first in terms of frequency (a wave property) then in terms of momentum (a particle property).

• **In quantum theory light has both wave and particle character.**

Putting the notion of the ultimate graininess of energy into the equations dealing with blackbody radiation gives predictions in accord with experiment. The quantum theory averts the ultraviolet catastrophe.

The quantum explanation of the photoelectric effect treats light as a stream of particles or packets or photons, each carrying a quantity of energy proportional to the frequency (not the amplitude) of the light:

$$E = h\nu$$

It requires a minimum energy to liberate a photoelectron against the forces holding it to its atom. This accounts for the observation or a minimum or threshold frequency for the liberation of photoelectrons. The **work function Φ** is the name for the minimum energy. Any extra energy from the photon can manifest itself as kinetic energy of the photoelectron:

$$KE_{\max} = h(\nu - \nu_0) = h\nu - \Phi$$

See **problems 13-19** and **13-45.**

Planck, Einstein, and Bohr

An important step in understanding the stability of the Rutherford nuclear atom was Niels Bohr's model of the hydrogen atom and ions of other elements that, like hydrogen, have only one electron.

Both atoms and molecules can emit and absorb light. They do so at various points in the electromagnetic spectrum, that is, at various wavelengths (various frequencies). Atoms emit and absorb at *lines of sharply defined, specific frequency.* Early spectroscopists found empirical formulas relating the observed frequencies of members of series of emission lines in the spectrum of atomic hydrogen. Bohr's model gave the same formulas entirely on a theoretical basis. Its postulates were:

1. The electron in the hydrogen atom (and one-electron ions like He^+, Li^{2+}, and so on) revolves around the nucleus in a circular orbit.

2. Coulomb (electrostatic) attraction between the negatively-charged electron and the positively-charged nucleus provides the force needed to sustain the circular orbital motion of the electron.

3. The only orbits allowed are those for which the angular momentum of the electron is an integer times the constant $h/2\pi$: That is, **angular momentum is quantized in units of $h/2\pi$:**

$$\text{angular momentum} = m_e v r = n\left(\frac{h}{2\pi}\right)$$

   In this equation n is the integer (or quantum number). The units of angular momentum are of course J s, the units of h. Since h is so small, the quantum of angular momentum is small. See **problem 13-49.**

4. Electrons make transitions from one orbit to another only by absorbing or releasing a quantity of energy equal to the energy difference between the two orbits. When this energy is absorbed or emitted in the form of light:

$$\Delta E = h\nu$$

These postulates, when combined with the ordinary rules of mechanics, give two major results:

- **The radius of the hydrogen atom is quantized:**

$$r = \frac{\epsilon_0 n^2 h^2}{\pi Z e^2 m_e}\frac{n^2}{Z} = \left(\frac{\epsilon_0 h^2}{\pi e^2 m_e}\right)\frac{n^2}{Z} = \frac{n^2}{Z}(0.529 \times 10^{-10}\text{ m}) = \frac{n^2}{Z}a_0$$

where m_e and e are the mass and charge of the electron, ϵ_0 is a constant called the permittivity of free space, n is the quantum number, a whole number greater than or equal to one, and Z is the atomic number of the nucleus. The grouping of constants in the large parentheses is called $\boldsymbol{a_0}$, the **Bohr radius**. It is a distance that occurs repeatedly in discussions of phenomena on the atomic level.

- **The energy of the hydrogen atom is quantized:**

$$E = -\frac{Z^2 e^4 m_e}{8\epsilon_0^2 n^2 h^2} = -\left(\frac{e^4 m_e}{8\epsilon_0^2 h^2}\right)\frac{Z^2}{n^2} = -\frac{Z^2}{n^2}(2.18 \times 10^{-18}\text{ J}) = -\mathcal{R}\frac{Z^2}{n^2}$$

The text identifies the collection of constants in parentheses in this equation as the **Rydberg constant $\mathcal{R}$.**

It is an excellent exercise to confirm the numerical values of a_0 and $\mathcal{R}$ by inserting all the constants in the preceding equation and making sure that the units cancel out. Note that ϵ_0 is 8.854×10^{-12} $C^2J^{-1}m^{-1}$. To study the equations further, set Z equal to 1. Then r has a minimum value of 0.529×10^{-10} m (a_0, the Bohr radius) when $n = 1$. The radius of the atom gets bigger without bound as n steps up to $2, 3, \ldots$. The energy E likewise has its minimum when $n = 1$, but this value is *less than zero;*. The state of lowest energy is the **ground state** of the atom. The energy rises toward a maximum of zero as n goes to infinity. Each different whole-number value of n greater than the ground-state value of 1 corresponds to a different **excited state** of the atom.

Infinite n corresponds to removal of the electron from the atom. The **ionization energy** of an atom is the minimum energy needed fully to remove an electron. Removing an electron from an atom of hydrogen in its ground-state means raising it from $n = 1$ to $n = \infty$. This requires 2.18×10^{-18} J of energy. An n of zero corresponds to having never put the electron in the atom! Only positive, integral values of n are allowed in the above equations. Such values generate the **allowed states** of the atom (or ion).

The atomic number Z is 1 for hydrogen, 2 for the one-electron He^+ ion, 3 for the Li^{2+} ion, 8 for O^{7+} (see **problem 13-57**) and so forth. **Problems 13-21** and 13-23 provide a comparison of the results of the Bohr model for hydrogen and an one-electron ion.

Waves, Particles and the Schrödinger Equation

Louis de Broglie suggested in 1924 that if light has both wave-like and particle-like properties, then perhaps matter (such as electrons, protons or even baseballs) has a wave aspect. The DeBroglie equation relates the wavelength of any moving particle to its momentum:

$$\lambda = \frac{h}{p} = \frac{h}{mv}$$

Because Planck's constant h is very small the wavelength of most moving objects is too short to measure. The wavelength only becomes important for subatomic particles which have small masses. See **problem 13-53.**

Heisenberg Indeterminacy Principle

The presence of a wave aspect in matter means that the positions and momenta of particles cannot be determined precisely. Instead there is a built-in indeterminacy. In the measurement of positions, the indeterminacy is on the order of the wavelength,

λ , of the particle. For momenta it is on the order of h / λ . More accurately:

$$(\Delta p)(\Delta x) \geq \frac{h}{4\pi}$$

where Δp and Δx are the indeterminacies in momentum and position respectively. **Problem 13-27** requires nothing more than substitution in this inequality and some attention to the units. **Problem 13-53** is a more complex calculation applying the indeterminacy principle to an important case.

The Schrödinger Equation

Erwin Schrödinger generalized the DeBroglie wave relation to apply to bound particles such as electrons held ("bound") in atoms. His contribution, a differential equation, is the fundamental equation of **quantum mechanics.** Its exact form is not as important as some key points about it and its solutions:

- Particles bound in different systems have different forms of the Schrödinger equation. The text considers only two of the many possible bound systems: the particle in a one-dimensional box and the hydrogen atom.

- Since the equation is a differential equation, its solutions are mathematical functions, not numbers. These solutions are **wave functions.** They are symbolized by ψ (the Greek letter psi).

- For any bound particle, there is a **family** of wave functions, not just one, satisfying the Schrödinger equation. All describe the particle with equal success. For example, in **problem 13-55,** substituting the positive integers for n in the function:

  $$\psi(x) = \sqrt{\frac{2}{L}} \sin\left(\frac{n\pi x}{L}\right) \qquad n = 1, 2, \ldots$$

  gives the (infinite) family of different solutions. The quantum number n generate the members of the family.

  Table 13-2 (text page 504) gives just *some* of the wave functions that satisfy the Schrödinger equation for the electron in one-electron species. As the table shows, solutions to the Schrödinger equation are often complex and tiresome to write out in full. They are frequently designated by a quantum number (or numbers) alone.

- For every different ψ that satisfies the Schrödinger equation in a given system, there is a corresponding energy ($E(\psi)$). These are the **allowed energies** of the system. Most energies are *not* allowed. For example, in one-electron atoms and ions the allowed energies are given by the formula:

$$E = -(2.178 \times 10^{-18}\ \text{J})\frac{Z^2}{n^2} = -(1312\ \text{kJ mol}^{-1})\frac{Z^2}{n^2}$$

When Z is 1, the species is the hydrogen atom. When n is 1, the energy of the hydrogen atom is -2.178×10^{-18} J. When n is 2, the energy of the H atom is -0.5445×10^{-18} J. All intermediate energies, such as -1.00×10^{-18} J, are impossible.

- The wave function that corresponds to the lowest E is the ground state. All others are **excited states** of a particle.

- The value of a wave function differs at different points in space, depending on the coordinates of the points. Regions of space in which ψ is positive have **positive phase.** Regions in which ψ is negative have **negative phase.** Regions where ψ passes through zero and changes sign are **nodes.** Finding a wave function's nodes helps in visualizing its shape. See **problem 13-39.**

- The square of the wave function of a bound particle, ψ^2, evaluated at some point, is the probability of finding the particle in a small volume about that point. A plot of ψ^2 against the space coordinates gives a picture of the **probability density distribution** of the bound particle.

The Particle in a Box

A simple case in which the Schrödinger equation can be solved is that of a particle confined in a one-dimensional box. The wave functions that describe this system are all sine functions. They are:

$$\psi(x) = \left(\frac{2}{L}\right)^{1/2} \sin\left(\frac{n\pi x}{L}\right) \quad n = 1, 2, \ldots$$

The energy of the particle is:

$$E_n = \frac{n^2 h^2}{8mL^2}$$

where n is an integer, m is the mass of the particle and L is the length of the box.

The particle in a box problem has slight practical application although in **problem 13-31** it does give a useful answer to a chemical question. Its main value is to illustrate how the wave property of a confined particle is accounted for. Key points are:

- In a one-dimensional box, the wave functions that satisfy the Schrödinger equation are characterized by *one* quantum number. In two-dimensional and three dimensional boxes the solutions are characterized by *two* and *three* quantum numbers, respectively.

- If the box is large (L large) or if the particle is massive (m large), the allowed energies of the particle are small. More importantly, the *differences* between neighboring values of allowed energy are small. When the allowed energies are closely spaced, the quantization of the energy is not apparent. This is the case with objects bigger than atoms.

- The quantum number alone is often used as a short-hand designation. Thus, the $n = 2$ wave function for a particle in a box is referred to as ψ_2 or simply "the $n = 2$ function".

The Hydrogen Atom

When the Schrödinger equation is set up and solved for the case of a single electron (charge e and mass m_e) moving in the vicinity of a central nucleus of charge $+Ze$, then a set of wave functions characterized by *three* quantum numbers (n, ℓ, m) arises. There are three quantum numbers because the hydrogen atom is three-dimensional. These wave functions are **orbitals.** They are very important in chemical bonding theory.

The solution of the hydrogen atom is in terms of **spherical coordinates, r, θ and ϕ**, with the origin at the nucleus, instead of x, y and z, the familiar Cartesian coordinates. Spherical coordinates simplify the mathematics of the solution. They are related to Cartesian coordinates by the equations:

$$x = r \sin\theta \cos\phi; \qquad y = r \sin\theta \sin\phi; \qquad z = r \cos\theta$$

See text Figure 13-17. The resultant wave functions can all be broken down into parts:

- A radial part, $R_{n\ell}(r)$, which depends only the distance from the nucleus, as given by the coordinate r. The form of $R(r)$ is determined by the first two quantum numbers, n and ℓ, which is why they appear as subscripts on the symbol for the function.

- An angular part $\chi(\theta\,\phi)$, which depends on the two angles θ and ϕ and is specified by the second and third quantum numbers, ℓ and m. See **problem 13-39** and text table 13-2.

The three quantum numbers, which arise naturally in the mathematics of the solution of the Schrödinger equation, occur in *sets* such as $\left(n = 3, \ell = 2, m = 0\right)$. Each set signifies a different wave function. For example, the set just listed specifies the function:

$$\psi(3,2,0) = R(3d)\chi(d_{z^2}) = \frac{4}{81\sqrt{30}}\left(\frac{Z}{a_0}\right)^{\frac{3}{2}} \sigma^2 \exp(-\sigma/3)\left(\frac{5}{16\pi}\right)^{\frac{1}{2}}(3\cos^2\theta - 1)$$

where the radial and angular functions on the right were copied from text Table 13-2. More importantly, each set tells the characteristic *energy and shape* of its wave function. Sets are *not* constructed by taking just any values for n, ℓ and m. Instead, possible values of the three are interlocked in a strict pattern that must be learned by memory:

- The principal quantum number n may have any integral value from +1 to infinity. The quantum number n gives the energy for a one-electron atom:

  $$E_n = -\frac{Z^2 e^4 m_e}{8\epsilon_0 n^2 h^2}$$

  It also tells the number of **nodes** that the wave function has:

  $$\text{Number of nodes} = n - 1$$

  The nodes of orbitals are a valuable physical reference for study. In the hydrogen atom wave functions, the nodes are always either **radial** or **angular.** Radial nodes are spheres of different radius with the nucleus at their center. They are called radial because their locations are specified by values of the coordinate r. Angular nodes are planes or curved surfaces. They are specified by values of the angular coordinates θ and ϕ. See **problem 13-39.** Angular nodes always contain the nucleus. Radial nodes encircle the nucleus but never contain it. Keeping track of the nodes helps to solve problems. See **problem 13-33.** *The more nodes an orbital has the higher its energy.*

- The angular momentum quantum number ℓ has allowed values ranging from 0 to $(n-1)$. The quantum number n imposes a strict ceiling on ℓ, which may *not* take on just any value. For example, when n is 1, then ℓ may equal 0 only.

  The value for ℓ always equals the number of angular nodes in the orbital. The presence of angular nodes lends a distinct shape to orbitals, so ℓ really tells the shape of the orbital. When $\ell = 0$, the orbital has no angular nodes. All the nodes are automatically radial, and all orbitals with $\ell = 0$ are therefore **spherically symmetrical.** When $\ell = 1$, the orbital has 1 angular node. When $\ell = 2$, it has 2 angular nodes, and so forth.

In referring to orbitals, the numerical value of ℓ is replaced by a letter, according to the code:

value of ℓ	0	1	2	3	4	...
code letter	s	p	d	f	g	rest of alphabet

An (n, ℓ) combination is referred to by prefixing the letter for the ℓ value with the integer equal to n. Examples: $2s$, $3d$.

All s orbitals are spherically symmetrical. All p orbitals have 1 angular node, a plane. All d orbitals have 2 angular nodes. The shapes of s, p and d orbitals are important in chemistry and should be memorized. Study Figures 13-19, 13-20, and 13-22, text page 505-7, noting the location of nodes especially.

- The magnetic quantum number m governs the behavior of the atom in external magnetic fields. It completes the description of an orbital. For a given value of ℓ, m may range from $-\ell$ through 0 and up to $+\ell$. The number of possible values for m is $(2\ell+1)$. For example, an f orbital ($\ell = 3$) has 7 different possible values of m: $-3, -2, -1, 0, 1, 2, 3$. There is only one s orbital (because $\ell = 0$), but there are 3 p orbitals: p_x, p_y, p_z, and 5 d orbitals: d_{xy}, d_{xz}, d_{yz}, $d_{x^2-y^2}$, d_{z^2}. In these designations, the subscripts are Cartesian coordinates that tell the orientation of the single angular node (in the p case) and the two angular nodes (in the d case). They should be learned. To help, notice that the first three labels comprise every possible combination of two of the three Cartesian coordinates. In the last two, all of the coordinates are squared. In the hydrogen atom, orbitals with the same n and ℓ are equivalent except in orientation. Thus, m, which distinguishes among the members of the set, can be interpreted as telling the orientation in space of the orbital's angular nodes. It is indeed sometimes called the space quantum number.

- The rules stating the possible values of the three quantum numbers give a pattern. As the following table shows, *the number of wave functions for a given n is n^2:*

n	Types of Orbitals					Total No. Orbitals
	s	p	d	f	g	
1	1	0	0	0	0	1
2	1	3	0	0	0	4
3	1	3	5	0	0	9
4	1	3	5	7	0	16
5	1	3	5	7	9	25
n						n^2

Studying the One-Electron Atom Wave Functions

The wave functions listed in text Table 12-2 are quite imposing. They can be understood by taking them apart and studying them piece by piece. Bear in mind that the wave functions *already* are in two parts. A complete wave function is the product of an angular part and a radial part.

Concentrate on one of the radial functions and prepare a table showing the values at different r of the several terms it contains. For example, take the $R(3s)$ function. Simplify matters by setting $Z = 1$, which means dealing with the hydrogen atom and not one of the one-electron ions. The function then becomes:

$$R(3s) = \frac{2}{81\sqrt{3}}\left(\frac{1}{a_0}\right)^{3/2}(27 - 18\sigma + 2\sigma^2)e^{-\sigma/3}$$

Work out the numerical value of the constants in the function:

$$R(3s) = (3.7051 \times 10^{13}\ \mathrm{m}^{-3/2})(27 - 18\sigma + 2\sigma^2)e^{-\sigma/3}$$

Now try out some values of r to see how the function behaves. The following table illustrates:

r (Å)	σ	$(27 - 18\sigma + 2\sigma^2)$	$e^{-\sigma/3}$	$R(r)$ ($\mathrm{m}^{-3/2}$)
0.0000	0.0000	27.00	1.0000	10.0×10^{14}
0.2645	0.5000	18.50	0.8465	5.80×10^{14}
0.5290	1.0000	11.00	0.7165	2.92×10^{14}
0.7935	1.5000	4.50	0.6065	1.01×10^{14}
1.0060	1.9019	0.00	0.5305	0.0
2.0000	3.7807	−12.46	0.2836	-1.31×10^{14}
3.0000	5.6711	−10.76	0.1510	-0.602×10^{14}
3.7548	7.0980	0.00	0.0938	0.0
10.000	18.904	401.4	1.834×10^{-3}	0.272×10^{14}
100.00	189.04	6.810×10^{4}	4.302×10^{-28}	1.09×10^{-9}

Note that the variable σ has no units. It is the distance r divided by the Bohr radius a_0, which is 0.529×10^{-10} m.

The term in the third column causes the overall function (last column) to become negative between $r = 1.006$ and 3.754 Å. The distances where $R(3s)$ changes sign (goes from positive to negative phase and vice versa) are the locations of the two radial nodes.

Despite the fact that the term in the third column gets large with increasing r, the exponential term in the fourth column gets so small that it forces the overall value of $R(r)$ toward zero at large r. Study the above numbers in combination with text Figure 13-21.

Finally, the unit of $R(r)$ is meters to the minus three-halves. It is hard to associate any physical meaning with this unit. When squared it becomes m^{-3}¿ If m^{-3} is multiplied by a volume, the result is a dimensionless number, a *probability*. (See next section.)

Note the similarity of the one electron wave functions and the Maxwell-Boltzmann distribution (page 55 of this Guide). It is worthwhile to graph the $3s$ and other wave functions.

Sizes and Shapes of Orbitals

The Heisenberg indeterminacy principle and the wave nature of the electron make it impossible to know exactly where an electron in an atom is located. Instead it is necessary to speak of probabilities.

The probability p of finding an electron at a point (r, θ, ϕ) in a hydrogen atom is:

$$p(r, \theta, \phi) = \left(\psi(r, \theta, \phi)\right)^2 = \left(R(r)\right)^2 \left(\chi(\theta, \phi)\right)^2$$

The quantities in parentheses are the coordinates of the point. Different ψ's give different probability distributions. Graphs of $R(r)^2$ and $\chi(\theta, \phi)^2$ convey the shape and size of orbitals. In studying these graphs, it is particularly important to note the number and orientation of the nodes of the orbital in question. The following generalizations hold:

- The size of orbitals of the same ℓ (for example all s orbitals) increases with increasing n. Thus, the $3p$ orbitals are bigger than the $2p$.

- An orbital with quantum numbers n and ℓ has ℓ angular nodes and $n - \ell - 1$ radial nodes.

- At very large r all orbitals go to zero.

- As r approaches zero, all orbitals go to zero except s orbitals. An electron in an s orbital has a finite probability of being found right at the nucleus.

- For a given value of n the orbital size *decreases* with increasing ℓ. The $3d$ is smaller than the $3p$, which is smaller than the $3s$.

Electron Spin

The electron behaves like a tiny magnet. In classical theory this magnetism is explained by imagining the electron to be a ball of charge spinning about its own axis like a top. Two directions of spin are possible. They are identified with two possible values of the **spin quantum number:**

$$m_s = +1/2 \quad \text{or} \quad -1/2$$

The electron spin quantum number adds a *fourth* quantum number to n, ℓ and m, the three quantum numbers that characterize the wave function of a bound electron. The existence of electron spin doubles the possible number of different quantum states of energy E_n for n^2 to $2n^2$.

Detailed Solutions to Odd-Numbered Problems

13-1 The speed of propagation of a wave equals the product of its frequency and wavelength. A wave-crest hits the beach once every 3.2 s, which means that slightly less than one-third of a wave reaches the beach per second: the frequency of the waves is the reciprocal of 3.2 s, which equals 0.3125 s^{-1}. Thus:

$$\text{speed} = \nu\lambda = \frac{1}{3.2 \text{ s}}(2.1 \text{ m}) = 0.66 \text{ m s}^{-1}$$

The time between the identical portions of a wave, the period of the wave, is always the reciprocal of the frequency of the wave.

13-3 The speed of propagation of electromagnetic radiation through a vacuum is c, a quantity that is very well known. As with all traveling waves, the speed of propagation of the FM signal equals the product of its wavelength and frequency: $c = \lambda\nu$. Hence:

$$\lambda = \frac{c}{\nu} = \frac{2.998 \times 10^8 \text{ m s}^{-1}}{9.86 \times 10^7 \text{ s}^{-1}} = 3.04 \text{ m}$$

13-5 a) Use the same relationship as in problem 13-3 except now the frequency must be solved for:

$$\nu = \frac{c}{\lambda} = \frac{2.998 \times 10^8 \text{ m s}^{-1}}{6.00 \times 10^2 \text{ m}} = 5.00 \times 10^5 \text{ s}^{-1}$$

b) The time for a wave to travel a distance d equals that distance divided by the speed of the wave. These electromagnetic waves advance at the known speed c. Hence:

$$t = \frac{d}{c} = \frac{8.0 \times 10^{10} \text{ m}}{3.00 \times 10^8 \text{ m s}^{-1}} = 2.7 \times 10^2 \text{ s} \qquad \text{which is 4.4 minutes}$$

13-7 The wavelength of the sound waves can be determined from the frequency and speed of propagation of the waves:

$$\lambda = \frac{\text{speed}}{\nu} = \frac{343.5 \text{ m s}^{-1}}{261.6 \text{ s}^{-1}} = 1.313 \text{ m}$$

The time to travel 30.0 m is:

$$t = \frac{d}{\text{speed}} = \frac{30.0 \text{ m}}{343.5 \text{ m s}^{-1}} = 0.0873 \text{ s}$$

13-9 Blue light has a higher frequency than green light (see Figure 13-4, text page 483). The photons of blue light are therefore more energetic than the photons of green light. Since the work function of the surface of the potassium is the same for both colors of light, the electrons ejected by the blue light have higher average kinetic energy.

13-11 The wavelength 671 nm is 6.71×10^{-7} m. Figure 13-4 (text page 483) shows that light of this wavelength is red.

13-13 The frequency corresponding to the transition energy is computed using a rearranged version of the Planck equation $E = h\nu$:

$$\nu = \frac{E}{h} = \frac{3.6 \times 10^{-19} \text{ J}}{6.626 \times 10^{-34} \text{ J s}} = 5.43 \times 10^{14} \text{ s}^{-1} = 5.4 \times 10^{14} \text{ s}^{-1}$$

The product of the wavelength and the frequency equals the speed of propagation. Hence:

$$\lambda = \frac{c}{\nu} = \frac{2.9979 \times 10^{8} \text{ m s}^{-1}}{5.43 \times 10^{14} \text{ s}^{-1}} = 550 \times 10^{-9} \text{ m}$$

This light is green (see Figure 13-4, text page 483).

13-15 a) The energy change of an atom of sodium and the wavelength λ of the radiation it emits are inversely related:

$$\Delta E = \frac{hc}{\lambda}$$

The λ is 589.3 nm (or 5.893×10^{-7} m). Substitution of h and c gives:

$$\Delta E = \frac{(6.626 \times 10^{-34} \text{ J s})(2.998 \times 10^{8} \text{ m s}^{-1})}{5.893 \times 10^{-7} \text{ m}} = 3.371 \times 10^{-19} \text{ J}$$

b) A mole of sodium atoms consists of Avogadro's number of sodium atoms. The energy change per mole is:

$$\left(3.371 \times 10^{-19} \frac{\text{J}}{\text{atom}}\right) \times \left(6.022 \times 10^{23} \frac{\text{atom}}{\text{mol}}\right) = 2.030 \times 10^{5} \text{ J mol}^{-1}$$

c) The sodium arc light puts out 1000 W (watt) of radiant energy; 1000 W is 1000 J s^{-1}. Assume that all of this energy is emitted at the sodium D-line. Then:

$$1.000 \times 10^3 \text{ J s}^{-1} \times \left(\frac{1 \text{ mol}}{2.030 \times 10^5 \text{ J}}\right) = 4.926 \times 10^{-3} \text{ mol s}^{-1}$$

13-17 Combine the relationships $c = \lambda\nu$ and $E = h\nu$ to compute the wavelength of the light that is just energetic enough to eject electrons from a surface of cesium metal:

$$\lambda = \frac{c}{\nu} = \frac{hc}{E} = \frac{(6.626 \times 10^{-34} \text{ J s})(3.00 \times 10^8 \text{ m s}^{-1})}{3.43 \times 10^{-19} \text{ J}} = 5.80 \times 10^{-7} \text{ m}$$

This light is yellow (see text Figure 13-4). Yellow light and light of any shorter wavelength (green, blue, violet, ultraviolet, etc.) will eject electrons from cesium in the photoelectric experiment.
For selenium, the work function is bigger (more energy is required to eject electrons):

$$\lambda = \frac{hc}{E} = \frac{(6.626 \times 10^{-34} \text{ J s})(3.00 \times 10^8 \text{ m s}^{-1})}{9.5 \times 10^{-19} \text{ J}} = 2.1 \times 10^{-7} \text{ m}$$

This wavelength is well into the ultraviolet.

13-19 a) The maximum kinetic energy of the ejected electrons equals the difference between the energy of the incident light and the work function of the chromium:

$$KE_{\text{max}} = h\nu - \Phi = \frac{hc}{\lambda} - \Phi$$

$$KE_{\text{max}} = \frac{(6.626 \times 10^{-34} \text{ J s})(3.00 \times 10^8 \text{ m s}^{-1})}{2.50 \times 10^{-7} \text{ m}} - 7.21 \times 10^{-19} \text{ J} = 0.74 \times 10^{-19} \text{ J}$$

b) The speed of a particle and its kinetic energy are related by $KE = 1/2mv^2$. Solving for v and substituting the kinetic energy computed in the previous part and the mass of the electron give:

$$v = \sqrt{\frac{2KE}{m}} = \sqrt{\frac{2(0.74 \times 10^{-19} \text{ J})}{9.109 \times 10^{-31} \text{ kg}}} = 4.0 \times 10^5 \text{ m s}^{-1}$$

13-21 The B^{4+} ion is a hydrogen-like ion. Like H, it has only one electron. Unlike H, its atomic number Z is 5. The questions about it can be answered by substitution into equations 13-7 and 13-8 on text pages 490 and 491:

$$r_n = \frac{n^2}{Z}a_0 \quad \text{and} \quad E_n = -\mathcal{R}\frac{Z^2}{n^2}$$

where a_0 is the Bohr radius and $\mathcal{R}$ is the Rydberg constant. The radius for the state $n = 3$ in an atom of B^{4+} ($Z = 5$) is:

$$r_3 = \frac{3^2}{5}(5.29 \times 10^{-11}\text{ m}) = 9.52 \times 10^{-11}\text{ m}$$

The energy for $n = 3$ and $Z = 5$ is:

$$E_3 = -\frac{5^2}{3^2}(2.18 \times 10^{-18}\text{ J}) = -6.06 \times 10^{-18}\text{ J}$$

The negative of this answer is the energy needed to strip the electron away from a single B^{4+} ion in the $n = 3$ state. It is $+6.06 \times 10^{-18}$ J. For a mole of B^{4+} ions, the energy is Avogadro's number times larger:

$$E = (6.022 \times 10^{23}\text{ mol}^{-1}) \times (6.06 \times 10^{-18}\text{ J}) = 3.65 \times 10^{6}\text{ J mol}^{-1}$$

The energy change in a B^{4+} ion undergoing a $3 \rightarrow 2$ transition is the difference between the energies of the two states. The two energies are:

$$E_3 = -\frac{5^2}{3^2}\mathcal{R} = -\frac{25}{9}\mathcal{R} \quad \text{and} \quad E_2 = -\frac{5^2}{2^2}\mathcal{R} = -\frac{25}{4}\mathcal{R}$$

The difference is the final energy minus the initial:

$$\Delta E = E_2 - E_3 = \left(\frac{-25}{4} - \frac{-25}{9}\right)(2.18 \times 10^{-18}\text{ J}) = -7.57 \times 10^{-18}\text{ J}$$

This is the energy change of the ion. The energy emitted in the surroundings in the form of a photon is the negative of this. It equals $+7.57 \times 10^{-18}$ J. Dividing this energy by h gives ν, the corresponding frequency:

$$\nu = \frac{E}{h} = \frac{7.57 \times 10^{-18}\text{ J}}{6.626 \times 10^{-34}\text{ J s}} = 1.14 \times 10^{16}\text{ s}^{-1}$$

The wavelength of the photon is also easily computed:

$$\lambda = \frac{hc}{E} = \frac{(6.626 \times 10^{-34}\text{ J s})(2.9979 \times 10^{8}\text{ m s}^{-1})}{7.57 \times 10^{-18}\text{ J}} = 2.63 \times 10^{-8}\text{ m}$$

13-23 The problem could be solved by substitution into the equations of the Bohr model. The fact that the $3 \rightarrow 2$ emission in hydrogen occurs at 656.1 nm is then not needed. A quicker solution uses the given fact. Energy levels in hydrogen-like atoms are characterized by the factor $-Z^2/n^2$ multiplied by a constant. The $3 \rightarrow 2$ transition for the neutral H atom ($Z = 1$) corresponds to an energy jump proportional to $(1^2/2^2 - 1^2/3^2)$, which equals 0.13889. For the Li^{2+} ion, $Z = 3$, and the $3 \rightarrow 2$ energy jump is proportional to $3^2/2^2 - 3^2/3^2$, or 1.2500. The energy jump is bigger in the Li^{2+} transition by the factor $(1.2500/0.13889) = 9.000$. Therefore the wavelength of the emitted light in the Li^{2+} transition is 9.000 times *shorter* than 656.1 nm. This is $656.1/9.000 = 72.90$ nm. Light of this wavelength is in the ultraviolet region.

13-25 The wave in a guitar string is a standing wave. Its allowed wavelength satisfies the equation:

$$\frac{n\lambda}{2} = L$$

where L is the length of the string and n is an integer.
a) The first harmonic has $n = 1$: Solving the above equation gives:

$$\lambda_1 = \frac{2L}{1} = \frac{2(50\ \text{cm})}{1} = 100\ \text{cm}$$

By similar substitution but with $n = 3$, the wavelength λ_3 of the third harmonic is 33 cm.
b) The number of nodes in a standing wave is always one less than the number of the harmonic. Thus the 1st harmonic has 0 nodes, the 2nd harmonic has 1 node, and the 3rd harmonic has 2 nodes.

13-27 The deBroglie wavelength of an object is related to its momentum p by the equation $\lambda = h/p$. The momentum of an object is its mass multiplied by its velocity.
a) For an electron moving at 1000 m s^{-1}:

$$\lambda = \frac{h}{p} = \frac{h}{m_e v} = \frac{6.626 \times 10^{-34}\ \text{J s}}{(9.11 \times 10^{-31}\ \text{kg})(1.00 \times 10^{3}\ \text{m s}^{-1})} = 7.27 \times 10^{-7}\ \text{m}$$

b) For a proton moving at the same velocity:

$$\lambda = \frac{h}{m_p v} = \frac{6.626 \times 10^{-34}\ \text{J s}}{(1.673 \times 10^{-27}\ \text{kg})(1.00 \times 10^{3}\ \text{m s}^{-1})} = 3.96 \times 10^{-10}\ \text{m}$$

c) A velocity of 75 km hr^{-1}, is equivalent to 20.8 m s^{-1} (multiply by 1000 m km^{-1} and then divide by 3600 s hr^{-1}). A 200.0-g baseball has a mass of 0.2000 kg. Then:

$$\lambda = \frac{h}{m_{\text{ball}} v} = \frac{6.626 \times 10^{-34}\ \text{J s}}{(0.2000\ \text{kg})(20.8\ \text{m s}^{-1})} = 1.6 \times 10^{-34}\ \text{m}$$

13-29 **a)** According to the Heisenberg indeterminacy principle:

$$(\Delta x)(\Delta p) \geq \frac{h}{4\pi} \quad \text{hence} \quad \Delta p_{\text{min}} = \frac{h}{4\pi}\left(\frac{1}{\Delta x}\right)$$

The minimum indeterminacy in the momentum is then:

$$\Delta p_{\text{min}} = \frac{6.626 \times 10^{-34} \text{ J s}}{4\pi(1.0 \times 10^{-9} \text{ m})} = 5.27 \times 10^{-26} \text{ kg m s}^{-1}$$

A joule is a kg m^2s^{-2}, which explains the way the units work out. The minimum indeterminacy in the velocity can be computed using the fact that the momentum is the product of the velocity and the mass of the electron:

$$\Delta v_{\text{min}} = \frac{\Delta p_{\text{min}}}{m} = \frac{(5.27 \times 10^{-26} \text{ kg m s}^{-1})}{9.11 \times 10^{-31} \text{ kg}} = 5.8 \times 10^{4} \text{ m s}^{-1}$$

b) The minimum uncertainty in the momentum is the same as in part a). The computation of the minimum indeterminacy of the velocity must use the mass of a helium atom (text table 12-1). Conversion from amu to kilograms gives the mass of the helium atom as 6.647×10^{-27} kg. Substitution gives:

$$\Delta v_{\text{min}} = \frac{\Delta p_{\text{min}}}{m_{\text{He}}} = \frac{5.27 \times 10^{-26} \text{ kg m s}^{-1}}{6.647 \times 10^{-27} \text{ kg}} = 7.9 \text{ m s}^{-1}$$

13-31 The allowed energies of a particle in a one-dimensional box are:

$$E_n = n^2 \left(\frac{h^2}{8mL^2} \right)$$

where the quantum number n appears in combination with Planck's constant h, with a quantity characterizing the particle (m), and with a quantity characterizing the box (L). The particle is an electron and the length of the box is 1.34 Å. If all quantities are expressed in SI units:

$$E_n = n^2 \left(\frac{(6.626 \times 10^{-34} \text{ J s})^2}{8(9.109 \times 10^{-31} \text{ kg})(1.34 \times 10^{-10} \text{ m})^2} \right) = n^2(3.36 \times 10^{-18} \text{ J})$$

Thus, the ground-state energy E_1 is 3.36×10^{-18} J, and the-first-excited-state energy E_2 is 2^2 times this value or 13.4×10^{-18} J. Similarly, E_3 is 3^2 times the constant, or 30.2×10^{-18} J. To excite an electron in this box from $n = 1$ to $n = 2$ requires energy equal to the difference between E_1 and E_2. This is 10.1×10^{-18} J. If this energy is supplied by one photon then:

$$10.1 \times 10^{-18} \text{ J} = \frac{hc}{\lambda}$$

Solving for λ and substitution of h and c gives λ equal to 1.97×10^{-8} m. This wavelength, 197 Å, occurs in the ultraviolet region of the spectrum.

13-33 Combination (a) is not allowed because ℓ must be less than n; it may never equal n. Combination (c) has $m > \ell$, which is not allowed. Combination (d) has $\ell < 0$, which is not allowed. Only combination (b) is allowed.
The rules are easy to apply when physical significance is attached. $(n-1)$ equals the total number of all nodes of the wave function. The quantum number ℓ tells the number of angular nodes. Since the number of angular nodes cannot exceed the total number of nodes, ℓ may equal $(n-1)$ at the most. Its *minimum* is zero because "-1 planar nodes" has no physical meaning. The third quantum number m is related to the orientation of the angular nodes and is allowed to be negative.

13-35 a) $4p$ **b)** $2s$ **c)** $6f$

13-37 a) $4p$: 2 radial nodes, 1 angular node. **b)** $2s$: 1 radial node, 0 angular nodes. **c)** $6f$: 2 radial nodes, 3 angular nodes.

13-39 The wave function $\psi(2p_z)$ is the product of a *radial* part, $R(2p)$ and an angular part $\chi(p_z)$. The two parts of the functions are given in Table 13-2 (text page 504). The wave function is the product of the two parts. It is:

$$\psi(2p_z) = \left(\frac{1}{32\pi}\right)^{1/2} \left(\frac{Z}{a_0}\right)^{3/2} \cos\theta \left(\frac{Zr}{a_0}\right) \exp\left(-Zr/2a_0\right)$$

The radial part of the function contains an exponential dependence on r, and the angular part contains a $\cos\theta$ dependence only. This particular orbital is not a function of the third spherical coordinate ϕ. The probability of finding an electron at a point defined by a set of coordinates (r,θ,ϕ) depends on the square of this function:

$$\text{probability}(r,\theta,\phi) = \frac{1}{32\pi}\left(\frac{Z}{a_0}\right)^3 \cos^2\theta \left(\frac{Z^2r^2}{a_0^2}\right) \exp\left(-Zr/a_0\right)$$

The question now becomes: what values of r and θ make the quantity on the right equal zero? Clearly, this happens when $r = 0$ (at the nucleus). It also happens when $\theta = \pi/2 = 90°$ (because $\cos^2\pi/2 = 0$), and when $\theta = 3\pi/2 = 270°$ (because $\cos^2 3\pi/2 = 0$). Consult Figure 13-17 (text page 502) to confirm that the coordinate θ starts at zero along the $+z$ axis and equals either $\pi/2$ or $3\pi/2$ at right angles to z, in the xy plane. Therefore, the probability equals zero at all points in the xy plane. This plane is a nodal plane.
Writing out and squaring the full wave function as in the previous discussion is not really necessary since the angular part alone controls the angular nodes of all the functions in Table 13-2. Thus, the square of the d_{xz} orbital has a $\sin^2\theta\cos^2\theta\cos^2\phi$ angular dependence. This product of trigonometric functions goes to zero whenever $\theta = \pi/2, 3\pi/2$ (in the xy plane) and whenever $\phi = \pi/2, 3\pi/2$ (in the yz plane). These two planes are the two nodal planes of the d_{xz} orbital.

The square of the $d_{x^2-y^2}$ orbital has a $\sin^4\theta\cos^2 2\phi$ angular dependence. This cluster of trigonometric functions goes to zero at these values of ϕ:

$$\phi = \pi/4\ (45°)\ \text{ and }\ 3\pi/4\ (135°)\ \text{ and }\ 5\pi/4\ (225°)\ \text{ and }\ 7\pi/4\ (315°)$$

These first and third values of ϕ define a plane containing the z-axis and half-way between the x and y axes. The second and fourth values of ϕ define a plane containing the z-axis and at right angles to the first plane. These are the nodal planes of the $d_{x^2-y^2}$ orbital. The function also goes to zero at $\theta = 0$. This happens only along the z axis, the line at the intersection of the two nodal planes just identified.

13-41 The wavelength is the speed of the wave divided by its frequency:

$$\lambda = \frac{\text{speed}}{\nu} = \frac{343\ \text{m s}^{-1}}{440\ \text{s}^{-1}} = 0.780\ \text{m}$$

Dividing the distance by the speed gives the time. It takes the sound wave 0.0292 s to travel 10 m.

13-43 As a blackbody is heated, the wavelength at which the maximum intensity is emitted becomes shorter, which explains the change from red to orange in the perceived color. Also, the intensity of the emitted radiation become larger at all wavelengths. A white-hot object emits considerable intensities at the wavelengths of all the colors. It emits strongly across a wide band of wavelength (see the shape of the curve plotted in Figure 13-5 on text page 448). As rising T shifts the wavelength of maximum intensity into the yellow and green, which are near the center of the visible range, the intensities at all other wavelengths are also so great than the object appears white.

13-45 The energy of the photon is sufficient to overcome the work function of the nickel surface (pry an electron loose) and to impart a kinetic energy as large as 7.04×10^{-19} J to the ejected electron. It is known that:

$$h\nu = \frac{hc}{\lambda} = \Phi + \frac{1}{2}mv^2$$

Hence:

$$\frac{(6.626 \times 10^{-34}\text{J s})(3.00 \times 10^{8}\ \text{m s}^{-1})}{131 \times 10^{-9}\ \text{m}} = \Phi + 7.04 \times 10^{-19}\ \text{J}$$

Solving gives $\Phi = 8.1 \times 10^{-19}$ J.

13-47 The Lyman series is emitted as hydrogen atoms undergo transitions from various excited states to the ground state. The energies of the emitted photons are:

$$E_n(\text{H}) = \mathcal{R}Z^2\left(\frac{1}{n_f^2} - \frac{1}{n_i^2}\right) = \mathcal{R}(1^2)\left(\frac{1}{1} - \frac{1}{n_i^2}\right)$$

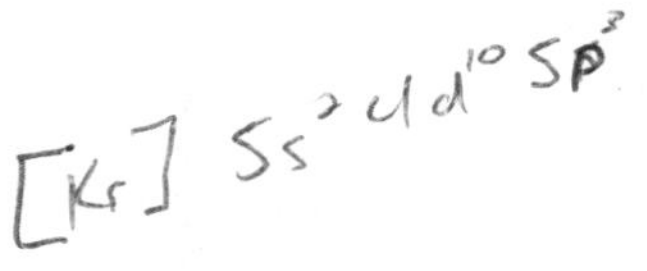

where $\mathcal{R}$ is the Rydberg constant and n_i is the quantum number of the excited state. To be absorbed by a ground-state He^+ ion, the energy of the incoming photon must exactly equal the energy it takes to raise the electron from the $n = 1$ state to the $n = 2, 3, 4 \ldots$ state. These energies are:

$$E_n(\text{He}^+) = \mathcal{R}Z^2\left(\frac{1}{n_i^2} - \frac{1}{n_f^2}\right) = \mathcal{R}(2^2)\left(\frac{1}{1} - \frac{1}{n_f^2}\right)$$

The question then becomes: is there any combination of positive whole-number values of n_i and n_f that makes E_n (H) and E_n (He^+) in the preceding two equations equal? To find out, take the difference of the two, set it equal to zero, and divide through by $\mathcal{R}$:

$$4\left(1 - \frac{1}{n_f^2}\right) - \left(1 - \frac{1}{n_i^2}\right) = 0 \text{ from which : } 3 + \frac{1}{n_i^2} = \frac{4}{n_f^2}$$

The left side of this expression is always between 3.25 and 3 as n_i takes on its possible values; the right side is always equal to or less than 1 as n_f takes on *its* possible values. It is clear that no suitable combination of n_i and n_f exists to make the equation valid. The answer to the question posed in the problem is therefore no.

13-49 The m of the earth and its v, and r in its orbit around the sun are all given in SI units. The angular momentum of the earth is the product of the three values. The angular momentum of the earth in its orbit around the sun is quantized in units of $h/2\pi$: $mvr = nh/2\pi$. Thus:

$$mvr = 2.7 \times 10^{40} \text{ kg m}^2\,\text{s}^{-2} = n\left(\frac{6.626 \times 10^{-34} \text{ J s}}{2\pi}\right)$$

Solving for n gives the desired answer: 2.6×10^{74}. Note that a J s is the same as kg $m^2\,s^{-2}$, so n is dimensionless. Since n is huge, ± 1 unit of angular momentum makes no meaningful change in the angular momentum.

13-51 First, compute the momentum of the photons as they approach the wall. The deBroglie relation gives p as a function of λ:

$$p = \frac{h}{\lambda} = \frac{6.626 \times 10^{-34} \text{ J s}}{550 \times 10^{-9} \text{ m}} = 1.205 \times 10^{-27} \text{ kg m s}^{-1}$$

Each photon bounces off the wall with momentum of the same magnitude but opposite sign. The change in momentum per collision is therefore:

$$\Delta p = p_2 - p_1 = (1.205 \times 10^{-27}) - (-1.205 \times 10^{-27}) = 2.410 \times 10^{-27} \text{ kg m s}^{-1}$$

This is the negative of the change in the momentum of the wall per collision of a photon. Hence:

$$\Delta p_{\text{wall}} = -2.410 \times 10^{-27} \text{ kg m s}^{-1}$$

The number of photons colliding per second equals the given power of the laser (1.0 watt = 1.0 J s^{-1}) divided by the energy delivered per photon. The energy delivered by each photon is:

$$E = \frac{hc}{\lambda} = \frac{(6.626 \times 10^{-34} \text{ J s})(3.00 \times 10^{8} \text{ m s}^{-1})}{550 \times 10^{-9} \text{ m}} = 3.614 \times 10^{-19} \text{ J}$$

Then:

$$\frac{\text{photons}}{\text{second}} = \left(\frac{1.0 \text{ J}}{1 \text{ s}}\right) \times \left(\frac{1 \text{ photon}}{3.614 \times 10^{-19} \text{ J}}\right) = 2.77 \times 10^{18} \text{ s}^{-1}$$

As developed in Section 3-5 (text page 105), the *total* force imparted to the wall is the change in momentum of the wall per collision multiplied by number of collisions per second:

$$F_{\text{wall}} = \frac{\Delta m_{\text{wall}}}{\Delta t} = (-2.41 \times 10^{-27} \text{ kg m s}^{-1}) \times (2.77 \times 10^{18} \text{ s}^{-1}) = -6.67 \times 10^{-9} \text{ kg m s}^{-2}$$

The negative sign simply tells the direction of the push on the wall and can be neglected. Pressure is defined as force divided by area. The area of the circular wall is πr^2 where r is its radius. Hence:

$$P = \frac{F_{\text{wall}}}{A} = \frac{F_{\text{wall}}}{\pi r^2} = \frac{6.67 \times 10^{-9} \text{ kg m s}^{-2}}{\pi (0.10 \times 10^{-3} \text{ m})^2} = 0.21 \text{ Pa}$$

This is a pressure of 2.1×10^{-6} atm.

13-53 a) The indeterminacy in the kinetic energy of the electron (call it ΔKE) is 0.02×10^{-19} J, the range of the values given in the problem. The indeterminacy in the momentum of the electron is related to that of the kinetic energy by:

$$\Delta p = \Delta(mv) = m\Delta v = m\sqrt{\frac{2(\Delta KE)}{m}}$$

The derivation of this equation employs $KE = 1/2mv^2$, the definition of kinetic energy. Compute Δp:

$$\Delta p = (9.11 \times 10^{-31} \text{ kg})\sqrt{\frac{2(0.02 \times 10^{-19} \text{ J})}{9.11 \times 10^{-31} \text{ kg}}} = 6 \times 10^{-26} \text{ kg m s}^{-1}$$

From the Heisenberg indeterminacy principle:

$$\Delta x \geq \frac{h/4\pi}{\Delta p}$$

it follows that the *minimum* indeterminacy in the position of the electron is:

$$\Delta x_{\rm min} = \frac{h/4\pi}{\Delta p} = \frac{(6.626 \times 10^{-34}\ \text{J s})/4\pi}{6 \times 10^{-26}\ \text{kg m s}^{-1}} = 9 \times 10^{-10}\ \text{m}$$

b) The mass of a helium atom is 6.647×10^{-27} kg (from text page 452 and problem 13-29). The indeterminacy in the momentum of helium atom with the same known range of energy as the electron in the previous part is

$$\Delta p = (6.647 \times 10^{-27}\ \text{kg})\sqrt{\frac{2(0.02 \times 10^{-19}\ \text{J})}{6.647 \times 10^{-27}\ \text{kg}}} = 5 \times 10^{-24}\ \text{kg m s}^{-1}$$

Note that the indeterminacy is larger than in the previous part because of the larger mass of the helium atom. The minimum indeterminacy in the position of the helium atom will be proportionately smaller. Thus:

$$\Delta x_{\rm min} = \frac{h/4\pi}{\Delta p} = \frac{6.626 \times 10^{-34}\ \text{J s}/4\pi}{5 \times 10^{-24}\ \text{kg m s}^{-1}} = 1 \times 10^{-11}\ \text{m}$$

13-55 The wave functions for the particle in a box are sketched in text Figure 13-16a (text page 500). The $n = 2$ wave function has exactly one node, half-way between $x = 0$ and $x = L$. This node results because $\sin(2\pi x/L)$ equals zero when x equals $L/2$. The function, on the other hand, has maxima at $x = L/4$ and $3L/4$. The figure clearly reveals the symmetry of the function. The probability of finding the particle at any point x between 0 and L is equal to ψ^2 evaluated at x. A plot of ψ^2 versus x has a zero at $x = L/2$ and symmetrical maxima at $x = L/4$ and $3L/4$ (see Figure 13-16b, text page 500). Passing from $x = 0$ to $x = L/4$ covers one-fourth of the area under the curve defined by ψ^2, the probability distribution function. This is evident from the symmetry of the function. The whole area under the curve scales to 1 since the particle must be in the box somewhere. The answer is therefore 1/4. The same answer comes from writing out the particle-in-a-box function for $n = 2$, squaring it, and integrating from $x = 0$ to $x = L/4$:

$$\text{probability} = \int_0^{L/4} \left(\sqrt{\frac{2}{L}} \sin \frac{2\pi x}{L}\right)^2 dx$$

In doing this it is helpful to know that

$$\int \sin^2 y dy = 1/2y - 1/4 \sin 2y$$

13-57 The $3d_{xy}$ orbital in O^{7+} ion is similar in shape to the $3d_{xy}$ orbital in an H atom; it has two nodal planes at right angles in both atoms. The orbital is much smaller in O^{7+} than in H because of the much larger nuclear charge attracting the electron in O^{7+}.

Chapter 14

Many-Electron Atoms and Chemical Bonding

Many-Electron Atoms and the Periodic Table

When an atom contains more than one electron, the Schrödinger equation becomes too complicated for easy use. The usual description of the **electronic structure** of many-electron atoms avoids the complexity of an exact solution by using the atomic orbitals of hydrogen as approximations of the actual orbitals. This is the reason for the heavy emphasis in chemistry on understanding hydrogen-atom wave functions. Hydrogen-atom orbitals are a guide to the electron configurations of all other atoms.

In the **orbital approximation** each electron in the many-electron atom is described by a set of four quantum numbers: n, ℓ, m, m_s. Certain rules govern how the orbitals accommodate electrons:

- The **Pauli Exclusion principle** states that no two electrons in an atom may have the same set of four quantum numbers. An orbital, which is by definition fully specified by a set of three quantum numbers, therefore holds at most two electrons, one with $m_s = +1/2$ and the other with $m_s = -1/2$.

- **Hund's rule** states that when electrons are added to a group of orbitals of equal energy one electron enters each separate orbital before the second electron enters any one of the set.

The ground-state electron configurations of all the elements in the periodic table are obtained by filling the H-atom orbitals in order of increasing energy while heeding the Pauli principle and Hund's rule. This embodies the **aufbau principle** or building-up principle. The following points are important in practicing aufbau:

- The order of filling the orbitals is:

$$1s\ \ 2s\ \ 2p\ \ 3s\ \ 3p\ \ 4s\ \ 3d\ \ 4p\ \ 5s\ \ 4d\ \ 5p\ \ 6s\ \ 4f\ \ 5d\ \ 6p\ \ 7s \ldots$$

In these designations, the numbers are n (the principal quantum number) and the letters tell the value of ℓ (the angular momentum quantum number) as explained on page 359 of this Guide. The s designations refer to just one orbital, but the p's refer to sets of 3 orbitals, and the d's and f's to sets of 5 and 7 orbitals because m is deliberately left unstated.

- A **shell** of orbitals in a many-electron atom is a set of orbitals with similar energies. A **subshell** is a group of orbitals with the same n and ℓ quantum numbers. Consider for example the $1s$, the $2s$, and $2p$ subshells in an atom of carbon. The $1s$ subshell, which contains a single orbital, is lower in energy than all other orbitals and is a distinct shell by itself. The $2s$ subshell (which contains one orbital) and the $2p$ subshell (which contains three orbitals) are relatively close together in energy and comprise a second shell.

- Orbitals of similar energy often are just orbitals with the same n. Sometimes however orbitals of different n are in the same shell. In cobalt the $4s$, $3d$ and $4p$ subshells are in the same shell.

- Electrons in the **outermost shell** of an atom determine the chemical behavior of that atom. These are the **valence electrons** of the atom. The valence electrons occupy the **valence orbitals.** Elements are classified as s-block, p-block, d-block, or f-block depending on whether the latest electrons added in aufbau went into s, p, d, or f orbitals.

- When two electrons in an atom have the same spin quantum number their spins are **parallel** and the two electrons are **unpaired.** Unpaired electrons must be in different orbitals, according to the Pauli principle. Two electrons in the same orbital must have spin quantum numbers $+1/2$ and $-1/2$. The two electrons are then **paired.** Electron spin is related to the magnetism of the electron. Paired electrons compensate for each other's magnetism. If all the electrons in an atom occur in pairs, then the atom is **diamagnetic.** If one or more electrons are not in pairs then their spins add together instead of cancelling each other out. An atom with one or more unpaired electrons is **paramagnetic.** Diamagnetism and paramagnetism are physically observable properties.

Learn to write a correct representation of the ground-state electron configuration of any atom or ion. See **problems 14-1** and **14-3.** In writing electron configurations:

1. The number of electrons in each subshell is shown as a right superscript (and is often mistaken for an exponent). The maximum superscript is 2 for an s subshell (1 s orbital), 6 for a p subshell (3 p-orbitals), 10 for a d subshell (5 d-orbitals, and 14 for an f subshell (7 f-orbitals).

2. Noble-gas electron configurations are abbreviated by writing the symbol of the noble gas in brackets as in **problems 14-1,** and **14-7.**

3. The order of filling of the subshells in the neutral atoms need not be memorized. It can be determined from the pattern of the different blocks of elements in the periodic table.

4. There are exceptions to the normal order for 19 elements, all of which are d-block are f-block elements. They are explained by attributing additional stability to half-filled and filled subshells. These are best memorized.

Another important exercise is writing the valence electron configuration of an element given its position in the periodic table. The number of the period that the element occupies gives the maximum value of n in such a configuration. For non-transition elements, the number of valence electrons equals the group number (number at head of the column in the periodic table).

Many problems are based on proposed violations of the rules that relate the quantum numbers of the electrons in many-electron atoms. See**problem 14-9,** and **14-15.**

Experimental Measures of Orbital Energies

Ionization Energy

The **ionization energy** (IE) of an atom is the minimum energy necessary to remove an electron from it:

$$X(g) \rightarrow X^{+}(g) + e^{-} \quad \text{Ionization Energy} = \Delta E$$

Ionization energies are measured by **photoelectron spectroscopy.** The positively charged ion produced by ionization may itself be able to lose an electron. Thus an atom with n electrons has a series of n ionization energies, each one larger than the previous: $IE_1, IE_2, IE_3 \ldots$

In general the ionization energy *increases* from left to right across a period (row) in the periodic table and *decreases* from top to bottom within a group (column). See **problem 14-11. Problem 14-15** concerns a hypothetical universe and tests not only an understanding of these trends but also the exceptions caused by the extra stability of half-filled subshells. For example, nitrogen, with a half-filled $2p$ subshell, has a higher IE_1 than the neighbor to the right, oxygen. The orbital description of many-electron atoms explains these trends and exceptions.

Electron Affinity

Energy is usually *released* when an electron is added to an isolated atom to form a negative ion. When a process releases energy it has a negative ΔE. The **electron affinity** *(EA)* of an atom is the energy *gained* by the system as the electron is added:

$$X(g) + e^- \rightarrow X^-(g) \quad \text{Electron Affinity} = -\Delta E$$

The use of the word "affinity," the tendency to gain, obliges the negative sign. A negative *EA* means that an atom repels the extra electron and the two must be forced together.

The periodic trends in *EA* parallel the changes in *IE*. Both atomic properties increase from left to right in a period and from bottom to top in a group.

If a second electron joins an originally neutral atom, there is a second electron affinity: EA_2 is *always* negative. The -1 ion formed by the atom's gain of the first electron repels additional negative charge.

Sizes of Atoms and Ions

Atomic size (the radii of ions and of metallic atoms) *decreases* with increasing atomic number across a period and *increases* going down a group in the periodic table. See **problem 1.** The *rate* of increase in size going down a group diminishes sharply once the 3*d* orbitals start to be filled. As Z increases across the transition metals, any one 3*d* electron is only imperfectly shielded from the attraction of the larger nuclear charge by the other electrons in the same subshell. Atomic size therefore contracts. The same effect occurs with *p* subshells and is responsible for the general contraction in size going across the table. The intervention of the first block of 10 elements in which *d*-orbitals are filled makes immediately subsequent elements smaller than they otherwise would have been.

Isoelectronic species have the same electron configuration but different nuclear charge. S^{2-}, Cl^-, Ar, and K^+ are isoelectronic. The smallest species among an isoelectronic group is the one with the highest Z.

Properties of the Chemical Bond

Different parameters characterize different aspects of bonding. It is important to understand their physical meaning and to get an idea of the units used to measure them. They include:

Bond Lengths. The distance from any nucleus in a molecule to any other is an **interatomic distance.** Interatomic distances have no upper limit because molecules can be arbitrarily large. Interatomic distances are labeled **bond**

lengths if the two atoms involved are connected by a chemical bond. Bond lengths are on the order of 1×10^{-10} m. This distance has a special name, the Ångstrom (Å).

$$\mathbf{1\ \text{Ångstrom} = 0.1\ \text{nanometer (nm)} = 100\ \text{picometer (pm)}}$$

It helps to get some typical bond lengths in mind: C-to-C bonds range from 1.20 to 1.54 Å; O-to-H bonds in various compounds range from 0.94 to 1.09 Å. Table 14-1 (text page 530) gives the bond length in many diatomic molecules. Bond lengths range up toward 3 Å. For example, the Sb-I distance in SbI_3 is 2.75 Å.

Bond lengths for a given type of bond do not change much from molecule to molecule. See Table 14-2, text page 530.

Bond lengths are symbolized R or R_e. The subscript (e for equilibrium) emphasizes that molecular vibrations can temporarily lengthen and shorten bonds.

Bond Enthalpies. Chemical bonds form because the atoms involved have a lower energy when close together than when far apart. Breaking a bond means moving the bonded atoms apart. It requires energy. This energy is ΔE_d, the **dissociation energy,** of the bond. It is a quantitative measure of the strength of a bond.

Bond *enthalpies* are ΔH values for breaking bonds. They are related to bond energies in the same way any enthalpy change is related to an energy change:

$$\Delta H_d = \Delta E_d + \Delta(PV) \quad \text{from which} \quad \Delta H_d = \Delta E_d + RT$$

because breaking a mole of bonds in a gaseous compound produces two moles of gaseous fragments.

Bond energies and enthalpies are always positive. They are measured in J mol^{-1} or kJ mol^{-1}. molar bond enthalpies are on the order of 100 to 1000 kJ mol^{-1}.

Bond enthalpies are fairly reproducible from one compound to another, but exact bond enthalpies depend on the exact surroundings of the bond in the molecule. The bond enthalpy of the first O—H bond in H_2O is 502 kJ mol^{-1}. The bond enthalpy of the O—H bond in the OH molecule left behind by removal of one H from H_2O is only 426 kJ mol^{-1}. Review the values in Table 8-3 (text page 296), which are *average* bond enthalpies. Such tables allow convenient estimates of the $\Delta H°$ of reactions. One imagines that a reaction proceeds by the complete disruption of all chemical bonds in the reactants to give isolated gaseous atoms followed by the recombination of these atoms to give the products. The total bond enthalpy of the reactants is the amount required to complete the disruption. It is the sum of the bond enthalpies of all the bonds the reactants have. The total bond enthalpy of the products is, similarly, the enthalpy to break all the bonds among the products. The *change* in bond enthalpy is the total bond enthalpy of the products minus the total bond enthalpy of the reactants. It is the negative of $\Delta H°$ of the reaction. See **problem 15-55b.**

The molar dissociation enthalpy of a diatomic molecule is *twice* the enthalpy required to produce 1 mol of atoms. Thus, it takes 243 kJ to dissociate 1 mol of $Cl_2(g)$. It takes only 141.5 kJ to produce 1 mol of $Cl(g)$ from $Cl_2(g)$. This factor of 2 can be a stumbling block in problems.

Bond Angles. Any three atoms in a molecule define an angle, measured in degrees. Indeed, three points in space define *two* angles, the second being equal to 360° minus the first. **Bond angles** are interatomic angles defined by bonded atoms. They are always taken in the sense that makes them less than or equal to 180°.

Bond Order. Sometimes the length and enthalpy of the bond between two specific kinds of atoms (for example C and N atoms) are sharply different from one compound to another. This breakdown in reproducibility is ascribed to the bonds have different **bond orders.** The most common bond orders are 1 (single bond, C—N) 2 (double bond C=N), and 3 (triple bond C≡N). Fractional bond orders are also possible.

Ionic and Covalent Bonds

Chemical bonds arise from the sharing or transfer of electrons between two or among several atoms. Bonding is **covalent** when sharing predominates; bonding is **ionic** when transfer predominates. Intermediate cases are **polar covalent** bonds.

Electronegativity

Atoms with low *IE*'s and low *EA*'s readily lose their own electrons and are poor at accepting new electrons. Such atoms lie to the lower left of the periodic table. Atoms with high *IE*'s and high *EA*'s strongly resist giving up their own electrons and are good at accepting new electrons. These atoms lie to the upper right of the periodic table. The first kind of atom is an electron donor and the second kind an electron acceptor.

A new atomic property quantifies the progression in the periodic table from good electron donor to good acceptor.

• The **electronegativity** of an atom is a measure of the capacity of the atom in a molecule to attract electrons to itself.

A numerical definition is:

$$\text{Electronegativity} \propto \frac{1}{2}(IE_1 + EA_1)$$

This definition is due to Robert Mulliken, and values attained with this formula are Mulliken electronegativities. Mulliken electronegativities are invariant atomic properties. The actual capacity of an atom to attract electrons in a molecule however depends on the surroundings of the atom. **Pauling electronegativities** are average electronegativities. A typical environment is assumed for each atom. Pauling electronegativities are versatile and widely used, but should not be thought of as precise experimental results. They run from a maximum of 3.98 for F to a minimum of 0.79 for Cs and have no units. Cesium is said to be a very **electropositive** element.

Chemical bonds are rarely entirely covalent or entirely ionic. Instead they have an intermediate character. Electronegativities help estimate just how covalent or ionic a bond is. The greater the difference in electronegativity between two atoms, then the more ionic is the bond between the two. The less the difference, then the more covalent is the bond. Atoms of the same element have the same atomic electronegativity. In compounds like H_2 and F_2, the bond is 100 percent covalent. If the difference in electronegativity exceeds about 2.0 for a pair of elements, then the chemical bond between them is mainly ionic. If it is less than about 0.4, then the bond is mainly covalent. In the intermediate range, the chemical bonds are covalent with partial ionic character.

The partial ionic character of a bond is approximately:

$$\text{Percent Ionic Character} = 16|\Delta| + 3.5\Delta^2$$

where Δ is the difference in electronegativity. A Δ of 2.0 corresponds to about 46 percent (about half) ionic character. **Problem 14-31** illustrates the use of this approximation.

Figure 14-12 (text page 533) lists a set of atomic electronegativities. The largest occur at the upper right of the periodic table, and the smallest χ's occur at the lower left of the table.

Ionic bonds

Ionic bonds arise from the transfer of electrons between atoms and the subsequent Coulomb attractions among the positive and negative ions that are formed.

To remove even one electron from a gaseous atom of any element always requires energy. This is the ionization energy, IE. Many atoms release energy, the electron affinity (EA), as they accept an electron. For every possible pair-wise combination of atoms:

$$\Delta E_\infty = (IE - EA) > 0$$

It always costs energy to transfer an electron from one atom to another.

Ionic bonds form because of the attraction between charges of unlike sign. The change in energy due to this attraction is:

$$\Delta E_{\text{coulomb}} = \frac{Z_1 Z_2 e^2}{4\pi\epsilon_0 R}$$

where R is the distance between the ions, the Z's are the charges of the ions in units of e, the charge on the electron, and ϵ_0 is a constant (8.854×10^{-12} $C^2\,J^{-1}m^{-1}$). According to this formula, the energy of a pair of unlike ions goes to negative infinity as R goes to zero. But ions are not point charges. At very short distances the electron clouds of the ions start to repel each other and the energy rises steeply. See Figure 14-13 (text page 535). The distance at which the short-range repulsions balance the Coulomb attraction is R_e.

The preceding formula applies to a pair-wise interaction between atoms A(g) and X(g) to form the molecule $A^+X^-(g)$. If the Coulomb energy is converted to a per mole basis then one must imagine a mole of widely separated A^+—$X^-(g)$ molecules. A collection of such ionic molecules held in the same region of space would, at ordinary temperatures, condense to form an ionic crystal, in which every ion interacts with all others. This is treated in Chapter 17 (text page 654-5).

Covalent Bonding

Covalent bonding arises from the sharing of electrons between atoms. Electrons have a negative charge. An electron shared by two nuclei spends most of its time *between* the two. It tends to lower the repulsion between the positively-charged nuclei because it is attracted to both.

One definition of a covalent bond is "a shared pair of electrons." This definition is over-simple (as proved by the covalent bond in the H_2^+ ion, which possesses a grand total of only one electron. Nevertheless it can sometimes be helpful.

Dipole Moment and Percent Ionic Character

Molecules in which the centers of positive and negative electrical charge do not coincide possess an electric **dipole moment.** In effect they have an electrically positive end and an electrically negative end.

If placed between a positively charged and a negatively charged plate (that is, in an electric field) a molecule with a dipole moment tends to align itself so that its negative end is nearer the positive plate and its positive end is nearer the negative plate. Molecules not aligned with the field experience a torque, or twist, which tends to align them. For a given electric field the size of the torque depends on the magnitude of the molecule's dipole moment. Numerically the dipole moment is:

$$\mu = QR$$

where R, is the distance separating charges of opposite sign and equal magnitude, Q.

Electric dipole moments are *vector* quantities. They have both magnitude and direction. Frequently they are represented by arrows. See Figure 14-20, text page 543. Larger dipoles are represented by longer arrows.

A difference in electronegativity between two bonded atoms means that there is a charge separation across the bond and consequently a dipole moment associated with the bond. For diatomic molecules, which have only one connection between atoms, this dipole moment is of course the dipole moment of the molecule.

In molecules which contain more than one bond the overall molecular dipole moment depends both on bond polarity and molecular geometry (see below).

Dipole moments can be experimentally measured. The magnitude of a dipole moment in a polyatomic molecule provide important information about the molecule's geometry. Absence of a dipole moment argues for some kind of symmetrical molecular structure. See **problem 14-43.**

Dipole moments are nearly always cited in a non-SI unit, the **Debye.** One Debye (D) equals 3.336×10^{-30} C m. The Coulomb meter (C m) is the SI unit for dipole moment. The Debye is more convenient for chemists because most molecular dipole moments are between 0 and 10 Debye.

- The Debye is the dipole moment of a +1 and −1 fundamental charge (charges of magnitude 1.602×10^{-19} C) held 0.2082 Å apart.

For diatomic molecules (and only for diatomic molecules), experimental dipole moments allow the estimation of the partial ionic character of the bond. The greater the dipole moment for a given bond distance, then the greater is the bond's ionic character. If δ is the fraction of a unit charge on each atom in a diatomic molecule then:

$$\delta = \frac{\mu}{eR}$$

When μ is in Debye, R in Ångstrom then:

$$\delta = 0.2082 \frac{\mu \text{ (in Debye)}}{R \text{ (in Å)}}$$

The quantity δ is the **fractional ionic character.** It ranges from 0.0 to 1.0. The fractional **covalent** character of a bond is $1.0 - \delta$. The equation is straightforward to use in problems. The only complication is the factor of 0.2082 which arises from the use of the non-SI unit for dipole moment. See **problem 14-29.**

The values for fractional ionic character estimated with the formula correspond fairly well with the values derived from electronegativity differences. See **problem 14-31.**

The Shapes of Molecules

Molecular Geometries

The **valence shell electron pair repulsion (VSEPR)** theory holds that valence electron pairs distribute themselves about atoms in such a way as to minimize electron pair repulsions. Electron pairs are either *lone pairs,* not directly involved in bonding, or *bonding pairs,* shared between atoms.

The **steric number** *(SN)* for an atom in a molecule is the sum of the number of atoms bonded to it and the number of lone pairs on it. The way to get the steric number of an atom in a molecule is first to draw a valid Lewis structure for the molecule and then just to count the atom's neighbors. If the molecule has several resonance Lewis structures, it does not matter which is used. A double or triple bond contributes only 1 the steric number of an atom.

In VSEPR theory the strength of the electron pair repulsions follows the order:

lone pair-lone pair > lone pair-bonding pair > bonding pair-bonding pair

Different geometrical shapes minimize the electron pair repulsions for the different steric numbers. The shapes are shown in text Figure 13-1. Knowing these shapes and employing the repulsion order just quoted, a table of the molecular shapes predicted by VSEPR theory for steric numbers from 2 to 6 is readily constructed. In this table, X is an atom bonded to the central atom A, and E stands for an electron pair.

SN of A	Molecular Type X is a bonded atom; E is a lone pair	Predicted Shape
2	AX_2	linear
3	AX_3	trigonal planar
3	AX_2E	bent
4	AX_4	tetrahedral
4	AX_3E	trigonal pyramidal
4	AX_2E_2	bent
5	AX_5	trigonal bipyramidal
5	AX_4E	distorted see-saw
5	AX_3E_2	distorted T
5	AX_2E_3	linear
6	AX_6	octahedral
6	AX_5E	square pyramidal
6	AX_4E_2	square planar

Problems 14-33 and 14-35 show the use of VSEPR. The theory can be extended to *SN*'s exceeding 6 (see **problem 14-67).**

In polyatomic molecules, the dipole moment of the molecule is the *vector sum* of the dipole moments of all of the bonds. For example, chlorine is more electronegative than carbon. The C–Cl bond is therefore polar. The CCl_4 molecule has four equal C—Cl bond dipoles. According to VSEPR theory the four bonds point from the central C atom to Cl atoms at the corners of a tetrahedron. This symmetrical arrangement means that the vector sum of the four bond dipoles is zero. Consult Figure 14-20d (text page 543). See **problem 14-39** for other examples.

Detailed Solutions to Odd-Numbered Problems

14-1 The ground-state configuration are:
a) C $1s^22s^22p^2$.
b)Se $1s^22s^22p^63s^23p^63d^{10}4s^24p^4$. The use of the bracketed symbol of a noble gas to represent the electron configuration of that noble gas shortens this to $[Ar]3d^{10}4s^24p^4$.

c) Fe $[Ar]3d^64s^2$.

14-3 The ground-state configuration of an ion is arrived at by writing the ground-state configuration of the atom and adding (in the case of a negative ion) electrons to available orbitals in order of energy. In the case of positive ions, electrons are subtracted starting with the highest-energy occupied orbitals:

Be^+	$1s^22s^1$	C^-	$1s^22s^22p^3$	Ne^{2+}	$1s^22s^22p^4$	Mg^+	$[Ne]3s^1$
P^{2+}	$[Ne]3s^23p^1$	Cl^-	$[Ne]3s^23p^6$	As^+	$[Ar]3d^{10}4s^24p^2$		
I^-	$[Kr]4d^{10}5s^25p^6$						

All of these electron configurations are *ground-state* (lowest energy) configurations. Be^+, C^-, Ne^{2+}, Mg^+, P^{2+} and As^+ all have at least one unpaired electron (they have incomplete subshells) and should be paramagnetic. The Cl^- and I^- ions are diamagnetic.

14-5 a) The atom has 49 electrons (36 represented by "[Kr]" and 13 more represented by superscripts). It is an indium atom.
b) The ion has 18 electrons, and a charge of −2. The atomic number of its nucleus must be 16. It is therefore S^{2-}.
c) The ion has 21 electrons, and a charge of +4. The atomic number of its nucleus must be 25. It is therefore Mn^{4+} ion.

14-7 As a halogen this element would have a ground-state configuration of the form $\ldots np^5$. The next p-subshell after the $6p$ (used in the sixth row of the periodic table) is the $7p$. Accordingly, the electron configuration of the element would be $[Rn]\,5f^{14}6d^{10}7s^27p^5$ where "[Rn]" stands for 86 electrons. Adding the other superscripts to 86 gives 117 electrons. Therefore Z equals 117.

14-9 If only one electron could occupy each orbital in many-electron atoms, then the configurations $1s^1$ and $1s^1 2s^1 2p^3$ and $1s^1 2s^1 2p^3 3s^1 3p^3$ would be closed-shell electron configurations. Atoms with $Z = 1, 5, 9$ respectively would have these ground state electron configurations. See also problem 14-15.

14-11 Use the periodic trends in ionization energy discussed on text pages 524-5. **a)** Sr has a higher first ionization energy than Rb. **b)** Rn has a higher IE_1 than Po. **c)** Xe has a higher IE_1 than Cs. **d)** Sr has a higher IE_1 than Ba.

14-13 Use the periodic trends in electron affinity discussed on text pages 525-6 and shown in Figure 14-9. **a)** Cs has a larger EA than Xe. **b)** F has a larger EA than Pm. **c)** K has a larger EA than Ca. **d)** At has a larger EA than Po.

14-15

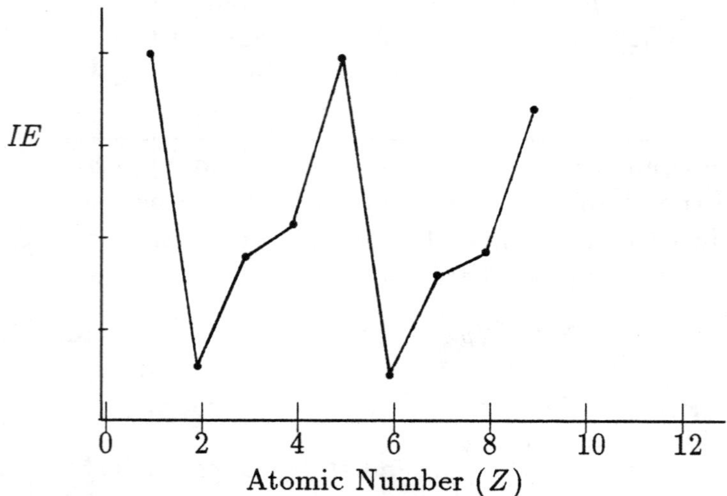

The exact differences in ionization energy from one "element" to the next in the preceding are of course completely conjectural. All that counts are the trends.

14-17 Convert the ionization energy of cesium from kilojoule per mole to joules per atom. This is done by multiplying it by 1000 J kJ^{-1} (to get to joules per mole) and then dividing by 6.022×10^{23} mol^{-1} (Avogadro's number). The answer is 6.239×10^{-19} J. Next, use the relationship $E = h\nu = hc/\lambda$ to compute the wavelength that corresponds to this energy:

$$\lambda = \frac{hc}{E} = \frac{(6.626 \times 10^{-34} \text{ J s})(2.9979 \times 10^{8} \text{m s}^{-1})}{6.239 \times 10^{-19} \text{ J}} = 3.184 \times 10^{-7} \text{ m} = 318.4 \text{ nm}$$

This wavelength is in the near ultraviolet region of the electromagnetic spectrum.

14-19 **a)** A ground-state K atom should have a larger radius than a ground-state Na atom. In K atoms the outermost electron occupies a $4s$ orbital, but in Na atoms the outermost electron occupies a closer $3s$ orbital.

b) The Cs atom is larger than the Cs^+ ion. As a Cs^+ ion gains an electron to produce a Cs atom, the electron is accommodated in the more distant n=6 shell.
c) The Rb^+ ion and the Kr atom are isoelectronic. The larger species is the one with smaller nuclear charge: Kr.
d) A Ca atom has two 4*s* electrons and a K atom has one 4*s* electron. The outermost electrons are in the same shell but Ca has a larger nuclear charge, contracting the electron cloud. Hence potassium is larger.
e) The Cl^- ion and the Ar atom are isoelectronic. The larger species is the one with smaller nuclear charge: Cl^-.

14-21 a) The S^{2-} ion should be larger than the O^- ion. Its outermost electrons occupy the $n = 3$ level whereas in O^- ion the outermost electrons are in the closer $n = 2$ level. **b)** The Ti^{2+} ion is larger than the Co^{2+} ion because the two have their outermost electrons in the same level, and Ti^{2+} has a smaller nuclear charge. **c)** The Mn^{2+} ion is larger than the Mn^{4+} ion because outermost electrons (those farthest away) are lost in going from the +2 to +4 ion. **d)** The Sr^{2+} ion is larger than the Ca^{2+} ion according to the trend to larger size going down the periodic table.

14-23 The As—H bond length should be intermediate between the 1.42 Å of P—H and the 1.71 Å of Sb—H. A length of 1.56 Å, (the average) is a reasonable guess. The experimental bond length is 1.519 Å. The X—H bond will be weakest in SbH_3, which has the longest bonds.

14-25 Binary ionic compounds have a large difference in electronegativity between their elements whereas binary covalent compounds have a small difference. High vapor pressure is associated with relatively weak intermolecular attractions and by extension with molecular (covalent) compounds. The correct answer in each case in the compound with the smaller difference in electronegativity: **a)** CI_4 **b)** OF_2 **c)** SiH_4.

14-27 The problem is similar to text Example 14-5. The gaseous KCl molecule is treated as two point charges separated by 2.67 Å. The Coulomb potential energy of a pair of charges separated by distance R_e is:

$$\Delta E_{\text{coulomb}} = \frac{Z_1 Z_2 e^2}{4\pi\epsilon_0 R_e}$$

where Z_1 and Z_2 are +1 (for K^+ ion) and −1 (for Cl^- ion), where ϵ_0 is 8.854×10^{-12} $C^2\,J^{-1}m^{-1}$, and where the unit of charge e is 1.602×10^{-19} C. Substitute $R_e = 2.67 \times 10^{-10}$ m and the various other values:

$$\Delta E_{\text{coulomb}} = \frac{(1)(-1)(1.602 \times 10^{-19}\ \text{C})^2}{4(3.1416)(8.854 \times 10^{-12}\ \text{C}^2\,\text{J}^{-1}\text{m}^{-1})(2.67 \times 10^{-10}\ \text{m})}$$

The answer is -8.64×10^{-19} J. This is the potential energy of a *single* K^+ to Cl^- attraction. A mole of such attractions has this ΔE multiplied by Avogadro's number. Thus $\Delta E_{\text{coulomb}}$ equals -520 kJ mol^{-1}. It would require $+520$ kJ to separate one mole of the KCl molecules, that is, to move all the Cl^- ions an infinite distance away from the K^+ ions to which they are bonded. The result would be a collection of $K^+(g)$ ions and a second collection of $Cl^-(g)$ ions. Removing the electrons from the $Cl^-(g)$ ions would *consume* 349 kJ mol^{-1}. This is the electron affinity of Cl(g) (see text Figure 14-9 or text Appendix F). Feeding the electrons into the $K^+(g)$ ions would release 419 kJ mol^{-1}. This value is the ionization energy of potassium. The energy change accompanying the electron transfer would be:

$$\Delta E_{\infty} = 419 - 349 = 70 \text{ kJ mol}^{-1}$$

When this *release* is subtracted from the 520 kJ mol^{-1} consumed in separating the ions, the answer is 450 kJ mol^{-1}, the dissociation energy.

14-29 In diatomic molecules, the fractional ionic character δ is:

$$\delta = (0.2082 \text{ Å D}^{-1})\left(\frac{\mu}{R}\right)$$

when the dipole moment μ is in Debye and the bond distance R is in Ångstrom. The percent ionic character is just 100 times the fractional ionic character. The results can be summarized in a table:

Compound	Bond Length (Å)	Dipole Moment (D)	(100 δ)
ClO	1.573	1.239	16
KI	3.051	10.82	74
TlCl	2.488	4.543	38
InCl	2.404	3.79	33

14-31 The results can be summarized in a table:

Compound	Δ	$16\lvert\Delta\rvert + 3.5\Delta^2$	Ionic Character
HF	1.80	40	41%
HCl	0.98	19	18
HBr	0.78	15	12
HI	0.48	8	6
CsF	3.19	87	70

The point of the problem is the generally good agreement between the values calculated from Δ and the dipole-moment-based values.

14-33 a) The molecule of CBr_4 is tetrahedral. The central C atom has *SN* 4. There are no lone pairs on the central carbon atom so this is an AX_4 case (see table on page 382 of this Guide).
b) In SO_3, the central S atom has *SN* 3 and no lone pairs. The molecule is trigonal planar with an O—S—O angle of 120°. The fact that one or more of the S-to-O bonds can be shown in a Lewis structure as a double bond does *not* affect the steric number.
c) In SeF_6, the central Se atom has *SN* 6. There are no lone pairs on the central Se atom, so the expected geometry of the F atoms about the Se is octahedral.
d) In ICl_3, the central I atom has *SN* 5. It is surrounded by 3 Cl atoms and 2 lone pairs. The geometry including the lone pairs is trigonal bipyramidal. The actual molecular geometry (which only counts atoms) of this AX_3E_2 case is a distorted T-shape.

14-35 a) The molecule-ion ICl_4^- is square planar. The central I atom has *SN* 6, which means the geometry of the electron pairs about the atom is octahedral. The 2 lone pairs lie opposite each other on the octahedral pattern, minimizing lone-pair to lone-pair interactions. The 4 Cl atoms surround the central I atom in a planar square.
b) In OF_2, the central O atom has *SN* 4. The molecule is of the type AX_2E_2. The molecule is *bent* to accommodate the steric requirements of the two lone pairs on the O atom. The F—O—F angle is less than 109.5°.
c) In BrO_3^-, the central Br atom has *SN* 4 and is of the type AX_3E. The single lone pair on the central Br atom occupies one corner of a tetrahedron about the Br atom. The resulting molecule is pyramidal. The presence of the lone pair forces the O atoms together slightly, so that the O—Br—O angle is less than 109.5°.
d) In CS_2, the central C atom has a *SN* of 2. Both of the C-to-S bonds are double bonds, but this plays no part in figuring this steric number on the C atom. The molecule, which is of the type AX_2, is linear.

14-37 a) Planar AB_3: BF_3, BH_3, SO_3. **b)** Pyramidal AB_3: NH_3, NF_3. **c)** Bent AB_2^-: ClO_2^-, NO_2^-. **d)** Planar AB_3^{2-}: CO_3^{2-}.

14-39 All of the *bonds* in all of the compounds are polar. The symmetry of certain molecular shapes makes the vector sum of the individual bond dipoles zero. Thus, SeF_6 (octahedral), CBr_4 (tetrahedral), and SO_3 (trigonal planar) are non-polar. The other two molecules, ICl_3 (distorted T) and $SOCl_2$ (isosceles planar), are less symmetrical and the vector sums of their bond dipoles are not zero. The VSEPR approach predicts the shapes.

14-41 The fact that the molecule is bent is unhelpful in deciding between the formulations :N̈=Ö—F̈: and :Ö=N̈—F̈: because both structures feature a central atom having *SN* 3 with 2 bonds and 1 lone pair. Theory predicts a bent molecule in both cases.

14-43 **a)** The resonance structures $:\ddot{N}{=}N{=}\ddot{O}:$ and $:N{\equiv}N{-}\ddot{\underset{..}{O}}:$ can be written for the NNO molecule. No matter which of the two resonance structures is considered, the *SN* of the central nitrogen atom is two. The predicted molecular geometry is linear. **b)** Linear geometry would allow a complete cancellation of the two bond dipoles if they were equal. They are not equal, so there is an observable net dipole moment. The polarity of the N=O bond is expected to exceed the polarity of the N=N bond because O is more electronegative than N. The N end of the molecule is therefore expected to have the positive partial charge.

14-45 The predicted maximum oxidation states are V (+5), P (+5), I (+7), Sr (+2). The predictions are made by looking at the group number of the element in the periodic table.

14-47 By comparison with other oxoacids of the general formula $XO(OH)_m$, $HCO(OH)_1$ should have K_a between 5×10^{-4} and 8×10^{-2}. The actual K_a is 2×10^{-4}. At pH 14, the formate ion $HCOO^-$ predominates in solutions of HCOOH(aq); Molecules of HCOOH are present only in very small concentration because the H atom attached to the O atom is so readily lost as an H^+ ion. There is however essentially no tendency for the H atom bonded to the C atom to be lost.

14-49 The four acids to be compared are $H_2B_4O_7$, H_3BO_3, $H_5B_3O_7$ and $H_6B_4O_9$. Rewrite their formulas as, respectively:

$$B_4O_5(OH)_2,\quad B(OH)_3,\quad B_3O_2(OH)_5,\quad B_4O_3(OH)_6.$$

The acid with the largest ratio of lone oxygens atoms to boron atoms should be the strongest. This ratios are 1.25, 0, 0.67, 0.75. The predicted order of acid strength is therefore:

$$B(OH)_3 < B_3O_2(OH)_5 < B_4O_3(OH)_6 < B_4O_5(OH)_2$$

14-51 This atom of sodium is in an excited state. It can lose energy in a variety of ways. to end up ultimately in its ground state, which is represented [Ne]$3s^1$.

14-53 In chromium(IV) oxide, the Cr^{4+} ion has the ground-state electron configuration: [Ar]$3d^2$. The neutral Cr atom has lost its $4s$ electron and three of its five $3d$ electrons. The two remaining $3d$ electrons are, at lowest energy, unpaired so CrO_2 has two unpaired spins per Cr atom.

14-55 The first ionization energy of Li(g) is 520×10^3 J mol^{-1}. Dividing this by N_0 converts it to an *IE* on a per atom basis. It is 8.635×10^{-19} J. The electronic energy of atoms in the one-electron approximation is given by the formula:

$$E = \left(-2.18 \times 10^{-18}\ \text{J}\right) \frac{Z_{\text{eff}}^2}{n^2}$$

where Z_{eff} is the effective nuclear charge of the approximated atom and n is the quantum number of the one electron being removed. The negative of this energy is what is required to excite the one electron to $n = \infty$. For lithium:

$$-(-2.18 \times 10^{-18}\ \text{J})\frac{Z_{\text{eff}}^2}{2^2} = IE = 8.635 \times 10^{-19}\ \text{J}$$

Solving gives $Z_{\text{eff}} = 1.26$. The actual Z of lithium is 3. The inner two electrons, which are located mainly between the 2s electron and the nucleus, reduce the influence of the nucleus considerably, but the screening is imperfect because Z_{eff} exceeds 1. For Na, n is 3 and the equation becomes:

$$-(-2.18 \times 10^{-18}\ \text{J})\frac{Z_{\text{eff}}^2}{3^2} = \frac{496 \times 10^3\ \text{J mol}^{-1}}{6.022 \times 10^{23}\ \text{mol}^{-1}} \quad \text{from which} \quad Z_{\text{eff}} = 1.84$$

For K, n is 4:

$$-(-2.18 \times 10^{-18}\ \text{J})\frac{Z_{\text{eff}}^2}{4^2} = \frac{419 \times 10^3\ \text{J mol}^{-1}}{6.022 \times 10^{23}\ \text{mol}^{-1}} \quad \text{from which} \quad Z_{\text{eff}} = 2.26$$

14-57 The smallest by far is the hydrogen-like Co^{25+} ion. The rest of the order follows from periodic trends (or text Figure 14-10)

$$Co^{25+} < F^+ < F < Br < K < Rb < Rb^-$$

14-59 As a general rule, the longer a bond is then the weaker it is. Since the C≡C bond in $Ni(CO)_4$ is longer than in free CO, the C≡O bond in $Ni(CO)_4$ is weaker.

14-61 The difference in electronegativities of the atoms in HF is 1.78; the difference in LiCl is 2.18. The two compounds differ greatly in their bonding according to the evidence of their physical properties. Lithium chloride is evidently a predominately ionic compound (high boiling, high melting); hydrogen fluoride is a predominately covalent compound (low melting, low boiling).

14-63 In general, large electronegativity is favored by small atomic size and a large effective nuclear charge. Therefore electronegativity usually decreases from the top to the bottom of a group in the periodic table. Such a trend is broken by Ge and Pb in the data given. The electronegativity of Ge is higher than expected because the intervention of the 3d elements in the build-up of the table makes the effective nuclear charge higher than expected. Similar reasoning explains the unexpectedly high electronegativity of Pb except that it is the intervention of the 4f series that is responsible.

14-65 a) At the critical distance R_c, the Coulomb energy between the $M^+(g)$ and $X^-(g)$ ion just compensates for the energy required to extract an electron from an isolated M(g) and place it on an isolated X(g). This energy (called ΔE_∞ on text page 534) is the difference between the *IE* of M(g) and the *EA* of X(g). Set it equal to the Coulomb potential energy:

$$\Delta E_\infty = IE - EA = \frac{Z_1 Z_2 e^2}{4\pi\epsilon_0 R_c}$$

In alkali halides, Z_1 and Z_2 are +1 and −1. Also, ϵ_0 is 8.854×10^{-12} C² J⁻¹m⁻¹, and e is 1.60218×10^{-19} C. Substitution and solution for R_c give:

$$R_c = \frac{(2.3071 \times 10^{-28}\ \text{J m})}{(IE - EA)\ \text{J}}$$

Appendix F and other sources give *IE*'s and *EA*'s on a per mole basis. Revise the preceding formula, which refers to a single pair of particles, to allow the use ionization energies and electron affinities that are given in joule per mole. Using such values multiplies the denominator by N_0. The required revision is to multiply the numerator by N_0 as follows:

$$R_c = \frac{(6.0221 \times 10^{23}\ \text{mol}^{-1})(2.3071 \times 10^{-28}\ \text{J m})}{(IE - EA)\ \text{J mol}^{-1}} = \frac{(1.3894 \times 10^{-4}\ \text{J m mol}^{-1})}{(IE - EA)\ \text{J mol}^{-1}}$$

b) For LiF, $(IE - EA)$ is $520.2 - 328.0 = 192.2 \times 10^3$ J mol⁻¹. The numbers come from Appendix F. Substituting in the preceding expression gives R_c equal to 7.229×10^{-10} m.

For KBr, $(IE - EA)$ is 94.1×10^3 J mol⁻¹ giving R_c equal to 14.8×10^{-10} m.

For NaCl, $(IE - EA)$ is 149.8 kJ mol⁻¹, making R_c equal to 9.275×10^{-10} m.

14-67 a) In $SbCl_5^{2-}$, the central Sb atom has *SN* 6. It is surrounded by five bonding pairs (single bonds to the five fluorine atoms) and one lone pair. As an AX_5E case (see table on page 382 of this Guide), the molecule-ion is square pyramidal.

b The central Sb atom in $SbCl_6^{3-}$ has *SN* 7. This steric number is rare. The required extension of VSEPR theory would be to include *SN* 7 and other higher *SN*'s. In fact, three geometries have been observed for *SN* 7: pentagonal bipyramidal, capped octahedral (in which the seventh atom occupies one face of an octahedron about the central atom) and capped trigonal prism (in which the seventh atom occupies one rectangular face of a trigonal prism about the central atom).

14-69 The central S atom in F_4SO has *SN* 5 and falls into the class AX_5. Its geometry should be trigonal bipyramidal. The real question is whether the oxygen atom is equatorial or axial. Putting the double-bonded oxygen atom at an equatorial position minimizes 90° interactions with the four fluorine atoms and should be preferred, according to VSEPR theory.

14-71 Set up a coordinate system and position the oxygen atom at its origin. Let the y-axis bisect the angle θ defined by H—O—H:

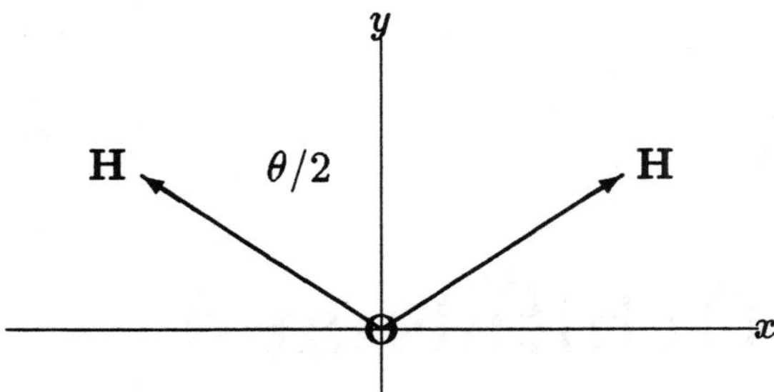

The dipole moments of the two O—H bonds are symbolized μ_{OH}. They are directed parallel to the two bonds. The x-components of these two vectors oppose each other and cancel. The y-components point in the same direction and add. The magnitude of the y-components is $\mu_{OH}\cos(\theta/2)$, and the sum of the two y-components equals the dipole moment of the molecule as a whole. Therefore:

$$\mu(H_2O) = 1.86\text{ D} = 2\mu_{OH}\cos\left(\frac{\theta}{2}\right) = 2\mu_{OH}\cos\left(\frac{104.5^\circ}{2}\right)$$

Solving for μ_{OH} gives 1.52 D.

14-73 Element 114 would lie below lead (Pb) in Group IV of the periodic table. It would therefore have four valence electrons. The maximum oxidation state is obtained when all valence electrons are removed, so the maximum oxidation state for element 114 would be +4. The trend down Group IV of the periodic table is toward greater importance for lower oxidation states and less importance for higher oxidation states. The most common oxidation state for element 114 would most likely be +2.

14-75 If the structure of the oxoacid H_3PO_2 were $HP(OH)_2$ (one hydrogen atom bonded directly to the phosphorus and the other two bonded to oxygen atoms), then the acid would be diprotic and much weaker than observed. The structure $H_2PO(OH)$ however goes well with the observed strength of the acid. According to this structure, the acid is monoprotic. Note that $H_2PO(OH)$ heads the column in the list of weak acids having one lone oxygen atom bonded to the central atom in Table 14-5 on text page 547.

Chapter 15

Molecular Orbitals and Spectroscopy

Atomic orbitals (abbreviated AO) *localize* the probability of finding their occupying electrons in regions close to the atom's nucleus. For example, the average distance of the $1s$ electron from the nucleus in hydrogen is only 0.8 Å. In contrast, **molecular orbitals** *delocalize* the probability of finding the electrons that occupy them over all the atoms of a molecule. Like atomic orbitals, molecular orbitals have a variety of shapes, sizes and energies. A molecular orbital (abbreviated MO) is constructed mathematically by mixing together the atomic orbitals of the atoms that make up the molecule. The process is **linear combination** of atomic orbitals, or **overlap** of atomic orbitals.

Diatomic Molecules

The simplest molecules are those containing just two atoms. The process of linear combination of atomic orbitals is fairly simple for *diatomic* molecules. A linear combination of two $1s$ atómic orbitals on a pair of neighboring atoms A and B is:

$$\psi(\text{bonding}) = C_{\text{A}}\psi_{1s}^{\text{A}} + C_{\text{B}}\psi_{1s}^{\text{B}}$$

$$\psi(\text{antibonding}) = C_{\text{A}}\psi_{1s}^{\text{A}} - C_{\text{B}}\psi_{1s}^{\text{B}}$$

where C_{A} and C_{B} are constants telling the degree of mixing, that is, the proportion of each parent's character that the daughter molecular orbital possesses. If the two parent atoms are identical, then C_{A} equals C_{B} on the basis of symmetry. If not, the constants differ. See **problem 15-51.** As the preceding equations indicate, two orbitals on two neighboring atoms can combine in two ways:

Bonding. The first combination places much electron probability density between the nuclei of the atoms. This tends to counteract internuclear repulsions and leads to bonding between the atoms. The energy of a daughter molecular orbital that is a bonding MO is *less* than the energy of either of its AO parents.

Antibonding. In contrast, the second combination of the atomic orbitals (the one with the minus sign) places a node (a region of zero electron probability density) between the atoms. Its energy is higher than the energy of the parent atomic orbitals. Electrons in an antibonding orbital are distributed in space in such a way as actively to oppose the continued association of the nuclei in a chemical entity.

There are several additional points concerning molecular orbitals:

1. Molecular orbitals are designated by Greek letters. The letters σ and π are analogous to the designations s and p used for atomic orbitals. A σ MO has no nodal plane containing the internuclear axis; a π MO has 1 nodal plane containing this axis. See Figures 15-4 and 15-5 (text pages 561 and 562). The parentage of the MO ($1s$ or $2p$ etc. atomic orbitals) is indicated by a subscript to the right of the Greek letter.

2. MO's may be non-bonding as well as bonding and antibonding. Antibonding orbitals are indicated with a superscript * to the right of the Greek letter. Non-bonding orbitals are indicated with a superscript *nb*. Bonding molecular orbitals have no special designation.

3. Occupancy of MO's determines bond order. The **bond order** of a molecule depends on the relative occupancy of the different kinds of MO's.

$$\text{Bond Order} = \frac{1}{2}(\text{No. } e^{-}\text{'s in bonding MO's}) - \frac{1}{2}(\text{No. } e^{-}\text{'s in antibonding MO's})$$

   See **problems 15-1, 15-3,** and **15-51.** Non-bonding electrons do not affect the bond order. Determining a bond order requires assigning all electrons to a bonding, an antibonding or a non-bonding orbital. Sharing electrons between atoms does not by itself guarantee bonding since the shared electrons could be in antibonding molecular orbitals.

4. Orbitals are conserved. The number of MO's formed in linear combination of AO's equals the number of AO's combined. If 10 atomic orbitals are mixed together, then 10 MO's result. In other words, in constructing molecular orbitals, the total number of orbitals is conserved.

5. Correlation diagrams show relative energies. Different MO's have different energies. A **correlation diagram** (text Figures 13-10, 13-12) tells the order of energy and the parentage of a molecule's molecular orbitals. These diagrams are difficult to derive. They should be taken as givens.

6. Electron configurations are possible for MO's. A molecular electron configuration is just like an atomic electron configuration except for using molecular orbitals instead of atomic orbitals.

To write a ground-state molecular electron configuration:

1. Consult the appropriate correlation diagram.

2. Put the available electrons into the MO's starting at lowest energy. Feed in electrons in accord with the Pauli principle and Hund's rule.

3. Write down each MO's designation and show the number of electrons occupying it with a superscript. See **problems 15-1** and **15-11**.

4. Treat only the *valence* electrons. Non-valence electrons are always equally distributed between bonding and antibonding orbitals.

Like atomic orbitals, MO's hold 0, 1 or 2 electrons. Like atomic orbitals, a single MO designation (example: π_{2p}) can refer to a *set* of two or more orbitals which have the same energy. Thus, there occurs notation like $(\pi_{2p})^4$ in **problem 15-9.** This particular example means that four electrons occupy a set of two π MO's of equal energy deriving from the overlap of four $2p$ atomic orbitals. It also appears in Table 15-2 (text page 565).

Polyatomic Molecules

The text describes bonding in polyatomic molecules by employing **valence-bond theory** for σ bonds and the molecular-orbital approach for π bonds.

In the valence bond theory, the valence atomic orbitals of atoms with two or more bonds are **hybridized** to form new atomic orbitals. Hybridization involves the same technique of linear combination that was used to construct molecular orbitals except: **all of the starting orbitals are on the same atom; all of the resultant orbitals are on that same atom.**

Hybridization is an attempt to explain the geometrical shapes that are matters of experimental fact in molecules. Different hybrid combinations give different geometrical shapes. For example, suppose that one $2s$ and three $2p$ orbitals ($2p_x$, $2p_y$, $2p_z$) on the same atom are hybridized. The linear combinations are:

$$\begin{aligned}
\psi_1 &= 1/2(s + p_x + p_y + p_z) \\
\psi_2 &= 1/2(s + p_x - p_y - p_z) \\
\psi_3 &= 1/2(s - p_x + p_y - p_z) \\
\psi_4 &= 1/2(s - p_x - p_y + p_z)
\end{aligned}$$

Taken together, the set of four hybrid daughter wave-functions ($\psi_1 \ldots \psi_4$) describes the same electron probability density as the four parents. This can be proved by squaring ψ_1, ψ_2, ψ_3, and ψ_4 and adding them together:

$$\psi_1^2 + \psi_2^2 + \psi_3^2 + \psi_4^2 = s^2 + p_x^2 + p_y^2 + p_z^2$$

However, unlike the four parents, the daughters in the new hybrid set of orbitals all have the same shape. They are exactly equivalent, except in their orientation. They are sp^3 orbitals and point toward the corners of a tetrahedron. The superscript in the symbol "sp^3" does not refer to the number of electrons occupying the orbital. Instead it refers to the number of parents of p character. An sp^3 orbital containing two electrons would be denoted by $(sp^3)^2$. This notation is a source of confusion but is deeply entrenched and must be learned.

Valence-bond theory invokes sp^3 hybridization when a central atom has a steric number (SN) of 4.

Other important linear combinations are:

$1\ s + 1\ p \rightarrow 2\ sp$ orbitals	Linear	(used when $SN = 2$)
$1\ s + 2\ p \rightarrow 3\ sp^2$ orbitals	Trigonal	(used when $SN = 3$)

A σ bond results from the end-to-end overlap of atomic orbitals or hybrid orbitals along a line connecting two atoms. In valence bond theory, a 109.5° angle is predicted between atoms bonded to a central atom using sp^3 hybrids. The predicted angles for sp^2 and sp hybrids are 120° and 180° respectively.

Once the framework of a molecule is set up using the valence-bond description, orbitals not involved in hybridization may, within limits imposed by their symmetry, mix together to form π bonds, The correlation diagrams (such as text Figure 15-17b) that result show only π orbitals. The main problem in using these diagrams is making sure that the right number of electrons is used. *All valence electrons do not necessarily take part in π bonding.* The number of electrons used in a π-system is the number of valence electrons minus the number used in σ bonding minus the number in lone pairs. See **problems 15-23** and **15-51.**

Bonding in Organic Compounds

The above bonding theory works well for the molecules of many organic compounds. These are compounds that involve carbon, hydrogen and other atoms from the first and second rows of the periodic table. The steps in the analysis of bonding in organic compounds are:

1. Write the Lewis electron-dot structure for the molecule. Take particular care to get the number of valence electrons right.

2. Determine the hybridization of every atom. To do this, use VSEPR theory (text page 539-42) to obtain the steric number of each atom. Then read the hybridization of the atoms from the following table:

Steric Number	Hybridization
2	sp
3	sp^2
4	sp^3

See **problem 15-17** for examples.

3. Place electron pairs in each localized (σ) molecular orbital. This constructs the single-bond framework of the molecule. In so doing, it uses two electrons between each pair of bonded atoms.

4. Identify the p orbitals that were not used in hybridization. Combine them to form π molecular orbitals.

5. Place the valence electrons that were not used in step 3 into the π molecular orbitals.

Interaction of Light with Molecules

As is the case in atoms, the energies of molecules are quantized. This means that molecules possess only certain allowed energy states. **Molecular spectroscopy** is the study of electromagnetic radiation (light) as it is absorbed or emitted by molecules passing from one allowed energy state to a second. The frequency of the light involved in such transitions is related to the difference in energy between the pair of energy levels (states):

$$|\Delta E| = h\nu$$

The absolute value is taken because ΔE is negative in emission and positive in absorption. The values of the energy in the different levels depends on molecular parameters such as bond distances and angles. Molecular spectroscopy thus reveals much about bonding.

The total energy of a molecule is approximately equal to the sum of contributions from three kinds of motion:

$$E(\text{total}) = E(\text{electronic}) + E(\text{vibrational}) + E(\text{rotational})$$

The three contributions are of different orders of magnitude and have quite different characteristics:

Electronic. Changes in the electron configuration of molecules generally require (or release) from 50 to 500 kJ mol^{-1}. Frequencies of light that supply (or take away) this amount of energy per photon range from 10^{14} to 10^{15} s^{-1}. The corresponding wavelength range is 2400 nm to 240 nm, right across the *visible* region of the spectrum into the ultra-violet. Electronic transitions are therefore associated with the colors of compounds. If a compound (or a solution of a compound) absorbs light of a visible wavelength (such as yellow), then the light that goes on through has the color that is the **complement** of the color absorbed (the complement of yellow is violet). Complementary colors lie opposite each other in the color wheel (Figure 15-23, text page 580). Molecules in excited electronic states often react more readily than ground-state molecules. **Photochemistry** is the study of reactions that follow the electronic excitation of molecules by light. For example, **photodissociation,** the breaking of bonds under the influence of light, may follow exposure of a compound to light of suitable frequency.

Vibrational. Chemical bonds can be likened to springs connecting the atoms of a molecule. Atoms in molecules move constantly about an average or home position. Changes in the relative positions of the atoms are molecular vibrations. Different vibrational states have different energies. and the energy of the vibrational states is quantized. Transitions between allowed vibrational states range in energy from about 2 to 40 kJ mol^{-1}. These energies correspond to the energy of light of frequencies between 5×10^{12} and 100×10^{12} s^{-1}. of the reaction.

When a chemical bond is stretched or compressed from its equilibrium length, there is a restoring force. The **force constant *k*** tells the strength of the restoring force when a bond is elongated or compressed. If a bond is like a spring, then the force constant is its stiffness. A typical molecular force constant is 500 newton per meter. If a bond with a k of 500 N m^{-1} is stretched 0.1 Åthe restoring force is 5×10^{-9} N. Experiments in molecular spectroscopy allow the computation of force constants (see **problem 15-37).** units of N m^{-1}. See **problem 15-37.** The force constant of the bonds of a molecule and the masses they connect determine the allowed vibrational frequencies of that molecule. In the case of a diatomic molecule, the relationship has a simple form. The stretching frequencies (symbol ν) depends on the force constant of the bond and the reduced mass:

$$\nu = \frac{1}{2\pi}\sqrt{\frac{k}{\mu}}$$

where k is the force constant, and the **reduced mass** μ of the diatomic molecule is:

$$\mu = \frac{m_1 m_2}{m_1 + m_2}$$

The wavelength of the light involved in transitions between vibrational states ranges from 60 000 nm to 3000 nm. This is in the infra-red region of the electromagnetic spectrum.

The allowed vibrational energies of a diatomic molecule are:

$$E(\text{vibrational}) = h\nu(v + 1/2) \qquad v = 0, 1, 2, \ldots$$

where v (the letter vee) is the vibrational quantum number. Unlike the electronic energy states, these vibrational energy states are uniformly spaced. Also, when $v = 0$, the molecule still has some energy, the **zero-point energy.** Sloppiness in writing the symbols ν (nu, a frequency) and v (a quantum number) can lead to confusion in using the preceding formula.

Rotational. Rotational transitions involve less energy yet. Typical energy differences range from 0.001 to 0.1 kJ mol^{-1}. The corresponding frequencies are 2.5×10^9 and 2.5×10^{11} s^{-1}. Molecules rotate with frequencies in this range. The corresponding wavelengths are 1200 mm to 1.2 mm. These wavelengths occur in the microwave and short-wave region of the electromagnetic spectrum.

Rotational energy depends on the molecular **moments of inertia.** Molecules rotate and tumble through space. A moment of inertia is related to the tendency of an object to persist in spinning about an axis. In general, molecules have three moments of inertia. Because they have a simple linear structure, diatomic molecules are a special case. There is only one non-zero moment of inertia, I:

$$I = \mu R_e^2 \qquad \text{diatomic molecule}$$

where μ is the **reduced mass** of the diatomic molecule.

The SI unit for moments of inertia is the kg m^2. Typical values for molecular moments of inertia are on the order of 10^{-45} kg m^2. See **problem 15-63.**

The quantization of the rotational motion of a *linear* molecule is in terms of a new quantum number J:

$$E(\text{rotational}) = \frac{h^2}{8\pi^2 I} J(J+1) \qquad J = 0, 1, 2, \ldots$$

The rotational spectrum of a diatomic molecule, which is of course always linear, consists of equally spaced lines. Each line corresponds to a transition from an initial state to a final state in which J is increased or decreased by exactly 1. Absorption of radiation does not occur for changes in when ΔJ exceeds 1.

In solving problems in molecular spectroscopy, it is important to:

- Pay heed to the units. Work exclusively in SI units and convert at the end as necessary. The units of moment of inertia (kg m^2) and reduced mass (kg) sometimes cause trouble; moments in amu Å^2 or g Å^2 and masses in amu or grams are used by mistake. Follow the pattern in **problem 15-63.**

- Distinguish between patterns in the **energy levels** of molecules and patterns in the **frequencies** of the transitions connecting those energy levels. Thus, in **problem 15-35** the observed spectroscopic lines in the pure rotational spectrum of a diatomic molecule are uniformly spaced by frequency. The rotational energy levels of this molecule are *not* uniformly spaced.

Atmospheric Photochemistry

Photochemical reactions in the upper atmosphere of the earth absorb much high-energy radiation from the sun. They thereby protect the surface from damaging effects of this radiation. In the outermost layer of the atmosphere, the **thermosphere,** photodissociation of oxygen ($O_2 + h\nu \to 2\,O$) removes most (harmful) radiation with wavelength less than 200 nm from incoming sunlight. In the **stratosphere,** atomic O formed by the preceding photodissociation reacts further to form ozone:

$$O_2 + O \to O_3^*, \quad \text{then}: O_3^* + M \to O_3 + M$$

at trace (10^{15} molecules per liter) concentration. Photodissociation of ozone in the stratosphere absorbs light of wavelength 200 to 350 nm. This is the ozone shield against harmful radiation. Free-radical pollutants in the stratosphere catalyze destruction of ozone. A radical is a species with an odd number of electrons. The radical NO (nitric oxide) from airplane exhausts catalyzes reactions: $NO + O_3 \to NO_2 + O_2$ and then $NO_2 + O \to NO + O_2$ to destroy ozone. The radical Cl from chlorofluorocarbons catalyzes similar reactions: $Cl + O_3 \to ClO + O_2$ and then $ClO + O \to Cl + O_2$ to destroy ozone.

Photochemical reactions in the layers of the atmosphere closest to the ground (the **troposphere**) create photochemical smog. "NO_x " stands for interconverting mixtures of nitrogen oxides NO, NO_2 and N_2O_4; these oxides are formed when air (N_2 plus O_2) is heated. Pollutant ozone is produced photochemically: $NO_2 + h\nu \to NO + O$, followed by $O + O_2 + M \to O_3 + M$. Ozone in the troposphere is harmful in itself and also oxidizes organic compounds to produce irritants.

Oxides of sulfur are a more wide-spread atmospheric pollutant. They arise from combustion of sulfur-containing impurities in fuels. The SO_2 that is immediately produced is eventually oxidized to SO_3; SO_3 reacts with water to form H_2SO_4.

Acid rain is precipitation (it includes fog and snow) made unduly acid by nitric acid (from NO_2 pollution) and sulfuric acid (from SO_2 pollution). Acid rain damages forests, lakes. Removal of SO_2 emissions essential control measure; straightforward acid-base chemistry suggests $CaO(s)$ to absorb SO_2.

Certain gases in the troposphere absorb infrared radiation emitted by warm surface of the earth. Since the IR radiation otherwise would go off into space, the effect is like than of a greenhouse. Water, carbon dioxide and methane are all effective absorbers; man-made increases in carbon dioxide and methane concentrations may lead to undesirable global warming through the **greenhouse effect.**

Detailed Solutions to Odd-Numbered Problems

15-1 a) Fluorine (F_2) is a homonuclear diatomic molecule with $Z = 9$. It has 14 valence electrons. The F_2^+ ion has only 13 valence electrons. Consult the correlation diagram in text Figure 15-6b to get the energetic order of valence molecular orbitals in both species. The MO's are then filled in order of increasing energy with the available valence electrons:

$$\begin{array}{ll} F_2 & (\sigma_{2s})^2(\sigma_{2s}^*)^2(\sigma_{2p})^2(\pi_{2p})^4(\pi_{2p}^*)^4 \\ F_2^+ & (\sigma_{2s})^2(\sigma_{2s}^*)^2(\sigma_{2p})^2(\pi_{2p})^4(\pi_{2p}^*)^3 \end{array}$$

b) The F_2 molecule has two more bonding than antibonding electrons. Its bond order is 1; F_2^+ ion has three more bonding than antibonding electrons. Its bond order is 3/2 .
c) The molecule of F_2 has no unpaired electrons; the substance should be diamagnetic. The F_2^+ ion has an odd number of electrons. One electron (a π_{2p}^* electron) is unpaired, and F_2^+ is paramagnetic.
d) The F_2^+ ion has a larger bond order and therefore requires more energy for dissociation than the F_2 molecule.

15-3 The valence-electron configuration of S_2 should be like that of O_2 except using $n = 3$ orbitals: $(\sigma_{3s})^2(\sigma_{3s}^*)^2(\sigma_{3p})^2(\pi_{3p})^4(\pi_{3p}^*)^2$. The bond order is 2, and the molecule should be paramagnetic (two unpaired electrons).

15-5 In each case determine the number of valence electrons. This result and the charge on the species identify the column of the periodic table in which the element is located. All the configurations involve MO's from the $n = 2$ shell and therefore involve elements in the second row of the periodic table. The bond order is one half the number of bonding electrons minus one half the number of antibonding electrons: **a)** F_2, bond order 1; **b)** N_2^+, bond order 5/2; **c)** O_2^-, bond order 3/2.

15-7 Check unpaired valence electrons: **a)** diamagnetic; **b)** paramagnetic; **c)** paramagnetic.

15-9 Nitrogen is more electronegative than carbon. The energies of its atomic orbitals are *lowered* in the correlation diagram (Figure 15-8, text page 566) relative to the energies of the corresponding orbitals of the carbon atom. A molecule of CN has 9 valence electrons. The valence electron configuration in the ground state is: $(\sigma_{2s})^2(\sigma^*_{2s})^2(\pi_{2p})^4(\sigma_{2p})^1$. The bond order of the molecule is 5/2, and it has a single unpaired electron, giving rise to paramagnetism.

15-11 The valence electron configurations are:
CF $(\sigma_{2s})^2(\sigma^*_{2s})^2(\sigma_{2p})^2(\pi_{2p})^4(\pi^*_{2p})^1$ and CF^+ $(\sigma_{2s})^2(\sigma^*_{2s})^2(\sigma_{2p})^2(\pi_{2p})^4(\pi^*_{2p})^0$. Removing an electron from the π^*_{2p} orbital in the CF molecule gives the CF^+ ion. The loss of an antibonding electron increases the bond order from 5/2 to 3.

15-13 The electron configuration for HeH^- would be $(\sigma_{1s})^2(\sigma^*_{1s})^2$. The ion has a bond order of zero and should not be stable.

15-15 The central N atom in NH_2^- is surrounded by eight valence electrons (5 from the N, 1 each from the H's and 1 for the overall negative charge). The valence orbitals of the N atom are sp^3 hybridized. Two of the hybrids overlap in σ bonds with $1s$ orbitals on the two H atoms. The other two contain lone pairs. The molecule-ion is bent with an H—N—H angle far less than 180°. In fact, the angle is 106.7° (even less than 109.5°.)

15-17 In all of these species, the hybridization on the central atom follows from the number of lone pairs plus bonded pairs surrounding the central atom (the steric number). The geometry depends on the hybridization, but is named only with reference to actual atoms, not actual atoms plus lone pairs.
a) The central C atom in CH_4 has *SN* 4. This atom is sp^3-hybridized, and the molecule is tetrahedral. **b)** The central C atom has *SN* 2 in CO_2 and is sp-hybridized. The molecule is linear. **c)** The central O atom has *SN* 4 in OF_2 and is sp^3-hybridized. Two of the hybrid orbitals on the O atom accommodate lone pairs of electrons, and two overlap with orbitals on the fluorine atoms. The molecule is bent. **d)** The central C atom in CH_3^- has *SN* 3 and is sp^3-hybridized. One of the four hybrid orbitals contains a lone pair of electrons. The other three overlap with $1s$ orbitals of the three hydrogen atoms. The molecule-ion is pyramidal. **e)** The central Be atom in BeH_2 has *SN* 2 and is sp-hybridized. The molecule is linear.

15-19 The ClO_3^+ and ClO_2^+ ions have 24 and 18 valence electrons respectively. The central Cl atom in ClO_3^+ has *SN* 3 and therefore three sp^2 hybrid orbitals overlapping with orbitals from the oxygen atoms. It has a trigonal planar geometry. The central Cl atom in ClO_2^+ likewise has a set of three sp^2 hybrid orbitals, but only two overlap with orbitals on oxygen atoms. The third sp^2 orbital contains a lone pair. The ClO_2^+ molecule-ion is bent. The central chlorine atoms in the following Lewis structures are shown with expanded octets; this minimizes formal charges.

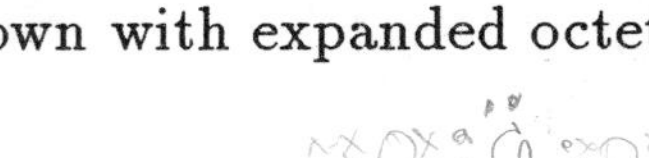

15-21 The central nitrogen atom in the orthonitrate ion can attain an octet by forming four single bonds, one to each of the four oxygen atoms. We expect sp^3 hybridization on the central N atom and a tetrahedral geometry. Notice that the Lewis structure of this ion puts a +3 formal charge on the central nitrogen atom! No wonder the ion is "unfamiliar."

15-23 Like the CO_2 molecule, the azide ion N_3^- has 16 valence electrons. The correlation diagram of Figure 15-16 (text page 573) applies. There are two N—N σ bonds, resulting from overlap of sp hybrid orbitals on the central N atom with $2p_z$ orbitals on the outer N atoms. These bonds use 4 electrons. Lone pairs in each of the $2s$ orbitals of the outer N atoms use another 4 electrons. The π system (text Figure 15-16) contains the remaining 8 valence electrons. Four of these are in bonding π orbitals, making two π bonds. The other four are in *non-bonding* π orbitals. Thus: $(\pi)^4(\pi^{nb})^4$. The overall bond order of the molecule is 4: (2 σ bonds plus 2 π bonds). The two N-to-N linkages are identical and have a bond order of 2.

The N_3 molecule has 15 valence electrons. It derives from N_3^- by the loss of an electron. Such loss comes from the highest energy molecular orbital which is (text Figure 15-16) a non-bonding MO. Thus N_3 has an overall bond order of 4, just like N_3^-. Unlike N_3^-, N_3 has an unpaired electron and is paramagnetic. The N_3^+ ion is a 14-valence-electron species derived from N_3^- by the loss of two non-bonding π electrons. The overall bond order of N_3^+ is therefore also 4. There are two unpaired electrons in the set of π^{nb} orbitals so N_3^+ is paramagnetic, too.

15-25 The Lewis structure of acetaldehyde is:

The hybridization on the —CH_3 carbon atom is sp^3; the hybridization on the other carbon atom is sp^2. In addition to the σ bonds from the overlap of these hybrid orbitals, two electrons in a π orbital bond the sp^2-hybridized C atom with the O atom. The three groups around the sp^2 C atom form an approximately trigonal planar structure with bond angles near 120°. The geometry at the other C atom is approximately tetrahedral, with all six H—C—H and H—C—C angles near 109.5°.

15-27 Draw the Lewis structure for NO_2^- and use VSEPR theory to determine the steric number and structure. There are two resonance forms:

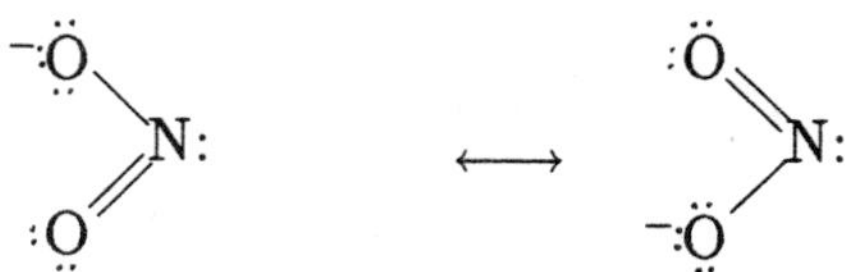

The VSEPR theory assigns a steric number of 3 for the nitrogen atom. The O atoms occupy two of the three sites, and the lone pair is at the third. The molecule-ion is therefore bent. The hybridization at the nitrogen atom is sp^2. Two of the three sp^2 hybrid orbitals form the σ-bonds to the oxygen atoms, and the third accommodates the lone pair. The remaining unhybridized $2p_z$ atomic orbital on the nitrogen atom is oriented perpendicular to the plane of the molecule and overlaps with $2p_z$ atomic orbitals of the oxygen to form a π system. In a localized-orbital scheme, two resonance forms are necessary to represent the bonding in NO_2^-. In the molecular orbital approach, the three $2p$ orbitals combine to form the three molecular orbitals shown in text Figure 15-17. There are four electrons in this π system. This leads to a π-bond order of 1 (the two electrons in the non-bonding orbital do not affect the bond order). Adding the σ MO's gives the overall bond order 3, which amounts to 3/2 for each bond.

15-29 The lowest unoccupied molecular orbital of ethylene is a π^* antibonding orbital. The electron that is added when the $C_2H_4^-$ ion is formed goes into this orbital. An additional antibonding electron means that the bond order in $C_2H_4^-$ is less than in C_2H_4.

15-31 The color of a substance is the complement of the color of the light that the substance absorbs. The complementary color of orange is blue. We would therefore expect absorption around 450 nm.

15-33 Conjugation of multiple bonds (π-delocalization) tends to increase the wavelength of absorbed light. Hence, we expect cyclohexene to absorb at shorter wavelengths than benzene because the π bonding is localized in cyclohexene and delocalized in benzene.

15-35 a) The spacing of the absorption lines in the pure rotational spectrum of a diatomic species is *uniform* with a frequency separation of $h/4\pi^2 I$ where I is the moment of inertia. From the data in the problem, the average frequency separation of the lines in the ^{12}C—^{16}O rotational spectrum is $1.155 \times 10^{11}\ s^{-1}$. Also, the frequency of the *first* rotational line is $1.15 \times 10^{11}\ s^{-1}$, as predicted by:

$$\nu = \left(\frac{h}{4\pi^2 I}\right)(J_i + 1)$$

with $J_i = 0$. The moment of inertia is computed by substitution in the equation:

$$1.155 \times 10^{11}\ s^{-1} = \left(\frac{h}{4\pi^2 I}\right) \quad \text{which gives:} \quad I = 1.45 \times 10^{-46}\ \text{kg m}^2$$

b) The *energy* of each rotational state is given by:

$$E(\text{rot}) = \left(\frac{h^2}{8\pi^2 I}\right) J(J+1) = \frac{h}{2}\left(\frac{h}{4\pi^2 I}\right) J(J+1)$$

The term $h/4\pi^2 I$ is 1.155×10^{11} s^{-1}, from the preceding part. Substitute it and also $h/2$ in the preceding:

$$E(\text{rot}) = \frac{6.626 \times 10^{-34} \text{ J s}}{2}(1.155 \times 10^{11} \text{ s}^{-1})J(J+1) = J(J+1)(3.826 \times 10^{-23} \text{ J})$$

Therefore, the rotational energy equals 7.65×10^{-23} J when J equals 1. It is 2.30×10^{-22} J when J equals 2, and 4.59×10^{-22} J when J equals 3.

c) The moment of inertia of the diatomic molecule is:

$$I = \frac{m_1 m_2}{m_1 + m_2} R_e^2$$

where R_e is the bond distance and the m's are the two atomic masses. The atomic mass of ^{12}C is 12.0000 amu, and the atomic mass of ^{16}O is 15.994915 amu. Dividing by 6.022137×10^{26} amu kg^{-1} converts the masses to kilograms. Substitution of these masses together with I (from part a) into the equation gives R_e equal 1.13×10^{-10} m or 1.13 Å.

15-37 The wavelength λ of the absorbed light is related to the difference in energy between two vibrational states of the $^7Li_2$ molecule:

$$\frac{hc}{\lambda} = \Delta E(\text{vib})$$

Absorption is allowed only for transitions between adjacent vibrational states; that is, the change in the vibrational quantum number v during absorption must equal +1. The change in vibrational energy is:

$$\Delta E(\text{vib}) = h\left(\frac{1}{2\pi}\right)\sqrt{\frac{k}{\mu}}\Big((v_2 + 1/2) - (v_1 + 1/2)\Big)$$

where k is the desired force constant, μ is the reduced mass of Li_2, and h and c are Planck's constant and the speed of light. The quantum-number part of the preceding equals 1 because Δv is 1. Hence:

$$\frac{hc}{\lambda} = \Delta E(\text{vib}) = h\left(\frac{1}{2\pi}\right)\sqrt{\frac{k}{\mu}}$$

The wavelength λ is quoted in the problem as 2.85×10^{-5} m. From Table 12-1 (text page 452), the mass of ^7Li is 7.016005 amu; the reduced mass of Li_2 is half of this or 3.508003 amu. Converting the reduced mass to kilograms gives 5.8253×10^{-27} kg. Insert this and the other quantities in the preceding equation, solve for k and complete the arithmetic:

$$k = \left(\frac{2\pi c}{\lambda}\right)^2 \mu = \left(\frac{2\pi(2.9979 \times 10^{10}\ \text{m s}^{-1})}{2.85 \times 10^{-5}\ \text{m}}\right)^2 (5.8253 \times 10^{-27}\ \text{kg}) = 25.4\ \text{kg s}^{-2}$$

This is the same as 25.4 N m^{-1}. Note that h cancels out of the calculation.

15-39 Dividing 440 kJ, the energy of a mole of bonds, by Avogadro's number gives the energy of just one bond. It is 7.31×10^{-19} J. Use the Planck equation $E = hc/\lambda$ to calculate the wavelength corresponding to the energy of the bond:

$$\lambda = \frac{hc}{E} = \frac{(6.626 \times 10^{-34}\ \text{J s})(2.9979 \times 10^{8}\ \text{m s}^{-1})}{7.31 \times 10^{-19}\ \text{J}} = 2.72 \times 10^{-7}\ \text{m}$$

15-41 The best Lewis structures are a resonance pair with one O=O double bond and one O—O single bond in each, differing only in which side has the double bond and which the single bond:

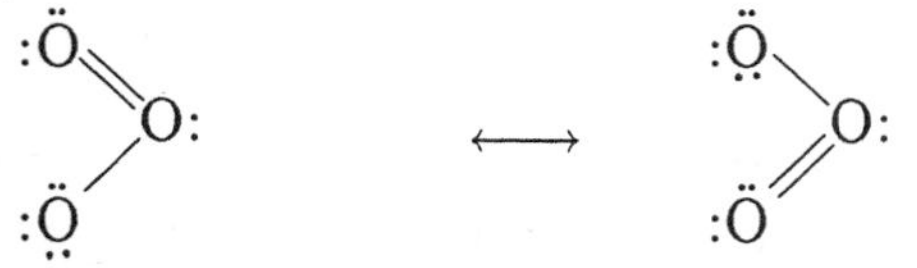

$$:\ddot{\text{O}}{=}\ddot{\text{O}}{-}\ddot{\underset{\cdot\cdot}{\text{O}}}: \quad \longleftrightarrow \quad :\ddot{\underset{\cdot\cdot}{\text{O}}}{-}\ddot{\text{O}}{=}\ddot{\text{O}}:$$

The central O atom has a lone pair in both structures. The VSEPR model assigns the central O atom *SN* 3 and thereby predicts an angle of (approximately) 120° at the central O atom. The ozone molecule is bent. There are two electrons in a bonding π orbital formed from the three $2p_z$ orbitals perpendicular to the molecular plane. The total bond order is 3, which is a bond order of 3/2 for each O-to-O linkage.

15-43 a) There are five occupied valence molecular orbitals in the N_2 molecule. The highest-energy occupied orbital is a σ_{p_z} MO, derived from $2p_z$-$2p_z$ overlap. Its shape is shown in Figure 15-4a (text page 561). The next two highest occupied MO's have equal energies and identical shapes, differing only in orientation. The shape is drawn in Figure 15-5a (text page 562). These are π bonding MO's. One is from $2p_x$-$2p_x$ overlap, and the other is from $2p_y$-$2p_y$ overlap. Then come a σ^* orbital derived from $2s$ orbitals of the nitrogen atoms and a σ orbital from the bonding overlap of the same pair of $2s$ orbitals.

b) Since the highest occupied molecular orbital of N_2 is a bonding orbital, the removal of one electron from N_2 will decrease the bond order, and lengthen the N-to-N bond.

15-45 The correlation diagram in Figure 15-6a (text page 563), gives the ground-state configuration $(\sigma_{2s})^2(\sigma^*_{2s})^2(\pi_{2p_x})^1(\pi_{2p_y})^1$ when applied to B_2 with its six valence electrons. There are two unpaired electrons in this configuration. The diagram in Figure

15-6b on the other hand implies the ground-state configuration $(\sigma_{2s})^2(\sigma^*_{2s})^2(\sigma_{2p_z})^2$. There are no unpaired electrons in this configuration. Diagram 15-6b is not consistent with the fact that B_2 is paramagnetic.

15-47 a) Consider the molecular orbital energy-level diagrams for H_2 and O_2 (text Figures 15-3 and 15-6 respectively). The ionization energy is the energy required to remove the highest energy electron from a gaseous molecule or atom. In the case of hydrogen, the 1*s* electron of the atom is higher in energy than a σ_{1s} electron of the H_2 molecule. Therefore, it requires less energy to remove the atomic 1*s* electron than a molecular σ_{1s} electron. In the case of oxygen, an atomic 2*p* electron lies lower in energy than a O_2 π^*_{2p} electron. Consequently, it requires more energy to ionize O than O_2.

b) The highest occupied molecular orbital of F_2 is the π^*_{2p} orbital (see text Figure 15-6). It is higher in energy than the atomic 2*p* orbital. Therefore, we expect the F_2 molecule to have a lower ionization energy than the F atom.

15-49 The molecular orbital and the square of the molecular orbital for the ground state of the heteronuclear molecule are:

$$\psi = C_A\psi_A + C_B\psi_B \quad \text{and} \quad \psi^2 = C_A^2\psi_A^2 + 2C_AC_B\psi_A\psi_B + C_B^2\psi_B^2$$

where the C's are constants. The *square* of the wave-function is given because it is the quantity that is related to the probability of finding the electron. Neglecting the overlap of the two orbitals means neglecting the cross-term in the squared wave-function:

$$\psi^2 \approx C_A^2\psi_A^2 + C_B^2\psi_B^2$$

If the electron spends 90 percent of its time in orbital ψ_A then $C_A^2 = 9C_B^2$. Also, the electron must be either on atom A or atom B so $C_A^2 + C_B^2 = 1$. Solution of the two simultaneous equations gives $C_A = 0.949$ and $C_B = 0.316$.

15-51 a) Nitramide has 24 valence electrons. It must have one double bond somewhere if the octet rule is obeyed (this can be confirmed using the procedure on text page 60). If the structure is non-planar, this double bond is strongly localized to the —NO_2 portion of the molecule. The two electrons occupy a π orbital constructed from $2p_z$ orbitals on the N atom and the two O atoms bonded to it. Inclusion in the π system of orbitals and electrons from the other N atom would require coplanarity of the H_2N— and —NO_2 portions of the molecule. If the two portions are not coplanar, then overlap and effective mixing of *p*-orbitals is not possible and the N—N bond order is 1.

b) If the nitramide molecule were planar, the four $2p_z$ orbitals present on the two nitrogen atoms and two oxygen atoms after completion of the σ bonding could overlap to form one π, two π^{nb}, and one π^* MO's. Four electrons would occupy this π system.

Two of the electrons would be in the bonding orbital, and the other two electrons would be in the non-bonding orbitals. The resulting π system would possess a total a net of two bonding electrons across the four atoms involved. The bond order of the N—N bond would be 1 (from the σ interaction) plus 1/3 (from the π system) or 4/3.

15-53 The Lewis structures for HCOOH and $HCOO^-$ are:

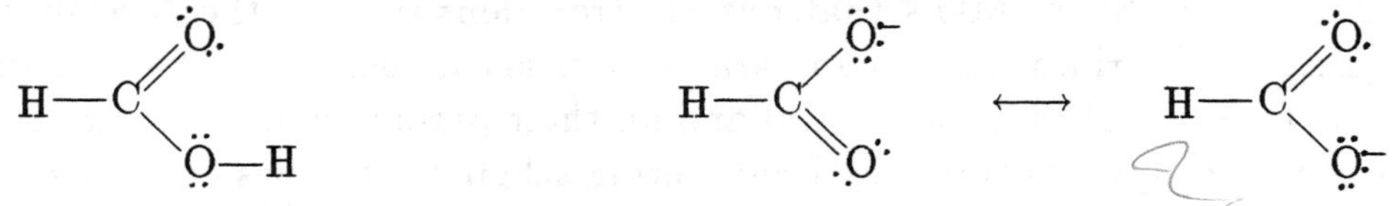

There is only one resonance form for HCOOH, but there are two for the formate anion $HCOO^-$. In formic acid, one oxygen atom is doubly bonded to the carbon atom, and the other is singly bonded. In the anion, there is partial double-bond character in both C—O bonds. The carbon atom in HCOOH is sp^2 hybridized (*SN* 3), and the OH oxygen atom is sp^3 hybridized (*SN* 4). The immediate surroundings of the carbon atom have trigonal planar geometry, and the C—O—H group is bent In the $HCOO^-$ ion, the carbon atom and both oxygen atoms are sp^2 hybridized (*SN* 3), possessing a three-center four-electron π system. In HCOOH, π overlap occurs between orbitals on the carbon atom and only one oxygen atom. The C—O bond lengths in the formate ion should lie somewhere between the value for the single bond (1.36 Å) and the value for the double bond (1.23 Å).

15-55 a) The $\Delta H°$ of the reaction $6\,C(g) + 6\,H(g) \rightarrow C_6H_6(g)$ equals the $\Delta H_f°$'s of the products minus the $\Delta H_f°$'s of the reactants:

$$\Delta H° = 1\underbrace{(82.93)}_{C_6H_6(g)} - 6\underbrace{(716.682)}_{C(g)} - 6\underbrace{(217.96)}_{H(g)} = -5524.92 \text{ kJ}$$

Neither of the $\Delta H_f°$'s of the reactants is zero because gaseous monatomic H and gaseous C are *not* the standard states of these two elements. The answer is an experimental result; it is based on experimental $\Delta H_f°$'s and Hess's law.

b) The formation of 1 mol of such a structure requires the formation of 6 mol of C—H bonds, 3 mol of C=C bonds, and 3 mol of C—C bonds. The negative of the sum of the bond dissociation enthalpies of these bonds is −5367 kJ.

(c) The resonance stabilization in 1 mol of benzene is the difference between the $\Delta H°$ expected and the $\Delta H°$ observed: $(-5367) - (-5524.92) = 158$ kJ.

15-57 The six π molecular orbitals of pyridine arise as combinations of the six p_z orbitals of the ring atoms, which are a nitrogen and five carbons atoms:

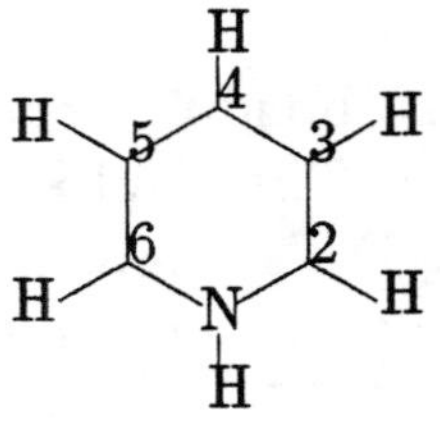

H
H 5 4 3 H
6 2
H N H
H

According to the problem, MO's that put electron density onto the N atom will be lower in energy in pyridine than comparable orbitals in benzene. Thus, molecular orbitals that have the N-atom p_z orbital among their parents will be lower in energy. Let the N atom occupy position 1 in a numbering scheme that goes counter-clockwise around the ring (see above). Also, refer to Figure 15-21, (text page 578), in which the six π MO's of benzene are sketched. The strongly bonding and strongly antibonding (the highest and lowest) MO's in Figure 15-21 both have parentage that includes the p_z orbital on atom 1. These two molecular orbitals are therefore both *lowered* in energy in pyridine relative to benzene. One of the two weakly bonding molecular orbitals in benzene has p_z(N) parentage, but the other does not. Instead, its parentage includes p_z-orbitals from carbon atoms 2 through 5. The first of the two weakly bonding MO's (located on the left in Figure 15-21) is therefore lowered in energy in pyridine relative to benzene, but the energy of the second is not affected. Similarly, the two weakly antibonding MO's in benzene are split in energy. The one that has some p_z(N) parentage (to the right in Figure 15-21) is lowered, but the other is unchanged. The final result is an energy-level diagram for pyridine with six different π-orbital energies, four lower than the corresponding benzene π-orbitals, and two unchanged in energy.

15-59 a) There are five C=C double bonds. The isomer to the left of the arrow has four *trans* C=C double bonds in the chain extending to the right from the six-membered ring. The double bond in the six-membered ring is also *trans* when the relative positions of the two largest groups, one of which is the long side-chain, are considered. The isomer to the right of the arrow is the same except that the second C=C double bond from the right end of the side chain is now *cis.*

b) The absorption maximum would shift to shorter wavelength. Loss of the ring and the —CHO group would reduce the range of delocalization of electrons in a system of alternating single and double bonds because the ring contains a C=C double bond, and the —CHO group contains a C=O double bond.

15-61 a) The carbon atom in formaldehyde is sp^2 hybridized.

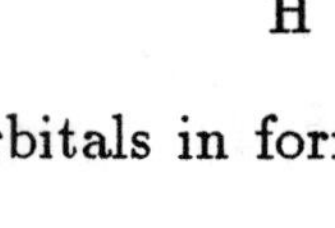

H
:Ö=C
H

b) There is a total of 10 valence orbitals in formaldehyde: three σ-orbitals formed from sp^2 orbitals of the C atom overlapping with $1s$ orbitals on the two H atoms and with the $2p$-orbital on the O atom that points toward the carbon; three empty σ^* orbitals with the same parents; two lone-pair $2s$ and $2p$ orbitals on the O atom; one occupied π (bonding) orbital derived from the two remaining $2p$ orbitals, which are directed perpendicular to the plane of the molecule; one empty π^* (antibonding) orbital derived from the same parents.

c) The weaker transition at lower frequency is probably due to excitation of an electron from a lone-pair $2p$ orbital on the oxygen atom to the π^*-orbital.

15-63 The moment of inertia of a diatomic molecule is $I = \mu R_e^2$ where R_e is the equilibrium bond distance and μ is the reduced mass:

$$\mu = \frac{m_1 m_2}{(m_1 + m_2)}$$

The two m's are the masses of the two atoms making up the molecule. Table 12-1 (text page 452) gives atomic masses. From those data μ for $^1H—^{19}F$ is 0.957055 amu, and μ for $^1H—^{81}Br$ is 0.99543 amu. Converting the two reduced masses to kg gives:

$$1.5893 \times 10^{-27} \text{ kg for HF} \quad \text{and} \quad 1.6530 \times 10^{-27} \text{ kg for HBr}$$

The equilibrium bond distances are: 0.926×10^{-10} m for HF and 1.424×10^{-10} m for HBr from Table 14-1 (text page 530).

Complete the substitution in the formula for I to obtain:

$$I = 1.363 \times 10^{-47} \text{ kg m}^2 \text{ for HF;} \qquad I = 3.352 \times 10^{-47} \text{ kg m}^2 \text{ for HBr}$$

The rotational spectra of these molecules consist of series of equally spaced lines separated by the frequency $h/4\pi^2 I$. This is the meaning of the formula:

$$\nu = \frac{\Delta E}{h} = \left(\frac{h}{4\pi^2 I}\right)(J_i + 1)$$

on text page 583. Compute the frequency spacing for the two molecules by substituting in this formula:

$$\frac{h}{4\pi^2 I} = 12.3 \times 10^{11} \text{ s}^{-1} \text{ for HF and } \frac{h}{4\pi^2 I} = 5.01 \times 10^{11} \text{ s}^{-1} \text{ for HBr}$$

The large change in mass between HF and HBr causes only a rather small change in the reduced mass of the diatomic molecule. To understand why, consider the behavior of the reduced mass as a mathematical quantity. If either m_1 or m_2 is much heavier than the other, then μ approaches the mass of the *lighter* atom. In these cases the reduced mass is already pretty close to the mass of ^1H in the case of HF and cannot get much closer even when F is replaced by HBr.

In the stratosphere, ozone (O_3) is created by irradiation of O_2 with sunlight. The trace of O_3 formed in this way absorbs a large fraction of the ultraviolet (UV) radiation from the sun, shielding the surface of the Earth from fluxes of UV that would otherwise interfere with the growth and reproduction of plants. In the stratosphere, nitrogen dioxide NO_2 gives NO which is part of a chain of reactions through which the destruction of O_3 is accelerated:

$$NO + O_3 \rightarrow NO_2 + O_2 \qquad NO_2 + O \rightarrow NO + O_2$$

In the troposphere, where the concentration of O_3 is small and the concentration of NO_2 is higher, NO_2 participates in the formation of O_3:

$$NO_2 + h\nu \rightarrow NO + O \qquad O + O_2 + M \rightarrow O_3 + M$$

Ozone is bad in the troposphere because of its high toxicity. Unfortunately, O_3 created in the troposphere is so reactive that it does not have a chance to diffuse up into the stratosphere, where it might do some good, before it reacts.

Chapter 16

Coordination Complexes

Coordination complexes contain transition metal atoms (the **central metal**) bonded to a small number of surrounding ions or neutral molecules (the **ligands**). The bonding can be thought of in acid-base terms. The ligands have unshared electron pairs which they donate, acting as Lewis bases. The metal atoms have vacant *d*-orbitals (as well as vacant *s*- and *p*-orbitals) and accept electron pairs, acting as Lewis acids. The resulting **coordinate** bonds are intermediate in character between ionic and covalent bonds. The bonded ligands are in the **coordination sphere** of the central metal.

The properties of the ligands (various ions and small molecules) and of the central metal are both altered by coordination. Coordination complexes persist in aqueous solution. The solutions have characteristic conductivities, colors, magnetism, and reactivities which are not the simple sums of the conductivities, colors, magnetism and reactivities of the parts. For example, coordinated Cl^- ion is often *not* precipitated by $Ag^+(aq)$. Instead it is held tightly in the coordination sphere. See **problem 16-45.**

The overall charge on a coordination complex is the algebraic sum of the charges of its coordinated parts. Ligands are often neutral (e.g. H_2O, NH_3, CO) or negatively charged (Cl^-, CN^-, NCO^-, and so forth). Thus, the compound $K_2[PtCl_6]$ contains the $[PtCl_6]^{2-}$ complex ion, which consists of a Pt(IV) atom and four Cl^- ions. Coordination complexes are often written enclosed in brackets.

Structure of Coordination Complexes

The number of bonds formed by the central atom to its surrounding ligands is its **coordination number (*CN*).** The geometrical structure of coordination complexes can be predicted from their coordination numbers using VSEPR theory with some modifications.

- Coordination numbers of 4 and 6 are more common than any others. When *CN* is 2, complexes are linear. Other *CN*'s (3, 5 and 7 or more) also occur.

- Most six-coordinated complexes have octahedral structures.

- Four-coordinated complexes are mostly either square-planar or tetrahedral. Tetrahedral coordination is the general rule for *CN* 4, but the square planar geometry is predominant for complexes of Pd(II) and Pt(II).

In reading the structural formulas of coordination complexes, remember that the coordination involves the central metal. Lines drawn from ligand to ligand (as in Figure 16-10b, text page 608 are *not* bonds but merely suggest the geometry of the coordination sphere. In text Figure 16-10b, there are exactly 4 coordinate bonds and 6 N—H covalent bonds within the two NH_3 ligands. There are no ligand-to-ligand bonds.

Chelation

Many ligands occupy only one position in the coordination sphere. When only **monodentate** ligands become coordinated, the number of ligands and the coordination number are equal. If a ligand has more than one **donor site,** then it can attach to the central metal in more than one place, provided that it can span the distance between the two points. Ligands which have more than one donor are **polydentate** ligands. Thus, $H_2\ddot{N}CH_2CH_2\ddot{N}H_2$ (ethylenediamine, "en") is a **bidentate** ligand. The dots represent electron pairs ready for donation by the two nitrogen atoms. Coordination complexes in which a polydentate ligand is coordinated at two or more donor sites are **chelates.** Chelates are more stable than non-chelate complexes with the same kind of donor atom. See **problem 16-41.**

Isomerism

Isomers are substances having the same number and kinds of atoms but arranged differently. Coordination complexes display many types of isomerism. In **geometrical isomerism,** the ligands are arrayed in different relative positions about the central metal atom. Two ligands which are near to each other, on the same side of a metal atom, are *cis.* Two ligands which are far apart (on opposite sides of a metal atom) are *trans.* In square-planar complexes, geometrical isomerism is possible when the formula is Ma_2b_2 or Ma_2bc where a, b, and c are different ligands. In octahedral complexes, any given site ("o" in the following figure) has one site that is *trans* to it (labeled "t") and four equivalent sites that are *cis* to it (labeled "c"). The c sites are equivalent although they do look different as usually sketched:

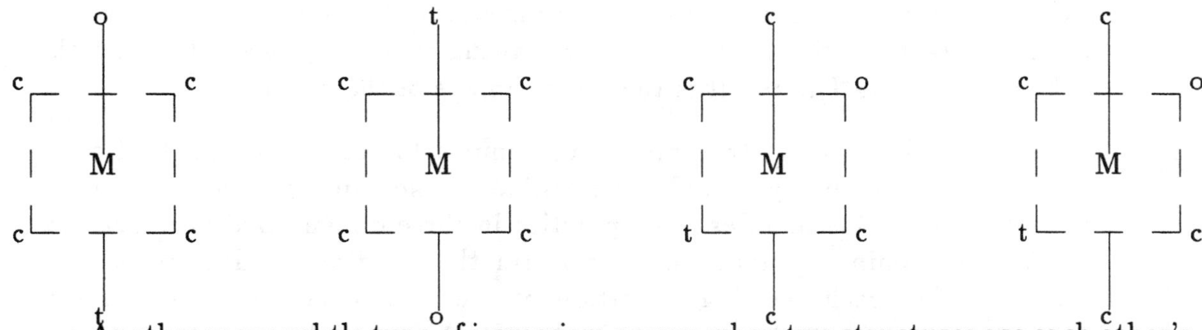

Another more subtle type of isomerism occurs when two structures are each other's mirror images. See **problem 16-17.** Non-superimposable mirror-image structures are **enantiomers.** Enantiomers have the same relationship to each other as the left and right hand. If the source of optical isomerism is the three-dimensional distribution of other, bonded atoms about a single atom, then that atom is a **chiral center.** Study the structure of the Co(III)-EDTA complex ion (Figure 16-13, text page 610), which has an enantiomer. The ligand EDTA has a backbone connecting the two N donors and four arms connecting the N donors to the four O donors. The ligand wraps around the central Co atom. The enantiomeric structure has one chelate arm leading from the front nitrogen atom to the *top* of the octahedron (instead of the bottom), and one chelate arm leading from the back nitrogen to the *bottom* of the octahedron (instead of the top). The wrapping of the chelate rings then has an opposite twist or chirality about the central cobalt atom, which is a chiral center.

In the tetrahedral geometry, optical isomerism with the metal as chiral center is possible only if there are four different ligands. See **problem 16-47.** In the square-planar geometry, optical isomerism with the central metal as the chiral center is not possible.

A common task is to sketch the geometry of a complex, including all isomers, given its formula. Both optical and geometrical isomerism occur in **problem 6-17** and **6-19.** Attack such problems as follows:

1. Determine the *CN* of the metal and the geometry of the complex.

2. Sketch a starting-point structure.

3. Test for geometrical isomerism by moving ligands around while systematically checking pair-wise relationships of all the ligands. Do the variant structures have the same *cis* and *trans* relationships as in the original sketch? If so, then the new structure is not a geometrical isomer of the original. If not, then it is.

4. Check for optical isomerism by drawing a vertical line next to the original sketch and using it as a mirror to create a mirror image of the original. Remember that atoms *near* the mirror line are still near it in the reflection on its other

side. Atoms far from the mirror remain far from it in the reflection. An effective alternative tactic is to turn the original drawing over and trace it through the back of the paper. The result is the mirror image of the original.

5. Check whether the new structure is superimposable on the original. Rotate it so as to put as many ligands as possible in the same positions as in the original sketch. This requires some practice in three-dimensional visualization. The best beginning approach is to perform the rotation by drawing several helper sketches each showing a portion of the total rotation. If the mirror image structure cannot be superimposed upon the original no matter how it is rotated, then the two are genuine enantiomers.

Nomenclature for Coordination Compounds

1. The cation is named before the anion.
2. Within the complex ion, the ligands are named first, followed by the metal ion. If more than one type of ligand is present, negatively charged ligands are listed first, followed by neutral ones. Within these categories ligands are listed in alphabetic order.
3. The names of anionic ligands end with the letter o, whereas neutral ligands are usually called by the names of the molecules. The exceptions are H_2O (aqua), CO (carbonyl), and NH_3 (ammine).
4. When several ligands of a particular kind are present, use the Greek prefixes di-, tri-, tetra-, penta-, and hexa-. Thus the ligands in $[Co(NH_3)_4Cl_2]^+$ are designated "dichlorotetraammine." If the ligand itself contains a Greek prefix, use the prefixes *bis* (2), *tris* (3), *tetrakis* (4) to indicate the number of ligands present. The ligand ethylenediamine already contains the term "di"; therefore "bis(ethylenediamine)" is used to indicate two ethylenediamine ligands.
5. If the complex is an anion, attach the suffix "-ate" to the name of the metal. Unfamiliar terms like "zincate", "cobaltate", "rhodate" result.
6. The oxidation number of the metal is written in Roman numerals following the name of the metal. Thus $Fe(CN)_6^{3-}$ is named hexacyanoferrate(III) ion.

Example: A coordination complex has the formula $[Co(NH_3)_5Cl]Cl_2$. Which atom is the central atom, what is the charge on the complex ion, what is the oxidation number of the central atom, and what is the name of the compound? **Solution:** The transition metal *cobalt* is the central atom and the ligands are ammonia and the chloride ion. Since two chloride ions (charge of -1 each) are needed to balance its charge, the complex ion (in brackets) must have a charge of +2. Cobalt is in the +3 oxidation state because there are three −1 chlorides in the formula to be countered electrically, and the ammonia molecules are neutral. The name of the compound is chloropentaamminecobalt(III) chloride.

Familiarity with nomenclature is tested in two ways: by giving names and asking for formulas and by giving formulas and asking for names. See **problems 16-5, 16-7,** and **16-43.**

Bonding in Coordination Complexes

A successful bonding theory must explain: a) the *geometry* of coordination complexes; b) the often striking *colors* of the complexes; c) the variation in the *magnetism* of the complexes (paramagnetism, caused by unpaired electrons, *versus* diamagnetism, when all electrons are paired).

The text presents two bonding theories. They are **crystal-field theory** and **ligand-field theory.** The "field" in these names refers to the electrostatic influence exerted by the ligands upon the central metal atom.

Crystal-Field Theory

In crystal-field theory the bonding between metal and ligands is modeled as ionic. The ligands are approximated by point charges situated at the proper distance from and in the proper geometry around the central metal atom. The ligands exert a negative electrostatic field which perturbs the *d*-orbitals of the central metal. If the symmetry of the ligands is **octahedral,** then the metal's five *nd* orbitals are split into two groups, the t_{2g} and e_g. The t_{2g} contains *three* of the *d*-orbitals (*t* for triple) and is lower in energy than the two orbitals of the e_g level. The degree of splitting is the **crystal-field splitting** and is symbolized Δ_o where the subscript refers to the octahedral field.

Splitting occurs because the 5 *d*-orbitals have different spatial distributions. Some orbital lobes point at the ligands and some point between them. **Problem 16-48** which tests understanding of the reasons for splitting among the *d*-orbitals by asking about what happens to the *p* orbitals in fields of various symmetries. An octahedral field gives one kind of splitting pattern. If the symmetry of the crystal field is tetrahedral or square planar instead of octahedral, then characteristic *different* patterns of splitting occur (see Figure 16-20, text page 618). Figures depicting these patterns are **crystal-field splitting diagrams.** The magnitudes of the crystal-field splittings are symbolized Δ_t (subscript t for tetrahedral) and Δ_1, Δ_2 and Δ_3 (for square planar). A square planar field creates three intervals of splitting because it is inherently less symmetrical than the tetrahedral or octahedral fields. A completely asymmetric arrangement of ligands would split the five *d*-orbitals to five different energies.

The magnitude of the splitting depends on the strength of the perturbing crystal-field which in turn depends on the identity of the ligands. The magnitude of the splitting varies considerably from ligand to ligand. It is on the order a few hundred kilojoules per mole.

• The crystal-field splittings are on the general order of the strength of chemical bonds.

The many colors exhibited by coordination complexes are interpretable in terms of this model. **Strong field** ligands split the *d*-orbitals far apart. Their Δ is big. It requires *more* energy to excite a *d*-electron from the low-lying set of *d*-orbitals to an orbital of higher energy. Complexes with strong field ligands therefore absorb light more toward the blue end of the visible spectrum. Complexes with weak-field ligands absorb more toward the red (low frequency) end of the spectrum. The **spectrochemical series** is an empirical ordering of common ligands according to the strength of the field that they exert:

weak-field strong-field

$$I^- < Br^- < Cl^- < F^- < OH^- < H_2O < NH_3 < \text{en} < CO \approx CN^-$$

To understand the visible spectra of coordination complexes, study closely the colors of the spectrum. From low-frequency to high-frequency they are:

Red Orange Yellow Green Blue Indigo Violet

The initials spell the mnemonic ROY G BIV. Figure 15-23 (text page 580) bends this array of colors into a circle, the color wheel.

• When a compound absorbs light of a given color, it removes those frequencies from the spectrum. The color perceived is the given color's **complement,** the opposite color on the color wheel (see Figure 15-23, text page 580). This idea in important in **problem 16-27.**

The splitting of the metal *d* orbitals in the crystal-field gives rise to two possible types of *d*-electron configurations. **Low-spin** configurations occur when Δ is large and the *d*-electrons remain paired in the orbitals that are split to lower energy. High-spin configurations occur when Δ is small, and the *d*-electrons remain unpaired because the energy it would require to pair them exceeds Δ, the energy required to push them up to occupy the *d*-orbitals that are split to higher energy. The correspondence in terminology is:

strong-field$\Longleftrightarrow$low-spin **weak-field$\Longleftrightarrow$high-spin**

The magnetic properties of complexes depend on the number of unpaired electrons and are successfully predicted by crystal-field theory.

Ligand-Field Theory

In this theory the ligands are no longer regarded as point negative charges exerting an electrostatic field as they surround the central metal atom. Instead, the ligands are allowed to have orbitals. Molecular orbitals are constructed from the metal valence orbitals and the ligand orbitals. Orbital correlation diagrams that are similar in principle to the correlation diagrams of Chapter 15 result. The ligand-field theory is intrinsically more realistic than crystal-field theory. It adds to crystal-field theory an understanding of the *variations* in field strength of the ligands. Also, the correlation diagrams of ligand-field theory include as a component the crystal-field splitting diagrams previously discussed. Hence, they do not contradict but include and expand upon crystal-field theory.

Ligands may, if they have properly arrayed orbitals, receive electrons from as well as give electrons to the metals. This is **π-back-bonding** and is associated with an increase in the field strength exerted by the ligand. Common ligands that can back-bond are the CN^- ion and the CO molecule.

Organometallic Compounds and Catalysis

Organometallic compounds have bonds between metal and carbon atoms. The carbon monoxide molecule (:C≡O:) has a lone pair at both ends. It bonds through its carbon end to many metals. It donates π electron probability density to the metal and accepts electron probability density back into its empty π^* (antibonding) orbitals.

Special stability occurs among organometallics when the central metal is surrounded by 18 valence electrons. See **problem 16-33.** Attaining 18 electrons means the metal has a closed shell electron configuration. The rule of 18 is thus conceptually like the rule of 8 (the octet rule) and similarly subject to violation. The special stability is associated with 18 valence electrons is as important for organometallics as the octet rule is for compounds of elements of the second and third row of the periodic table. The rule of 18 explains dimeric structures of $Mn_2(CO)_{10}$ and $Co_2(CO)_8$ in contrast to monomeric $Ni(CO)_4$.

Sandwich compounds are organometallic compounds in which a metal atom or ion is situated between two planar aromatic rings as in $Fe(C_5H_5)_2$. Ligands of the type found in this compound can donate more than one electron pair (see **problem 16-55).** The rule of 18 still works in sandwich compounds.

Many transition metals catalyze processes giving valuable organic compounds. Catalysis proceeds with intermediate formation of metal-carbon bonds. An example is the **Monsanto process** for the production of acetic acid from methanol and carbon monoxide. The process involves an organometallic complex of rhodium (see Figure 16-23 text page 620).

Coordination Complexes in Biology

Many biologically important compounds contain some derivative of **porphine,** a cyclic organic compound in which four N-donors lie in a plane, surrounding and coordinating a metal ion. The heme portion of hemoglobin contains an iron ion coordinated by a porphine-derived tetradentate ligand. Chlorophyll contains a different porphine derivative coordinated to Mg^{2+} ion. Vitamin B_{12} contains a corrin ring (somewhat similar to porphine) coordinated to a central cobalt atom.

Detailed Solutions to Odd-Numbered Problems

16-1 Methylamine is a monodentate ligand that binds to a central metal ion by donating a lone pair of electrons from the N atom. This is the only lone pair in the molecule.

16-3 a) $V(NH_3)_4Cl_2$: the V atom is in the +2 oxidation state. **b)** $[Mo_2Cl_8]^{4-}$: the Mo atom is in the +2 oxidation state. **c)** $[Co(H_2O)_2(NH_3)Cl_3]^-$: the Co atom is in the +2 oxidation state. **d)** $Ni(CO)_4$: the Ni atom is in the 0 oxidation state.

16-5 a) $Na_2[Zn(OH)_4]$ **b)** $[Co(H_2NCH_2CH_2NH_2)_2Cl_2]NO_3$
c) $[PtBr(H_2O)_3]Cl$ **d)** $[Pt(NH_3)_4(NO_2)_2]Br_2$

16-7 a) Ammonium diamminetetraisothiocyanatochromate(III). **b)** Pentacarbonyltechnetium(I) iodide.
c) Potassium pentacyanomanganate(IV). **d)** Tetraammineaquachlorocobalt(III) bromide.

16-9 The color perceived in a solution is the complement of the color of light absorbed. If an ion (like $[Zn(H_2O)_6]^{2+}$) is colorless, then it does not absorb a significant amount of visible light.

16-11 If the ligands of a complex ion can be rapidly substituted for by other ligands, the complex ion is labile. A ligand that is thermodynamically stable and labile will persist in solution and will also undergo rapid substitution reactions with other ligands. The fact that the complex undergoes substitution reactions can be verified by isotopic labeling experiments.

16-13 The reaction is: $[Cu(NH_3)_4]^{2+}(aq) + 4\,H_2O(l) \rightarrow [Cu(H_2O)_4]^{2+}(aq) + 4\,NH_3(aq)$. Another way to represent $[Cu(H_2O)_4]^{2+}(aq)$ is $Cu^{2+}(aq)$. Thus the equation can be rewritten: $[Cu(NH_3)_4]^{2+}(aq) \rightarrow Cu^{2+}(aq) + 4\,NH_3(aq)$.
Taking data from Appendix D:

$$\Delta G^\circ = 1\underbrace{(65.49)}_{Cu^{2+}(aq)} + 4\underbrace{(-26.50)}_{NH_3(aq)} - 1\underbrace{(-111.07)}_{[Cu(NH_3)_4]^{2+}(aq)} = +70.56 \text{ kJ}$$

Because ΔG° is greater than zero, the reaction is not spontaneous as written. The tetraamminecopper(II) ion is therefore thermodynamically stable with respect to this reaction. Under acidic conditions, the reaction is represented:

$$[Cu(NH_3)_4]^{2+}(aq) + 4\,H_3O^+(aq) \rightarrow Cu^{2+}(aq) + 4\,NH_4^+(aq) + 4\,H_2O(l)$$

$$\Delta G^\circ = 1\underbrace{(65.49)}_{Cu^{2+}(aq)} + 4\underbrace{(-79.31)}_{NH_4^+(aq)} + 4\underbrace{(-237.18)}_{H_2O(l)} - 1\underbrace{(-111.07)}_{[Cu(NH_3)_4]^{2+}(aq)} - 4\underbrace{(-237.18)}_{H_3O^+(aq)}$$

In this case, ΔG° is -140.68 kJ. Hence $[Cu(NH_3)_4]^{2+}(aq)$ *does* tend spontaneously to react with the solvent in aqueous solution at a pH of 0.

16-15 The four substances all dissolve in water to make 0.010 M solutions. The more ions per mole of solute then the greater the conductivity of the solution at a given concentration of solute. Hence in order of increasing conductivity:

$[Cu(NH_3)_2Cl_2]$	$< KNO_3$	$< Na_2[PtCl_6]$	$< [Co(NH_3)_6]Cl_3$
0 ions	2 ions	3 ions	4 ion

16-17 a) There are two isomers of $[Pt(NH_3)_2BrCl]$. Neither is optically active:

```
Br ┌ ─ ─ ┐ NH3          H3N ┌ ─ ─ ┐ Br
   |  Pt  |                 |  Pt  |
Cl └ ─ ─ ┘ NH3          Cl  └ ─ ─ ┘ NH3
```

b) There are three possible isomers of $[Co(CN)_3(H_2O)_2Cl]^{2-}$. None of the three isomers is optically active.

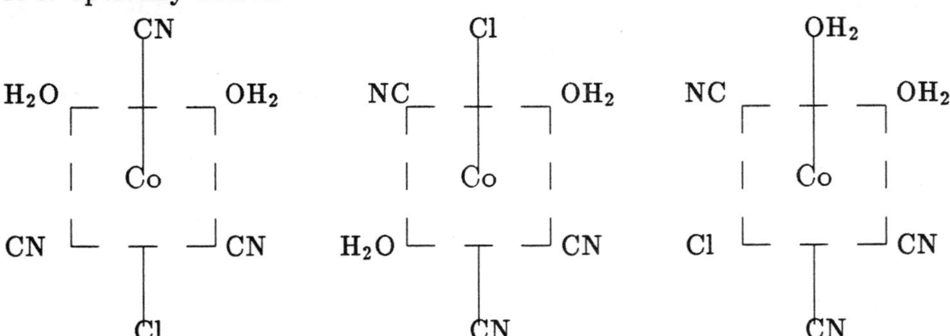

c) $V(C_2O_4)_3^{3-}$ is optically active with two possible optical isomers. In the following the oxalato ligand (^-OOC—COO^-) is abbreviated is "O—·—·—O":

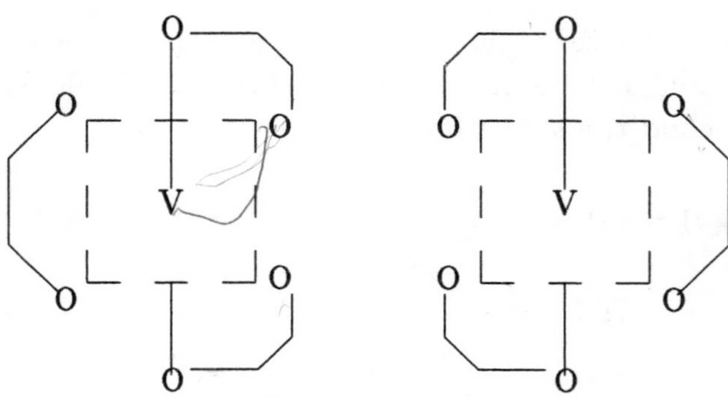

16-19 There are *three* isomeric $[Fe(en)_2Cl_2^+]$ complexes. They are the *trans*-dichlorobis(ethylenediamine)iron(III) ion (left in the following) and the two mirror-image *cis*-dichlorobis(ethylenediamine)iron(III) ions. All involve octahedral coordination about the Fe atom. The en is a bidentate ligand, coordinating through its two $-NH_2$ groups. It can span an edge of the Fe octahedron but not opposite corners. In the following the en-ligand (see Figure 16-2, text page 599) is "N—·—·—N":

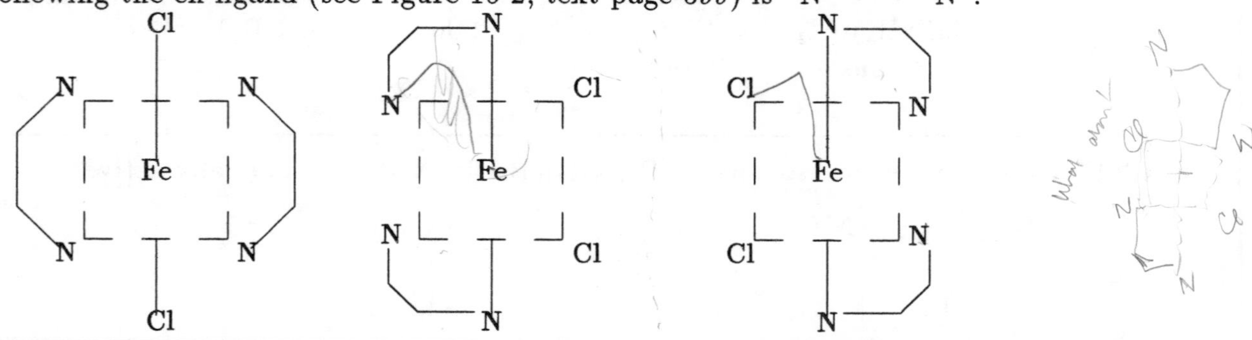

16-21 Strong-field octahedral complexes have a large splitting between the t_{2g} and e_g sets of orbitals; weak-field complexes have a small splitting between the t_{2g} and e_g orbitals. When the energy by which the levels are split exceeds the pairing energy of the electrons, electrons pair up in the t_{2g} level and fill it completely before occupying the e_g level. Otherwise, electrons remain unpaired as long as possible. The number of unpaired electrons is given in the following occupancy diagrams:

a) The electron configuration of Mn^{2+} is $[Ar]3d^5$. It has 1 unpaired electron in a strong field and 5 unpaired electrons in a weak field: Strong Field

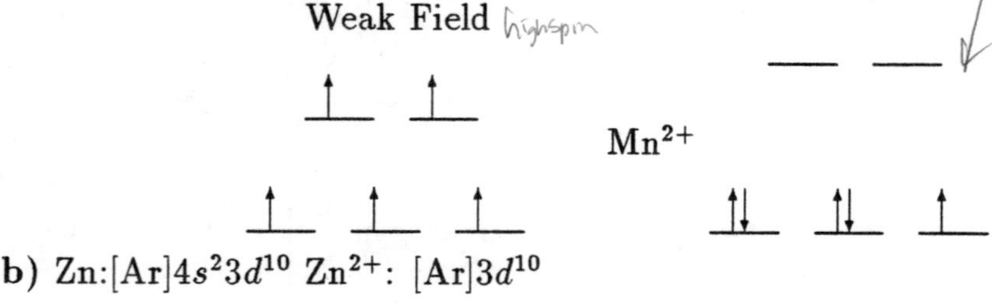

b) Zn:$[Ar]4s^23d^{10}$ Zn^{2+}: $[Ar]3d^{10}$

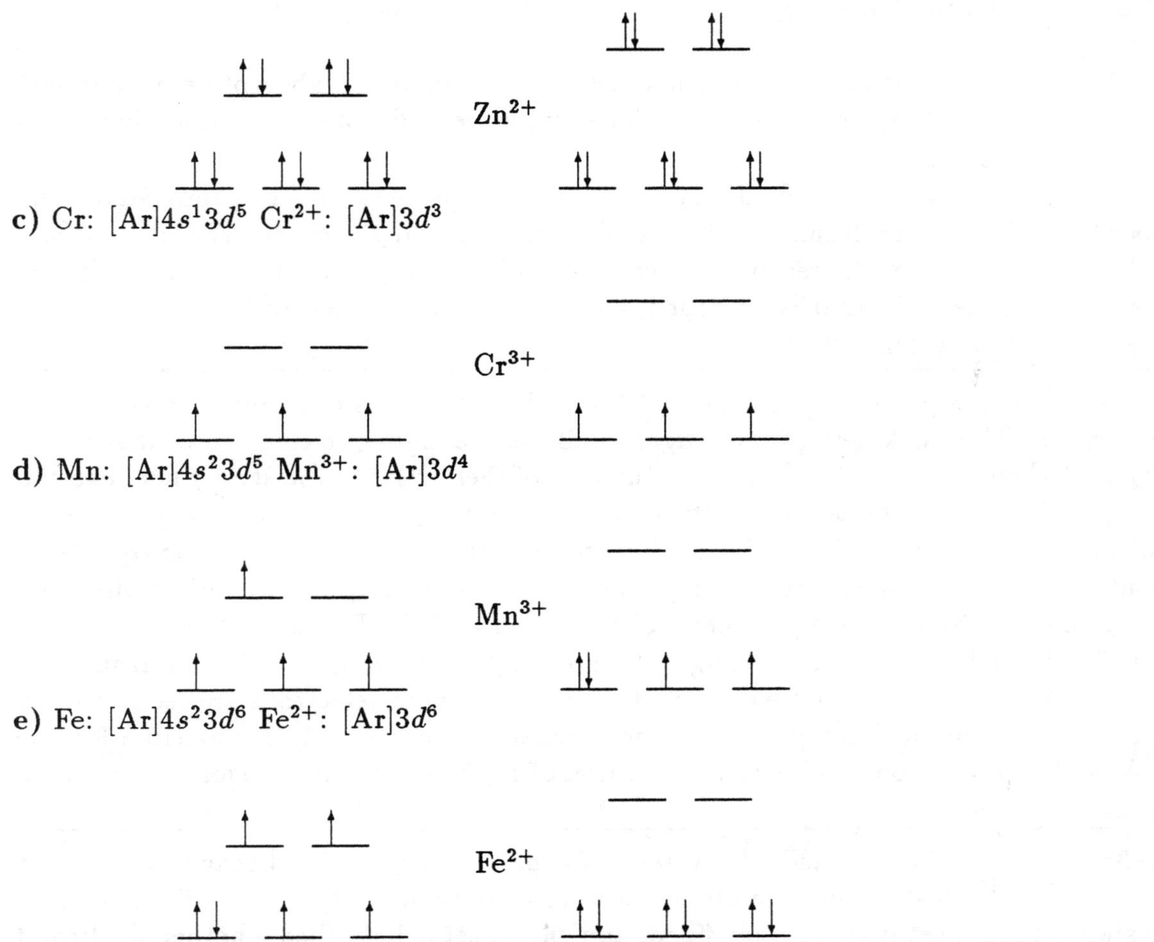

c) Cr: [Ar]$4s^1 3d^5$ Cr^{2+}: [Ar]$3d^3$

d) Mn: [Ar]$4s^2 3d^5$ Mn^{3+}: [Ar]$3d^4$

e) Fe: [Ar]$4s^2 3d^6$ Fe^{2+}: [Ar]$3d^6$

16-23 The ground-state Fe^{3+} ion has 5 *d*-electrons. In the strong octahedral field exerted by six CN^- ligands, the *d*-electron configuration is $(t_{2g})^5(e_g)^0$. All of the *d*-electrons are in the three t_{2g} orbitals; only one of these electrons can be unpaired. In the weak field exerted by six H_2O ligands, the *d*-electron configuration is $(t_{2g})^3(e_g)^2$. All five of the *d*-electrons remain unpaired.

16-25 A solution of $[Fe(CN)_6]^{3-}$ ion transmits red light. Assuming that it absorbs any visible light, the complex ion must absorb light in the green portion of the spectrum. See Figure 15-23 (text page 580). According to Figure 13-4 (text page 483), green light has a wavelength of about 5×10^{-7} m. Using the relationship $E = hc/\lambda$ with Planck's constant and the speed of light in the proper units gives an energy of 4×10^{-19} J. This is equivalent to 2.4×10^5 J mol^{-1}. Assume that absorption of the light excites a single electron from a low-lying t_{2g} orbital to an e_g level. The separation in energy between the two levels, which equals Δ_o, is then also about 2.4×10^5 J mol^{-1}.

16-27 a) The complementary color of blue-violet is orange-yellow (see Figure 15-23 (text page 580).
b) The absorbed light is orange-yellow with a λ of maximum absorption around 600 nm (see Figure 13-4, text page 483). The experimentally observed transition turns out to be at 575 nm.
c) Cyanide ion is a strong-field ligand and water is an weak-field ligand. Replacing coordinated water molecules with cyanide ions accordingly causes the crystal-field splitting to increase. Increasing the crystal-field splitting causes an increase in the frequency of the light that is absorbed, and therefore a decrease in the wavelength of maximum absorption.

16-29 a) In an aqueous solution of $Fe(NO_3)_3$, the Fe^{3+} ion is coordinated to six water molecules. The weak field of these ligands allows the high-spin electron configuration $(t_{2g})^3(e_g)^2$ on the central Fe^{3+} ion. In the case of $[Fe(CN)_6]^{3-}$, the strong field exerted by the CN^- ligands forces the electron configuration $(t_{2g})^5(e_g)^0$. Replacing the weak-field ligand water with the weak-field ligand fluoride should not change the $(t_{2g})^3(e_g)^2$ configuration. The absorption of light by the fluoride complex ion should therefore resemble that by the aqua complex. The solution of $K_3[FeF_6]$ should be pale.
b) The ground-state electron configuration of Hg^{2+} ion is $[Xe]4f^{14}5d^{10}$. A full subshell of 10 d-electrons means that electronic transitions in which electrons are redistributed among d-orbitals are not possible. Such transitions are mostly responsible for the colors of coordination complexes. A solution of $K_2[HgI_4]$ would therefore be expected to be colorless.

16-31 In an octahedral field, d^3 systems are particularly stable because the t_{2g} set is half filled whether the ligands are strong-field or weak-field. In d^8 octahedral systems, the t_{2g} set is completely filled, and the e_g set is half filled whether the ligand is strong-field or weak-field. This also promotes stability. Octahedral d^5 systems would be expected to be particularly stable in complexes of weak-field ligands. The configuration of the d-electron system is then $(t_{2g})^3(e_g)^2$, in which the t_{2g} and e_g sets of orbitals are half filled. Octahedral d^6 systems would be expected to be particularly stable in complexes of strong-field ligands. The configuration of the d-electron system is then $(t_{2g})^6(e_g)^0$ in which the t_{2g} orbitals are a filled set.

16-33 In general, ligands donate a pair of electrons to the central metal atom of a complex except that hydrogen donates only a single electron. The number of valence electrons which the central metal atom provides is given by its position in the periodic table. The compound $[Cr(CO)_4]$ has a total of 14 valence electrons, 6 from the Cr atom and 8 from the CO ligands. The compound $[Os(CO)_5]$ has a total of 18 valence electrons: 8 from the Os atom and 10 from the ligands. The compound $[H_2Fe(CO)_4]$ has a total of 18 valence electrons. The compound $K_3[Fe(CN)_5CO]$ contains the $[Fe(CN)_5CO]^{3-}$ ion. The Fe(II) atom contributes 6 valence electrons, and the six ligands contribute 12 for a total of 18 valence electrons about the Fe atom.

16-35 In 1.00 mol of vitamin B_{12s} there is 1.00 mol of Co. A mole of cobalt weighs 58.93 g. but this is only 4.43% of the mass of a mole of the vitamin. The mass of one mole of the vitamin is therefore larger than that of a mole of cobalt of the factor 100/4.43:

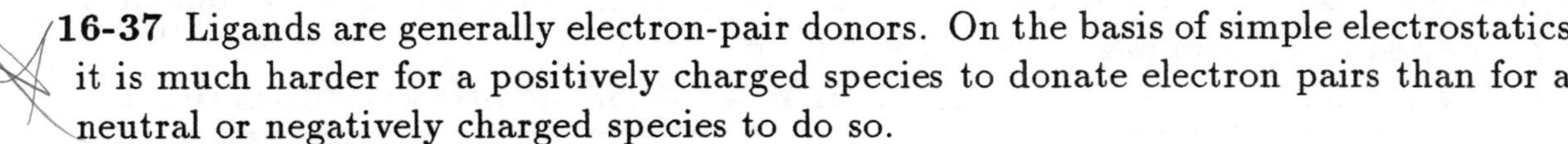

$$\left(\frac{100}{4.43}\right) \times 58.93 \text{ g mol}^{-1} = 1330 \text{ g mol}^{-1}$$

16-37 Ligands are generally electron-pair donors. On the basis of simple electrostatics it is much harder for a positively charged species to donate electron pairs than for a neutral or negatively charged species to do so.

16-39 In $[Ru_2(NH_3)_6Br_3](ClO_4)_2$, all six ammonia ligands are neutral, each bromide ion has a -1 oxidation state and each perchlorate ion has a -1 charge. Because the sum of the oxidation states must equal zero (the overall charge of the complex), the oxidation state of the ruthenium is 2.5. It is possible that one of the ruthenium atoms is in the $+2$ oxidation state, and the other is in the $+3$ oxidation state.

16-41 The problem requires the comparison of the two reactions:

$$\begin{aligned} [Cd(NH_2CH_3)_4]^{2+}(aq) &\rightarrow Cd^{2+}(aq) + 4\,NH_2CH_3 \quad (1) \\ [Cd(en)_2]^{2+}(aq) &\rightarrow Cd^{2+}(aq) + 2\,\text{en} \quad (2) \end{aligned}$$

where en stands for ethylenediamine $H_2NCH_2CH_2NH_2$. Judging from the number of particles formed among the products, ΔS_1° is a larger positive number than ΔS_2°. For the two reactions:

$$-RT \ln K_1 = \Delta H_1^\circ - T\Delta S_1^\circ \quad \text{and} \quad -RT \ln K_2 = \Delta H_2^\circ - T\Delta S_2^\circ$$

Subtracting the first equation from the second gives:

$$RT \ln K_1 - RT \ln K_2 = \Delta H_2^\circ - \Delta H_1^\circ + T\Delta S_1^\circ - T\Delta S_2^\circ \approx T\Delta S_1^\circ - T\Delta S_2^\circ$$

because the ΔH°'s of the two reactions are about the same, according to the problem. Thus:

$$R \ln \left(\frac{K_1}{K_2}\right) \approx \Delta S_1^\circ - \Delta S_2^\circ$$

Since ΔS_1° exceeds ΔS_2°, K_1 exceeds K_2. The first reaction lies farther to the right at equilibrium because of its larger K, that is, the instability constant of $[Cd(NH_2CH_3)_4]^{2+}$ ion is larger than the instability constant of $[Cd(en)_2]^{2+}$ ion. The chelate effect stabilizes the $[Cd(en)_2]^{2+}$ complex.

16-43 a) In the preferred formulation, the Pt atom in the $[Pt(en)_2(SCN)_2]^{2+}$ ion is in the +4 oxidation state, and the Pt atom in the $[PtBr_2(SCN)_2]^{2-}$ ion is in the +2 oxidation state. These complexes form a 1-to-1 ionic compound with the correct empirical formula. The Pt^{4+} ion has 6 *d*-electrons, and is surrounded by 6 donors. If all 6 *d*-electrons are in the t_{2g} level (strong-field), the ion has no unpaired electrons and is diamagnetic. The Pt^{2+} ion has 8 *d*-electrons and is surrounded by 4 donors. The anion can therefore also have all of its electrons paired and be diamagnetic. By contrast, if the molecular formula were $[PtBr(en)(SCN)_2]$, then all Pt atoms would be in the +3 oxidation state and would have 7 *d*-electrons. The compound would then be expected to be paramagnetic, which it is not.

b) Bis(ethylenediamine)dithiocyanatoplatinum(IV) dibromodithiocyanatoplatinate(II).

16-45 Water coordinated to the central Cr ion is tightly bonded and more difficult for a dehydrating agent to remove than water that is loosely held in the solid as water of crystallization. Compound 1 loses two moles of H_2O per mole; it has therefore two waters of crystallization. The other water is in the coordination sphere: $[Cr(H_2O)_4Cl_2]Cl{\cdot}2H_2O$. This ion has octahedral coordination about the central Cr atom. Compound 2 loses only one mole of H_2O per mole so it has only one water of hydration: $[Cr(H_2O)_5Cl]Cl_2{\cdot}H_2O$. Compound 3 loses no water of hydration, so it must have all six water molecule coordinated: $[Cr(H_2O)_6]Cl_3$. Solutions of silver nitrate precipitate AgCl only with chloride ions that are not in the coordination sphere. Therefore, compound 1 gives one mole of AgCl per mole of complex; compound 2 gives two moles of AgCl per mole of complex; compound 3 gives three moles of AgCl per mole of complex. For compound 1:

$$100.0\text{ g} \times \left(\frac{1\text{ mol}}{266.44\text{ g}}\right) \times \left(\frac{1\text{ mol AgCl}}{1\text{ mol}}\right) \times \left(\frac{143.32\text{ g AgCl}}{1\text{ mol AgCl}}\right) = 53.79\text{ g AgCl}$$

Similar 100.0 g samples of compounds 2 and 3, which have the same molar mass, give respectively twice and three times as much AgCl (107.6 g and 161.4 g).

16-47 Tetrahedral structures never display *cis-trans* isomerism because each corner of a tetrahedron is the same distance from the other three corners. In this case:

N N Co Cl Cl

Tetrahedral structures exhibit optical isomerism if they have four different atoms attached to the central atom; the complex $[CoCl_2(en)]$ has two identical ligands and cannot exhibit optical isomerism. The complex $[CoClBr(en)]$ also lacks *cis-trans* isomers, as explained above. It has no optical isomers because the two ends of the en ligand are the same.

16-49 The cyanide ligand has a -1 charge. This means that in $[Mn(CN)_6]^{5-}$ the oxidation state of the Mn must be $+1$: a $+1$ added to $6(-1)$ gives the observed -5 charge on the complex ion. Similarly, the oxidation states of Mn in $[Mn(CN)_6]^{4-}$ and $[Mn(CN)_6]^{3-}$ are $+2$ and $+3$, respectively. The ground-state electron configurations of the manganese ions are:

$$Mn^{+1} : [Ar]3d^6 \qquad Mn^{+2} : [Ar]3d^5 \qquad Mn^{+3} : [Ar]3d^4$$

The low-spin (strong-field) d-orbital occupancy diagrams for each complex are:

Mn(I)	Mn(II)	Mn(III)
__ __	__ __	__ __
↑↓ ↑↓ ↑↓	↑↓ ↑↓ ↑	↑↓ ↑ ↑

16-51 The cesium ions each have a $+1$ charge, and the fluoride ions each have a -1 charge. Thus, copper is in the $+4$ oxidation state. There are six monodentate ligands attached to the Cu^{4+}, so the most likely geometry about the central metal atom will be octahedral. The ground-state electron configuration of Cu^{4+} ion is $[Ar]3d^7$. In the weak octahedral field, the d-electron configuration would become $(t_{2g})^5(e_g)^2$. This high-spin configuration is far more likely than the low-spin $(t_{2g})^6(e_g)^1$ configuration because F^- ligands exert only a weak field.

16-53 Minimum radii occur for the ions with 3 and 8 d-electrons. These are V^{3+} and Ni^{2+} ions. Presumably the oxides are high-field, low-spin complexes in which the 4th and 9th electrons are accommodated in a higher energy orbital, increasing the radius.

16-55 The cyclopentadienyl ion ($C_5H_5^-$) has a -1 charge and is a six-electron donor. The neutral C_6H_6 molecule is also a six-electron donor. The other ligands are all two-electron donors.

The Co atom in $[Co(C_5H_5)_2]^{2+}$ has 6 $3d$ electrons and shares 12 more from the two ligands. It sees a total of 18 valence electrons.

The Fe atom in $[Fe(C_5H_5)(CO)_2Cl]$ also sees a total of 18 valence electrons: the Fe^{2+} ion starts with 6, the $C_5H_5^-$ ion donates 6, the Cl^- ion and each CO molecule donate 2.

The Mo atom in $[Mo(C_5H_5)_2]Cl_2$ sees a total of 14 valence electrons: the Mo^{4+} ion starts with 2 and each $C_5H_5^-$ ion contributes 6. The Cl^- ions are not coordinated to the Mo atom, according to the formula, and donate no electrons to the Mo atom. If the two Cl^- ions do coordinate to the Mo atom, they donate 2 electrons each, and the Mo atom sees a total of 18 electrons.

The Mn atom in the complex $[Mn(C_5H_5)(C_6H_6)]$ sees a total of 18 valence electrons: the Mn^+ ion has 6, the $C_5H_5^-$ ion contributes 6, and the C_6H_6 molecule also contributes 6.

16-57 The oxidation state of W in $[WH_2(C_5H_5)_2]$ is +4; the ground-state electron configuration of W^{4+} is $[Xe]4f^{14}5d^2$. The total number of valence electrons about the W in $[WH_2(C_5H_5)_2]$ is 18: W^{4+} has 2 in its $5d$ orbitals, each $C_5H_5^-$ contributes 6, and each H^- contributes 2 electrons. There is a lone pair of valence electrons on the W atom that can be donated to an acid. In $[TaH_3(C_5H_5)_2]$, the tantalum is in the +5 state. and has the ground-state electron configuration $[Xe]4f^{14}5d^0$. There are 18 valence electrons about the Ta in $[TaH_3(C_5H_5)_2]$, but there are no lone pairs on the Ta atom (or elsewhere). Hence the complex is not a base.

Chapter 17

Solids and Liquids

Crystal Structure and Symmetry

Crystals are classified according to the number and kind of their **symmetry elements.** Elements of symmetry include:

Rotational axes. If the rotation of a crystal about an axis leaves it superimposed upon its original appearance then the axis is an axis of symmetry. A C_2 symmetry axis attains the original appearance at 180° intervals; a C_4 axis attains the original appearance at 90° intervals; a C_n axis does so at $360°/n$ intervals.

Mirror planes. Imagine a plane slicing through a crystal. If reflection across the plane, exchanging left for right and right for left, leaves the crystal unchanged in appearance then the plane is a mirror plane.

Centers of inversion. A crystal has a center of inversion if an imagined operation of "pulling itself through" its central point leaves it unchanged in appearance. If the central point is taken as the origin $(0,0,0)$ a crystal with a center of inversion has an identical feature at $(-x,-y,-z)$ for every feature (x,y,z).

Symmetry operations may exist in any geometrical object, including molecules. See **problem 17-1.**

The long-range order in a crystal means that sites with identical surroundings recur in a regular pattern. The array of all such points (called **lattice points**) in a given crystal is the **crystal lattice** of that crystal. Like physical crystals themselves, crystal lattices, which are mathematical constructions, can be characterized by their symmetry. There are only seven different possible combinations of symmetry elements in crystals lattices, the seven **crystal systems.** Each crystal system has an essential or minimum symmetry and a descriptive name: **hexagonal, cubic, tetragonal, trigonal, orthorhombic, monoclinic,** and **triclinic.** Crystal systems have more symmetry that just the essential minimum. It is very important to study text Table 17-1 closely and visualize the relationship of the symmetry elements in the different crystal systems. The text identifies *all* of the symmetry elements for the cubic system (Figure 17-3, text page 630). An obvious exercise is to make such a listing for the other six crystal systems.

A crystal on the microscopic level can be thought of as built from blocks of a unit structure, its **unit cell,** stacked side by side in three-dimensional space. Unit cells always have three pairs of parallel faces so that they can stack snugly top to bottom, side to side, and front to back. There are of course no little cells walls inside actual crystals because unit cells are constructions of the mind. As such, unit cells are chosen as **the smallest units that retain all the symmetry of the crystal system.** A primitive unit cell contains one lattice point. Unit cells containing two or more lattice points are **non-primitive.**

Usually, the walls of a unit cell will slice through one or more atoms. When portions of atoms are excluded just outside one wall, the loss is compensated for inside the unit cell since a similar portion is automatically included just *inside* the opposite wall.

There are seven types of unit cell, one for each of the crystal systems. Each type has the same symmetry as one of the seven crystal systems. A set of six **cell constants,** three edge lengths ($\boldsymbol{a}$**,** $\boldsymbol{b}$**,** and $\boldsymbol{c}$) and three angles ($\boldsymbol{\alpha}$**,** $\boldsymbol{\beta}$**,** and $\boldsymbol{\gamma}$) between the edges specify the size and shape of a unit cell. The angles are *not* necessarily 90°. Unit cells with higher symmetry have conditions relating the values of a, b and c and restricting the angles α, β, and γ to certain values, usually 90°. The conditions linking the cell constants in the seven crystal systems should be memorized. See Table 17-1, text page 632.

The volume V_c of a unit cell can be computed from its six cell constants:

$$\boldsymbol{V_c = abc\sqrt{1 - \cos^2\alpha - \cos^2\beta - \cos^2\gamma + 2\cos\alpha\cos\beta\cos\gamma}}$$

When one or more of the three angles equals 90°, the above formula is drastically simplified, since $\cos 90° = 0$. See **problem 17-53.** Unit cells have axial lengths starting at about 4 Å and can be exceedingly large (see **problem 17-3).** The structure of the entire crystal can be understood in terms of its unit cell.

Real crystals are huge compared to unit cells. Most unit cells are buried deep in the crystalline interior. In the interior, every unit cell is just like the next only moved over in one of three directions by the length of one of the three cell edges. Therefore, the surroundings of the eight corners of every unit cell are identical. These corners are the lattice points. They are at the intersections of a regularly-spaced three dimensional lattice, or grid, of lines running along the edges of the unit cells.

- Every lattice point has identical surroundings. In many problems and applications it is important to know how many lattice points there are per unit cell. A unit cell has eight corners but shares each one with seven other cells. Its corners alone contribute $1/8 \times 8 = 1$ lattice point.

Some lattices have more than one lattice point per unit cell and still conform to the symmetry of one of the seven crystal systems. The additional lattice points appear:

- At the interior center of the unit cell. This is **body-centering.** A body-centered cell has 2 lattice points. One stems from the 8 corners and the second is the point at the body center.

- At the centers of all six faces. This is **face-centering.** A face-centered unit cell has 4 lattice points. Understanding the face-centered structure is essential in many problems such as **problem 17-65.**

- At the centers of just one of the three sets of parallel faces. This is **side-centering.** An side-centered unit cell has 2 lattice points.

Unit cell constants are also sometimes called **lattice parameters** because they tell the spacing of the lattice points.

Crystal Structure of Simple Substances

The cubic crystal system is particularly important to understand because it is the system chosen by nature to underlie the solid-state structure of many simple substances. Three lattices have cubic symmetry: the **simple cubic (s.c.)**, the **face-centered cubic (f.c.c.)**, and the **body-centered cubic (b.c.c.).** Some geometrical facts about the cube are useful:

1. One parameter, the length of an edge a, fully characterizes a cube. This is the same as saying that one parameter, the length of a unit cell edge, fully characterizes a cubic lattice.

2. The volume of a cube is equal to its edge cubed (a^3). Similarly the volume of a cubic unit cell is a^3.

3. A line drawn diagonally across one face of a cube is the **face diagonal** $\boldsymbol{f}$**.** The length of f is $\sqrt{2}a$.

4. A line connecting the opposite corners of a cube is the **body diagonal** $\boldsymbol{b}$**.** The length of b is $\sqrt{3}a$.

The density of a crystal is equal to its mass divided by its volume. In terms of the unit cell, the density of a crystal is the mass of the contents of the cell divided by the volume of the cell:

$$\text{density} = \rho = \frac{m \text{ (contents of cell)}}{V_c}$$

This leads to the useful expression:

$$\rho = \frac{n_c \mathcal{M}}{N_o V_c}$$

in which n_c, which is always an integer, is the number of formula units in the unit cell, $\mathcal{M}$ is the molar mass of the formula units, N_0 is Avogadro's number, and V_c is the volume of the unit cell. If $\mathcal{M}$ is in g mol^{-1} and V_c is in cm^3, then the density comes out in g cm^{-3}. This equation is the basis for answers to **problems 17-5, 17-7, 17-9,** and **17-53.** One error in its use is failure to compute V_c properly. If cell constants are given in Å, then cell volume will come out in $Å^3$. Remember that 1 cm^3 is 10^{24} $Å^3$. Another error in computing densities is to omit N_0. One way to understand the inclusion of N_0 in the formula is to reason that $\mathcal{M}$ has units of g mol^{-1} and something has to be done to remove the mol^{-1} part because densities do not have moles in their units. **Problem 17-5** gives additional perspective on why N_0 appears.

Even in simple substances, all of the atoms in a unit cell do *not* automatically sit at lattice points. Atoms can be anywhere in the cell. Their locations are given in terms of a set of three **fractional coordinates.** One of the corners of the unit cell serves as the origin and the three cell edges which intersect at the origin serve as axes. Any point inside the cell has coordinates (x, y, z) where x tells how far away from the origin the point is parallel to the first edge in units of a, that edge's length and y and z do the same for the second and third edges. For example, an atom in a cubic cell might have the fractional coordinates (0.40, 0.15, 0.35). If the three cell edges were all equal to 4.0 Å, this atom would lie 1.60 Å out from the origin in a direction parallel to the a cell edge, 0.60 Å out parallel to b, and 1.40 Å out parallel to c.

The eight corners of the unit cell have the fractional coordinates:

$$(0,0,0)\ (1,0,0)\ (0,1,0)\ (0,0,1)\ (1,1,1)\ (0,1,1)\ (1,0,1)\ (1,1,0)$$

These eight corners are exactly equivalent. In fractional coordinates, adding or subtracting 1 to any member of the triple does not create new or unique positions but instead refers to the *same* relative location in the *adjoining* unit cell.

Example: What are the fractional coordinates of atoms located at the face-centers of a cubic unit cell? **Solution:** The face centers all have one fractional coordinate equal to 0 and the other two equal to 1/2. The correct answer:

$$(1/2,1/2,0)\ (1/2,0,1/2)\ (0,1/2,1/2)$$

The trouble with this answer is that everybody "knows" that a cube has six face centers because it has six faces. The centers of the other three faces are:

$$(1/2,1/2,1)\ (1/2,1,1/2)\ (1,1/2,1/2)$$

These coordinates do *not* specify distinct points. They are instead for points that are equivalent to the first three but translated by the length of a cell edge along z, y and x respectively. A cubic unit cell possesses as it own only three face centers because each of its six faces is shared 50:50 with a neighboring unit cell.

Fractional coordinates are used to advantage in **problem 17-25.**

Atomic Packing in Crystals

In elemental metals there is only one kind of atom in the crystal. This makes for simplicity in the crystal structure. Many metals crystallize in cubic lattices in which the atoms are in contact with their neighbors and are located at each point of the lattice:

- In a simple cubic lattice, the distance between neighboring lattice points is a. The radii of identical metal atoms sitting at these points and in contact is $1/2a$.
- In a face-centered cubic (f.c.c.) lattice, nearest-neighbor identical atoms situated at the lattice points touch along a face diagonal. The distance between their centers is $\sqrt{2}a/2$, and the radius of the metal atoms is $\sqrt{2}a/4$. This statement applies only with all the atoms in the structure are identical. See **problem 17-65** for a contrasting case.
- In a body-centered cubic (b.c.c.) lattice, nearest-neighbor identical atoms touch along a body diagonal. The distance between their centers is $\sqrt{3}a/2$; the radius of the metal atoms is accordingly $\sqrt{3}a/4$.

If atoms are hard spheres, then it is impossible not to leave gaps between them when packing them together in a crystal. A **close-packed** structure is a structure in which identical spheres (atoms) occupy the greatest possible fraction of the total space. The two most efficient methods for packing identical spheres are **cubic close-packing (c.c.p.)** and **hexagonal close-packing (h.c.p.)**. In both, the atoms occupy 74.0 percent of the volume. The other 26.0 percent of the volume of the unit cell is empty space. Cubic close-packing corresponds to setting down layers of identical spheres in the sequence *abcabcabc*..., where the letters refer to the offset in position of the higher layers relative to the first. The recurrence of a in the fourth position means that atoms in the fourth layer are positioned exactly above atoms in the first layer. Cubic close-packing gives rise to a face-centered cubic lattice. Hexagonal close-packing corresponds to the sequence of layers *ababab*....

The gaps between close-packed atoms are **interstitial sites.** Octahedral interstitial sites are surrounded by 6 atoms. Tetrahedral interstitial sites are smaller and surrounded by only 4 atoms. In the face-centered cubic lattice there are 4 octahedral sites and 8 tetrahedral sites per unit cell.

In other kinds of lattices there are other kinds of interstitial sites. An examination problem might refer to the geometry of a **cubic** site (surrounded by 8 atoms) to test understanding.

Scattering of X-rays by Crystals

X-rays have wavelengths λ on the order of the distances between layers of atoms in crystals. When a beam of x-rays strikes a layer of atoms in a crystal, it is **scattered** in all directions by interaction with the electrons of the atoms in the layer. The atoms lie in a regular array. Some of the scattered x-rays from one layer **constructively interfere** with some of the scattered x-rays from another layer. The result is a large number of scattered beamlets shooting out from the crystal in all directions. If x-rays were visible the effect might resemble what is seen at dance-halls when a spotlight is played on a large sphere which is tiled with small flat mirrors. The reason there are so many beamlets is that many different sets of layers of atoms exist simultaneously in the crystal.

Each scattered beamlet is called a **reflection.** The **Bragg law** is the criterion for scattering:

$$\boldsymbol{n\lambda = 2d \sin\theta}$$

In this important equation, n is an integer, λ is the wavelength of the x-rays, d is the perpendicular distance between the layers of atoms, and θ is one-half of the angle between the incident beam of x-rays and the reflected beamlet in question. The angle θ is the **Bragg angle.**

In common applications, λ is known and 2θ (and thus θ) is measured. The order of the reflection is often given explicitly. See **problems 17-13** and **17-15.**

Example: Suppose that x-rays of wavelength 1.54 Å are scattered from a set of evenly-spaced layers of atoms in a crystal with a spacing of 4.62 Å. Compute the Bragg angle of all the resultant reflections. **Solution:** From the Bragg equation with $\lambda = 1.54$ Å and $d = 4.62$ Å:

$$\theta = \sin^{-1}\left(\frac{n\lambda}{2d}\right) = \sin^{-1}(n/6.00)$$

Systematic substitution for the integer n gives:

$$\begin{array}{llll} n=1 & \theta = 9.59,\ 170.41^\circ & n=4 & \theta = 41.81,\ 138.19^\circ \\ n=2 & \theta = 19.47,\ 160.53 & n=5 & \theta = 56.44,\ 123.56 \\ n=3 & \theta = 30.00,\ 150.00 & n=6 & \theta = 90.00,\ 90.00 \end{array}$$

The listing ends at $n = 6$ because higher n's give $\sin\theta > 1.0$, but numbers exceeding 1.0 do not have an inverse sine. There are two values of θ for each n because the inverse sine function gives both an angle and its supplement. The value between 90 and 180° is often simply neglected. The solution to **problem 17-15** shows what happens when it is not neglected. Every reflection in the list is found scattered at an angle 2θ relative to the incident beam. For $n = 6$ the angle 2θ is 180°. The $n = 6$ reflection scatters from the crystal right back into the incident beam.

X-rays are not unique in being scattered by matter. Both neutrons and electrons have wave-like properties, as predicted by the DeBroglie equation (Chapter 12):

$$\lambda = \frac{h}{p} = \frac{h}{mv}$$

where p is the momentum of the particle. When beams of neutrons or electrons of the proper momentum impinge on crystals, they are scattered according to the Bragg equation. **See problem 17-65,** which is like **problem 17-9** except that the proper λ has to be calculated (using the DeBroglie equation) before applying the Bragg equation.

Intermolecular Forces

Electromagnetic forces cause all interactions among molecules. If two electrical charges are at rest, then **Coulomb's law** gives the electrostatic force between them:

$$F = \frac{-Q_1 Q_2}{4\pi\epsilon_0 R^2}$$

where Q_1 and Q_2 are the charges, R is their separation and ϵ_0 is a constant (8.854×10^{-12} C^2 J^{-1} m^{-1}) to make the units come out right. The Q's are signed quantities. Charges of opposite sign attract each other. With the negative sign in the above equation a positive Coulomb force means an attraction between the charges; a negative force is a repulsion.

A deceptively similar but distinct formula gives the Coulomb potential energy of interaction between two charges. In this equation R is not squared and the negative sign does not appear:

$$V(R) = \frac{Q_1 Q_2}{4\pi\epsilon_0 R}$$

When the charges have opposite signs they attract each other, and the energy of interaction is negative. If the two charges are $Z_1 e$ and $Z_2 e$ where e stands for the charge on the electron, then the potential energy of interaction is:

$$V(R) = \frac{Z_1 Z_2 e^2}{4\pi\epsilon_0 R}$$

This equation is identical to the equation for the interaction energy between two ions that was introduced in the discussion of ionic bonding on text page 534.

Dipole-Dipole Interactions

Electrostatic forces act between electrically neutral molecules, as well as between ions, if the molecules have a non-spherical distribution of electrical charge. Such molecules are **electric dipoles.** They have a positive end and a negative end separated by a distance R. The computation of the Coulomb energy of a system of two or more dipolar molecules can become quite complicated. The Coulomb energy depends on the relative *orientation* of the dipoles are well as the magnitude of the charges on their ends and the distances between them. A good procedure is:

1. Count all of the pair-wise intermolecular combinations of charge centers. Every dipole has two centers of charge. Two molecules have four combinations. An arrangement of three molecules would have eight. Combinations *within* molecules are not counted.

2. Figure out the distances between the centers of charge in each pair. Calculating these distances in an arbitrary arrangement of dipoles can involve much drawing and solving of triangles, a procedure that is subject to error. **Problem 17-61** illustrates the assignment of Cartesian coordinates to each charge to facilitate the calculation of the distance between them.

3. Determine the magnitude and sign of every charge. This information may have to be obtained from the dipole moments of the molecules. Recall that the dipole moment μ equals QR, the product of the distance separating charges of opposite sign and magnitude Q.

4. Compute the total Coulomb energy, the sum of all of the pair-wise interactions. Adding up pair-wise interactions is what is meant by the subscripts and summation signs in the formula:

$$V = \sum_i \sum_j \frac{Q_i Q_j}{4\pi\epsilon_o r_{ij}}$$

   See **problem 17-21.** If the sum is negative, then the interaction is an attractive one. If it is positive, then the interaction is a repulsive one.

Dispersion Forces

Dispersion forces between molecules arise from the correlation of the motions of electrons. They are induced-dipole to induced-dipole attractions. The electron probability density on one molecule experiences a momentary fluctuation. This induces a correlated fluctuation in the probability density on the second. A build-up of negative charge on one side of the first atom induces a build-up of positive charge on the near side of the second atom. The result is an attraction between the molecules.

Pauli Repulsive Forces

Pauli repulsive forces operate *in addition to* electrostatic repulsive forces as two atoms are pushed close together. The electron probability densities of the two distort to maintain the Pauli principle in the new, two-atom system. This distortion is a rearrangement of the electron probability density of the atoms to a situation of higher energy. The Pauli principle forbids the simple overlapping of the electrons in two closely neighboring atoms. The repulsive effect caused by the operation of the principle rises very rapidly at short distance. This fact allows assignment of an atomic radius, the **van der Waals** radius for every atom. High energies are required to bring two atoms closer together than the sum of their van der Waals radii.

Potential Energy Curves

As two atoms (or molecules) approach each other they usually (depending on the orientation of any dipoles, see **problem 17-21**) attract each other at first. The potential energy of the system drops. This continues until short-range repulsive forces become important. Then the potential starts to rise again. The result is a characteristic hook-shaped potential energy curve (Figure 17-17, text page 644). The minimum of the curve is the bottom of a "potential-energy well" of the system. The depth of the well varies enormously. When two atoms form a chemical bond, the well is hundreds of kJ mol^{-1} deep. When atoms experience only dipole-dipole attractions or dispersion forces it may be only a few tenths of a kJ mol^{-1} deep.

The **Lennard-Jones potential** approximates the shape of the potential energy curve for non-bonding interactions between two identical atoms and molecules. It consists of a repulsion term (the term involving the twelfth power) and an attraction term:

$$V_{\mathrm{LJ}} = 4\epsilon\left[\left(\frac{\sigma}{R}\right)^{12} - \left(\frac{\sigma}{R}\right)^{6}\right]$$

where ϵ and σ are constants that differ for various atoms and molecules (see Table 17-3, text page 645). This ϵ is *not* the same as the ϵ_0 used earlier in the chapter. It is instructive to graph the intermolecular potential energy curve for one or two different gases, given ϵ and σ. Thus, the parameters for argon in Table 17-3 (text page 645) can be used to draw the curve labeled Ar + Ar in Figure 17-17 (text page 644). **Problem 17-63** lends additional physical meaning to the Lennard-Jones parameters by relating them to the van der Waals constants.

Cohesion in Solids and Liquids

The strength of the forces that hold crystalline substances together varies enormously. Crystals are classified according to these forces. The categories are: **molecular, ionic, metallic** and **covalent.**

Molecular crystals

Molecular crystals are held together by van der Waals or dipole forces. These forces are relatively weak. The building-block molecules in the crystal are only slightly affected in their internal geometry by the interaction with their neighbors. Molecular crystals are soft, low-melting, and generally poor conductors of electricity.

Ionic crystals

Compounds composed of elements that differ considerably in electronegativity occur as ionic crystals. Such crystals behave to a good approximation as if constructed of hard, charged spheres held together *in contact* by electrostatic attractions. Such attractions are strong but *nondirectional* in character.

To maximize the attractions (between ions of unlike charge) and to minimize repulsions (between ions of like charge), the lattice of an ionic crystal has every positive ion surrounded by as many negative ions as possible and vice versa. Since fixed bond angles play no part in electrostatic attractions, the ionic crystal favors densely packed structures. Each positive ion occupies an interstitial site in a lattice of negative ions and each negative ion occupies an interstitial site in a lattice of positive ions. The shape (octahedral, tetrahedral or other) and proportion of occupied sites depend on the relative size and charge of the ions.

In ionic crystals, the negative ions are nearly always larger than the positive ions. The cation/anion **radius ratio,** a number less than 1, determines the structure which charged hard spheres adopt in a lattice. As the radius ratio gets larger, more anions can surround the cation and just touch it without bumping into each other. See **problem 17-55.** Important structures for ions of equal charge are the **zinc blende** structure (when the radius ratio is less than 0.414), the **rock salt** structure (when the radius ratio is between 0.414 and 0.732), and the **cesium chloride** structure (when the radius ratio is greater than 0.732. Learn these structures, which are diagrammed in text Figures 17-21, 17-19, and 17-20, respectively.

Metallic Crystals

In a metallic crystal, the attractive force holding the crystal together is the metallic bond. Metallic bonds are *nondirectional* in character and involve wide-spread *delocalization* of electrons. They involve the sharing of electrons among all the atoms in the crystal. The electrons occupy molecular orbitals arising from the overlap of atomic orbitals of similar energy on all of the atoms. For example, in 1 mol of sodium metal, there are 6.02×10^{23} $3s$ orbitals. These orbitals overlap to make a **band** of molecular orbitals. The molecular orbitals comprise a band because they are so numerous and their energies are so nearly the same. The band has room for $2 \times 6.02 \times 10^{23}$ electrons and therefore is only half-full, under the Pauli principle.

The excellent electrical and thermal conductivities of metals are explained by the great mobility of electrons at the top of the sea of occupied levels. The uppermost occupied or half-occupied molecular orbital in the band is the **Fermi level.** Electrons at or near the Fermi level require only slight amounts of energy to excite them to occupy levels lying above.

Covalent Crystals

In covalent crystals, the atoms are linked by covalent bonds. These bonds are *strong* and *directional.* Diamond, discussed in **problem 17-51** and diagrammed in Figure 17-25 (text page 651) is a typical case. Its crystal structure consists of a face-centered cubic array of C atoms with additional C atoms occupying every other tetrahedral interstitial site. In this arrangement, each C atom is surrounded by 4 nearest neighbors situated perfectly for σ overlap between sp^3 hybrid orbitals.

The structures of the elements provide examples of three of the four classes of crystalline solids. Atoms of nonmetallic elements at the top of the periodic table tend to use up all their bonding capacity in intramolecular bonds, thereby mainly forming molecular crystals. Structures trend away from the formation of multiple bonds and toward formation of chains and rings of atoms moving down the table.

Lattice Energies of Crystals

The **lattice energy** of a crystal is the energy decrease that results when the constituent atoms, molecules or ions condense from a gas to form the crystal at 0 K.

Lattice energies of molecular crystals can be estimated from the Lennard-Jones parameters of the molecules. They are much smaller than the lattice energies of other types of crystals. For example, compare the answer to **problem 17-29a** with the answer to **problem 17-67a.**

In ionic crystals, Coulomb forces are the predominant type of interaction between the constituent particles, which are ions. The Coulomb contribution to the lattice energy of an ionic crystal is:

$$V = \frac{MZ_1Z_2e^2N_0}{4\pi\epsilon_0R_0}$$

where the Z's are the charges on the two ions in the lattice in units of e, where e is 1.6021×10^{-19} C, where ϵ_0 is 8.854×10^{-12} $C^2\ J^{-1}m^{-1}$, where R_0 is the *minimum* distance between the two ions, and where M is a special constant, the **Madelung constant.** The Madelung constant accounts for the fact that the electrostatic energy of a crystal is the sum of all possible pair-wise attractions and repulsions in the crystal. The value of the Madelung constant depends on the type of crystal lattice. The constant can be computed, using the methods of infinite series, for any lattice (see Table 17-6, text page 655 for the results).

Calculations using the above formula give *theoretical* estimates of lattice energies. Experimental lattice energies are obtained using the **Born-Haber cycle,** an application of the first law of thermodynamics. It is impossible directly to decompose an ionic crystal into its constituent ions in their gaseous states. Fortunately, ΔE for this process, which defines the lattice energy, can be determined as the sum of several measurable ΔE's for steps that combine to give this reaction. The solution to **problem 17-29** shows the use of the Born-Haber cycle. Form an idea of the size of the different energies involved in the Born-Haber cycle. This will allow a common-sense approach to checking lattice energy calculations. Difficulties in working with the energetics of crystals come from:

- Carelessness with the sign. Either Z_1 and Z_2 in the above formula must be negative. All of the other quantities are positive. Consequently the Coulomb energy of a crystal is always negative.

- Confusion between lattice energy and Coulomb energy. The lattice energy of an ionic crystal is an experimental number. The Coulomb energy is a calculated, or theoretical quantity. The Coulomb energy approximates the lattice energy in magnitude but is always 10 to 15 percent larger. It errs as an estimate of the lattice energy because the hard-sphere picture of the lattice is an oversimplification.

- Confusion between ΔH and ΔE. As defined, the lattice energy is an energy change, not an enthalpy change. The Born-Haber cycle must use ΔE values for all of its steps to compute a lattice energy properly. In practice, the difference between ΔH and ΔE is usually small. See **problem 17-29.**

Many problems require some creative use of a Born-Haber cycle.

Example: Estimate the enthalpy of formation of the hypothetical compound $CaCl(s)$. Why does $CaCl_2(s)$ always form when $Ca(s)$ and $Cl_2(g)$ react? **Solution:** The enthalpy of formation of $CaCl(s)$ can only be estimated because the compound does not exist, and direct measurements are not possible. The key insight is to use the lattice energy of KCl to approximate the lattice energy of the hypothetical compound. First, either look up or calculate this lattice energy. Then apply the Born-Haber cycle to the formation of the hypothetical $CaCl(s)$. The answer is a ΔH_f° of about -150 kJ mol^{-1}. The compound is favored thermodynamically relative to the elements!

Why does it not form? The answer is in the second part of the question. The real compound, $CaCl_2(s)$, has a ΔH_f° of -795.8 kJ mol^{-1} (Appendix D). It is even more favored.

Defects and Amorphous Solids

Real crystals are imperfect. In a real crystal some lattice sites are unoccupied. Sometimes the missing atom is entirely lost, and sometimes it is just displaced to a near-by interstitial site. The first case is a **Schottky defect.** This second is a **Frenkel defect.**

The existence of defects explains why many solid compounds have stoichiometries differing from strict whole-number ratios. If some proportion of the positive ions in an ionic lattice is oxidized (meaning that some ions lose additional electrons), then there can be Schottky defects at some of the positive ion sites. The result is a ratio of positive ion to negative ion that is slightly *less* than stoichiometric. The fraction of positive ions that is oxidized is related to the number of defects and the degree of nonstoichiometry in the compound. See **problems 17-33** and **17-71.** Both Frenkel and Schottky defects in crystals are mobile. They jump from one lattice sit to another. Diffusion in crystalline solids is due mainly to the presence and mobility of defects.

An **F-center**, which is also called a **color center** occurs in an ionic crystal when an anion site is occupied by an electron alone (see text Figure 17-31).

Amorphous Solids and Glasses

When a solid has so many defects that crystalline order is destroyed, then it is an **amorphous** solid or a **glass.** A strong tendency to form glasses is associated with the presence in a solid of long or irregularly shaped molecules than can easily become tangled. In principle however, any solid that can be liquefied can be prepared in an amorphous state, usually by cooling it very rapidly. Glasses soften and flow when heated instead of melting sharply at a defined temperature. This plasticity makes glasses very easy to fabricate into desired shapes.

Liquids

In a gas, even at high temperatures, attractive intermolecular forces temporarily hold molecules together in pairs or in small clusters. These associations quickly break apart under the random jostling of collisions from other molecules. As the temperature is lowered, the collisions from outside become weaker. Instead of breaking a small cluster apart, impinging molecules tend to join it. Finally, when the temperature gets down to the boiling point, a single large cluster forms and grows rapidly. The result is a liquid.

Structure of Liquids

In liquids there is short-range, local order in the arrangement of the molecules, but no long-range order. A predictable, repeating pattern holds temporarily within a cluster of molecules. The pattern is frayed at its edges by holes or defects. With time it disintegrates completely. Another region of order pops up. The relative positions of the molecules constantly change as regions of order form and decay. From another point of view, ordered regions migrate from place to place, changing their membership as they go.

The average structure of a liquid is given by a **radial distribution function,** abbreviated $g(R)$. To understand this function, consider the surroundings of a *test molecule* first in an ideal gas and then in a liquid. Let ρ represent the average density over time of the gas or liquid. This average density has the units of molecules per unit volume. In the ideal gas there is no order, either at short or long range, among the molecules, which are points. Therefore, ρ is the same up close to the test molecule as it is far away from it. In the liquid, the test molecule influences its surroundings. It exerts intermolecular attractions and also has a small volume of its own. As a result, the time-average density near the test molecule depends on R, the distance from the test molecule:

$$\text{Average Density} = \rho g(R)$$

where the dependence on R has been separated off into the function $g(R)$, the radial distribution function. This function is a unitless multiplier that modifies ρ. For an ideal gas, $g(R) = 1.0$. In a liquid, the function $g(R)$ is very small at short distances because the intermolecular forces are strongly repulsive at short distances. At intermediate distances, $g(R)$ has one or more maxima (see Figure 17-35, text page 660). It then goes to 1.0 at distances around 10 to 15 Å because the order in the liquid is only short-range.

To find the average number of neighbors that the test molecule has at distance R, imagine a spherical shell of radius R englobing the test molecule and having thickness ΔR. The volume of this shell is $4\pi R^2 \Delta R$. Multiplying this volume by $\rho g(R)$, which tells the time-average density of molecules at the distance R, gives the desired result.

Example: Estimate the average number of neighbors at a distance of 3.1 ± 0.6 Å from a typical molecule in liquid argon (density 1.4 g cm^{-3}), if the liquid has the radial distribution function graphed in Figure 17-35. **Solution:** The number of neighbors within the specified range is:

$$N(\text{neighbors}) = 4\pi^2 R^2 \rho g(R) \Delta R$$

ΔR is 1.2 Å, and R is 3.1 Å. The graph of $g(R)$ uses Ångstroms as the unit of R so it is sensible to convert the density of the liquid from g cm^{-3} to a unit involving Å^3. The molar mass of Ar is 39.95 g mol^{-1} so the number density (see Chapter 3) of the atoms in liquid Ar is 0.021 $Å^{-3}$. Figure 17-35 shows that as R goes from 2.5 to 3.7 Å, $g(R)$ goes from 1.0 to a maximum near 2.5 and then returns to about 1.0. It has an average value of around 1.3 (estimated by eye). Substituting these values gives:

$$N \text{ (neighbors)} = 4\pi^2(3.1 \text{ Å})^2(0.021 \text{ Å}^{-3})(1.3)(1.2 \text{ Å}) \approx 12$$

Cohesion in Liquids

As a gas condenses to a liquid at its boiling temperature T_b, the enthalpy of the system *decreases.* The internal potential energy in the liquid is much less than in the vapor. The kinetic energy is about the same as long as the temperature is still T_b. The total energy of the liquid is therefore less and the total enthalpy is closely related. See **problem 17-73.**

On the other hand, vaporizing the liquid requires enthalpy from outside to pry the molecules away from each other. In this reverse process, the enthalpy of the system increases.

During condensation of a gas to a liquid, the entropy of the system decreases because there are fewer microstates available to the molecules in the product liquid than in the gas (see text page 326, Chapter 9). Obviously, in the reverse process the entropy of the system increases.

Whether approached by lowering the temperature to condense gas, or by raising the temperature to vaporize liquid, T_b is the temperature at which liquid and gas phases are in equilibrium. At equilibrium:

$$\Delta G = \Delta H_{vap} - T_b \Delta S_{vap} = 0 \quad \text{from which} \quad \Delta S_{vap} = \frac{\Delta H_{vap}}{T_b}$$

The subscripts could as well indicate condensation.

The magnitude of the ΔH of vaporization depends on the strength of the intermolecular forces in the liquid. If the forces are strong, then ΔH is large. Stronger attractive forces also make for a higher boiling temperature. The upshot is that ΔS_{vap} is roughly constant for a wide variety of liquids. This is **Trouton's rule:**

$$\boldsymbol{\Delta S_{vap} \approx 88 \text{ J mol}^{-1}\text{K}^{-1}}$$

Dynamics of Liquids

Molecules in a liquid diffuse according to the same law given for gases:

$$\Delta r_{rms} = \sqrt{6Dt}$$

where Δr_{rms} is the-root-mean-square displacement of a molecule after time t, and D is the diffusion coefficient of the liquid. An important difference between diffusion in liquids and gases is that the diffusion coefficient for typical liquids is about 10 000 times smaller than for gases at STP, about 10^{-9} m^2 s^{-1}. Compare **problem 17-41** to **problem 3-76b.** Diffusion in gases is already slow. In liquids it is orders of magnitude slower.

The motion of a molecule in a liquid consists of random-walk diffusion upon which is superimposed a rattling motion as the molecule is trapped by intermolecular attractions in a temporary cage of its neighbors. An **encounter** between two molecules consists of a series of collisions within such a cage over a long period of time. Liquid phase encounters are much more productive of chemical reaction than simple gas phase collisions because the candidates for reaction have many more chances to interact just right. Moreover, there are plenty of near neighbors to carry away any excess energy. For **diffusion-controlled reactions** in liquids, the slow step in reaction is the step of diffusing together. See **problem 17-77**.

Hydrogen Bonding in Water and Ice

Hydrogen bonds are the strongest type of non-bonded intermolecular forces. They are found only in compounds containing hydrogen chemically bonded to nitrogen, oxygen or fluorine. The latter three elements are all strongly electronegative. The hydrogen bond is a strong dipole-dipole interaction linking the hydrogen atom of one molecule to an N, O, or F atom of a neighboring molecule. The strength of hydrogen bonds depends on the identity of the H-donor, the atom chemically bonded to the hydrogen, and the H-acceptor, the atom with lone pairs of electrons which interact with the hydrogen. Hydrogen bonds range in strength from 5 to 25 kJ mol^{-1}. This is roughly 10 percent of the strength of regular chemical bonds.

Hydrogen bonds cause many anomalies in the properties of substances in which they are prevalent. They are responsible for the abnormally open, low-density structure of ice (see **problem 17-81**) and for the abnormally high molar heat capacity of liquid water. These properties are profoundly influential in the biosphere.

Liquid Crystals

In many substances, particularly organic substances, the solid-to-liquid transition is not a single clear-cut change but rather a series of changes through intermediate phases. Materials in such intermediate states are **liquid crystals.** Liquid crystals flow like liquids, but there is a degree of statistical order in the distribution of their molecules. Liquid crystals usually form in substances having molecules with elongated rod-like shapes. They are solid-like in showing orientational ordering of these molecules, but liquid-like in the random distribution of the centers of the molecules. In the **nematic phase,** there is a preferred orientation for the long axis of the molecules.

In **smectic phases** there are other forms of microscopic order.

Detailed Solutions to Odd-Numbered Problems

17-1 The CCl_2F_2 molecule has two mirror planes. The first is defined by the two Cl atoms and the central C atom, and the second is defined by the two F atoms and the central C atom. The intersection of the two mirror planes coincides with a 2-fold axis of rotation. This axis passes through the central C atom and bisects the angles defined Cl—C—Cl and F—C—F.

17-3 a) The cell angles are all 90°, as required by the orthorhombic crystal system. The volume of the cell V_c is then just the product of the lengths of the three edges. It is 5.675×10^6 Å^3.
b) The volume of the box-shaped crystal is likewise the product of the lengths of the three edges. It is 3 mm^3—small, but still easily visible with the unaided eye. For comparison, convert V_c from the previous part to mm^3 by multiplying by 10^{-21} mm^3/Å^3. The result is 5.675×10^{-15} mm^3. The ratio of the volume of the crystal to the volume of the unit cell is the number of units cells in the crystal. It is 5×10^{14}, or 500 trillion.

17-5 a) The volume of the cubical unit cell in elemental silicon is just the edge of the cell cubed. It equals (5.431 Å)^3, which is 160.19 Å^3. There are 10^8 Å in a centimeter and consequently 10^{24} Å^3 in a cubic centimeter. Hence:

$$\left(\frac{160.19\ \text{Å}^3}{1\ \text{unit cell}}\right) \times \left(\frac{1\ \text{cm}^3}{10^{24}\ \text{Å}^3}\right) = 1.602 \times 10^{-22}\,\frac{\text{cm}^3}{\text{unit cell}}$$

b) The mass of the contents of the unit cell of silicon is the volume of the unit cell multiplied by its density:

$$\left(\frac{1.602 \times 10^{-22}\ \text{cm}^3}{\text{unit cell}}\right) \times \left(\frac{2.328\ \text{g Si}}{1\ \text{cm}^3}\right) = \frac{3.729 \times 10^{-22}\ \text{g Si}}{\text{unit cell}}$$

c) The unit cell contains eight atoms of silicon, which have the total mass just computed. Consequently, a single atom has a mass of 4.662×10^{-23} g.
d) One mole of silicon contains Avogadro's number of atoms of silicon. The molar mass of silicon is 28.0855 g mol^{-1}. Divide this by the mass of a single atom of silicon to obtain Avogadro's number:

$$\frac{28.0855\ \text{g mol}^{-1}}{4.662 \times 10^{-23}\ \text{g}} = 6.025 \times 10^{23}\ \text{mol}^{-1} = N_0$$

This is only about 0.05% larger than the accepted value.

17-7 The volume V_c of a unit cell of sodium sulfate equals the product of the three different cell edges. It is 708.47 Å^3, which is 7.0847×10^{-22} cm^3. The volume of a mole of unit cells of sodium sulfate is Avogadro's number times the volume of one cell:

$$(7.0847 \times 10^{-22}\ \text{cm}^3) \times (6.022 \times 10^{23}\ \text{mol}^{-1}) = 426.6\ \text{cm}^3\ \text{mol}^{-1}$$

The density of a unit cell equals the density of the substance itself since a crystal is just many unit cells stacked side by side. Multiplying the volume of a mole of unit cells by the density of the substance gives the mass of a mole of unit cells:

$$\left(\frac{426.6\ \text{cm}^3}{1\ \text{mol}}\right) \times \left(\frac{2.663\ \text{g}}{1\ \text{cm}^3}\right) = 1136.1\ \text{g mol}^{-1}$$

The molar mass corresponding to the formula Na_2SO_4 is only 142.04 g mol^{-1}, far short of the mass of a mole of unit cells (1136.1 g mol^{-1}) just obtained. There must be several formula units per unit cell. Because 142.04 is almost exactly 1/8th of 1136.1, it follows that there are 8 Na_2SO_4 formula units in every unit cell.

17-9 a) The body-centered cubic structure means 2 Fe atoms per unit cell, one in the center of the cell and 1 at each of the 8 corners of the cell (each corner atom is shared by 7 neighboring cells). Such Fe atoms are in contact along the body diagonal of the cell. They do not touch along the cell edges. The volume of the unit cell is the mass of its contents divided by its density. The density, as given in the problem, is 7.86 g cm^{-3}. The mass of the contents is 2×55.847 amu or 111.694 amu. Convert this mass from amu to grams by dividing by the factor 6.02214×10^{23} amu g^{-1}. The result, 1.8547×10^{-22} g, is the mass of the iron in one unit cell. The volume of the cell is its mass divided by its density. The volume comes out to 2.36×10^{-23} cm^3. The edge a of the cubic unit cell is the cube root of the volume. It is 2.87×10^{-8} cm, which is 2.87 Å. The nearest-neighbor distance is one-half the body diagonal b of the unit cell. The body diagonal is related to the edge as follows:

$$b = \sqrt{3}a = \sqrt{3}(2.87\ \text{Å}) = 4.97\ \text{Å}$$

Hence nearest neighbors are 2.48 Å apart.
b) The lattice parameter is 2.87 Å, the cubic cell's edge. See above.

17-11 a) A body-centered cubic lattice has two lattice points per unit cell. In metallic sodium, a single Na atom is associated with each lattice point so there are two Na atoms per cell.
b) Let r equal the radius of the Na atom. In the crystal, Na atoms touch along the body diagonal b of the cubic cell, which has atoms exactly at its corners and center. This means $4r = b$. But b is $\sqrt{3}$ times the edge of the cell. Hence $4r = \sqrt{3}a$. Cubing the equation gives:

$$64r^3 = 3\sqrt{3}a^3$$

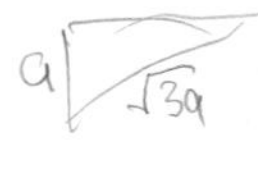

The volume of the cell V_c is a^3. The volume of a single Na atom is $4/3\pi r^3$, and, obviously, two Na atoms have twice this volume:

$$V_c = a^3 \quad \text{and} \quad V_{2\text{Na}} = 2 \times \left(\frac{4\pi r^3}{3}\right)$$

Solving these equations for a^3 and r^3 respectively and substituting into the equation that precedes them gives:

$$64\left(\frac{3V_{2\text{Na}}}{8\pi}\right) = 3\sqrt{3}V_c$$

Solving for the ratio of the two volumes gives:

$$\left(\frac{V_{2\text{Na}}}{V_c}\right) = \frac{(3\sqrt{3})(8\pi)}{3(64)} = 0.6802$$

17-13 The Bragg law $n\lambda = 2d\sin\theta$ becomes in this case:

$$2(1.660\ \text{Å}) = 2d\sin\left(\frac{54.70°}{2}\right) \quad \text{from which} \quad d = \frac{1.660\ \text{Å}}{\sin 27.35°} = 3.613\ \text{Å}$$

17-15 The Bragg law $n\lambda = 2d\sin\theta$ becomes:

$$4(1.936\ \text{Å}) = 2(4.950\ \text{Å})\sin\theta$$

where $n = 4$ follows from the specification of fourth-order diffraction and 4.950 Å is the interplanar spacing. Solving gives θ equal to 51.46° and 2θ equal to 102.9°. Students of trigonometry will recall that $\theta = 128.54°$ $(180 - 51.46°)$ also fulfills the above equation. This gives a second solution for 2θ: 257.1°, which is equivalent to −102.9°.

17-17 Solve the Bragg law for 2θ and substitute the values given for this case of diffraction by crystalline LiCl:

$$2\theta = 2\sin^{-1}\left(\frac{n\lambda}{2d}\right) = 2\sin^{-1}\left(\frac{n\,2.167\ \text{Å}}{22.570\ \text{Å}}\right) = 2\sin^{-1}(0.4216n)$$

Inserting integers for n gives 2θ equal to $\pm 49.87°$ for $n = 1$ and 2θ equal to $\pm 115.0°$ for $n = 2$. Higher values of n lead to arguments of $\sin^{-1}$ that exceed 1.00. The inverse sine function is not defined in such cases. Consequently 2θ has only the values listed.

17-19 **a)** Ion-ion, induced dipole and dispersion forces all contribute to the interactions in KF. Ion-ion forces predominate.

b) Dipole-dipole and dispersion forces contribute to interactions in HI; dipole-dipole forces predominate.

c) Dispersion forces only cause interactions in Rn.

d) Dispersion forces only cause interactions in N_2.

17-21 The H—F bond in hydrogen fluoride is polar. The charge distribution approximates that of a point charge, $-e\delta$ at the fluorine and another point charge, $+e\delta$ at the hydrogen atom where δ is 0.41, and e is the charge on the electron. When two HF molecules approach each other, there are four interactions: two attractive (H to F) and two repulsive (H to H and F to F). The pairs *within* the molecule are not counted.

Let R_1 and R_2 represent the H-to-H and F-to-F distances respectively, and let R_3 and R_4 represent the two H-to-F distances. The potential energy of the interaction between two HF molecules is the sum of the pair-wise Coulomb energies of interaction:

$$V = \left(\frac{e^2\delta^2}{4\pi\epsilon_0}\right)\left[\frac{1}{R_1} + \frac{1}{R_2} - \frac{1}{R_3} - \frac{1}{R_4}\right]$$

Attractions *lower* V and repulsions *raise* it. This explains the signs of the four terms in the brackets. Substitution of $e = 1.602 \times 10^{-19}$ C, $\epsilon_0 = 8.854 \times 10^{-12}$ C^2 J^{-1}m^{-1} and $\delta = 0.41$ gives:

$$V = \left(3.8774 \times 10^{-29} \text{ J m}\right)\left[\frac{1}{R_1} + \frac{1}{R_2} - \frac{1}{R_3} - \frac{1}{R_4}\right]$$

In the geometry of Figure 17-15a, the H-to-H and F-to-F distances (R_1 and R_2) are both 3.00 Å (3.00×10^{-10} m). The two H to F distances (R_3 and R_4) are 3.138×10^{-10} m. The latter result comes by applying the Pythagorean theorem to the right triangle defined by three of the four atoms. Substituting these R values gives V equal to $+1.1 \times 10^{-20}$ J. When the two HF molecules have the relative orientation shown in text Figure 17-15c, R_1 is 1.16×10^{-10} m, R_2 is 3.00×10^{-10} m, and R_3 and R_4 are both 2.08×10^{-10} m. These four distances are easily read from a sketch of the geometry. The term in brackets is $+0.2338 \times 10^{10}$ m^{-1}, and V is $+9.1 \times 10^{-20}$ J.

17-23 **a)** $BaCl_2$–ionic **b)** SiC–covalent **c)** CO–molecular **d)** Co–metallic.

17-25 In the simple cubic CsCl lattice, the positive ion has 8 Cl^- ions as nearest neighbors. Next comes a set of 6 more distant Cs^+ ions and then a set of 12 yet more distant Cs^+ ions. To obtain this answer, imagine a Cs^+ ion at the center of a cube with Cl^- ions on its eight corners. These are the nearest-neighbor Cl^- ions. The cube has 6 faces and 12 edges. The 6 second-nearest-neighbors are the Cs^+ ions at the centers of the 6 face-adjoining unit cells. The 12 Cs^+ ions are at the centers of the 12 edge-adjoining cells. One danger in this approach is getting bogged down using messy sketches to count neighbors and to decide which neighbors are nearer. A better way uses the formalism of *fractional coordinates.* Define an origin (0,0,0) at the Cs^+ ion. Then express the location of any neighbor in terms of its fractional coordinates. In fractional coordinates the unit of length is always taken to be the length of the cell edge. For instance there is an ion (a Cl^-) at (1/2,1/2,1/2). The cubic symmetry means that the x, y and z coordinates are equivalent and that the plus and minus directions on each coordinate are equivalent, too. Therefore permuting the fractional coordinates and changing the signs of the fractional coordinates generate equivalent locations. For $(1/2, 1/2, 1/2)$ the following sets of coordinates result from these operations:

$$\left(+\tfrac{1}{2},+\tfrac{1}{2},+\tfrac{1}{2}\right)\quad\left(-\tfrac{1}{2},+\tfrac{1}{2},+\tfrac{1}{2}\right)\quad\left(+\tfrac{1}{2},-\tfrac{1}{2},+\tfrac{1}{2}\right)\quad\left(+\tfrac{1}{2},+\tfrac{1}{2},-\tfrac{1}{2}\right)$$
$$\left(-\tfrac{1}{2},-\tfrac{1}{2},-\tfrac{1}{2}\right)\quad\left(+\tfrac{1}{2},-\tfrac{1}{2},-\tfrac{1}{2}\right)\quad\left(-\tfrac{1}{2},+\tfrac{1}{2},-\tfrac{1}{2}\right)\quad\left(-\tfrac{1}{2},-\tfrac{1}{2},+\tfrac{1}{2}\right)$$

The eight locations are the eight equivalent nearest-neighbors of the Cs^+ ion previously identified. A bonus in this approach is that it is easy to compute interionic distances. The distance of the 8 equivalent Cl^- ions from the central Cs^+ ion is the square root of the sum of the squares of the three coordinates. In this case the distance comes out $\sqrt{3}/2$ times a, the edge of the unit cell.

17-27 Substitute the values specific to the case of $RbCl(s)$ into the formula for the Coulomb energy of a crystal:

$$V = -\frac{MN_0e^2}{4\pi\epsilon_0 R_0}$$

This formula is implied on text page 655. In it where M is the Madelung constant and R_0 is the distance between neighboring Rb^+ and Cl^- ions. The M that is given in the problem is for the rock-salt structure which is the structure adopted by $RbCl(s)$; R_0 is approximately the sum of the radii of the two ions, which equals 3.29 Å, or 3.29×10^{-10} m. The result is:

$$V = -\frac{(1.7476)(1.602 \times 10^{-19}\ \text{C})^2(6.022 \times 10^{23}\ \text{mol}^{-1})}{4\pi(8.854 \times 10^{-12}\ \text{C}^2\,\text{J}^{-1}\text{m}^{-1})(3.29 \times 10^{-10}\ \text{m})} = -738 \times 10^3\ \text{J mol}^{-1}$$

Reducing the magnitude of the Coulomb energy by 10% gives -664 kJ mol^{-1} for the lattice energy of $RbCl(s)$. The dissociation energy is the negative of the lattice energy. Hence $+664$ kJ are required to dissociate one mole of $RbCl(s)$ to gaseous ions. The experimental value is just slightly larger (680 kJ).

17-29 a) The lattice energy of $LiF(s)$ is the energy change of the reaction:

$$Li^+(g) + Cl^-(g) \rightarrow LiF(s) \qquad \text{lattice energy} = \Delta E$$

Direct experimental measurement of this ΔE is not possible. The Born-Haber cycle is a series of lesser steps taking place at 25°C that add up to the above change. The ΔE of each step *can* be measured. The steps are: **1,** the transfer of electrons from $F^-(g)$ ions to $Li^+(g)$ ions to give neutral gaseous atoms: **2,** the condensation of $Li(g)$ to $Li(s)$ and association of $F(g)$ to $F_2(g)$; **3,** the reaction of $Li(s)$ and $F_2(g)$ in their standard states to give LiF(s). Applying the first law of thermodynamics to this cycle:

$$\Delta E = \Delta E_1 + \Delta E_2 + \Delta E_3$$

Begin with ΔE_1, which is the electron affinity of F(g) *minus* the first ionization energy of Li(g):

$$\Delta E_1 = 328 - 520 = -192 \text{ kJ mol}^{-1}$$

The negative result means that energy is released to the surroundings by the electron transfer. Next, ΔE_2 is the energy change accompanying the condensation of Li(g) to Li(s) plus the energy change accompanying the association of F(g) to $F_2(g)$. The $\Delta H°$'s for these processes (text Appendix D) are -159.37 kJ mol^{-1} and -78.99 kJ mol^{-1}, respectively. Although the tabulated $\Delta H°$'s are not the same as the desired ΔE's, they are related as follows:

$$\Delta E = \Delta H° - RT\Delta n_{\text{g}} = \Delta H° - 2.48 \text{ kJ mol}^{-1}\Delta n_{\text{g}} \qquad (\text{at } T = 298 \text{ K})$$

For Li(g) $\rightarrow$ Li(s), Δn_{g} is -1. It follows that:

$$\Delta E_{\text{condense}} = -159.37 \text{ kJ mol}^{-1} - (2.48)(-1) = (-159.37 + 2.48) \text{ kJ mol}^{-1}$$

For the reaction F(g) $\rightarrow$ 1/2 $F_2(g)$, Δn_{g} is $-1/2$. Therefore, for this change:

$$\Delta E_{\text{react}} = -78.99 \text{ kJ mol}^{-1} - (2.48)(-1/2) = (-78.99 + 1.24) \text{ kJ mol}^{-1}$$

Adding the two parts of step 2:

$$\Delta E_2 = (-159.37 + 2.48) + (-78.99 + 1.24) = -234.64 \text{ kJ mol}^{-1}$$

The third step in the cycle is the reaction of one mole of Li(s) and one-half mole of $F_2(g)$ to give one mole of LiF(s). In Appendix D, the $\Delta H°$ for this reaction is listed as the standard enthalpy of formation of LiF(s) and equals -615.97 kJ mol^{-1}. Again, there is a small correction to obtain ΔE. This time, Δn_{g} is $-1/2$:

$$\Delta E_3 = \Delta H_3^\circ - RT\Delta n_{\text{g}} = -615.97 - (2.48)(-1/2) = -614.73 \text{ kJ mol}^{-1}$$

At this point, ΔE values for all three steps in the Born-Haber cycle are available. Add them up:

$$\Delta E_{\text{cycle}} = -192 - 234.64 - 614.73 = -1041 \text{ kJ mol}^{-1}$$

If the $\Delta n_{\text{g}}RT$ corrections in step 2 and 3 are omitted, the answer comes out to be $2RT$ more negative. It then equals -1046 kJ mol^{-1}. The omission causes less than a 1% difference. In view of the experimental uncertainty of many ΔE values, taking ΔE to equal ΔH is readily defensible. The lattice energy is -1041 kJ mol^{-1}.

b) The computation of the Coulomb energy for LiF follows the pattern of problem 17-27:

$$V = -\frac{MN_0e^2}{4\pi\epsilon_0 R_0}$$

where M is the Madelung constant and R_0 is the distance between neighboring positive and negative ions. For lithium fluoride R_0 is 2.014 Å, or 2.014×10^{-10} m. The ionic radii of Li^+ and F^- ions are 0.68 and 1.33 (text Appendix F) The ratio of these radii is 0.51, which confirms that LiF(s) adopts the rock-salt structure. Choose the rock-salt M from Table 17-6 (text page 655). Then:

$$V = -\frac{(1.7476)(1.602 \times 10^{-19}\ \text{C})^2(6.022 \times 10^{23}\ \text{mol}^{-1})}{4\pi(8.854 \times 10^{-12}\ \text{C}^2\,\text{J}^{-1}\text{m}^{-1})(2.014 \times 10^{-10}\ \text{m})} = -1205 \times 10^3\ \text{J mol}^{-1}$$

The calculated Coulomb energy is about 15 percent larger than the experimental (Born-Haber) lattice energy. The discrepancy arises because the Coulomb calculation ignores non-ionic interactions.

17-31 The presence of Frenkel defects will not change the density of a crystal by a significant amount, because the vacancies at lattice sites are compensated for by interstitial atoms. In large numbers Frenkel defects might cause a small bulging of the crystal and consequent decrease in its density.

17-33 A sample of, say, 100 g of this sample of iron(II) oxide contains 76.55 g of Fe and 23.45 g of O. This corresponds to 1.3707 mol of Fe and 1.4657 mol of O. Dividing one by the other gives the formulas $FeO_{1.0693}$ or $Fe_{0.9352}O$. It is improper to round off to the stoichiometric formula FeO. The experimental analysis is precise to four significant figures, and the chemical formula should have the same precision.

b) Let a equal the fraction of sites occupied by Fe^{3+} ions and b equal the fraction of sites occupied by Fe^{2+} ions. The Fe^{3+} ions that occur in Fe^{2+} sites compensate with their extra charge for missing Fe^{2+} ions elsewhere and make the compound as a whole electrically neutral. The average positive charge per site must be 2. Also the sum of a and b is 0.9352, as shown by the empirical formula. In equation form this means:

$$3a + 2b = 2 \quad \text{and} \quad a + b = 0.9352$$

Solution of these simultaneous equations gives $a = 0.1296$. This is the fraction of sites occupied by Fe^{3+} ions. The fraction of the iron in the +3 state is the fraction of sites having +3 iron divided by the fraction having iron of either kind: $0.1296/0.9352 = 0.1386$.

17-35 Dispersion forces increase with molar mass in a given series; the stronger forces in bromine make it condense at a higher temperature than fluorine or chlorine. Iodine condenses at a still higher temperature and is a solid at room temperature.

17-37 Applying the statement in the problem, $SnBr_4$ is expected to have a lower boiling point than $SnBr_2$. The molecular substance Br_2 should have the lowest boiling point of the three.

17-39 Substances with the strongest intermolecular forces require the highest temperature to make them boil. Liquid RbCl has strong electrostatic forces holding its ions together. It has the highest boiling point. Liquid NH_3 has dipole-dipole attractions as does liquid NO. In NH_3 these are particularly strong. They are hydrogen bonds. In NO the dipole-dipole attractions are weaker. Liquid NH_3 boils at a higher temperature than liquid NO. Weak dispersion (van der Waals) forces are the only intermolecular attraction in liquid neon. It has the lowest boiling point of all: $Ne < NO < NH_3 < RbCl$

17-41 **a)** The root-mean-square displacement by diffusion after time t is $r_{rms} = \sqrt{6Dt}$. The diffusion constant D for water is given as 2.3×10^{-9} m^2 s^{-1}. Substitution of t equal to 1 s, gives r_{rms} equal to 1.2×10^{-4} m.
b) After 1 day (8.64×10^4 s), r_{rms} is 0.035 m. The calculation shows just how slow diffusion is in liquids. After a day at room temperature an average water molecule moves less than 1 1/2 inches from its starting position.

17-43 The fact that the molecule is a discrete tetramer suggests that its structure is cyclic. Why would a straight chain of molecules stop at exactly four links? The ring is probably formed by hydrogen bonding of the hydroxide hydrogen atom with the oxygen atom of a neighboring molecule. It would consist of 8 alternating H and O atoms.

17-45 We expect hydrazine to have a higher boiling point, higher heat capacity, and higher viscosity as a liquid than acetylene. The two have comparable molar masses, but there is considerable hydrogen bonding in hydrazine (N—H···N), and none in acetylene.

17-47 Ethane will be the least soluble in water because its molecule is non-polar and there is no possible hydrogen bonding to the solvent molecules. Ethanol would be the most soluble with abundant $—OH\cdots OH_2$ hydrogen bonds. Chloroform would be intermediate in its solubility. Its molecule is polar, but it does not form hydrogen bonds.

17-49 The entropy of the isotropic liquid phase exceeds the entropy in the smectic liquid crystal phase of a substance. Compare the degrees of order apparent in text Figure 17-40a and 17-40b. The enthalpy of the isotropic liquid phase exceeds that of the smectic phase because heating the smectic phase converts it to the isotropic phase.

17-51 In diamond the C—C bond distance will be the distance between any two nearest-neighbor atoms. Reviewing the list of coordinates given in the problem shows one such pair of atoms is the C at $(0,0,0)$ and the C at $(1/4,1/4,1/4)$. This is also clear in Figure 17-25 in the text. Other pairs of carbons are equally close but none is closer. These carbons are separated by one-fourth of the body diagonal of the unit cell. The body diagonal is $\sqrt{3}$ times the edge of the cell or $3.57\sqrt{3}$ Å. The bond distance is 1/4 of this or 1.55 Å.

17-53 a) The cell is monoclinic so two of the three cell angles are automatically equal to 90°.

b) The volume of the cell is:

$$V_c = abc\sqrt{1 - \cos^2\alpha - \cos^2\beta - \cos^2\gamma + 2\cos\alpha\cos\beta\cos\gamma}$$

Because both α and γ are 90°, this becomes (with the use of the trigonometric identity $\sin^2\beta + \cos^2\beta = 1$):

$$V_c = abc\sqrt{1 - \cos^2\beta} = abc\sin\beta = (11.04)(10.98)(10.92)\sin 96.73° = 1314.6\ \text{Å}^3$$

The volume equals 1.3146×10^{-21} cm^3. The density is then computed as follows:

$$\rho = \frac{n_c\mathcal{M}}{N_0 V_c} = \frac{48(32.066\ \text{g mol}^{-1})}{(6.022 \times 10^{23})(1.3146 \times 10^{-21}\ \text{cm}^3)} = 1.944\ \text{g cm}^{-3}$$

17-55 Any tetrahedral interstitial site can be viewed as occupying the center of a cube that has every other one of its eight corners occupied by spherical atoms of radius r_1. Let the edge a of such a cube have length 1. Then the face diagonal f has length $\sqrt{2}$ and the body diagonal b has length $\sqrt{3}$. The four atoms at the alternate corners surround the center and touch each other along the face diagonals of the cube. Therefore:

$$2r_1 = \sqrt{2}$$

Let r_2 be the radius of a spherical atom placed at the interstitial site, the center of the cube. The largest such atom will just touch all four atoms at the corners. The body diagonal in that case equals the sum of the diameters of the two atoms:

$$2r_1 + 2r_2 = b = \sqrt{3}$$

Dividing the second equation by the first gives:

$$\frac{(r_1 + r_2)}{r_1} = \frac{\sqrt{3}}{\sqrt{2}} \quad \text{hence} \quad 1 + \frac{r_2}{r_1} = 1.225$$

Since r_2/r_1 is 0.225, the largest possible value for r_2 is 0.225 times r_1.

17-57 The Bragg law $n\lambda = 2d\sin\theta$ becomes, in this case of first-order diffraction of water waves:

$$1(3.00 \text{ m}) = 2(5.00 \text{ m}) \sin\theta$$

Solving for θ gives 17.46° and 162.54°, so 2θ is ±35°.

17-59 a) The unit-cell volume V_c of NaCl is the cell edge cubed or 179.43 Å^3.
b) The volume V_p of the primitive unit cell of NaCl is one-fourth of the volume of the conventional unit cell or 44.856 Å^3.
c) For relatively long wavelength (2.2896 Å) x-rays:

$$N_{\text{beams}} = \frac{4}{3}\pi\left(\frac{2}{\lambda}\right)^3 V_p = \frac{4}{3}\pi\left(\frac{2}{2.2896 \text{ Å}}\right)^3 (44.856 \text{ Å}^3) = 125$$

d) For short wavelength (0.7093 Å) x-rays there are far more diffracted beams.

$$N_{\text{beams}} = \frac{4}{3}\pi\left(\frac{2}{\lambda}\right)^3 V_p = \frac{4}{3}\pi\left(\frac{2}{0.7093 \text{ Å}}\right)^3 (44.856 \text{ Å}^3) = 4212$$

17-61 In a diatomic molecule the dipole moment and the polarity of the bond are simply related (see text page 537):

$$\mu \text{ (D)} = \frac{R \text{ (Å)}}{(0.2082 \text{ Å D}^{-1})}\delta$$

where δ is the dipole moment in debye, R is the internuclear distance in Å, and δ is the fraction of the charge of the electron residing at each atom. In HBr, R is 1.414 Å and μ is 0.78 D, making δ equal to 0.115. The problem shows two HBr molecules in a geometry in which the bond axis of one molecule is cocked by 30° relative to the bond axis of the other. Number the atoms of the two molecules 1 (H_1), 2 (Br_1), 3 (H_2), and 4 (Br_2). The distance between the centers of the molecules is 5.000 Å. The H_1-to-H_2 distance exceeds 5.000 Å, and the Br_1-to-Br_2 distance is less than 5.000 Å by an equal amount. There are many ways to get the four R_{ij}, the distances from all the atoms of the first molecule to all the atoms of the second. One is to construct a Cartesian coordinate system with its origin midway between the molecular centers and with its x axis parallel to the bond in one molecule. The (x, y) coordinates (in Å) of the atoms in such a system are:

H_1 (−0.7070, −2.500) Br_1 (0.7070, −2.500)
H_2 (−0.6123, 2.8535) Br_2 (0.6123, 2.1465)

Note that 0.6123 is cos 30° times one-half the HBr bond distance. Also, the average of 2.1465 and 2.8535 is 2.5000. The numerical coordinates thus reflect the symmetry of the layout. The distance between any two atoms is:

$$R = \sqrt{(x_2 - x_1)^2 + (y_2 - y_1)^2}$$

Four interatomic distances (R_{13}, R_{14}, R_{23}, and R_{24}) are needed. They come out:

R_{14} (H_1–Br_2)	4.830 Å	R_{13} (H_1–H_2)	5.354
R_{24} (Br_1–Br_2)	4.647 Å	R_{23} (Br_1–H_2)	5.514

The Coulomb energy of interaction of the two HBr molecules is now:

$$V = \frac{e^2\delta^2}{4\pi\epsilon_0}\left(-\frac{1}{R_{14}} + \frac{1}{R_{13}} + \frac{1}{R_{24}} - \frac{1}{R_{23}}\right)$$

This is the highlighted equation on text page 642 of the text specialized to this case. The δ is 0.115; e and ϵ_0 are 1.602×10^{-19} C and 8.854×10^{-12} C^2 J^{-1} m^{-1} respectively. Inserting these values and the four R_{ij} (expressed in meters) gives:

$$V = 3.05 \times 10^{-30} \text{ J m}\left(-\frac{10^{10}}{4.830 \text{ m}} + \frac{10^{10}}{5.354 \text{ m}} + \frac{10^{10}}{4.647 \text{ m}} - \frac{10^{10}}{5.514 \text{ m}}\right) = 3.1 \times 10^{-20} \text{ J}$$

The positive V means that in this geometry the two molecules repel each other. If the second molecule were cocked by 30° the *other* way, so that the H atoms of the molecules were close, V would be the same. If either molecule were turned end for end, then V would be negative, and the two would attract.

17-63 a) The data for the plot are in the fourth and fifth columns of the table in the next part. The values for b are directly from Table 3-3 (text page 116). The values for $N_0\sigma^3$ have been converted from units of Å^3 mol^{-1}, which are obtained from direct combination of the data in Table 17-3 (text page 645), to L mol^{-1} for better comparison. The required conversion is division by 10^{27} Å^3 per liter. Study of the two sets of data reveals a strong correlation between b and $N_0\sigma^3$: the ratio of b to $N_0\sigma^3$ lies in a range between 1.17 to 1.36, except for the case of H_2.

b) The usual unit of the van der Waals constant a is atm L^2 mol^{-2}. Rewrite this as L atm L mol^{-2}. Both the joule and the L atm are units of energy. In fact, 1 L atm equals 101.325 J. Also, 1 L equals 10^{-3} m^3. Therefore, one atm L^{2}mol^{-2} equals 0.101325 J m^3 mol^{-2}. The values of a have been converted to this unit in the following table. Combine the two Lennard-Jones constants to form the product $\epsilon\sigma^3$. The motivation for doing this is that the units of this particular combination can be converted from J Å^3 to J m^3. Then, multiplication by N_0^2 (Avogadro's number squared) to give $\epsilon\sigma^3 N_0^2$ furnishes a quantity having the units J m^3 mol^{-2}, the same units as those of a. The data from Table 3-3 and Table 17-3 are converted to these units in the second and third columns in the following table:

Gas	a (J m^3 mol^{-2})	$\epsilon\sigma^3 N_0^2$ (J m^{3}mol^{-2})	b (L mol^{-1})	$N_0\sigma^3$ (L mol^{-1})
Ar	0.13628	0.0236	0.03219	0.0237
H_2	0.02476	0.00466	0.02661	0.0151
CH_4	0.22829	0.0414	0.04278	0.0336
N_2	0.14084	0.0241	0.03913	0.0305
O_2	0.13780	0.0270	0.03183	0.0273

The two sets of data are *not* the same, but there is a strong correlation between a and $\epsilon\sigma^3 N_0^2$ in their variation from one gas to the next. The ratio of the two stays within a fairly narrow range (from 5.1 to 5.8) for the five gases.

17-65 a) Use the deBroglie relation to obtain the wavelength of the neutrons:

$$\lambda = \frac{h}{mv} = \frac{6.626 \times 10^{-34} \text{ J s}}{(1.6750 \times 10^{-27} \text{ kg})(2.639 \times 10^{3} \text{ m s}^{-1})} = 1.499 \times 10^{-10} \text{ m}$$

b) The edge length of the unit cell is the interplanar spacing of the planes doing the scattering. Compute it by solving the Bragg law and d and substituting:

$$a = d = \frac{n\lambda}{2\sin\theta} = \frac{2(1.499 \times 10^{-10} \text{ m})}{2\sin(36.26°/2)} = 4.817 \times 10^{-10} \text{ m} = 4.817 \text{ Å}$$

c) In the rock-salt structure, the Na^+ ions and the H^- ions touch along the edges of the unit cell. The Na^+ ions occupy five sites on each face of the cell, forming a pattern like the pattern of five dots on the face of die. Four H^- ions also lie in each face. They are on edges exactly in between the Na^+ ions at the four corners. The distance from the center of an Na^+ ion to the center of the nearest neighbor H^- is therefore one-half of the edge of the unit cell. This equals 2.409 Å.
d) As established in slightly different words in the preceding, the edge a of the unit cell is the sum of two Na^+ radii and two H^- radii:

$$2r_{H^-} + 2r_{Na^+} = a = 4.817 \text{ Å}$$

Substitution of r_{Na^+} equal to 0.98 Å gives $r_{H^-} = 1.43$ Å.

17-67 According to the equations developed on text page 654, the energy and intermolecular distance in a face-centered-cubic molecular crystal depend on the Lennard-Jones parameters for the molecules comprising the crystal:

$$R_0 \approx 1.09\sigma \quad \text{and} \quad V_{tot} \approx -8.61\epsilon N_0$$

where σ is the first Lennard-Jones parameter, R_0 is the equilibrium spacing (at 0 K), V_{tot} is the total potential energy of the lattice, and ϵ is the other Lennard-Jones parameter. For N_2, σ is 3.70 Å. (Table 17-3). Therefore, R_0 is about 4.03 Å.
For N_2, ϵN_0 is 0.790 kJ mol^{-1}. The potential energy of the lattice is accordingly -6.80 kJ mol^{-1}. This is a reasonable estimate of the lattice energy.
b) The density of a crystal is related to the volume of its unit cell by:

$$\rho = \frac{n_c \mathcal{M}}{N_0 V_c}$$

For $N_2(s)$, ρ is 1.026 g cm^{-3}. The crystal has four N_2 molecules per unit cell and each molecule has a mass of 28.014 g mol^{-1}. Solve the preceding for V_c and substitute:

$$V_c = \frac{n_c\mathcal{M}}{N_0\rho} = \frac{4(28.014 \text{ g mol}^{-1})}{6.022 \times 10^{23} \text{ mol}^{-1}(1.026 \text{ g cm}^{-3})} = 181.36 \times 10^{-24} \text{ cm}^3$$

The edge of the cubic cell is the cube root of the volume of the cell. It equals 5.660×10^{-8} cm (5.660 Å). In a face-centered cubic lattice there is a nitrogen molecule at the center of every face of the unit cell and at every corner. The face diagonal is $5.660\sqrt{2}$ Å or 8.005 Å long. One-half of this is the distance from an N_2 at a face center to an N_2 at a face corner. This, the intermolecular distance, is 4.002 Å. This result is only about 0.7 percent less than the distance computed using the Lennard-Jones parameter. The agreement tends to confirm the analysis in text section 17-4.

17-69 Sodium chloride is an ionic solid. If there are Schottky defects, a fraction of the Na^+ sites is vacant. To maintain electrical neutrality an equal fraction of the Cl^- sites must be vacant. The density of defect-free NaCl is 2.165 g cm^{-3}. Introducing 0.0015 mole fraction of Schottky defects reduces the chemical amount of NaCl per cm^3 to 0.9985 of what had been. Therefore, the mass of NaCl per cm^3 is 0.9985 of what it had been, or 2.162 g cm^{-3}.
Frenkel defects involve displacement from a regular lattice site to an interstitial site. No mass is removed from the crystal, so the density stays at 2.165 g cm^{-3} as long as the volume of the crystal is not changed.

17-71 a) The compound is 28.31 percent O and therefore 71.69 percent Ti. In 100 g of the compound there is 1.4973 mol of Ti and 1.7694 mol of O. The formula is $Ti_{0.8462}O$ where the subscript is the ratio of 1.4973 to 1.7694.
b) Only 0.8462 of the stoichiometric quantity of Ti is present; 0.1538 of the total Ti^{2+} sites then must be vacant. Let a equal the fraction of Ti^{2+} sites with a Ti^{3+} occupying them, and b the fraction of sites with a Ti^{2+}. Clearly: $a + b = 0.8462$. The net positive charge per oxygen must be +2. Each Ti^{3+} contributes +3 and each Ti^{2+} contributes +2. Electrical neutrality requires $3a + 2b = 2$. Solution of the equations in a and b gives b equal to 0.5386 and a equal to 0.3026. Just under 31 percent of the Ti^{2+} sites contain a Ti^{3+} ion.

17-73 Imagine a quantity of a monatomic liquid of the type mentioned in the problem boiling at the temperature T_b. Assume that the vapors formed in contact with the liquid are ideal. According to Trouton's rule, the ΔH of the vaporization is $\Delta H_{vap} = 88T_b$ J mol^{-1}. In vaporization, Δn_g is +1. Insert this and the previous relation into the equation $\Delta H = \Delta E + \Delta n_g RT$ to obtain:

$$\Delta E = 88T_b - RT_b$$

The quantity ΔE is the total energy of the liquid subtracted from the total energy of the gas. For each phase, the total energy is the sum of the kinetic energy, KE, and the potential energy, V. Therefore:

$$\Delta E = \big(KE(g) + V(g)\big) - \big(KE(l) + V(l)\big)$$

The kinetic energies of the gas and the liquid are both equal to $3/2RT_\text{b}$ (see text page 106), since the Maxwell-Boltzmann distribution of molecular kinetic energies applies to both phases. Inserting this into the previous equation and taking the potential energy of the gas, $V(g)$, to be zero gives:

$$-V(l) = \Delta E = 88T_\text{b} - RT_\text{b}$$

The $KE(l)$ is, as just stated, equal to $3/2RT_\text{b}$. The ratio of the potential energy of the liquid to its kinetic energy is:

$$\frac{V(l)}{KE(l)} = \frac{-(88T_\text{b} - RT_\text{b})}{3/2\,RT_\text{b}} = \frac{-(88 - R)}{3/2\,R}$$

Taking R as 8.315 J $\text{mol}^{-1}\text{K}^{-1}$ (so that its units cancel the units of the Trouton constant) gives -6.4 as the numerical value for the ratio.

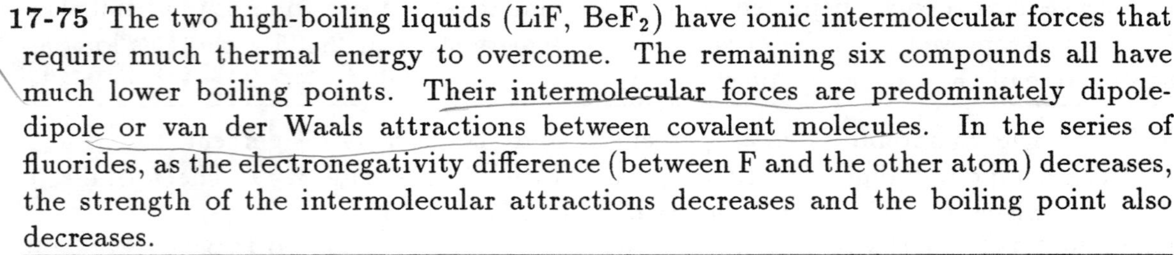

17-75 The two high-boiling liquids (LiF, BeF_2) have ionic intermolecular forces that require much thermal energy to overcome. The remaining six compounds all have much lower boiling points. Their intermolecular forces are predominately dipole-dipole or van der Waals attractions between covalent molecules. In the series of fluorides, as the electronegativity difference (between F and the other atom) decreases, the strength of the intermolecular attractions decreases and the boiling point also decreases.

17-77 Suppose that the concentrations of $H_3O^+(aq)$ and of $OH^-(aq)$ are 1.0 M. Then there are 6.022×10^{23} H_3O^+ ions and an equal number of OH^- ions in a liter of solution. A liter is 10^{-3} m^3 so there are $2 \times 6.022 \times 10^{26}$ ions per cubic meter of solution. It follows that, on the average, every ion "owns" a volume of 8.303×10^{-28} m^3. If this volume is spherical, the radius of the sphere is 5.831×10^{-10} m (computed using $V = 4/3\pi r^3$). The mean separation of H_3O^+ and OH^- ions is clearly the sum of the radii of the spheres surrounding each or 11.7×10^{-10} m. This is 11.7×10^{-8} cm. Make the suggested approximation that this distance is the root-mean-square distance r_rms that the ions traverse by diffusion to collide. It is known in liquid water that:

$$r_\text{rms} = \sqrt{6Dt} = \sqrt{6(2.3 \times 10^{-5}\ \text{cm}^2\,\text{s}^{-1})t}$$

Inserting r_rms and solving for t gives the root-mean-square time initially required for diffusion before collision. It is 9.9×10^{-11} s. In the instant that the reaction starts, the number of collisions per second by an average pair of ions is the reciprocal of this diffusion time or 1.0×10^{10} s^{-1}. There is a mole of pairs in the solution. Assume that every collision is a productive reaction event. Then, as the reaction starts, H_3O^+ and OH^- pairs diffuse toward each other throughout the bulk of the 1 L of solution, producing water at the initial rate of 1.0×10^{10} mol L^{-1}s^{-1}. The reaction is second-order:

$$\text{rate} = k[H_3O^+][OH^-]$$

Substitution of the initial values of the rate and of the two concentrations gives $k = 1.0 \times 10^{10}$ L mol^{-1}s^{-1}.

17-79 **a)** Among the Group IV hydrides the enthalpy change of sublimation, ΔH_{subl}, increases by about 4.3 kJ mol^{-1} with each step going down the periodic table:

CH_4, 8.4; SiH_4 ?; GeH_4, 17; SnH_4, 21 (all in kJ mol^{-1})

The expected value for SiH_4 would be about 12.7 kJ mol^{-1}. The two Group VI hydrides H_2Te and H_2Se have ΔH_{subl} 5 to 7 kJ mol^{-1} larger than the Group IV elements in the same row. Furthermore, the Group VI ΔH_{subl}'s increase going down the table by 6 kJ mol^{-1} from H_2Se to H_2Te. Extrapolating this trend upward would give a ΔH_{subl} of 10 kJ mol^{-1} for H_2O. Extrapolating across the chart (from CH_4) would give ΔH_{subl} of 14 kJ mol^{-1}. Estimate water's hypothetical ΔH_{subl} as midway between these two: 12 kJ mol^{-1}.

b) The experimental ΔH_{subl} of ice is 51 kJ mol^{-1}. Presumably, 39 kJ mol^{-1} of this is due to hydrogen bonds. Each O atom in a mole of ice is hydrogen-bonded to 2 other O atoms through its two hydrogens. The ratio is 2 mol of H-bonds per mole of ice. The estimated energy of an H-bond in ice is therefore half of 39, that is, about 20 kJ mol^{-1}.

17-81 Applying the rule of thumb assigns each water molecule a volume of 18 Å^3. The mass of a water molecule is 18.02 amu so the density of water would be 18.02 amu/18 Å^3, which is 1.0 amu Å^{-3}. Convert this density to g cm^{-3}:

$$\frac{1.0 \text{ amu}}{1 \text{ Å}^3} \times \left(\frac{1 \text{ g}}{6.022 \times 10^{23} \text{ amu}}\right) \times \left(\frac{10^{24} \text{ Å}^3}{1 \text{ cm}^3}\right) = 1.7 \text{ g cm}^{-3}$$

The density based on the rule of thumb is much higher than the actual density of solid water (0.90 g cm^{-3}). The rule of thumb fails in this case because hydrogen bonding in ice maintains a very open structure.

Chapter 18

Chemical Processes

The Chemical Industry

In a **process-based** approach to chemistry, the interaction of the complete range of chemical principles is considered with respect to the operation of practical industrial processes that transform starting materials to desired products. Chemical processes required careful development from their original laboratory scale through a series of **pilot-plant** stages to a final production status. This also often involves a changeover from a **batch** to a **continuous** process, if the latter is feasible and more economical. See **problem 18-1.** Starting materials should be easily accessible and transportable. Most starting materials have low free energies, a fact that gives particular important to the existence of high-free-energy materials like hydrocarbons from petroleum. The best sources of the most important elements should be learned (see **problem 18-3).** Desired reactions for which ΔG is positive can be made to proceed by linking them to other reactions for which ΔG is negative.

An ideal chemical process gives the desired product in 100% yield, at an acceptable and controllable rate, and in acceptable purity or in a form that is easily purified. The proper disposal of waste is an integral part of the design of chemical processes. Chemical processes often involve many steps and intermediate chemical compounds. Most large-volume chemicals are little used by consumers, but they are essential in making the products that consumers do use.

The products of a practical process must be obtained at a sufficient yield, at a sufficient rate and either pure enough to use immediately or readily purified. Thermodynamics imposes fundamental limitations on the yield of chemical reactions. Pressure and temperature must be selected accordingly. But practical processes may not proceed arbitrarily slowly. Compromises sometimes must be made between the best conditions based on thermodynamics and conditions under which the process will go at an acceptable rate. Much effort goes into designing catalysts to speed up thermodynamically favored processes enough to make them practical. Finally, if side-reactions contaminate the desired product with by-products that are dangerous to handle or hard to remove or both, the conditions of the process may again have to be modified to eliminate the troublesome by-products.

Geochemical and Biochemical Processes

There are lessons for the chemical industry in a study of natural geochemical and biochemical processes. Geochemical processes are characteristically slow because they are frequently solid-state reactions Biochemical processes are notable for their speed and selectivity. Biochemical catalysts, called **enzymes**, are unmatched for their ability to advance specific reactions, and their study provides insights into the design of industrial catalysts. Important **natural products** are now often made in chemical processes, as well as in processes that deliberately use specialized living agents like bacteria. Probably the most important chemical raw materials that come from biological activity are the fossil fuels coal, petroleum, and natural gas. Deposits of phosphate rock and some deposits of sulfur also have their origins in biological activity.

Hydrogen

The production and uses of hydrogen exemplify process-based chemistry. The element is abundant in water, but the free energy of water is too low for economically direct conversion to gaseous hydrogen (with oxygen as by-product. Instead, the major source is the **reforming reaction** between water and methane (or another hydrocarbon) from natural gas (see **problem 18-7).** This reaction produces **synthesis gas,** which is a mixture of hydrogen and carbon monoxide. Synthesis gas can be used as a mixture for a variety of syntheses. Further reaction of the carbon monoxide in synthesis gas with water in the exothermic **shift reaction** gives more hydrogen and some carbon dioxide, which finds uses of its own. See **problem 18-8.** Many of the uses of hydrogen are **captive**—that is, occurring directly after production. Hydrogen is used to make ammonia, methanol, or (in direct-reduction processes), to convert iron ore to iron. Another use is in the **hydrogenation** of **unsaturated** hydrocarbons, in which carbon-carbon double bonds are reduced to single bonds, and the compound is thereby saturated. Hydrogenation is applied to petroleum products as well as to edible oils (see **problem 18-11).**

Detailed Solutions to Odd-Numbered Problems

18-1 A batch process is one in which a quantity of reactants is mixed and allowed to react. The products are then removed. A continuous process is one in which the reactants are continuously added to the reaction vessel and the products are continuously removed. A continuous process requires more than one opening in the reaction vessel; a batch process can take place in a vessel with only one opening.

18-3 a) Sulfur is found in useful amounts in salt domes under the surface of the earth. The deposits were produced when calcium sulfate reacted with natural gas and carbon dioxide. Interestingly, the reaction is catalyzed by enzymes that are secreted by bacteria: $CaSO_4(s) + CH_4(g) + CO_2(g) \rightarrow CaCO_3(s) + S(s) + 6\,H_2O(l)$.
b) A useful source of carbon is bituminous coal. Coal is formed from plant matter by bacterial reduction, heat, and pressure.
c) The most useful source of phosphorus is in phosphate rock, which is nearly pure $Ca_5(PO_4)_3F$.
d) A good source of calcium is in limestone or marble, both of which are principally composed of calcium carbonate.

18-5 A good source of small amounts of pure hydrogen is the reaction of elemental zinc with a strong acid: $Zn(s) + 2\,H_3O^+(aq) \rightarrow H_2(g) + Zn^{2+} + 2\,H_2O(l)$. Zinc is far too expensive to use in large-scale preparation of hydrogen.

18-7 Use the relationship $\Delta G = \Delta H^\circ - T\Delta S^\circ$. If ΔG is equal to zero, then:

$$T = \Delta H^\circ / \Delta S^\circ$$

Appendix D supplies the standard entropies and standard enthalpies of formation of the reactants and products of the reaction:

$$CH_4(g) + H_2O(g) \rightarrow CO(g) + 3\,H_2(g)$$

Combine the data to get the ΔH° and ΔS° of the reaction:

$$\Delta H^\circ = 3\underbrace{(0)}_{H_2(g)} + 1\underbrace{(-110.52)}_{CO(g)} - 1\underbrace{(-74.81)}_{CH_4(g)} - 1\underbrace{(-241.82)}_{H_2O(g)} = +206.11 \text{ kJ}$$

$$\Delta S^\circ = 3\underbrace{(130.57)}_{H_2(g)} + 1\underbrace{(197.56)}_{CO(g)} - 1\underbrace{(186.15)}_{CH_4(g)} - 1\underbrace{(188.72)}_{H_2O(g)} = +214.40 \text{ J K}^{-1}$$

$$T = \frac{\Delta H^\circ}{\Delta S^\circ} = \frac{206.11 \times 10^3 \text{ J}}{214.40 \text{ J K}^{-1}} = 960 \text{ K}$$

18-9 a) The equation is $3\,Fe(s) + 4\,H_2O(g) \rightarrow Fe_3O_4(s) + 4\,H_2(g)$.
b) Use the values of ΔH_f° in Appendix D to calculate ΔH° for the above reaction:

$$\Delta H^\circ = 1\underbrace{(-1118.4)}_{Fe_3O_4(s)} + 4\underbrace{(0)}_{H_2(g)} - 4\underbrace{(-241.82)}_{H_2O(g)} - 3\underbrace{(0)}_{Fe(s)} = -151.1 \text{ kJ}$$

A negative ΔH° means that the reaction is exothermic. Increasing the temperature reduces K and shifts the reaction from right to left. Thus, the yield of the reaction at equilibrium would decrease, although an elevated temperature is needed for the reaction to proceed rapidly.

18-11 The balanced chemical equation is: $C_{18}H_{32}O_2 + 2\,H_2(g) \rightarrow C_{18}H_{36}O_2$. Recall that at STP 1 mol of any gas will occupy a volume of 22.4 L so the volume of H_2 at STP is:

$$500.0 \text{ g } C_{18}H_{32}O_2 \times \left(\frac{1 \text{ mol}}{280.45 \text{ g } C_{18}H_{32}O_2}\right) \times \left(\frac{2 \text{ mol } H_2}{1 \text{ mol } C_{18}H_{32}O_2}\right) \times \left(\frac{22.4 \text{ L } H_2}{1 \text{ mol } H_2}\right) = 79.9 \text{ L}$$

18-13 Very reactive chemicals are always chemicals of high free energy. All systems at constant T and P tend toward states of minimum free energy. A compound of high free energy will tend to react spontaneously with another compounds. This is why, for example, there is very little elemental aluminum in the Earth's crust. Any elemental aluminum initially present has long since reacted to form minerals like Al_2O_3 or $Al_2Si_2O_5{\cdot}(OH)_4$ which have lower free energies.

a) Coal, hydrogen and air do not spontaneously react to form aspirin. This means that ΔG exceeds zero for this reaction. Therefore, the aspirin has a higher free energy than its constituent carbon (from coal) hydrogen and oxygen (from air).

b) Pure water, sodium chloride and magnesium carbonate react spontaneously to form seawater. Thus, seawater is in a lower free-energy state.

c) Carbon dioxide and water do not spontaneously react to form vitamin A and oxygen. This means that the vitamin A and oxygen must lie in higher free-energy states than carbon dioxide and water.

18-15 The reaction of iron or zinc with sulfuric acid is not a useful large-scale preparation for hydrogen because zinc and iron are relatively expensive starting materials. They require a large input of energy to free them from their ores. A economical and useful large-scale preparation must use energetically cheaper starting materials.

18-17

$$63 \times 10^9 \text{ bottle} \times \frac{0.355 \text{ L}}{1 \text{ bottle}} \times \frac{0.15 \text{ mol } CO_2}{1 \text{ L}} \times \frac{0.0440 \text{ kg } CO_2}{1 \text{ mol } CO_2} = 1.5 \times 10^8 \text{ kg } CO_2$$

18-19 The problem review the classification of reactions established in Section 2-5 (text page 69).

a) This reaction is a redox reaction. Sulfur is reduced from the +6 oxidation state (in $CaSO_4$) to the 0 oxidation state (in S). Methane is oxidized.

b) This reaction is a redox reaction. Hydrogen is oxidized from the 0 oxidation state (in H_2) to the +1 oxidation state (in CH_3OH). The reduction that accompanies this oxidation is the conversion of CO (C in the +2 oxidation state) to CH_3OH (C in the −2 oxidation state).

c) This reaction (called a hydrogenation) is a redox reaction. Hydrogen is oxidized from the 0 oxidation state (in H_2) to the +1 oxidation state (in H_3CCH_3). The reduction is the conversion of $H_2C{=}CH_2$ (C in the −2 oxidation state) to $H_3C{-}CH_3$ (C in the −3 oxidation state).

d) This reaction is a redox reaction. Iron is oxidized from the 0 oxidation state (in Fe) to the +2 oxidation state (in $FeSO_4$). The reduction which accompanies this oxidation involves converting H_2SO_4 (H in the +1 oxidation state) to H_2 (H in the 0 oxidation state).

Chapter 19

The Lithosphere

The earth (approximate radius 6370 km) consists of the three distinct parts: **core** (thickness 3470 km), **mantle,** (2880 km) and **crust** (17 km). The thin crust is the **lithosphere.** The lithosphere is the region of contact between the cool atmosphere and the warm interior of the earth. It is in a state of continual chemical transformation.

Silicates

Compounds containing silicon and oxygen make up much of the earth's crust. The bond between Si and O is the basis for a large class of minerals called **silicates.** The orthosilicate anion, SiO_4^{4-} is the structural unit of these minerals. The molar ratio of Si to O in this anion is 1:4. Orthosilicate groups can link to each other. As several SiO_4^{4-} tetrahedra link together by sharing corners, the number of oxygen atoms per silicon diminishes. **Example:** Determine the charge on the cyclic trisilicate ion in Figure 19-1c, text page 690. **Solution:** A trisilicate must contain three Si atoms. There are three links between SiO_4 tetrahedra in this cyclic structure. This reduces the number of oxygen atoms from 12, the number in $(SiO_4)_3$, to 9. In silicates the oxidation number of Si is +4 and of O −2. Hence, the ion is $Si_3O_9^{6-}$. Another approach notes that silicon atoms have a zero formal charge in silicates, that oxygen atoms with just one bond in a silicate always have a −1 formal charge, and that oxygen atoms that are shared between silicon atoms have a zero formal charge. There are three oxygen atoms shared between the SiO_4 tetrahedra in text Figure 19-1c, but six oxygen atoms on the perimeter of the structure. The sum of the formal charges of all atoms is −6, hence the net charge on the ion is −6. Compare to **problem 19-1.**

A wide variety of silicate structures occurs in nature, ranging from single tetrahedra to pairs, rings, chains, sheets, and three-dimensional networks of linked tetrahedra, with cations interspersed to balance the total charge in the crystal. The ratio of the number silicon to oxygen atoms provides a means to classify silicate structures.

The criteria for classification are summarized in Table 19-2 (text page 689). There are two complications:

- Not all of the oxygen atoms in a mineral are part of the silicate structural system. Some may be present in water of crystallization, for example. Such casual oxygen atoms are not counted in determining the Si:O ratio. See **problem 19-3.**
- Aluminum can substitute for silicon in the silicate structures. If it does, it must be counted as an Si atom in predicting silicate structures. See **problem 19-5.** Note that aluminum in silicate minerals does not always replace silicon. Sometimes it is present merely to maintain electrical neutrality.

Aluminosilicates

When an Al atom replaces an Si atom in tetrahedral coordination in a silicate mineral, an **aluminosilicate** results. Because Al contributes only three valence electrons instead of four, aluminosilicate structures always contain a positive metal ion nearby to balance the change. **Feldspars** and **micas** are two important kinds of aluminosilicate minerals.

Clays and Zeolites

In **clay** minerals, water can be absorbed between sheets of bonded atoms and cause swelling. These materials are used to seal bore holes in the drilling of oil wells, in construction, and in pottery making. Zeolites are aluminosilicates with three-dimensional network structures that are open enough to allow the inclusion of small molecules and ions. Every time an aluminum(III) replaces a silicon(IV) in a silicate structure, the resulting aluminosilicate gains a negative charge. There must be cations nearby to maintain electrical neutrality. These cations reside in the openings in the zeolite framework.

Zeolites are used for ion-exchange. An equilibrium is established as one type of positive ion becomes lodged in the pores and tunnels of the aluminosilicate structure and another is released. In addition, small neutral molecules like water can be taken up and held in the internal cavities of zeolites. Finally zeolites can hold small molecules in favorable orientations for reaction, thus serving as catalysts.

Geochemistry

Geochemistry is the study of the chemistry of the earth. Understanding earth science requires use of nearly all the principles covered in the first half of the text. Thus,

problem 19-9 could come from Chapter 9, **problem 19-33** from Chapter 2, and **problem 19-37** from Chapter 7.

The **geochemical cycle** starts with the crystallization of a **magma,** a molten silicate fluid, to form **igneous** rocks. **Sedimentary** rocks comes from the weathering, dissolution, and subsequent precipitation of minerals. Many of these minerals show strong **differentiation** based on chemical factors, from the original material. Another type of change is the **metamorphosis** of sedimentary rocks under heat and pressure. High-grade metamorphism leads to remelting and magma formation once again

The text introduces a new device for displaying chemical data. It is the **potential versus pH diagram** (Figure 19-8, text page 697). The key points in understanding and using these diagrams are:

- Inconsistencies in plotting and scaling give confusing results. Always plot *reduction* potentials *(not* oxidation potentials) on the vertical axis. Then, the top of the diagram is favored in oxidizing environments, and the bottom in reducing environments. Remember that as pH increases (to the right on the horizontal scale) the concentration of $H_3O^+(aq)$ becomes less, (not more).

- Half-reactions in which H_3O^+ or OH^- do not occur explicitly have no dependence of potential upon pH.

- Breaks in potential versus pH lines occur when one half-reaction supersedes another. Thus, in **problem 19-7,** the species $Co^{2+}(aq)$ is reduced in acid solution, but $Co(OH)_2(s)$ is reduced in basic solution. A bend in the potential versus pH curve marks the cross-over.

- The slope of the reduction potential versus pH line derives from the logarithm term in the Nernst equation. At 25°C, the slope of this line is some multiple or submultiple of 0.0592 V per pH unit.

- The potential versus pH diagrams identify *stability regions* for the elements in different oxidation states. They help to decide which minerals will form under certain conditions (oxidizing versus reducing and acidic versus basic).

Clapeyron Equation

The Clapeyron equation states the effect of pressure on a chemical equilibrium:

$$\frac{dP}{dT} = \frac{\Delta H}{T\Delta V}$$

This equation gives the slopes of phase coexistence lines in the phase diagrams discussed in Chapter 4.

In using the Clapeyron equation, remember that it is valid *only* at combinations of temperature and pressure for which the system is at equilibrium. A common error is to substitute a temperature which does *not* fit the equation:

$$\Delta G = 0 = \Delta H - T\Delta S$$

and expect the Clapeyron equation to work. See **problem 19-11d.** As long as ΔG is 0 for a change taking place at constant temperature and pressure, the quotient $(\Delta H/T)$ equals ΔS. An alternate version of the Clapeyron equation then is:

$$\frac{dP}{dT} = \frac{\Delta S}{\Delta V}$$

This form of the equation is used in **problems 19-9 and 19-39.**

Difficulties arise with the Clapeyron equation because it concerns equilibrium at constant temperature and pressure, yet the term dP/dT clearly involves changes in both temperature and pressure. Imagine that a given system is brought to equilibrium under constant temperature and pressure in many separate experiments at many different (P, T) combinations. It would certainly be possible to plot these sets of values on a graph with P on the vertical axis and T on the horizontal axis. The term dP/dT is the slope of the resulting line, telling how the observed pressures and temperatures interrelate.

A second difficulty in applying the Clapeyron equation occurs because it has constantly been assumed in problems that the ΔS of a change does not depend strongly on the temperature. This is true, at least across limited ranges of temperature. Unfortunately, it encourages the incorrect conclusion that ΔS values are also only weakly dependent upon the *pressure*. If the change in the number of moles of gas Δn_g is non-zero in a process, then the ΔS of the process is *strongly* dependent upon pressure **problem 19-9.**

The units of dP/dT are Pa K^{-1} (pascal per kelvin) in the SI system. These will be the units of the answer if ΔS (or $\Delta H/T$) is in J K^{-1} and ΔV is in m^3. Volume changes however nearly always derive from densities, which rarely are quoted with m^3 as part of their unit. Do not forget to convert if ΔV is in liters or cubic centimeters or if $(\Delta H/T)$ is in kJ K^{-1}. The unfamiliar Pa K^{-1} unit can readily be converted to atm K^{-1}, if desired (see **problem 19-11.**

Extractive Metallurgy

Extractive metallurgy concerns the winning of useful metals from ores. Ores contain metals in positive oxidation states in compounds (sulfides, oxides) of low free energy. Reductions of ore to metal plus oxygen (or sulfur) are nonspontaneous (example: Fe_2O_3 to give Fe and O_2). Therefore, compensating spontaneous reactions (such as

O_2 plus C to give CO_2) are run to furnish the needed free energy. **Pyrometallurgy** refers to recovery of metals at high temperature.

The extraction of copper from its sulfide ores requires ore benefaction and high-temperature reduction. The iron that is usually present in copper ores must also be removed. Froth-flotation of copper ores uses water, oil, and a detergent: oil-wetted ore particles rise in air-churned froth to separate from water-wetted rock. Roasting of enriched copper ores converts iron that is present to oxide; leaves copper as copper(I) sulfide. Further reaction of Cu_2S with oxygen gives metallic Cu.

The **hydrometallurgical** method of separating copper from iron in ore uses aqueous redox chemistry and thereby avoids production of the pollutant $SO_2(g)$.

Iron and Steel

The winning of iron from ores is also a high-temperature reduction. Iron ores are reduced in a blast furnace by $CO(g)$ from coke ($C(s)$). Alumina (Al_2O_3) and silica (SiO_2) impurities are taken up to form slag upon addition of limestone ($CaCO_3$) as flux. A blast furnace has a charge of iron ore, coke and limestone, the charge meets a blast of pre-heated air from nozzles (tuyeres). Ensuing reactions produce impure iron (pig iron) and slag. Steel (alloy of iron with lower carbon content than pig iron) is made from pig iron by the Bessemer process or by the open-hearth process. In the **Bessemer process,** a stream of air forced through molten pig iron with a flux to remove most carbon as CO, to oxidize Si and P, and transfer the oxides to a slag phase. This creates steel. The **basic oxygen process** is a Bessemer process but using blast of oxygen instead of air. The open-hearth process is another oxidative extraction process to remove excess C and impurities from pig iron and make steel. In the electric-arc process, open-hearth steel is melted by an electric arc. This method is used to make high-tensile-strength steels. It avoids fuel-borne impurities and lowers the concentrations of oxygen, sulfur and hydrogen impurities in the steel.

Electrometallurgy

The production of aluminum requires the reduction of aluminum ores in an electrolytic cell. Magnesium metal is produced by the electrochemical reduction of $MgCl_2(l)$. Metals are refined as well as produced by electrometallurgical methods. Metals can be plated out electrochemically as coatings on other metals.

Detailed Solutions to Odd-Numbered Problems

19-1 The structure of $Si_2O_7^{6-}$ is:

```
           ..          ..
          :O:⁻        :O:⁻
     ..    |           |    ..
  ⁻:O—————Si————O————Si————O:⁻
     ..    |           |    ..
          :O:⁻        :O:⁻
           ..          ..
```

The six O atoms on the perimeter of the structure each have three lone pairs and a single bond to an Si atom. All six perimeter O atoms have a formal charge of -1. The O atom between the Si atoms has a formal charge of zero. The Si atoms also have formal charges of zero. There are 56 valence electrons in the Lewis structure. The $P_2O_7^{4-}$ and $S_2O_7^{2-}$ ions also have 56 valence electrons and have the same Lewis structure except for the change in the central atoms. Both P atoms in the Lewis structure of $P_2O_7^{4-}$ have a $+1$ formal charge. Both S atoms have a $+2$ formal charge in the $S_2O_7^{2-}$ structure. The analogous compound of chlorine is Cl_2O_7 in which both Cl atoms have a formal charge of $+3$.

If the octet rule is broken for the Si (or P or S) atoms in these structures, then many additional resonance structures are possible. Such structures would have one or more double bonds between the surrounding O atoms and the central atoms. They would increase the average bond order and would make the formal charge on the central atoms less positive.

19-3 In each example, determine the Si:O ratio for the network. Ignore oxygen atoms found in, for example (OH) groups. Then use Table 19-2.

a) Tetrahedra. Ca, $+2$; Fe, $+3$; Si, $+4$; O, -2. **b)** Infinite sheets. Na, $+1$; Zr, $+2$; Si, $+4$; O, -2. **c)** Pairs of tetrahedra. Ca, $+2$; Zn, $+2$; Si; $+4$; O, -2. **d)** Infinite sheets. Mg, $+2$; Si, $+4$; O, -2; H, $+1$.

19-5 The problem is exactly like the preceding except that Al atoms grouped in the formulas with the Si atoms are counted as Si atoms in determining the Si:O ratio.

a) Infinite network. Li, $+1$; Si, $+4$; Al, $+3$; O, -2. **b)** Infinite sheets. K, $+1$; Al, $+3$; Si, $+4$; O, -2; H, $+1$. **c)** Closed rings or infinite single chains. Al, $+3$; Mg, $+2$; Si, $+4$; O, -2.

19-7 The standard reduction potential for the half-reaction $Co^{2+}(aq) + 2\,e^- \rightarrow Co(s)$ is -0.28 V at pH 0. As the pH rises the potential remains constant until, after a brief intermediate regime, cobalt(II) hydroxide precipitates, and the half-reaction $Co(OH)_2(s) + 2e^- \rightarrow Co(s) + 2\,OH^-(aq)$ more accurately describes what is going on. The standard potential for this half reaction is -0.73 V. The Nernst equation for the new half-reaction is:

$$\mathcal{E} = \mathcal{E}^\circ - \frac{0.0592}{2}\log[OH^-]^2 = -0.73 - 0.0592\log[OH^-]$$

The term "$-0.0592 \log[OH^-]$" equals $+0.0592$ pOH by the definition of pOH. It therefore also equals $0.0592(14 - pH)$. Thus:

$$\mathcal{E} = -0.73 \text{ V} + (0.0592 \text{ V}(14 - \text{pH}) = (0.0988 - 0.0592\,\text{pH}) \text{ V}$$

This equation tells how the half-cell potential varies with pH. A plot of $\mathcal{E}$ versus pH is a straight line with a slope of -0.0592. If the line were extended back to pH 0, it would intercept the voltage axis at 0.0988 V. Of course the actual behavior at low pH is governed by the -0.28 V half-reaction.

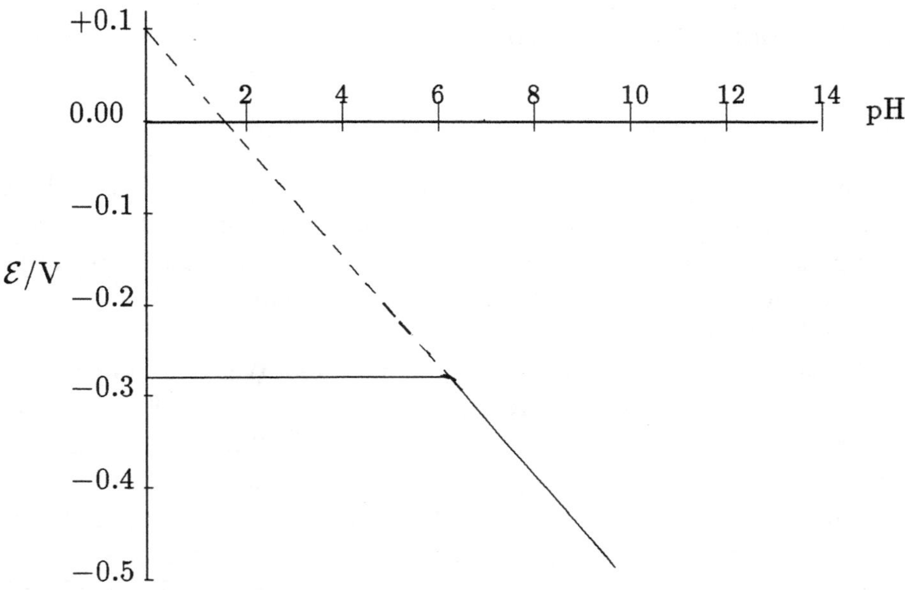

19-9 a) $CaCO_3(s) + SiO_2(s) \rightarrow CaSiO_3(s) + CO_2(g)$.
b) The results are obtained using data from Appendix D in the usual way:

$$\Delta H^\circ = 1\underbrace{(-1634.94)}_{CaSiO_3(s)} + 1\underbrace{(-393.51)}_{CO_2(g)} - 1\underbrace{(-1206.92)}_{CaCO_3(s)} - 1\underbrace{(-910.94)}_{SiO_2(s)} = +89.41 \text{ kJ}$$

$$\Delta S^\circ = 1\underbrace{(81.92)}_{CaSiO_3(s)} + 1\underbrace{(213.63)}_{CO_2(g)} - 1\underbrace{(92.9)}_{CaCO_3(s)} - 1\underbrace{(41.84)}_{SiO_2(s)} = +160.8 \text{ J K}^{-1}$$

c) Assume that ΔH° and ΔS° are independent of temperature. Then, when ΔG° is zero:

$$T = \frac{\Delta H^\circ}{\Delta S^\circ} = \frac{89.41 \times 10^3 \text{ J}}{160.8 \text{ J K}^{-1}} = 556 \text{ K}$$

At this temperature the equilibrium constant for the above reaction equals 1.

d) Although ΔH° for the process is nearly entirely independent of the pressure, ΔS° *does* depend on pressure. Gaseous CO_2, if released at a higher P, has less entropy. Hence ΔS for this reaction at 500 atm is less than ΔS°. Compressing 1 mol of $CO_2(g)$ from 1 atm to 500 atm changes the entropy by $-R\ln(500/1)$ which is -51.67 J K^{-1}. This formula is discussed in Chapter 9 (page 239 of this Guide). Therefore, replace $\Delta S^\circ = 160.8$ J K^{-1} in part b) above with a ΔS of $160.8 + (-51.67)$ or 109.1 J K^{-1}. Then set ΔG equal to zero and compute T. as in part c). It is 819 K. At high pressure, it takes a *higher* temperature to make the equilibrium constant of the reaction equal 1. This fits with the prediction of LeChatelier's principle. Higher pressure favors the reactants.

19-11 a) The two forms of calcium carbonate are calcite and aragonite. Represent the conversion from the first to the second as:

$$1 \text{ mol calcite} \longrightarrow 1 \text{ mol aragonite}$$

The ΔH° for this change is the difference between the ΔH_f°'s (Appendix D) of calcite and aragonite. The ΔH° of the reaction is -0.21 kJ. The ΔS° is the difference in the absolute entropies (Appendix D) of the two substances at 298.15 K. It equals -4.2 J K^{-1}. The ΔG° of the reaction is $+1.04$ kJ, by a similar calculation with ΔG_f°'s, using the same source for data.
b) The ΔG° for the calcite$\rightarrow$aragonite conversion is positive. This means that calcite is thermodynamically favored over aragonite at 298.15 K and 1.00 atm.
c) Assume that ΔH° and ΔS° depend only weakly on T. Then: $\Delta G \approx \Delta H^\circ - T\Delta S^\circ$. When the two forms are equally favored, ΔG° (which equals $-RT\ln K$) is 0. Setting ΔG equal to zero gives $T = 50$ K. Below 50 K, aragonite is favored. At temperatures above 50 K, calcite is favored (as long as the pressure is 1.00 atm).
d) Higher pressure favors the more dense form. The molar volume of aragonite is 34.16 cm^3 mol^{-1}. Aragonite is more dense than calcite, which occupies 36.94 cm^3 mol^{-1}. Aragonite is favored at high pressure.
e) The Clapeyron equation gives the slope of the coexistence line for two phases in a P versus T phase diagram:

$$\frac{dP}{dT} = \frac{\Delta H}{T\Delta V}$$

For the transition of 1 mol of calcite to 1 mol of aragonite ΔH° equals -210 J. The ΔV for this transition is $(34.16 - 36.94)$ cm^3 or -2.78 cm^3 which is 2.78×10^{-6} m^3. The coexistence line for the two solids passes through the point defined by the (P, T) combination: $(P = 1$ atm, $T = 50$ K). See part c) above. Assume that the ΔH and ΔV values are good at this low temperature on the P versus T graph. Then the slope of the line at this point is:

$$\frac{dP}{dT} = \frac{-210 \text{ J}}{(50 \text{ K})(2.78 \times 10^{-6} \text{ m}^3)} = 1.5 \times 10^6 \text{ Pa K}^{-1}$$

Appendix B (text page A-11) confirms that a J m^{-3} equals a pascal (Pa). One pascal is 101 325 atm, so an equivalent answer is 15 atm K^{-1}.
The result tells the rate of change of pressure with temperature at $P = 1$ atm and $T = 50$ K. As long as ΔH and ΔV are both independent of T and P, the coexistence line is a straight line with this slope.

19-13 The balanced equation for the reduction of pyrolusite to manganese by aluminum is:

$$3\,\mathrm{MnO_2}(s) + 4\,\mathrm{Al}(s) \rightarrow 2\,\mathrm{Al_2O_3}(s) + 3\,\mathrm{Mn}(s)$$

Combine the standard enthalpies of formation and standard free energies of formation of the two reactants and two products (Appendix D) to get the standard enthalpy change and standard free energy change of the reaction:

$$\Delta H^\circ = 2(-1675.7) + 3(0) - 3(-520.03) - 4(0) = -1791.3 \text{ kJ}$$
$$\Delta G^\circ = 2(-1582.3) + 3(0) - 3(-465.17) - 4(0) = -1769.1 \text{ kJ}$$

These changes in enthalpy and free energy apply to the production of 3 mol of Mn. The values per mole of Mn are only one-third as large: $\Delta H^\circ = -597.1$ kJ mol^{-1} and $\Delta G^\circ = -589.7$ kJ mol^{-1}.

19-15 The balanced equation for the decomposition of mercury(II) oxide to this elements is:

$$2\,\mathrm{HgO}(s) \rightarrow 2\,\mathrm{Hg}(l) + \mathrm{O_2}(g)$$

Combine the standard enthalpies of formation and absolute entropies of the reactants and products at 25° (Appendix D) to get the standard enthalpy change and standard entropy change of the reaction:

$$\Delta H^\circ = 2(0) + 1(0) - 2(-90.83) = +181.66 \text{ kJ}$$
$$\Delta S^\circ = 1(+205.03) + 2(76.02) - 2(70.29) = +216.49 \text{ J K}^{-1}$$

If ΔH° and ΔS° are independent of temperature, ΔG for this reaction becomes equal to zero at the temperature fulfilling the equation:

$$T = \frac{\Delta H^\circ}{\Delta S^\circ} = \frac{+181.66 \times 10^3 \text{ J}}{+216.49 \text{ J K}^{-1}} = 839 \text{ K}$$

At temperatures exceeding 839 K, ΔG is negative.

19-17 The equation for the reduction of $WO_3(s)$ to $W(s)$ with gaseous hydrogen is:

$$\mathrm{WO_3}(s) + 3\,\mathrm{H_2}(g) \rightarrow \mathrm{W}(s) + 3\,\mathrm{H_2O}(g)$$

By combining the proper ΔH_f°'s and S°'s (Appendix D), it can be confirmed that the ΔH° of this reaction is 117.41 kJ, and ΔS° is 131.19 J K^{-1}. The reaction is endothermic. High temperatures favor the products in such a case. When K is one, then ΔG is zero. The temperature that makes ΔG equal to zero (if ΔH° and ΔS° are temperature-independent) is:

$$T = \frac{\Delta H^\circ}{\Delta S^\circ} = \frac{+117.41 \times 10^3 \text{ J}}{+131.19 \text{ J K}^{-1}} = 895 \text{ K}$$

19-19 According to its formula, chalcopyrite is 34.62% Cu. Similarly, covellite is 66.46% copper by mass, and bornite is 63.61% copper by mass. The mass of copper per 100 g of this ore is:

$$\left(\frac{34.62 \text{ g Cu}}{100 \text{ g chalcopyrite}} \times \frac{1.1 \text{ g chalcopyrite}}{100 \text{ g ore}}\right) + \left(\frac{66.46 \text{ g Cu}}{100 \text{ g covellite}} \times \frac{0.42 \text{ g covellite}}{100 \text{ g ore}}\right)$$

$$+ \left(\frac{63.61 \text{ g Cu}}{100 \text{ g bornite}} \times \frac{0.51 \text{ g bornite}}{100 \text{ g ore}}\right) = \frac{0.984 \text{ g Cu}}{100 \text{ g ore}}$$

A metric ton of ore is 1000 kg of ore. This is 10^4 times more than 100 g of ore. Accordingly there is 0.98×10^4 g of copper per metric ton of ore, or 9.8 kg of copper per metric ton of ore.

19-21 a) The balanced equations are:

$$\begin{array}{ll} (1) & Fe_2O_3(s) + 3\,CO(g) \rightarrow 2\,Fe(s) + 3\,CO_2(g) \\ (2) & Fe_3O_4(s) + 4\,CO(g) \rightarrow 3\,Fe(s) + 4\,CO_2(g) \\ (3) & FeCO_3(s) + CO(g) \rightarrow Fe(s) + 2\,CO_2(g) \end{array}$$

b) Each reaction involves 1 mol of an iron compound. For each reaction, ΔG° is the ΔG_f° of the reactants minus the ΔG_f° of the products. Taking free energies of formation from Appendix D, perform calculations using Hess's law:

Equation No.	ΔG°	ΔG° per mol Fe
(1)	−29.4 kJ	−14.7 kJ mol^{-1}
(2)	−13.3	−4.4
(3)	+15.1	+15.1

The first reaction (reduction of Fe_2O_3) has the most negative ΔG°. Iron(III) oxide is the easiest compound to reduce with CO from the standpoint of thermodynamics. Which compound is easiest to reduce in practice depends on factors such as ease of handling and rate of reduction.

19-23 At the anode, chloride ion in the molten NaCl is oxidized to gaseous chlorine; at the cathode, sodium ion is reduced to metallic sodium:

$$Cl^-(NaCl) \rightarrow 1/2\,Cl_2(g) + e^- \quad \text{and} \quad Na^+(NaCl) + e^- \rightarrow Na(l)$$

19-25 A period of 24 hours is 8.64×10^4 s. A steady current of 55,000 A for this period means that 4.75×10^9 C passes through a single cell. Dividing by the Faraday constant gives the chemical amount of electricity passing through the cell. It is 4.93×10^4 mol. It takes 3 mol of electrons to deposit 1 mol of Al. The theoretical yield of Al is therefore 1.64×10^4 mol, which is 4.43×10^5 g of Al per cell. There are 100 cells, so the total theoretical yield of Al is 100 times larger than for a single cell. It is 4.4×10^7 g.

19-27 The Kroll process used the reaction:

$$\text{TiCl}_4(l) + 2\,\text{Mg}(s) \rightarrow \text{Ti}(s) + 2\,\text{MgCl}_2(s)$$

The minimum mass of magnesium to produce 100 kg of titanium by the Kroll process is:

$$100\text{ kg Ti} \times \left(\frac{1\text{ kmol Ti}}{47.88\text{ kg Ti}}\right) \times \left(\frac{2\text{ kmol Mg}}{1\text{ kmol Ti}}\right) \times \left(\frac{24.305\text{ kg Mg}}{1\text{ kmol Mg}}\right) = 102\text{ kg Mg}$$

19-29 The problem states that 7.32 g of zinc is to be coated to the steel garbage can. This is 0.112 mol of Zn. Each mole of Zn requires 2 mol of electrons to plate it out, and a mole of electrons is 96,485 coulombs. The total charge passed through the cell is therefore 2.161×10^4 C. A current of 8.50 A means that 8.50 C passes through the cell every second. The time required to pass the required charge is:

$$t = \frac{Q}{I} = \frac{2.161 \times 10^4\text{ C}}{8.50\text{ C s}^{-1}} = 2.54 \times 10^3\text{ s} = 42.4\text{ min}$$

19-31 a) Apophyllite contains infinite sheets of silicate units. The Si and O are in the +4 and −2 oxidation states respectively. The F is −1, the K is +1 and the Ca is +2 in oxidation state. The water of hydration in the mineral has H in the +1 and O in the −2 oxidation states.

b) Rhodonite contains infinite single chains of SiO_4 units. The Ca and Mn are both have +2 oxidation states.

c) Margarite is an aluminosilicate. It contains infinite sheets of aluminosilicate groups. The Ca and non-infinite-sheet Al are in the +2 and +3 oxidation states, respectively. The Si, Al, and O in the aluminosilicate framework are in the +4, +3 and −2 states; the hydroxide H and O are in the +1 and −2 states respectively.

19-33 Let x equal the oxidation number of the Fe and write an equation to express the electrical neutrality of the compound:

$$\underbrace{2.36(2)}_{\text{Mg}} + \underbrace{0.48x}_{\text{Fe}} + \underbrace{0.16(3)}_{\text{Al}} + \underbrace{2.72(4)}_{\text{Si}} + \underbrace{1.28(3)}_{\text{Al}} + \underbrace{10(-2)}_{\text{O}} + \underbrace{2(-1)}_{\text{OH}} + \underbrace{0.32(2)}_{\text{Mg}} = 0$$

Solving gives $x = 3$.

19-35 $CaAl_2Si_2O_8(s) + CO_2(aq) + 2\,H_2O(l) \rightarrow CaCO_3(s) + Al_2Si_2O_5(OH)_4(s)$. Lowering the pH will solubilize the product $CaCO_3$, which dissolves readily in acid (see problem 19-37), and thereby increase the extent of weathering, by LeChatelier's principle.

19-37 **a)** The dissolution of $CaCO_3(s)$ in water involves the equilibria:

$$CaCO_3(s) \rightleftharpoons Ca^{2+}(aq) + CO_3^{2-}(aq) \qquad K_{sp} = 7.6 \times 10^{-9}$$
$$CO_3^{2-}(aq) + H_2O(l) \rightleftharpoons HCO_3^-(aq) + OH^-(aq) \qquad K_{b1} = 2.08 \times 10^{-4}$$
$$HCO_3^-(aq) + H_2O(l) \rightleftharpoons H_2CO_3(aq) + OH^-(aq) \qquad K_{b2} = 2.32 \times 10^{-8}$$

The mass-action expressions for the two acid-base equilibria are:

$$K_{b1} = \frac{[OH^-][HCO_3^-]}{[CO_3^{2-}]} \quad \text{and} \quad K_{b2} = \frac{[OH^-][H_2CO_3]}{[HCO_3^-]}$$

Note that K_{b1} is K_w divided by the K_{a2} for H_2CO_3 (text Table 6-2), and K_{b2} is K_w divided by K_{a1} for H_2CO_3. The K_{sp} for $CaCO_3$ is from text Table 7-2.
Since the water is at pH 7, $[OH^-]$ is 1.0×10^{-7} M. Substituting this value into the two K_b expressions and rearranging gives:

$$[HCO_3^-] = (2080)[CO_3^{2-}] \quad \text{and} \quad [H_2CO_3] = (482)[CO_3^{2-}]$$

Let S equal the equilibrium solubility of the $CaCO_3(s)$. Then S is equal to $[Ca^{2+}]$ and is also equal to the sum of the concentrations of the three carbon-containing species. This statement is a material-balance condition (text page 232) for the carbonate. Expressed mathematically:

$$S = [Ca^{2+}] = [CO_3^{2-}] + [HCO_3^-] + [H_2CO_3]$$

Substituting the independent expressions for the bicarbonate and carbonic acid concentrations into this equation gives:

$$S = [CO_3^{2-}]\,(1 + 2080 + 482) = [CO_3^{2-}](2560)$$

Combining this equation with the K_{sp} expression for $CaCO_3(s)$ allows the calculation of S:

$$8.7 \times 10^{-9} = [Ca^{2+}][CO_3^{2-}] = S\frac{S}{2560}\,; \qquad S = 0.0047 \text{ M}$$

One error to avoid in this problem is the unjustifiable assumption that $HCO_3^-(aq)$ is the only carbon-containing species present in significant concentration. This corresponds to leaving out the first and third terms in the parentheses in the preceding equation. Committing this error yields an S equal to 0.0042 M, which is 10 percent less than the correct answer. A worse error is to ignore the acid-base interaction of the carbonate ion with the water entirely. This corresponds to omitting the second and third terms within the parentheses and gives an S of 9.3×10^{-5} M, about 50 times too low.

b) Decreasing the pH will increase the solubility of $CaCO_3(s)$ in water. More carbonate ion is converted to bicarbonate ion or carbonic acid at lower pH.

c) The river's annual flow is 8.8×10^{12} L. This much water would, at equilibrium, dissolve 4.1×10^{10} mol of $CaCO_3$ ($\mathcal{M} = 100$ g mol^{-1}) which is 4.1×10^6 metric tons.

19-39 The balanced equation is:

$$Mg_3Si_4O_{10}(OH)_2(s) + Mg_2SiO_4(s) \rightarrow 5\,MgSiO_3(s) + H_2O(g)$$

Trial-and-error balancing gives this answer. Another approach is to note that Mg_2SiO_4 loses O^{2-} ions on a per silicon basis, and $Mg_3Si_4O_{10}(OH)_2$ gains O^{2-} ions on the same basis. Hence, O^{2-} ion in this reaction plays a role similar to that of the electron in oxidation-reduction reactions. Arranging the gain of O^{2-} to equal the loss of O^{2-} rapidly gives a balanced equation:

$$O^{2-}\text{ loss:}\quad Mg_2SiO_4(s) \rightarrow MgSiO_3(s) + O^{2-} + Mg^{2+}$$
$$O^{2-}\text{ gain:}\quad O^{2-} + Mg^{2+} + Mg_3Si_4O_{10}(OH)_2(s) \rightarrow 4\,MgSiO_3(s) + H_2O(g)$$

b) Use LeChatelier's principle. Increasing the total pressure increases the activity of $H_2O(g)$ and shifts the reaction to the left. The products are disfavored.

c) The slope of the coexistence curve is given by the Clapeyron equation:

$$\frac{dP}{dT} = \frac{\Delta H}{T\Delta V} = \frac{\Delta S}{\Delta V}$$

For the reaction as written above, ΔV is clearly positive because the products include 1 mol of gas, which has a large volume, but the reactants include no gas. The ΔS of the reaction is positive, for the same reason. The slope of the curve is accordingly positive.

19-41 The equilibrium $CO_2(g) + C(s) \rightarrow 2\,CO(g)$ is crucial in the reduction of iron ore in a blast furnace. At room temperature it lies far to the left, but at 1875° (2148 K) it is very much shifted to the right. To answer the question, compute $\Delta H°$ and $\Delta S°$ of this reaction from the data in Appendix D and then use the relationship:

$$-RT\ln K = \Delta H° - T\Delta S° \text{ which gives } \quad \ln K = \frac{-\Delta H°}{RT} + \frac{\Delta S°}{R}$$

to solve for K at both of the temperatures. It is found that $\Delta H°$ is 172.47 kJ and $\Delta S°$ is 175.75 J K^{-1}. The answers are $K_{298} = 9.2 \times 10^{-22}$ and $K_{2148} = 9.7 \times 10^4$. The second K is a rough approximation because $\Delta H°$ and $\Delta S°$ do change between 298 and 2148 K. Note that the answer to this problem is also given on page 708 of the text.

19-43 Iron ore is Fe_2O_3 admixed with silica, alumina and minor impurities; copper ore is usually a sulfide. Iron ore is reduced with $CO(g)$ in a blast furnace by reaction of coke (carbon) with a limited amount of air. Fluxes like CaO are added to make a fluid slag from the silica and alumina impurities. Blast furnace slag has often been used as road ballast, and also finds use in making portland cement. The roasting of copper ores gives off sulfur dioxide, which must not be vented into the atmosphere and is often processed into sulfuric acid. It reduces the copper only partially (to the copper(I) sulfide) and oxidizes impurities. Roasting of copper ores takes place at relatively low temperatures. Removal of impurities goes on at higher temperature (about 1100°C) but still below the 1535°C at which iron is reduced in a blast furnace. The flux in this step of copper recovery is $CaCO_3$, and the slag contains impurity iron. The copper is still combined chemically in Cu_2S. This compound is decomposed into impure metallic copper by heating it in the air, which generates more $SO_2(g)$.

19-45 Both aluminum and magnesium are produced electrochemically. The starting material for the production of aluminum is a mixture of cryolite with bauxite. The mixture is melted and electrolyzed; aluminum is produced at the cathode. The commercial source of magnesium is seawater. Magnesium hydroxide is precipitated from seawater by treating it with a cheap base such as calcined dolomite $CaO{\cdot}MgO$. The $Mg(OH)_2$ is then converted to $MgCl_2$ by reaction with $HCl(aq)$. The $MgCl_2$ is melted and electrolyzed, producing chlorine gas at the anode and metallic magnesium at the cathode.

19-47 a) The balanced equation is:

$$4\,FeCr_2O_4(s) + 8\,Na_2CO_3(l) + 7\,O_2(g) \rightarrow 2\,Fe_2O_3(s) + 8\,Na_2CrO_4(l) + 8\,CO_2(g)$$

b) The standard enthalpies of formation (in kJ mol^{-1}) of these substances are:

$FeCr_2O_4(s)$	$Na_2CO_3(l)$	$O_2(g)$	$Fe_2O_3(s)$	$Na_2CrO_4(l)$	$CO_2(g)$
-1445	$(-1330.7 + \Delta H_{fus})$	0	-824.2	$(-1342 + \Delta H_{fus})$	-393.5

The entries for the two liquids require comment. The ΔH_f°'s for the two exceed the ΔH_f°'s for their solids by an amount equal to the enthalpy of fusion of the solid. According to the problem, the two enthalpies of fusion are about the same. The two therefore cancel out of the computation of ΔH° by Hess's law:

$$\Delta H^\circ = 2(-824.2) + 8(-1342 + \Delta H_{fus}) + 8(-393.5) - 8(1130.7 + \Delta H_{fus}) - 4(-1445)$$

The answer is 2590 kJ.

c) Leach the soluble sodium chromate away from the insoluble by-products by treating the product mixture with water. Evaporate the water. Heat the resulting solid sodium chromate with carbon (charcoal) to reduce it to chromium. Other products would include $CO_2(g)$ and $Na_2O(s)$, which could be taken up in a suitable acidic flux.

19-49 First compute the volume of zinc that will be coated onto the steel. It is the thickness of the coating times the width of the steel times the length times two (for the two sides). It equals 0.0500 m^3. Next comes the mass of the zinc. Appendix F provides the density of zinc (7.133 g cm^{-3}). Convert the volume of zinc to cm^3 and multiply it by this density to get a mass of 3.566×10^5 g of Zn. This is 5.454×10^3 mol of Zn. Each mole of Zn requires 2 mol of electrons to plate it out, and a mole of electrons is 96 485 C. The total amount of electricity passing the cell is therefore 1.053×10^9 C. The energy used is this amount of charge multiplied by the voltage, which is 3.5 V. The energy is thus 3.68×10^9 V C which is 3.68×10^9 J. Divide this energy by 0.9 because the galvanizing is only 90% efficient. This raises the energy consumption to 4.09×10^9 J, equivalent to 1.14×10^3 kW-hr, which costs 114 dollars. It is not necessary to know the current.

Chapter 20

Ceramics and Semiconductors

The Properties of Ceramics

Ceramics are synthetic materials that have as their essential components inorganic nonmetallic materials. They are able to resist high temperatures and corrosion, and are stiff and hard, but also brittle, making them liable to failure when stressed. They are subject to **thermal shock.**

Composition and Structure of Ceramics

Ceramics employ a wide range of compounds. Useful ceramic bodies are nearly always a mixture of compounds. The classification of ceramics includes **silicate ceramics,** which all contain the SiO_4 group; **oxide ceramics,** in which some elements other than silicon combines with oxygen; and **nonoxide ceramics,** in which oxygen is replaced by another nonmetal as a principal component. Nonoxide ceramics are based on compounds like Si_3N_4 and SiC.

A **ceramic phase** is any portion of the whole ceramic body that is physically homogeneous and bounded by a surface that separates it from other parts. The **microstructure** of a ceramic piece can be quite complex, with grains of different phases, voids, and pores. Microstructure has a significant effect on the properties of the finished products. Ceramics are made by **firing** to a high temperature, which leads to partial melting, crystal phase changes, and to **sintering,** in which grains grow together and the ceramic body shrinks or **densifies.** The properties of a ceramic piece depend on its microstructure which in turn depends markedly on the exact conditions under which the piece was formed and fired. The biggest problem with ceramics as structural materials is inconsistent quality.

Silicate Ceramics

Many **silicate ceramics** use as their starting material natural clay minerals such as kaolin. Clays are aluminosilicate minerals that contain hydrated cations between the layers of an infinite sheet structure. Firing an object formed of clay drives away the water and, among other reactions and phase changes, leads to the formation of mullite ($Al_6Si_2O_{13}$), which lends strength to the ceramic. Compare the reaction:

$$3\,Al_2Si_2O_5(OH)_4(s) \rightarrow 3\,Al_6Si_2O_{13}(s)\ (\text{mullite}) + 4\,SiO_2(s) + 6\,H_2O(g)$$

to the reaction in **problem 20-1.** A **glaze** is a thin, glassy layer that coats the object, giving it strength and also often decorating it.

Glass

Glasses are amorphous solids of variable composition. Different compositions of silicate glass are used for different applications. See **problem 20-5.** Pure silica (SiO_2) forms a glass that has a high melting point. The melting point is lowered by the addition of sodium oxide and calcium oxide to give the **soda-lime** glass used in most applications. Incorporating boron oxide in silicate glass yields a glass with better resistance to thermal shock because it has a smaller coefficient of linear thermal expansion. Soda-lime glass has the approximate formula $Na_2O{\cdot}CaO{\cdot}6SiO_2$. Its structure consists of SiO_4 tetrahedra linked in a random three dimensional network. The cations (Na^+ and Ca^{2+}) occupy voids in the network and maintain electrical neutrality. Glasses are **isotropic** in their physical properties (the properties are the same in all directions). They soften over a range of temperatures rather than melt sharply. This makes it possible to work glass and to **anneal** it. Annealing reduces internal stresses that freeze in when a glass body hardens rapidly.

Cements

Portland cement is another silicate ceramic. It is a complex mixture of compounds produced from oxides of calcium, aluminum, silicon, and iron, to which calcium sulfate is added. Portland cement is made from ground limestone mixed with aluminosilicates. Firing the mixture in a cement kiln causes many reactions from which the major product is tricalcium silicate (Ca_3SiO_5), and the minor product is tricalcium aluminate ($Ca_3Al_2O_6$). This mixed material, in the form of *clinkers,* is mixed with a small percentage of gypsum. Cement hardens shortly after being mixed with water because of the hydration of the tricalcium silicate. Successive reactions take place that give off heat and incorporate the water in microscopic needle-shaped hydrated crystals, which interlock, solidify the cement, and give it impressive strength. **Mortar** is sand mixed with portland cement. **Concrete** is a mixture of sand and an aggregate (crushed stone or pebbles).

Nonsilicate Ceramics

Oxide ceramics do not contain major amounts of silicon but instead have a metal in combination with oxygen. Their high melting points and resistance to corrosion make them good **refractories** for lining furnaces and chemical tank reactors. **High-density alumina** Al_2O_3, a material fabricated with minimal pore size, is useful in cutting tools and armor plating. **Magnesia** MgO combines very high thermal conductivity and heat resistance with very low electrical conductivity.

Some oxide ceramics have been found to be **superconducting,** losing all resistance to the flow of an electric current at or below a certain temperature. This temperature exceeds the boiling point of liquids nitrogen for certain formulations. The **1-2-3** compound $YBa_2Cu_3O_{(9-x)}$ is one such high-temperature superconductor. It is a mixed-valence compound that is a nonstoichiometric solid, with oxygen atoms missing from many sites in the ideal **perovskite** structure.

Nonoxide ceramics have strong covalently bonded network structures. They include **silicon nitride** Si_3N_4, which can be made by reaction of silicon with nitrogen at high temperatures, or by reaction of $SiCl_4$ with NH_3. **Boron nitride** BN occurs in different structures that are related to the structure of graphite and diamond. It is used for high-temperature ceramic vessels. **Silicon carbide** SiC is an abrasive that is now used to reinforce other ceramics. Such **composite ceramics** are important new structural materials.

Silicon and Semiconductors

In the elemental state, silicon forms a covalently bonded network solid in which all the electrons are localized between pairs of atoms, filling the **valence band** and leaving the conduction band virtually empty. The result is a low electrical conductivity, which can be increased by heating the solid or illuminating it (see **problem 20-15)** thereby exciting electrons across the **band gap.** Other **intrinsic semiconductors** are made by combining elements to give crystals with the same number of valence electrons per atom as silicon (four). An important example is gallium arsenide (compare with **problem 20-19).**

An **insulator** such as diamond has an even larger band gap than a semiconductor, so that its conduction band is unoccupied by electrons at ordinary temperatures (see **problem 20-17).** By contrast, the conduction band in a metal is partly filled and electrons can change levels with only minimal input of energy. The conductivities of silicon and other intrinsic semiconductors can be increased by **doping,** the deliberate addition of impurity atoms of selected types. If a Group V element is added to pure silicon, the excess electrons occupy levels just below the conduction band, and the conductivity in increased 1000-fold, giving a ***n*-type semiconductor.** If atoms of a Group III element substitute for silicon atoms, they make a ***p*-type semiconductor,** in which electrical conduction occurs by the hopping of positive holes in the valence band when a voltage is applied (see **problem 20-35**).

Phosphors are wide band-gap materials with dopants selected to create new energy levels such that particular colors of light are emitted when elements previously excited to these levels fall back to the ground state (see **problem 20-21**).

Detailed Solutions to Odd-Numbered Problems

20-1 $Mg_3Si_4O_{10}(OH)_2(s) \rightarrow 3\,MgSiO_3(s) + SiO_2(s) + H_2O(g)$

20-3 Preparation of glass of the composition cited on text page 727 proceeds:

$$Na_2CO_3(s) + CaCO_3(s) + 6\,SiO_2(s) \rightarrow Na_2O{\cdot}CaO{\cdot}(SiO_2)_6(s) + 2\,CO_2(g)$$

The problem now becomes a routine calculation in stoichiometry using the molar ratios established in the equation. Recall that 1.00 mol an ideal gas occupies 22.4 L at STP. The molar mass of the glass is 479 g mol^{-1}, which equals 0.479 kg mol^{-1}. Then:

$$2.50\text{ kg glass} \times \left(\frac{1\text{ mol glass}}{0.479\text{ kg glass}}\right) \times \left(\frac{2\text{ mol }CO_2(g)}{1\text{ mol glass}}\right) \times \left(\frac{22.4\text{ L }CO_2(g)}{1\text{ mol }CO_2(g)}\right) = 234\text{ L }CO_2$$

20-5 Assume a sample of exactly 100 g of the soda-lime glass and calculate the chemical amount of each element which is present. This requires use of the molar masses of the several binary oxides. The following table summarizes the results:

Mass of Oxide	$\mathcal{M}$(g mol^{-1})	Amounts of Elements	
72.4 g SiO_2	60.08	1.205 mol Si	2.410 mol O
18.1 g Na_2O	61.98	0.5841 mol Na	0.2920 mol O
8.10 g CaO	56.07	0.1444 mol Ca	0.1444 mol O
1.00 g Al_2O_3	101.96	0.01962 mol Al	0.02942 mol O
0.20 g MgO	40.304	0.004962 mol Mg	0.004962 mol O
0.20 g BaO	153.33	0.001304 mol Ba	0.001304 mol O

The total chemical amount of oxygen in the sample is the sum of all the listings for O in the right-most column. It is 2.882 mol. The chemical amounts of the various elements per mole of oxygen are the amounts in the above table divided by 2.882. They are: 0.418 mol Si, 0.203 mol Na, 0.0501 mol Ca, 0.00681 mol Al, 0.00172 mol Mg, 0.000452 mol Ba.

20-7 The reaction for the production of tricalcium silicate is:

$$SiO_2(s, \mathit{quartz}) + 3\,CaO(s) \rightarrow (CaO)_3{\cdot}SiO_2(s)$$

The enthalpy of the reaction is the sum of the enthalpies of formation of the products minus the sum of the enthalpies of formation of the reactants. Use Appendix D to obtain the ΔH_f°'s for $SiO_2(s, quartz)$ and $CaO(s)$.

$$\Delta H^\circ = 1\underbrace{(-2929.2)}_{(CaO)_3SiO_2(s)} - 1\underbrace{(-910.94)}_{SiO_2(s)} - 3\underbrace{(-635.09)}_{CaO(s)} = -113.0 \text{ kJ}$$

20-9 The sum of the oxidations numbers of the atoms in the compound must equal zero. Assign the oxidation numbers -2 to oxygen, $+2$ to Ba, and $+3$ to Y. A simple equation then shows that the copper must have an oxidation number of 7/3 to bring the sum to zero.

20-11 a) $SiO_2(s, quartz) + 3\,C(s, graphite) \rightarrow SiC(s) + 2\,CO(g)$.
b) Refer to Appendix D for the necessary ΔH_f°'s. Then:

$$\Delta H^\circ = 1\underbrace{(-65.3)}_{SiC(s)} + 2\underbrace{(-110.52)}_{CO(g)} - 1\underbrace{(-910.94)}_{SiO_2(s)} - 3\underbrace{(0)}_{C(s)} = -624.6 \text{ kJ}$$

c) Silicon carbide should resemble diamond in its properties: hard, high melting, and a poor conductor of electricity.

20-13 The reaction is: $SiC(s) + 2\,O_2(g) \rightarrow SiO_2(s, quartz) + CO_2(g)$. Refer to Appendix D for the necessary values of ΔG_f°:

$$\Delta G^\circ = 1\underbrace{(-394.36)}_{CO_2(g)} + 1\underbrace{(-856.67)}_{SiO_2(s)} - 1\underbrace{(-62.8)}_{SiC(s)} - 2\underbrace{(0)}_{O_2(g)} = -1188.2 \text{ kJ}$$

The ΔG° for this reaction is less than zero so the reaction is spontaneous. Thus, $SiC(s)$ tends spontaneously to react with oxygen. The rate of this reaction is vanishingly slow at room temperature.

20-15 Use Planck's relation to compute the energy that corresponds to a wavelength of 920 nm:

$$E = h\nu = \frac{hc}{\lambda} = \frac{(6.626 \times 10^{-34} \text{ J s})(2.9979 \times 10^{8} \text{ m s}^{-1})}{920 \times 10^{-9} \text{ m}} = 2.16 \times 10^{-19} \text{ J}$$

This energy is the band gap energy in InP.

20-17 The problem requires substitution in the formula for the number of electrons excited. The gap energy E_g is 8.7×10^{-19} J, which is equivalent to 5.24×10^5 J mol^{-1}. With T at 300 K and R equal to 8.315 J K^{-1}mol^{-1} the formula becomes:

$$n_e = (4.8 \times 10^{15} \text{ cm}^{-3}\,\text{K}^{-3/2})(300 \text{ K})^{3/2} \exp\left(\frac{-5.24 \times 10^5 \text{ J mol}^{-1}}{2(8.315 \text{ J K}^{-1}\text{mol}^{-1})(300 \text{ K})}\right)$$

The answer is 6.1×10^{-27} cm^{-3}, which is quite small. In a one cm^3 diamond (pretty big for a diamond) at room temperature, only 6.1×10^{-27} electrons are excited to the conduction band; there are essentially no electrons in the conduction band.

20-19 a) Phosphorus-doped silicon is an n-type semiconductor because substitution of a P (five valence electrons) at an Si (four valence electrons) site populates the conduction band. The carriers of electric current are mobile electrons.

b) Zinc-doped indium antimonide is a p-type semiconductor. Mobile holes are the charge carriers in this material.

20-21 Use the Planck equation to compute the wavelength that corresponds to an energy of 2.9×10^{-19} J:

$$\lambda = \frac{hc}{E} = \frac{(6.626 \times 10^{-34} \text{ J s})(2.9979 \times 10^{8} \text{ m s}^{-1})}{2.9 \times 10^{-19} \text{ J}} = 6.8 \times 10^{-7} \text{ m}$$

Light of this wavelength is red.

20-23 At room temperature, zinc white does not absorb in the visible region, although it does absorb ultraviolet light. It appears white. When it is heated, the absorption in the UV is shifted into the blue end of the visible region. Yellow is the complement of blue so the absorption of blue light makes the substance appear yellow. The shift into the blue from the UV is a shift to lower frequency and indicates a decrease in band gap.

20-25 To impart a red color to the pot, the oxidation state of the iron must be high. The iron in iron oxides will be in a high oxidation state if bound to many oxide anions. Thus, an air-rich (oxygen-rich) atmosphere should be employed. To impart a black color to the pot, the oxidation state of the iron must be low. A smoky fuel-rich atmosphere has little oxygen in it. In such an atmosphere the iron is not oxidized. To make a red pot use an air-rich atmosphere; to make a black pot use a smoky atmosphere.

20-27 According to Table 20-1 (text page 727), leaded glass contains 56% SiO_2, 29% PbO, 9% K_2O, 4% Na_2O, and 2% Al_2O_3 by mass. Assume 100 g of leaded glass. There is then 56 g of SiO_2 and 29 g of PbO. Use the molar masses of SiO_2 ($\mathcal{M}$ = 60.08 g mol^{-1}) and of PbO ($\mathcal{M}$ = 223 g mol^{-1}) to calculate the number of moles of Si and Pb that are present. The answer is 0.93 mol of Si and 0.13 mol of Pb. The ratio of the number of lead atoms to silicon atoms is 0.14.

20-29 Consider a sample of exactly 100 g of pure dolomite. This sample contains 45.7 g of $MgCO_3$ ($\mathcal{M}$ = 84.31 g mol^{-1}) and 54.3 g of $CaCO_3$ ($\mathcal{M}$ = 100.08 g mol^{-1}). Dividing the masses of the compounds by their molar masses shows that the chemical amounts of the two compounds are both equal to 0.542 mol. If the two compounds are present in equal chemical amount the formula is "$(MgCO_3)_1 \cdot (CaCO_3)_1$" which is better written $MgCO_3 \cdot CaCO_3$ or $MgCaC_2O_6$, or still better $MgCa(CO_3)_2$.

20-31 $Si_3N_4(s) + 12\,HF(aq) \rightarrow 3\,SiF_4(g) + 4\,NH_3(aq)$.

20-33 The hybridization of silicon atoms in Si(s) is sp^3, giving rise to a 3-dimensional network of tetrahedral silicon atoms that is just like the diamond structure of carbon. Graphite consists of parallel sheets of hexagonally arrayed σ bonded carbon atoms. Less directional bonds join the sheets. The result is highly electrically conductive in a direction parallel to the sheets as a result of extensive electron delocalization in the out-of-plane π-system . If silicon were to adopt the graphite structure, we would expect high electrical conductivity.

20-35 The empty seat will appear to move to the right at a rate of one seat position per five minutes. The analogy with hole motion in p-type semiconductors is evident. The empty seat is the hole, and the people are the electrons. Each seat corresponds to a lattice site.

Chapter 21

Sulfur, Nitrogen, and Phosphorus

Sulfuric Acids and Its Uses

Sulfuric acid plays a central role in modern industrial society. Historically, it was produced in quantity in the **lead-chamber** process, in which oxides of nitrogen serve as oxygen carriers for the oxidation of SO_2 to SO_3. Currently, the **contact process,** which uses the direct oxidation of SO_2 to SO_3 in the presence of a catalyst, followed by absorption of the SOL_3 in previously produced H_2SO_4, is the main synthetic route to sulfuric acid. Sulfur for sulfuric acid comes from metal sulfides or from deposits of elemental sulfur that are mined by the **Frasch process.** In this technique, superheated water is pumped underground to melt the sulfur, which is then forced up from the sulfur-bearing formation with heated compressed air. Sulfur present as sulfide (mainly hydrogen sulfide) in natural gas and petroleum is another source of sulfur. These impurities must be removed in any case before combustion of the fuels to avoid objectionable smells and the release of SO_2 to the atmosphere. The sulfides are acidic and are extracted from the fuel by scrubbing it with a basic solution. Then, in the **Claus process,** the hydrogen sulfide reacts with SO_2 to yield elemental sulfur.

Sulfuric acid is an intermediate in a large number of chemical processes. Most of its uses are indirect in that sulfur from the acid does not become a part of the product, but ends up as a sulfate waste or as **spent acid.** The important chemical sodium sulfate, however, is synthesized from SO_2, NaCl, and O_2 in the **Hargreaves process** and is a by-product of rayon manufacture. An important application of sulfur chemistry is the processing of wood to wood pulp for paper and cardboard. In the **sulfate**, or **Kraft, process,** an alkaline digestion liquor containing NaOH and Na_2S breaks down the lignin in the wood and allows separation of the cellulose. This pulp is further processed into paper and the digestion liquor is recycled. In the **sulfite process,** wood chips are digested in an acidic solution containing the hydrogen sulfite ion to free the cellulosic pulp.

Phosphorus Chemistry

Elemental phosphorus exists in several allotropic forms. White phosphorus consists of tetrahedral P_4 molecules. In red and black phosphorus the element is extensively bonded in network structures.

The important oxides of phosphorus are P_4O_6 and P_4O_{10}. They are both acidic oxides and react with excess water to give H_3PO_3 (phosphorous acid) and H_3PO_4 (phosphoric acid) respectively.

The single most important use of sulfuric acid is in the production of phosphate fertilizers. Phosphate rock (principally $Ca_5(PO_4)_3F$) is treated with $H_@SO_4$ to give **superphosphate** fertilizer, a mixture of calcium dihydrogen phosphate ($Ca(H_2PO_4)_2{\cdot}H_2O$) and gypsum ($CaSO_4{\cdot}2H_2O$). The same type of reaction using an excess of sulfuric acid produces **wet-process** phosphoric acid. The main use of wet-process phosphoric acid is in reaction with further phosphate rock to generate **triple superphosphate** fertilizer (see **problem 21-43).** A higher grade of phosphoric acid than the wet-process product is afforded by the **furnace process,** in which phosphate rock is first reduced to elemental phosphorus. The P_4 is burned in air to P_4O_{10}, which reacts with water to give H_3PO_4. The reaction of phosphoric acid with ammonia gives ammonium phosphate, the major phosphorus-containing fertilizer in current use. Sodium phosphates from the neutralization of phosphoric acid by sodium hydroxide are used in cleaning products and as builders in detergents.

Phosphate ions link together to form polyanions. Tetrahedral PO_4^{3-} groups join at their corners to give chains that are analogous to the polysilicates. **See problem 19-1.**

Example: Phosphoric acid (H_3PO_4) condenses to give the dimer pyrophosphoric acid ($H_4P_2O_7$) and the trimer tripolyphosphoric acid ($H_5P_3O_{10}$). Predict the formula of the polyphosphoric acid with four phosphates linked in a straight chain. Predict the formula of the polyphosphoric acid with four phosphates linked in a ring. **Solution:** Each upward step corresponds to adding H_3PO_4 and subtracting H_2O. This is equivalent to adding HPO_3. The next formula is $H_6P_4O_{13}$. Closing a tetraphosphate chain into a ring would involve one more condensation step and the concomitant loss of one more H_2O. The cyclic tetraphosphoric acid would have the formula $H_4P_4O_{12}$. This example is related to **problem 19-1.**

The existence of condensed phosphates is very important in biochemistry. The crucial biological molecules adenosine diphosphate and adenosine triphosphate, ADP and ATP, are derivatives of pyrophosphoric acid and triphosphoric acid respectively, just as adenosine monophosphate, AMP, is a derivative of phosphoric acid. See Figure 23-31, text page 843.

Nitrogen Fixation

Nitrogen is an essential nutrient for plants and animals. Although the element is abundant at or near the surface of the earth, most of the supply is chemically inaccessible because of the great stability of the triple bond in the N_2 ,molecule. The formation of compounds between molecular nitrogen and other elements is called **nitrogen fixation.** In nature, atmospheric nitrogen is fixed in the form of NO (which reacts immediately to form NO_2) by the intense heat in lightning flashes. Nitrogen is also fixed by the action of certain bacteria that grow in association with the roots of some plants. Large supplies of nitrogen fertilizers, which are essential to high crop yields, required an economical means for the fixing of atmospheric nitrogen. An early method, the **electric-arc process,** fixed nitrogen by electrical discharge, but was expensive to operation and hard to regulate. In the **cyanamide,** or **Frank-Caro, process,** calcium carbide (CaC_2) was generated by heating the inexpensive materials lime (CaO) and coke (C) in an electric furnace. Calcium carbide then reacted with nitrogen to give calcium cyanamide ($CaCN_2$), which was used directly as fertilizer or treated with water to give ammonia. The major source of industrial fixed nitrogen today is the **Haber-Bosch process,** the direct combination of N_2 and H_2 to yield NH_3 (see **problems 5-27** and **5-65).** The practical process requires a compromise between low temperature, which favors high yield of ammonia at equilibrium, and high temperature, which increases the rate of attainment of equilibrium. The best operating temperature is between 700 and 900 K. A catalyst speeds up the reaction, and high pressure favors higher equilibrium yields of ammonia. The Haber-Bosch process is energy intensive and requires large capital investment to build large structures strong enough to withstand high pressure.

Once nitrogen is fixed in the form of ammonia, a variety f products are easily formed by acid-base reactions for used as fertilizers. The oxidation of ammonia by the **Ostwald process** produces nitric acid (HNO_3. This conversion is performed in stepwise fashion, first to NO, and then to NO_2. The reaction of the NO_2 with water gives HNO_3 and NO, which is cycled back into the process. The major use of nitric acid is in reaction with ammonia to give ammonium nitrate (NH_3NO_3), which is used for fertilizer and as an industrial explosive.

Hydrazine (N_2H_4) is made from aqueous ammonia by oxidation with the aqueous hypochlorite ion ($ClO^-(aq)$) in the **Raschig process.** Note that its formula is twice the formula of ammonia with H_2 subtracted. Accordingly, the oxidation number of nitrogen in hydrazine is -2, *versus* -3 in ammonia. Ammonia is NH_3, and its conjugate acid, NH_4^+, is the ammonium ion. Similarly, H_2NNH_2 is hydrazine and its conjugate acid is the hydrazinium ion. In fact, hydrazine can gain *two* hydrogen ions; both nitrogen atoms are basic.

Explosives are compounds that are thermodynamically unstable, but kinetically stable.s Under proper circumstances they detonate, releasing stored chemical energy at a rapid rate. Primary explosives detonate when shocked or heated; secondary explosives are less sensitive and require a primary explosion to induce detonation.

Nitrogen forms 6 oxides, in which its oxidation number ranges from +1 to +5. Nitrogen oxide (NO) and nitrogen dioxide (NO_2) both have an odd number of valence electrons. Much of the chemistry of these gases can be rationalized on the basis of their tendency to react to form products with even numbers of valence electrons. Thus, $NO(g)$ rather easily loses an electron to form the nitrosyl ion (NO^+), and NO_2 dimerizes to N_2O_4. Both NO_2 and N_2O_4 are fairly strong oxidizing agents in aqueous solution. N_2O_5 is the acid anhydride of nitric acid.

Note the distinction between the thermodynamic stability of a compound and its kinetic stability. This sounds an important theme in descriptive chemistry. Thus, NO_2^- is thermodynamically unstable in basic aqueous solution with respect to disproportionation but such solutions can be kept indefinitely. Similarly, the oxidation of ammonia with oxygen is strongly favored thermodynamically at room temperature but does not proceed at any perceptible rate because no effective kinetic pathway is available. Addition of a Pt catalyst speeds the reaction.

Detailed Solutions to Odd-Numbered Problems

21-1 The reaction is $SO_2(g) + 1/2\,O_2(g) \rightleftharpoons SO_3(g)$. Increasing the pressure will drive this equilibrium to the right. Decreasing the temperature favors the products (the reaction is exothermic). Finally, continuous removal of the SO_3 would favor the production of more product to replace it.

21-3 The oxidation of the pollutant SO_2 by H_2O_2 is represented:

$$SO_2(g) + H_2O_2(aq) \rightarrow HSO_4^-(aq) + H^+(aq)$$

Taking data from Appendix D to compute ΔG° gives:

$$\Delta G^\circ = 1\underbrace{(0)}_{H^+(aq)} + 1\underbrace{(-755.91)}_{HSO_4^-(aq)} - 1\underbrace{(-134.03)}_{H_2O_2(aq)} - 1\underbrace{(-300.19)}_{SO_2(g)} = -321.69 \text{ kJ}$$

The reaction is therefore spontaneous and would be thermodynamically feasible.

21-5 $ZnS(s) + 2\,O_2(g) + H_2O(l) \rightarrow ZnO(s) + H_2SO_4(l)$.

21-7 Sodium sulfide is prepared by heating sodium sulfate with carbon:

$$Na_2SO_4(s) + 2\,C(s) \rightarrow Na_2S(s) + CO_2(g)$$

The required sodium sulfate is cheaply obtained by neutralization of sulfuric acid with sodium hydroxide: $H_2SO_4(l) + 2\,NaOH(s) \rightarrow Na_2SO_4(s) + 2\,H_2O(l)$.

21-9 $5\,H_2SO_4(aq) + 2\,P(s) \rightarrow 5\,SO_2(g) + 2\,H_3PO_4(aq) + 2\,H_2O(l)$.

21-11 If *pairs* of phosphoric acid molecules react and split out water, the reaction must be $2\,H_3PO_4(l) \rightarrow H_4P_2O_7(aq) + H_2O(l)$. The structure of $H_4P_2O_7$ is:

```
    :Ö—H  :Ö—H
     |     |
 Ö=P—O—P=Ö
     |     |
    :Ö—H  :Ö—H
```

21-13 This is a straightforward unit-conversion problem:

$$2 \times 10^{11}\ \text{g N} \times \frac{1\ \text{mol N}}{14.0\ \text{g N}} \times \frac{6.02 \times 10^{23}\ \text{atom N}}{1\ \text{mol N}} \times \frac{1\ \text{year}}{3.15 \times 10^{7}\ \text{s}} = \frac{3 \times 10^{26}\ \text{atom N}}{\text{s}}$$

21-15 The first step is the thermal decomposition of calcium carbonate (limestone) to calcium oxide. The calcium oxide is then converted to calcium carbide by reaction with carbon (coke). Next the calcium carbide is reacted at high temperature with nitrogen to form calcium cyanamide. This is the actual nitrogen-fixing step. Finally the calcium cyanamide is treated with water to liberate ammonia. The four steps and their sum are:

$$CaCO_3(s) \rightarrow CaO(s) + CO_2(g)$$
$$CaO(s) + 3\,C(s) \rightarrow CaC_2(s) + CO(g)$$
$$CaC_2(s) + N_2(g) \rightarrow CaNCN(s) + C(s)$$
$$CaNCN(s) + 4\,H_2O(l) \rightarrow Ca(OH)_2(s) + CO_2(g) + 2\,NH_3(g)$$

$$CaCO_3(s) + 2\,C(s) + N_2(g) + 4\,H_2O(l) \rightarrow Ca(OH)_2(s) + 2\,CO_2(g) + CO(g) + 2\,NH_3(g)$$

The standard enthalpy change of this reaction is obtained by Hess's law combination of appropriate ΔH_f°'s from Appendix D. Assume the calcium carbonate to be in the form of calcite. The answer is +374.4 kJ. Therefore the ΔH° is +187.2 kJ per mole of $NH_3(g)$.

21-17 The advantage of working at higher pressures is the increase in yield. The equilibrium, $N_2(g) + 3\,H_2(g) \rightleftharpoons 2\,NH_3(g)$ will shift to the right with increasing pressure. The disadvantage of higher pressure is that stronger equipment must be constructed. This means more costly vessels, valves, and pipes. The standard enthalpy change for the reaction is negative (−92.2 kJ) indicating that the reaction is exothermic. Therefore, equilibrium yield is also increased by lower temperature. The general rule is that reaction rates increase with increasing temperature, and that gas-phase reactions speed up with increasing pressure. At fixed yield a higher temperature can be used if the pressure is higher, giving a faster reaction.

21-19 The three steps in the Ostwald process are:

$$4\,NH_3(g) + 5\,O_2(g) \rightarrow 4\,NO(g) + 6\,H_2O(g) \quad \Delta H^\circ = -905.48 \text{ kJ}$$
$$2\,NO(g) + O_2(g) \rightarrow 2\,NO_2(g) \quad \Delta H^\circ = -114.14 \text{ kJ}$$
$$3\,NO_2(g) + H_2O(l) \rightarrow 2\,HNO_3(l) + NO(g) \quad \Delta H^\circ = -71.66 \text{ kJ}$$

All three steps are exothermic. Hence, for all three steps, a high equilibrium yield will be favored by low temperature.

21-21

$$3\,HNO_3(aq) + Cr(s) + 3\,H_3O^+(aq) \rightarrow 3\,NO_2(g) + Cr^{3+}(aq) + 6\,H_2O(l)$$

$$3\,HNO_3(aq) + Fe(s) + 3\,H_3O^+(aq) \rightarrow 3\,NO_2(g) + Fe^{3+}(aq) + 6\,H_2O(l)$$

21-23 The compound has 24 valence electrons: $H_2N{-}N{=}N{-}NH_2$

21-25 One way to obtain the equation for the disproportionation reaction is to write one of the half-equations as an oxidation and combine it with the reduction:

reduction	$2\,N_2H_4 + 4\,H_2O + 4\,e^- \rightarrow 4\,NH_3 + 4\,OH^-$
oxidation	$N_2H_4 + 4\,OH^- \rightarrow N_2 + 4\,H_2O + 4\,e^-$
total	$3\,N_2H_4 \rightarrow N_2 + 4\,NH_3$

Note that the reduction half-equation has been multiplied by 2 to equalize the electron count. The disproportionation reaction does not involve OH^- or H_3O^+ ion. Its potential is consequently independent of pH. Combining the two half-cell potentials listed in the problem gives the potential difference of the reaction. It is +1.26 V, indicating that 1 M aqueous hydrazine is thermodynamically unstable under standard conditions with respect to disproportionation to nitrogen and ammonia. It does not follow that the reaction occurs at any appreciable rate.

21-27 The reaction is: $5\,N_2H_4(l) + 4\,HNO_3(l) \rightarrow 7\,N_2(g) + 12\,H_2O(l)$.
Combining ΔH_f°'s: $\Delta H^\circ = 12(-285.83) + 7(0) - 5(50.63) - 4(-174.10)] = -2986.71$ kJ.

21-29 The more electronegative the hydrogen-replacing group, the more the nitrogen lone pair is drawn toward it, and the poorer the electron-donating ability of the nitrogen atom. Hydroxylamine should be less basic than ammonia because O is more electronegative than H. Experiment bears out this prediction.

21-31 A good tactic in working out a possible synthetic route to a target compound, is to work backwards. N_2O_4 is in equilibrium with NO_2. This equilibrium represents the last step in the synthesis:

$$2\,NO_2(g) \rightleftharpoons N_2O_4(s)$$

This exothermic reaction can be shifted to the right by cooling. Also, N_2O_4, a gas at room temperature, freezes out at $-11°C$. The gaseous NO_2 is prepared from ammonia as part of the Ostwald process for nitric acid (see problem 25-7):

$$4\,NH_3(g) + 5\,O_2(g) \rightarrow 4\,NO(g) + 6\,H_2O(g)$$
$$2\,NO(g) + O_2(g) \rightarrow 2\,NO_2(g)$$

Ammonia can be prepared directly from the elements: $N_2(g) + 3\,H_2(g) \rightarrow 2\,NH_3(g)$. Thus, the final four-step route to N_2O_4 from the elements:

$$N_2(g) + 3\,H_2(g) \rightarrow 2\,NH_3(g) \quad 4\,NH_3(g) + 5\,O_2(g) \rightarrow 4\,NO(g) + 6\,H_2O(g)$$
$$2\,NO(g) + O_2(g) \rightarrow 2\,NO_2(g) \qquad 2\,NO_2(g) \rightarrow N_2O_4(s)$$

21-33 In addition to its value as a propaganda ploy, this practice added a good reducing agent to the explosive and made the explosion more energetic. The reaction $2\,Al(s) + 3/2\,O_2(g) \rightarrow Al_2O_3(s)$ has $\Delta H°$ equal to -1675.7 kJ. The aluminum served the same function as fuel oil does in its explosive mixtures with ammonium nitrate.

21-35 Compute the standard free-energy changes of the two reactions using the ΔG_f° data in Appendix D. The $\Delta G°$ of the oxidation of SO_2 by $O_2(g)$ according to the first equation in the problem is -70.89 kJ. The $\Delta G°$ for the oxidation of SO_2 by $NO(g)$ according to the second equation in the problem is -35.63 kJ. The -oxidation by O_2 has a larger thermodynamic driving force, but the oxidation by NO is used in practice. It is faster.

21-37 The reaction in the electrolysis unit combines the oxidation of sulfate ion and the reduction of hydronium ion:

$$2\,SO_4^{2-}(aq) \rightarrow S_2O_8^{2-}(aq) + 2\,e^- \qquad \text{oxidation}$$
$$2\,H_3O^+(aq) + 2\,e^- \rightarrow H_2(g) + 2\,H_2O(l) \qquad \text{reduction}$$
$$2\,H_3O^+ + 2\,SO_4^{2-}(aq) \rightarrow H_2(g) + S_2O_8^{2-}(aq) + 2\,H_2O(l) \qquad \text{overall}$$

The potential difference is:

$$\Delta\mathcal{E}° = \mathcal{E}°(\text{reduction}) - \mathcal{E}°(\text{oxidation}) = 0.0 - 2.0 = -2.0 \text{ V}$$

The Nernst equation for this reaction (at 25°C) is:

$$\Delta\mathcal{E} = -2.0 \text{ V} - \frac{0.0592 \text{ V}}{2} \log\left(\frac{P_{H_2}[S_2O_8^{2-}]}{[H_3O^+]^2[SO_4^{2-}]^2}\right) = -2.0 - \frac{0.0592}{2}\log\left(\frac{(0.1)(0.5)}{(1.0)^2(1.0)^2}\right)$$

Completion of the arithmetic gives a potential difference of -1.96 V. The electrolysis unit requires a minimum voltage of $+1.96$ V to force the reaction to run.

21-39 a) $Ba^{2+}(aq) + H_2SO_4(aq) + 2\,H_2O(l) \rightarrow BaSO_4(s) + 2\,H_3O^+(aq)$
b) $2\,NH_3(aq) + H_2SO_4(aq) \rightarrow 2\,NH_4^+(aq) + SO_4^{2-}(aq)$
c) $Cu(s) + 2\,H_2SO_4(aq) \rightarrow Cu^{2+}(aq) + SO_2(g) + SO_4^{2-}(aq) + 2\,H_2O(l)$

21-41 a) The ΔG at 1000 K for this reaction can be estimated by assuming that the ΔH and ΔS at 1000 K remain unchanged from their values at 298 K and writing:

$$\Delta G = \Delta H - T\Delta S = \Delta H^\circ - T\Delta S^\circ = \Delta H^\circ - (1000\text{ K})\Delta S^\circ$$

The ΔH° and ΔS° of the reaction are computed from the data in Appendix D:

$$\Delta H^\circ = 2\underbrace{(-241.82)}_{H_2O(l)} + 4\underbrace{(-296.83)}_{SO_2(g)} + 1\underbrace{(-393.51)}_{CO_2(g)} - 4\underbrace{(-395.72)}_{SO_3(g)} - 1\underbrace{(-74.81)}_{CH_4(g)} = -406.78\text{ kJ}$$

$$\Delta S^\circ = 2\underbrace{(188.72)}_{H_2O(l)} + 4\underbrace{(248.11)}_{SO_2(g)} + 1\underbrace{(213.63)}_{CO_2(g)} - 4\underbrace{(256.65)}_{SO_3(g)} - 1\underbrace{(186.15)}_{CH_4(g)} = 370.76\text{ J K}^{-1}$$

Inserting these numbers in the previous equation gives:

$$\Delta G = \Delta H^\circ - T\Delta S^\circ = -406.78\text{ kJ} - (1000\text{ K})(0.37076\text{ kJ K}^{-1}) = -780\text{ kJ}$$

b) The ΔH° and ΔS° for this reaction are computed just as in the previous:

$$\Delta H^\circ = 1\underbrace{(-241.82)}_{H_2O(l)} + 1\underbrace{(-20.63)}_{H_2S(g)} + 1\underbrace{(-393.51)}_{CO_2(g)} - 1\underbrace{(-395.72)}_{SO_3(g)} - 1\underbrace{(-74.81)}_{CH_4(g)} = -185.43\text{ kJ}$$

$$\Delta S^\circ = 1\underbrace{(188.72)}_{H_2O(l)}\,1\underbrace{(205.68)}_{H_2S(g)} + 1\underbrace{(213.63)}_{CO_2(g)} - 1\underbrace{(256.65)}_{SO_3(g)} - 1\underbrace{(186.15)}_{CH_4(g)} = 165.23\text{ J K}^{-1}$$

From which: $\Delta G = -185.43\text{ kJ} - (1000\text{ K})(0.16523\text{ kJ K}^{-1}) = -351\text{ kJ}$.
c) Both reaction are spontaneous at 1000 K. Thermodynamics cannot predict which, if either, will occur preferentially. Although the first is more favored than the second, the second could still predominate if it were faster and if the reaction:

$$H_2S + 3\,SO_3 \rightarrow 4\,SO_2 + H_2O$$

were slow. Note that this reaction added to the second reaction gives the first. The best way to settle the issue is to run the reaction and analyze the product gases.

21-43 Superphosphate is $(Ca(H_2PO_4)_2{\cdot}H_2O)_3\,(CaSO_4{\cdot}2H_2O)_7$. This compound has a molar mass of 1961.4 g mol^{-1}. The mass percentage of P_2O_5 in superphosphate is:

$$\frac{3 \times 141.94 \text{ g mol}^{-1}}{1961.4 \text{ g mol}^{-1}} \times 100\% = 21.7\%$$

This is about one-third the percentage of P_2O_5 in triple superphosphate.

21-45 The formation of nitrogen compounds from available starting materials frequently requires the input of energy. The Haber-Bosch process requires high pressures of $N_2(g)$ and $H_2(g)$. It takes energy to produce these pressures. The first step in the Ostwald process for nitric acid requires heating to 800°C. In the 1970's, when the price of petroleum rose, so did the price of energy, and consequently, the cost of manufacturing ammonia, nitric acid and from them fertilizer.

21-47 a) $6\,NO_2(g) + 4\,(NH_2)_2CO(aq) \rightarrow 7\,N_2(g) + 4\,CO_2(g) + 8\,H_2O(l)$
b) $6\,NO(g) + 2\,(NH_2)_2CO(aq) \rightarrow 5\,N_2(g) + 2\,CO_2(g) + 4\,H_2O(l)$

21-49 In the Raschig process hydrazine is produced from aqueous hypochlorite ion and ammonia:

$$2\,NH_3(aq) + OCl^-(aq) \rightarrow N_2H_4(aq) + H_2O(l) + Cl^-(aq)$$

If we assume NH_2Cl to be an intermediate in the two-step mechanism for the above reaction, then one possibility is:

$$NH_3(g) + OCl^-(aq) \rightarrow NH_2Cl(aq) + OH^-(aq)$$
$$NH_2Cl(aq) + NH_3(aq) + OH^-(aq) \rightarrow N_2H_4(aq) + H_2O(l) + Cl^-(aq)$$

21-51 The problem concerns a 1.00 M aqueous solution of hydrazine (NH_2NH_2) with a pH of 0.00. At this pH, $[H_3O^+]$ is 1.0 M and $[OH^-]$ is 1.0×10^{-14} M. Hydrazine is a base in water:

$$N_2H_4(aq) + H_2O(l) \rightleftharpoons N_2H_5^+(aq) + OH^-(aq) \quad K_b = \frac{K_w}{K_a} = \frac{1.0 \times 10^{-14}}{1.17 \times 10^{-8}} = 8.5 \times 10^{-7}$$

Substitution in the equilibrium expression

$$\frac{[OH^-][N_2H_5^+]}{N_2H_4]} = 8.5 \times 10^{-7} \text{ gives } \frac{[N_2H_5^+]}{N_2H_4]} = 8.5 \times 10^{7}$$

when $[OH^-] = 1.0 \times 10^{-14}$ M This large ratio means that effectively no $N_2H_4(aq)$ is present at equilibrium at this pH. It is all converted to $N_2H_5^+$ by the high concentration of H_3O^+. Substitute 1.00 M for the concentration of $N_2H_5^+$ in the preceding equilibrium equation. Solving gives $[N_2H_4] = 1.17 \times 10^{-8}$ M.

21-53 a) The molar free energies of formation (in kJ mol^{-1}) of the reactants and products are:

	$6\,Fe_2O_3(s)$	$+\ N_2H_4(aq)$	$\rightarrow 4\,Fe_3O_4(s)$	$+\ N_2(g)$	$+2\,H_2O(l)$
ΔG_f°	-742.2	128.1	-1015.5	0	-237.18

Combine these in the usual way to obtain ΔG° of the reaction:

$$\Delta G^\circ = 4(-1015.5) + 1(0) + 2(-237.18) - 6(-742.2) - 1(128.1) = -211.3 \text{ kJ}$$

b) The proposal is no good. When hydrazine reacts with rust it acts as a reducing agent. Under highly acidic conditions, hydrazine gains H^+ to form the hydrazinium ion $N_2H_5^+$ which is a good oxidizing agent. The suggested improvement would actually probably aggravate the rusting problem.

Chapter 22

The Halogen Family and the Noble Gases

The halogens include chlorine, bromine, iodine and fluorine, and comprise Group VII of the periodic table. The noble gases are helium, neon, argon, krypton, and radon, and are Group VIII.

Chemicals from Salt

Glassmaking requires sodium carbonate (soda ash), and soapmaking requires a cheap base that is more soluble than calcium hydroxide (slaked lime). A third important process, the bleaching of cloth or fibers, requires a quick-acting bleach. Suitable chemicals for all these purposes come from sodium chloride, by a variety of production methods.

In the **Leblanc process** for sodium carbonate, sodium chloride was heated with sulfuric acid to form sodium sulfate. Heating sodium sulfate with a mixture of carbon (from coal) and limestone ($CaCO_3$) gives a mixture from which the water-soluble sodium carbonate could be extracted. The process, which is no longer used, had two noxious by-products—CaS and HCl. Oxidation of the HCl from the Leblanc process to Cl_2 either by reaction with the mineral pyrolusite (MnO_2) or by the **Deacon process** changed the HCl from a nuisance to an asset because chlorine, in the form of **bleaching powder** CaCl(OCl) makes an excellent bleach for textiles.

The **Solvay process** for sodium carbonate supplanted the Leblanc process. In it, the net reaction is the conversion of NaCl and $CaCO_3$ to Na_2CO_3 and the by-product $CaCl_2$. Ammonia and ammonium chloride are important intermediates that are recycled through this continuous process. Much Na_2CO_3 is now obtained by mining **trona,** and sodium hydroxide (formerly, made by reacting Solvay Na_2CO_3 with $Ca(OH)_2$ is now largely produced electrolytically.

Sodium hydroxide and chlorine are produced from salt by the electrolysis of concentrated aqueous solutions. The by-product is hydrogen, which has many uses. In the **diaphragm cell,** for the electrolysis of brine, the anode is made of titanium coated with a noble metal, and the cathode is steel. The two are separated by a diaphragm that prevents the mixing of the product gases Cl_2 (anode) and H_2 (cathode) but allows migration of ions to preserve electrical neutrality. In the **mercury cell,** the electrolysis produces chlorine at the anode, and a solution of sodium in mercury at the cathode. The dissolved sodium is subsequently converted to NaOH by reaction with water. Sodium hydroxide, a strong soluble base, has many uses in chemical and materials processing.

The Chemistry of Chlorine, Bromine and Iodine

This section treats the properties, reactions and compounds of the elements of Group VII, the halogens, with the exception of fluorine. One format for studying such material is to outline it under the following headings:

1. Occurrence and preparation of the element
2. Characteristic chemical and physical properties
3. Important compounds and types of compounds
4. Uses in society

The halogens exhibit many regular trends in their properties in accord with the predictions of the periodic law. All are reactive non-metals that serve as oxidizing agents. Their oxidizing strength decreases regularly going down the group. The solution chemistry of the halogens can in large part be summarized in a set of reduction potential diagrams. (see page 278 of this Guide). Many of the potentials come from Appendix E.

Reduction Potential Diagrams of the Halogens
Acidic Solution (pH 0)

$$F_2 \xrightarrow{2.87} F^-$$

$$ClO_4^- \xrightarrow{1.20} ClO_3^- \xrightarrow{1.18} HClO_2 \xrightarrow{1.70} HClO \xrightarrow{1.63} Cl_2 \xrightarrow{1.36} Cl^-$$

$$BrO_4^- \xrightarrow{1.85} BrO_3^- \xrightarrow{1.45} HBrO \xrightarrow{1.60} Br_2 \xrightarrow{1.06} Br^-$$

$$H_5IO_6 \xrightarrow{1.60} IO_3^- \xrightarrow{1.13} HIO \xrightarrow{1.44} I_2 \xrightarrow{0.53} I^-$$

Basic Solution (pH 14)

$$F_2 \xrightarrow{2.87} F^-$$

$$ClO_4^- \xrightarrow{0.37} ClO_3^- \xrightarrow{0.29} ClO_2^- \xrightarrow{0.59} ClO^- \xrightarrow{0.42} Cl_2 \xrightarrow{1.36} Cl^-$$

$$BrO_4^- \xrightarrow{1.03} BrO_3^- \xrightarrow{0.49} BrO^- \xrightarrow{0.45} Br_2 \xrightarrow{1.06} Br^-$$

$$H_3IO_6^{2-} \xrightarrow{0.65} IO_3^- \xrightarrow{0.15} IO^- \xrightarrow{0.42} I_2 \xrightarrow{0.53} I^-$$

Observe how much these diagrams can convey. They show, for example:

- The high reduction potential of iodine(VII) in acid solution. Iodine(VII) ion can oxidize $Mn^{2+}(aq)$ to $MnO_4^-(aq)$ because the reduction potential of the MnO_4^- to Mn^{2+} couple is only about 1.5 V and lower than 1.6 V.

- That F_2 oxidizes Cl^-. In addition the diagrams show that each elemental halogen will oxidize all halides (X^- ions) in compounds of the halogens beneath it in the periodic table. **See problem 22-43.**

- That the chemistry of periodic acid involves complications not present in perbromic and perchloric acids. The formulas of the major compound of iodine(VII) both in acid and base show the expansion of the iodine coordination sphere to include additional water molecules.

Fluorine and Its Compounds

The chemistry of fluorine is discussed under a separate heading because of the extreme differences between it and the other halogens. These include:

- Fluorine exhibits only the 0 and -1 oxidation states unlike the other halogens, all of which can be oxidized to positive oxidation states ($+1$, $+3$, $+5$, and $+7$).

- Only fluorine of all the halogens has been shown to form compounds with any of the noble gases (Group VIII).

- The dissociation energy of the F_2 molecule is abnormally low.

- The oxidizing ability of fluorine is extremely high. This is related to its high electronegativity, its low dissociation energy, and the strong bonds it forms with other elements.

In studying the descriptive material here, note how the theoretical concepts of previous chapters are put to work in systematizing great quantities of chemical information. For example, fluorine forms compounds with nearly every other element. It might be possible to memorize the ΔH_f° of all of these compounds. But linking the data to the periodic table allows useful *general* conclusions. Look for valid generalizations and avoid memorizing minutiae.

Fluorine atoms form weak bonds to other fluorine atoms but particularly strong bonds to most other kinds of atoms. These facts underlie the extraordinary chemical reactivity of F_2 and the difficulties encountered in the first attempts to prepare it in elemental form. Modern methods for preparing fluorine are based on the original 1886 electrolytic process, and employ KF dissolved in HF, both fluorides derived from natural sources in fluoroapatite or fluorite minerals. Fluorine displaces other halogens from their compounds and oxidizes elements in compounds exposed to it. Elemental fluorine is used in uranium production and isotope separation, and in the manufacture of sulfur hexafluoride. Calcium fluoride is used in the steel industry, and hydrogen fluoride is used to make synthetic cryolite for aluminum production.

Most fluorine compounds are made directly or indirectly from hydrogen fluoride, a colorless hydrogen-bonded liquid. The compounds of fluorine with metals in their lower oxidation states are salts with strong ionic character. The higher oxidation states of many of the same elements yield volatile fluorides with significant covalent character (see **problem 22-29**). Compounds of fluorine with the nonmetals possess a wide range of reactivities and molecular geometries. Some compounds, such as SF_6, are quite inert toward chemical reaction; others such as BF_3 are strong Lewis acids (see **problem 22-27**) that readily accept a share in electron pairs from fluoride ions or other Lewis bases; still others, such as SbF_5, are strong fluorinating agents. One of the strongest of these agents PtF_6 opened up modern research on the compounds of the noble gases by its reaction with xenon; xenon and krypton fluorides are made by direct reaction with fluorine or with other fluorinating agents. Xenon fluorides react with water to yield xenon trioxide.

Fluorinated hydrocarbons arise by the substitution of fluorine for hydrogen in hydrocarbons and often have greater thermal and chemical stability than the parent hydrocarbon. The release of chlorine upon eventual decomposition of **chlorofluorocarbons** in the outer atmosphere leads to depletion of the ozone in that region. The unique properties of fluorine shape the properties of the important fluorinated polymer Teflon (see **problem 22-55**).

Compounds of Fluorine and the Noble Gases

The noble gases are in Group 0 (VIII) of the periodic table. Their principal uses come from their physical properties. The unreactive character of these elements prevents their entry into chemical compounds with most other elements. Fluorine and oxygen form several compounds with Xe. Note that the ΔH_f°'s of all of the xenon fluorides are negative (see **problem 22-35**). The compounds are thermodynamically stable with respect to their elements.

• Problems in Chapter 22 all involve the application of previously studied principles to the particular cases of the halogens and noble gases.

Detailed Solutions to Odd-numbered Problems

22-1 Use ΔG_f° data (Appendix D) to compute ΔG° of this reaction: The set-up is:

$$\Delta G^\circ = 2\underbrace{(-36.8)}_{OCl^-(aq)} + 1\underbrace{(-237.18)}_{H_2O(l)} - 2\underbrace{(-157.24)}_{OH^-(aq)} - 1\underbrace{(97.9)}_{Cl_2O(g)} = -94.2 \text{ kJ}$$

Substitute this value into the equation $\Delta G^\circ = -RT \ln K$. Use $T = 298.15$ K and R $= 8.315$ J mol^{-1}K^{-1}. Be sure to convert from the units of ΔG° from kJ to J. The $\ln K$ is 38.0, and K is 3.2×10^{16}.

22-3 The overall reaction in the Leblanc process is:

$$2\,NaCl(s) + H_2SO_4(l) + 2\,C(s) + CaCO_3(s) \longrightarrow Na_2CO_3(s) + 2\,CO_2(g) + CaS(s) + 2\,HCl(g)$$

Use the enthalpies of formation from Appendix D in the usual way to compute the ΔH° of this process:

$$\Delta H^\circ = 1\underbrace{(-1130.68)}_{Na_2CO_3(s)} + 2\underbrace{(-393.51)}_{CO_2(g)} + 1\underbrace{(-482.41)}_{CaS(s)} + 2\underbrace{(-92.31)}_{HCl(g)}$$
$$-2\underbrace{(-411.15)}_{NaCl(s)} - 1\underbrace{(-813.99)}_{H_2SO_4(l)} - 2\underbrace{(00)}_{C(s)} - 1\underbrace{(-1206.92)}_{CaCO_3(s)} = +258.48 \text{ kJ}$$

The process is endothermic because ΔH° is greater than zero.

22-5 The production of Cl_2 is tied to the production of NaOH by the stoichiometry of the reaction: $2\,NaCl + 2\,H_2O \longrightarrow Cl_2 + 2\,NaOH + H_2$. According to this equation, 2 mol of NaOH is produced per 1 mol of Cl_2. The mass of 2 mol of NaOH is 80.0 g, and the mass of 1 mol of Cl_2 is 70.9 g. If there were no losses of either material during production, the ratio of their masses would always equal 1.13 to 1. Evidently, more NaOH is lost in practice than Cl_2, reducing this ratio to the 1.05 quoted in the problem.

22-7 The problem is a reversal on the more common task of determining an equilibrium constant from ΔG data. It gives an equilibrium constant (in slightly disguised form) and asks for an estimate of ΔG°. The reaction is: $Br_2(l) \rightleftharpoons Br_2(aq)$. At 25°C

(298.15 K), 33.6 g of bromine dissolves in one liter of water. Once the solution is saturated, the above equilibrium exists. The concentration of $Br_2(aq)$ at the point of saturation is 0.210 M (taking the molar mass of Br_2 as 159.82 g mol^{-1}). The equilibrium constant expression for the reaction is:

$$K = [Br_2](aq)$$

so K is 0.210. The equilibrium constant and $\Delta G°$ are related:

$$\Delta G° = -RT \ln K$$

Substitution of $R = 8.315$ J $mol^{-1}K^{-1}$ and $T = 298.15$ K gives $\Delta G°$ equal to +3.87 kJ for the reaction as written, which is the reaction giving 1 mol of $Br_2(aq)$. The Appendix D value of ΔG_f° for one mole of $Br_2(aq)$ is +3.93 kJ. Note that *some* Br_2 does dissolve despite the fact that $\Delta G°$ is positive.

22-9 The $\Delta G°$ of the reaction: $2\,Br^-(aq) + Cl_2(g) \rightleftharpoons Br_2(g) + 2\,Cl^-(aq)$ is:

$$\Delta G° = 2(-131.23) + 3.14 - 2(-103.96) \text{ kJ} = -51.4 \text{ kJ}$$

where the numerical data come from Appendix D. This $\Delta G°$ is related to K by the equation:

$$\Delta G° = -RT \ln K$$

Substituting $R = 8.315$ J $mol^{-1}K^{-1}$ and $T = 298.15$ K gives $\ln K = 20.7$, from which K is 1.0×10^9. The same value is obtained using the standard potential difference of the reaction in the highlighted equation on text page 373. The K is large enough to make the extraction of $Br_2(g)$ from sea water by oxidation with $Cl_2(g)$ thermodynamically feasible. If the chlorine is cheap and the reaction is fast, the process might also be economically feasible.

22-11 a) The balanced equation is: $Br^-(aq) + H_3PO_4(aq) \rightarrow H_2PO_4^-(aq) + HBr(g)$.

Other versions are possible (involving NaBr as a reactant and giving PO_4^{3-} or Na_3PO_4 or other forms of phosphate as a product). This ionic equation is best.
b) The 100 mL of solution contains 0.050 mol $L^{-1} \times 0.100$ L = 0.0050 mol of $Br^-(aq)$. According to the balanced equation, 0.0050 mol of $HBr(g)$ will form. At standard temperature and pressure, gaseous HBr, if ideal, occupies 22.4 L mol^{-1}. Hence, this portion of $HBr(g)$ occupies 0.11 L.

22-13 A reaction involving reactants and products in their standard states is spontaneous at constant T and P if $\Delta G°$ is less than zero, Because $\Delta G° = -nF\Delta\mathcal{E}°$, a reaction is also spontaneous if $\Delta\mathcal{E}°$ is greater than zero. If the first half-reaction in the problem runs as a reduction (at the cathode) and the second as an oxidation (at the anode), then

$$\Delta\mathcal{E}^\circ = \mathcal{E}^\circ \text{ (cathode)} - \mathcal{E}^\circ \text{ (anode)} = 0.954 - (-0.25) = 1.20 \text{ V} > 0$$

the reaction is spontaneous as written. The corresponding reaction is:

$$2\,ClO_2 + 2\,OH^- \rightarrow ClO_2^- + ClO_3^- + H_2O$$

Thus, ClO_2 is tends to disproportionate in basic solution.

22-15 The standard reduction potential diagram for iodine in strong base shows that $I_2(s)$ disproportionates spontaneously to $IO^-(aq)$ and $I^-(aq)$:

$$I_2(s) + 2\,OH^-(aq) \rightarrow I^-(aq) + IO^-(aq) + H_2O(l)$$

This is because 0.535 V, the standard reduction potential for $I_2(s)$ to $I^-(aq)$, exceeds 0.45 V, the potential for $IO^-(aq)$ to $I_2(aq)$. But $IO^-(aq)$ itself disproportionates spontaneously to $IO_3^-(aq)$ and $I^-(aq)$:

$$3\,IO^-(aq) \rightarrow IO_3^-(aq) + 2\,I^-(aq)$$

as shown by a comparison between the $IO_3^-(aq) \rightarrow IO^-(aq)$ and $IO^-(aq) \rightarrow I^-(aq)$ reduction potentials. The overall balanced reaction for the disproportionation is:

$$6\,OH^-(aq) + 3\,I_2(aq) \rightarrow IO_3^-(aq) + 5\,I^-(aq) + 3\,H_2O(l) \quad \Delta\mathcal{E}^\circ = 0.49 - 0.14 = 0.35 \text{ V}$$

b) The half-reaction: $IO_3^-(aq) + 3\,H_2O(l) + 6\,e^- \rightarrow I^-(aq) + 6\,OH^-(aq)$ is the sum of:

$$IO_3^- + 2\,H_2O + 4\,e^- \rightarrow IO^- + 4\,OH^- \quad \mathcal{E}^\circ = 0.14 \text{ V}$$
$$IO^- + H_2O + 2\,e^- \rightarrow I^- + 2\,OH^- \quad \mathcal{E}^\circ = 0.49 \text{ V}$$

The first half-reaction transfers 4 electrons, and the second transfers 2 electrons for an overall transfer of 6 e^-'s:

$$\mathcal{E}^\circ = \frac{(4 \times 0.14) + (2 \times 0.49)}{6} = 0.26 \text{ V}$$

22-17 a) $2\,F_2(g) + 2\,SrO(s) \rightarrow 2\,SrF_2(s) + O_2(g)$
b) $2\,F_2(g) + O_2(g) \rightarrow 2\,OF_2(g)$ **c)** $UF_4(s) + F_2(g) \rightarrow UF_6(g)$

22-19 Calculate the relative number of moles of Na and F. The empirical formula is NaF·2HF.

22-21 The compound is 38.35% Cl and 61.65% F, which corresponds to the empirical formula ClF_3. See problem 1-41. The molecular formula is then Cl_nF_{3n} where n is an integer. To obtain the molecular formula, determine the molar mass. The molar mass of an ideal gas can be calculated from its measured vapor density ρ at any conditions of T and P. It can be shown from the ideal-gas equation and the definition of molar volume (text page 25) that:

$$\mathcal{M}_{\text{vap}} = \frac{\rho_{\text{vap}} RT}{P}$$

Substitution of the P and T values in this case along with the vapor density give a molar mass of 92.5 g mol^{-1}. When n is one, the above molecular formula gives exactly this molar mass. Hence the molecular formula is CF_3.

22-23 The shapes of all of these compounds can be predicted using the valence shell electron pair repulsion (VSEPR) theory. See page 382 in this Guide.

a) OF_2 has a central O atom with a steric number of 3. This includes the two F atoms and one lone pair of electrons. The molecule is bent.

b) BF_3 has a central B atom with *SN* 3. The molecule is trigonal planar.

c) In BrF_3, the central Br has *SN* of 5. The molecule has a distorted T shape.

d) In BrF_5, the central Br is surrounded by five F atoms and one lone pair for a *SN* of 6. The molecule has a square pyramidal shape.

e) In IF_7, the central I is surrounded by seven F atoms. One arrangement that minimizes repulsions among 7 electron pairs distributed about a center is a pentagonal bipyramid. See the solution to problem 14-67b in this Guide. This is the structure of IF_7.

f) In SeF_6, the central Se is surrounded by six F atoms and no lone pairs. The *SN* is 6 and the predicted geometry is octahedral.

22-25 a) Molecules of the two compounds will have the same geometry because the central atoms of the two have both the same number of valence electrons and the same number of bonded atoms. Use the VSEPR model to predict the molecular geometry. The central atoms of the molecules have *SN* 4 with 1 lone pair and 3 bonded atoms. The predicted molecular geometry is pyramidal.

b) Recall that Lewis bases are electron-pair donors and Lewis acids are electron-pair acceptors. In the reaction between SOF_2 and BF_3, the SOF_2 molecule donates its lone pair of electrons to the BF_3 molecule. Thus, SOF_2 is acting as a Lewis base.

22-27 Fluorine chemists often think of fluoride-ion donors as Lewis bases and fluoride-ion acceptors as Lewis acids. The equilibrium is:

$$2\,BrF_3(l) \rightleftharpoons BrF_4^-(solv) + BrF_2^+(solv)$$

In this reaction BrF_3 is amphoteric, acting as both a Lewis acid and a Lewis base. The BrF_2^+ is a Lewis acid because it can accept a fluoride ion (from BrF_4^-); BrF_4^- is a Lewis base because it can donate a fluoride ion (to BrF_2^+).

22-29 The molecule with the least ionic character, and smallest intermolecular interactions is expected to have the lowest boiling point. The ranking would therefore be $PtF_6 < PtF_4 < CaF_2$. In fact, PtF_4 decomposes upon melting, and so does not possess a normal boiling point.

22-31 The reaction of interest is: $C_2F_4(g) \rightarrow C(s) + CF_4(g)$. Combine the ΔH_f°'s to obtain ΔH°:

$$\Delta H^\circ = 1\underbrace{(-925)}_{CF_4(g)} + 1\underbrace{(0)}_{C(s)} - 1\underbrace{(-651)}_{CF_4(g)} = -274 \text{ kJ}$$

Thus, if one mol of C_2F_2 were to explode, 274 kJ of heat would be released. The molar mass of C_2F_2 is 100 g mol^{-1}. The 1.00 kg of C_2F_4 therefore is 10.0 mol. Its explosion would release about 2740 kJ.

22-33 The redox reaction: $XeF_6(g) + 3\,H_2(g) \rightarrow Xe(g) + 6\,HF(g)$ has ΔH° equal to the sum of the ΔH_f°'s of the products minus the sum of the ΔH_f°'s of the reactants:

$$\Delta H^\circ = 1\underbrace{\Delta H_f^\circ}_{Xe(g)} + 6\underbrace{\Delta H_f^\circ}_{HF(g)} - 3\underbrace{\Delta H_f^\circ}_{H_2(g)} - 1\underbrace{\Delta H_f^\circ}_{XeF_6(g)}$$

Inserting data from the problem and Appendix D gives:

$$-1282 \text{ kJ} = 1\underbrace{(0)}_{Xe(g)} + 6\underbrace{(-271.1)}_{HF(g)} - 3\underbrace{(0)}_{H_2(g)} - 1\underbrace{(\Delta H_f^\circ)}_{XeF_6(g)}$$

Solving gives the ΔH_f° for $XeF_6(g)$. It is -345 kJ mol^{-1}.
b) From part a) and Appendix D we have:

$$XeF_6(g) \rightarrow Xe(g) + 3\,F_2(g) \quad \Delta H^\circ = +345 \text{ kJ}$$
$$1/2\,F_2(g) \rightarrow F(g) \quad \Delta H^\circ = 78.99 \text{ kJ}$$

Multiplying the second equation by 6 and adding it to the first gives:

$$XeF_6(g) \rightarrow Xe(g) + 6\,F(g) \quad \Delta H^\circ = 818.94 \text{ kJ}$$

The ΔH° here is the enthalpy change to atomize 1 mol of $XeF_6(g)$. The average bond enthalpy of the Xe—F bond is 1/6 of this or 136 kJ mol^{-1}.

22-35 The ΔH° for the hydrolysis:

$$XeF_6(g) + H_2O(l) \rightarrow XeOF_4(l) + 2\,HF(g)$$

is calculated by combining ΔH_f° values according to Hess's law. The first and third in the following are given in the problem; the others are from Appendix D:

$$\Delta H^\circ = 1\underbrace{(148)}_{XeOF_4(l)} + 2\underbrace{(-271)}_{HF(g)} - 1\underbrace{(-298)}_{XeF_6(g)} - 1\underbrace{(-285.83)}_{H_2O(l)} = 190 \text{ kJ}$$

The standard enthalpy change for hydrolysis of $XeF_6(g)$ is 190 kJ mol^{-1}.

22-37 The balanced half-equations are:

$$Mn^{2+}(aq) + 12\,H_2O(l) \rightarrow MnO_4^-(aq) + 8\,H_3O^+(aq) + 5\,e^-$$
$$XeO_6^{4-}(aq) + 12\,H_3O^+(aq) + 8\,e^- \rightarrow Xe(g) + 18\,H_2O(l)$$

The oxidation of the Mn^{2+} releases 5 e^- per ion and the reduction of the XeO_6^{4-} consumes 8 e^- per ion. The overall equation is accordingly:

$$8\,Mn^{2+}(aq) + 5\,XeO_6^{4-}(aq) + 6\,H_2O(l) \rightarrow 8\,MnO_4^-(aq) + 5\,Xe(g) + 4\,H_3O^+(aq)$$

22-39 Gaseous chlorine is a stronger oxidizing agent than $Br_2(l)$ because the reduction potential for Cl_2/Cl^- is more positive than the reduction potential for Br_2/Br^-. The stronger oxidizing agent will be the stronger bleach.

22-41 (a) A current of 100 000 A corresponds to the passage of 100 000 C s^{-1}. There are 86 400 seconds in a day, so 8.64×10^9 C passes through each of the 250 cells daily. This equals 8.95×10^4 mol of electrons per cell (computed using the Faraday constant ($\mathcal{F}$ = 96 485 C mol^{-1}). As each mole of electrons that passes through a cell, 1/2 mol of Cl_2 is produced, according to the half-reaction $Cl^- \rightarrow Cl_2 + 2\,e^-$. Therefore, 4.48×10^4 mol of Cl_2 is produced by one cell in a day. This is 3.17×10^6 g of Cl_2 per cell. The total production is 7.94×10^8 g of Cl_2 per day for the 250-cell plant. This is 794 metric tons of Cl_2.

b) The power consumption in watts is the operating potential in volts multiplied by the current. Each cell uses 3.5 V × 100 000 A = 350 000 W. The set of 250 cells use 250 times this: 8.75×10^7 W. Multiplying this power by the 24 hours in a day gives 2.10×10^6 kilowatt-hour as the daily energy consumption of the plant. Since a kWh is 3.600×10^6 J (see problem 10-73), the daily energy consumption is 7.56×10^{12} J.

c) The electric bill for one day at this plant is $105,000.

22-43 a) The trend in acid strength among the hydrogen halides acid strength runs: HI > HBr > HCl. Therefore HAt should be an even stronger acid than HI.

b) The question concerns the standard reduction potential for the half-reaction:

$$At_2(s) + 2\,e^- \rightarrow 2\,At^-(aq)$$

Zn(*s*) reduces $At_2(s)$ and in the process forms $Zn^{2+}(aq)$. The standard reduction potential for the $Zn^{2+}(aq)/Zn(s)$ half-reaction is −0.763 V. Therefore, $\mathcal{E}°$ for the $At_2(s)/At^-(aq)$ reduction exceeds −0.763 V. On the other hand, At_2 does not react with $Fe^{2+}(aq)$. The standard reduction potential for reduction of $Fe^{3+}(aq)$ to $Fe^{2+}(aq)$ is 0.770 V. Therefore $\mathcal{E}°$ for the reduction of $At_2(s)$ is less than 0.770 V. Moreover, on the basis of the trends in Group VII, the standard reduction potential for $At_2(s)$ should be less than 0.535 V, the value for $I_2(s)$.

c) The $Cl_2(g)$ will easily oxidize $At^-(aq)$: $2\,At^-(aq) + Cl_2(g) \rightarrow 2\,Cl^-(g) + At_2(s)$.
d) Solid At_2 should have an even greater tendency to disproportionate in aqueous base than $I_2(s)$. Thus: $6\,At_2(s) + 12\,OH^-(aq) \rightarrow 2\,AtO_3^-(aq) + 10\,At^-(aq) + 6\,H_2O(l)$.

22-45 Fluorine chemists often think of fluoride-ion acceptors as Lewis acids and fluoride-ion donors as Lewis bases. The reaction is:

$$K_2[NiF_6](s) + TiF_4(s) \rightarrow K_2[TiF_6](s) + NiF_2(s) + F_2(g)$$

In this reaction, TiF_4 accepts F^- ions to form TiF_6^{2-}. Thus, TiF_4 is acting as a Lewis acid.

22-47 The standard reduction potential $\mathcal{E}°$ for the reduction of $F_2(g)$ to $F^-(aq)$ is 2.87 V. This is much more positive than the 0.815 V potential for the reduction of $O_2(g)$ to 10^{-7} M $OH^-(aq)$ ion. Because 1 M $F^-(aq)$ has a very much smaller tendency to be oxidized than 10^7 M $OH^-(aq)$, the product at the anode in the electrolysis of almost any concentration of F^- ion is $O_2(g)$, not $F_2(g)$.

22-49 The balanced equation: $4\,HF(aq) + SiO_2(s) \rightarrow SiF_4(g) + 2\,H_2O(l)$ shows that 4 mol of $HF(aq)$ is needed to dissolve 1 mol of quartz rock. The rock contains mere traces of gold (1 part in 100 000 by mass). Suppose enough rock is dissolved to recover 1 troy ounce (31.3 g) of gold. This is 10^5 times 31.3 g of rock. The mass of HF required is:

$$31.3 \times 10^5 \text{ g rock} \times \left(\frac{1 \text{ mol rock}}{60.1 \text{ g}}\right) \times \left(\frac{4 \text{ mol HF}}{1 \text{ mol rock}}\right) \times \left(\frac{20.0 \text{ g HF}}{1 \text{ mol HF}}\right) = 4.17 \times 10^6 \text{ g HF}$$

The cost of this HF, which comes in the form of a solution, is:

$$4.17 \times 10^6 \text{ g HF} \times \left(\frac{2 \text{ g HF sol'n}}{1 \text{ g HF}}\right) \times \left(\frac{1 \text{ L sol'n}}{1170 \text{ g sol'n}}\right) \times \left(\frac{\$0.25}{1 \text{ L sol'n}}\right) = \$1780$$

The solvent cost more that the \$350 the troy ounce of gold would bring. The method is a losing proposition. It breaks even on materials when the ore is 5.1 times richer in gold (5.1 is the ratio of \$1780 to \$350). For true feasibility the cost of labor and equipment would have to be covered, too.

22-51 Follow the rules for writing Lewis dot structures outlined in Chapter 2. See Problem 3-37 for examples. A good Lewis dot structure of FOOF is:

$$:\ddot{\underset{..}{F}}—\ddot{\underset{..}{O}}—\ddot{\underset{..}{O}}—\ddot{\underset{..}{F}}:$$

The bond order of the O—F bond is 1, and the bond order of the O—O bond is 1. Each F—O—O bond angle is roughly 109° reflecting the steric number of four of each O atom. The analogous compound of oxygen and hydrogen is hydrogen peroxide.

22-53 Use the VSEPR model (text page 539) to predict the molecular geometry. The Lewis structure of the SF_5^- ion is:

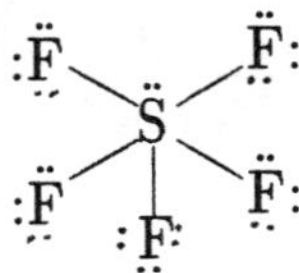

The central S atom in this ion has *SN* 6 with 1 lone pair and 5 bonded atoms; the molecular geometry is square pyramidal.

22-55 Teflon has the formula $(CF_2)_n$. It consists of many CF_2 monomer units linked together in a chain. The mass percentage of fluorine in the polymer is the same as the mass percentage of fluorine in the simplest repeating unit. The first value of mass percentage of fluorine was quite low compared to the actual mass percentage giving the apparent formula $CF_{0.59}$. There should be no chlorine in Teflon, but under the circumstances it was wise to check.

22-57 The reaction of interest is: $Xe(g) + 2\,F_2(g) \rightarrow XeF_4(s)$. The given free energy of formation applies directly to this reaction. Using the relationship $\Delta G^\circ = -RT \ln K$ we have:

$$\ln K = \frac{\Delta G^\circ}{-RT} = \frac{-134 \times 10^3 \text{ J}}{(8.315 \text{ J K}^{-1}\text{mol}^{-1})(298.15 \text{ K})} = 54.1 \quad \text{from which } K = 3 \times 10^{23}$$

Chapter 23

Organic Chemistry and Biochemistry

Organic chemistry is the study of the reactions and properties of compounds containing carbon. Carbon has an intermediate electronegativity and its atoms form strong covalent bonds to other carbon atoms and to hydrogen atoms **Biochemistry** is the study of compounds and reactions involved in life processes. Most biochemically important compounds are organic compounds.

The text first considers **hydrocarbons,** which contain only carbon and hydrogen. It then takes up derivatives of hydrocarbons in which other atoms attach as a **functional group** to the hydrocarbon skeleton.

Petroleum Refining and the Hydrocarbons

The primary source of hydrocarbons at this time is petroleum. The processing of petroleum yields fuels such as gasoline and starting materials for the petrochemical industry, which creates a large variety of useful products.

Hydrocarbons fall into four categories: **alkanes, alkenes, alkynes,** and the **aromatic** hydrocarbons. Carbon's unique propensity for forming stable chains allows the existence of straight chain, branched chain and **cyclic,** or ring structures in all four categories of hydrocarbon compound. Hydrocarbons lacking any rings are **acyclic.** All aromatic hydrocarbons contain at least one ring.

Alkanes

Alkanes are **saturated hydrocarbons** because all the carbon atoms in the molecule are combined with as many hydrogen atoms as possible. All C-to-C (and C-to-H) bonds in alkanes are *single* bonds. This gives alkanes the general formula C_nH_{2n+2}.

When "n" precedes the name of an alkane, it means the carbon atoms are linked in straight, unbranched chains. The ***n*-alkanes** are the major constituents of petroleum.

Branched-chain alkanes have the same general formulas as normal alkanes, but now the chain of carbon atoms has branches so that at least one carbon atom is bound to three other carbon atoms. The names of the alkanes having from 1 to 15 C atoms are listed in textbook Table 23-1.

• Memorize the names of at least the first ten alkanes. Note that after $n = 4$ the names employ familiar stems followed by the suffix "-ane."

There is free rotation about C—C single bonds and therefore many **conformations** for acyclic alkanes as various segments of the chain rotate into proximity with each other. Different conformations of a given carbon chain look different but are the same compound as long as they have the same sequence of atoms.

Compounds that have the same chemical formula but *different* atom-to-atom bonding sequences are structural **isomers.** No amount of free rotation can convert one structural isomer into another. Instead, bonds must be broken and re-formed. Normal and branched-chain alkanes with the same formula are isomers.

The standard (IUPAC) system of naming casts light on the nature of isomers by it way it deals with the problem of naming them. Follow these steps:

1. Identify the longest continuous chain of C atoms. Chain length can be concealed by writing structures in zig-zags on the page. Do not be deceived. What counts is *sequence.* Follow the chain around corners.

2. Number the atoms in the chain, starting from the end nearer any branches. The aim is to give the lowest possible number to the positions of side-groups. **Substituent groups** (or side-groups) are branches from the main chain. Alkyl side-groups are named by removing the suffix "-ane" from the name of the alkane and adding "-yl."

3. Write the name of the compound using a number for each side-group to tell where it is attached, followed by the name of each side-group and finally by the name for the main chain.

4. Side-groups may be named either in alphabetic order or in order of increasing complexity. The presence of identical side-groups at different locations on the chain is indicated by the appropriate multiplying prefix. See **problem 23-9.**

The above procedure does not deal with every possible naming situation in the alkanes, but it does work with most common compounds. **Example:** Name the following alkane:

```
    CH3 H  CH3 H  H
     |  |   |  |  |
H — C — C — C — C — C — CH3
     |  |   |  |  |
     H CH3 CH3 H  H
```

Solution: The name is 2,3,3-trimethylheptane. One common wrong answer is "1,2,3,3-tetramethylhexane." The methyl group that is apparently a side-group on the left-most carbon in the main line of the structure is really part of the chain. Once the longest chain is identified, there is the question of which direction to use for numbering. Numbering this chain from the left to the right, instead of the reverse, gives *smaller* prefix numerals in the name. That is, "2,3,3-" is preferable to "4,4,5-".

An increased fraction of branched-chain alkanes in a gasoline improves the smoothness of its combustion, a property that is measured by **octane number.**

Alkenes and Alkynes

Alkenes are hydrocarbons that contain at least one carbon-carbon double bond. They are **unsaturated** because they can take up H_2 to form alkanes. Acyclic alkenes have the generic formula C_nH_{2n}. Note that such alkenes are isomers of cycloalkanes. If there are two double bonds in an alkene, it is a **diene**. and has the generic formula $C_nH2n - 2$. Trienes have the formula C_nH_{2n-4}, etc. Hydrocarbons with multiple double bonds are **polyenes.** There is no free rotation about the C=C double bond. This allows the existence of *cis-trans* isomers (review text pages 575 and 608). *Cis* and *trans* isomers have their atoms connected in the same sequence but lying in different geometrical relationships. See **problem 23-7.**

Alkenes are named similarly to alkanes. Always pick the longest chain *that includes the double bond.* The position of the double bond is signified in the name by the number of the carbon atom that is followed by the double bond in the structural formula. Number the atoms of the chain and designate side-groups so that this number is as small as possible. Then comes the root name of the parent hydrocarbon followed by the ending "-ene" to indicate the alkene. See **problem 23-9.** The trivial names "ethylene" and propylene" for ethene and propene respectively must be memorized. Alkenes do not occur in petroleum, but they can be produced from it by **cracking** reactions. These include **catalytic cracking** and **thermal cracking.** The two methods produce shorter-chain alkenes from the alkanes in the starting petroleum distillate (see **problem 23-3**).

Alkynes contain C-to-C triple bonds. The simplest alkyne is ethyne (acetylene). Again, the chain selected for naming must include the triple bond. In naming these compounds, the position of the triple bond is indicated by a number, and the ending "-yne" is attached. When there is no ambiguity the number is omitted. For example,

"1-propyne" and "2-propyne" are the same compound, propyne. The same policy of simplification applies to alkenes (above). There is only one possible propene, so numbers are not needed in its name.

Cyclic alkenes, polyenes, and alkynes and polyalkynes are all possible. These hydrocarbons all contain at least one ring. They are named by using the prefix "cyclo-" in conjunction with the appropriate root name for the number of carbons found in the ring. Thus, propane is C_3H_8, an alkane or saturated hydrocarbon, but, cyclopropane is C_3H_6. It has the same formula as propene. Cyclopropene is C_3H_4. A cycloalkane is the subject of **problem 23-1.** Naming cyclic alkanes and alkenes is part of **problems 23-5, 23-6,** and **23-49.** Recall **problem 15-33**, which concerns a cyclic alkene, and the several cyclic hydrocarbons in text Figure 23-10. Formation of a ring from an alkane reduces the number of H atoms by 2.

Do not forget to include cyclic structures when listing the possible isomers of a hydrocarbon which is, by its formula, unsaturated.

Aromatic Hydrocarbons

The **aromatic hydrocarbons** contain rings of carbon atoms in which delocalized π-electrons significantly increase the molecular stability. Benzene C_6H_6 is the archetypal aromatic hydrocarbon. It has an unsaturated six-carbon ring with a system of six π-electrons. See text page 576. Each C-to-C bond is effectively a 3/2 bond. The increase in molecular stability caused by this aromatic system in benzene is computed in **problem 15-55.** It is a **resonance stabilization energy.** Petroleum contains useful quantities of benzene.

Substituent groups may be attached to the benzene ring. It is customary to omit the hydrogen atoms when drawing the structure of the benzene ring. Avoid the errors of not counting these H atoms in molecular formulas or forgetting to subtract them from molecular formulas when they are replaced by substituent groups. Thus, C_6H_5—CH_3 is methylbenzene. One of the six hydrogen atoms of benzene is replaced by a methyl group. This compound is more commonly called **toluene.**

If two side-groups are attached to the benzene ring, *three* different isomers are possible. Relative positions are distinguished by numbers. The 1,2-, 1,3-, and 1,4- di-substitution patterns are named *ortho, meta and para,* respectively. This means that there are three dimethylbenzenes. These isomeric compounds are called **xylenes.**

The **BTX aromatics** are a mixture of benzene, toluene, and *ortho-*, *meta-*, and *para-*xylene. These compounds are very important in gasoline because they significantly increase the octane number of the gasoline. They are made by **reforming reactions** from straight-chain alkanes separated from petroleum by distillation. Toluene and the xylenes can be converted to benzene by **hydrodealkylation** reactions. In such reactions an alkyl side-group is removed (hence "-dealkyl") and replaced by hydrogen.

Functional Groups and Organic Synthesis

The insertion of non-carbon atoms in a hydrocarbon chain or their attachment to such a chain creates sites of greater reactivity that are called **functional groups.** The structures of the common functional groups are given in text Table 23-3 (text page 812).

Many functional groups can be regarded as derived from a simple inorganic molecule by substitution of an alkyl group. The general symbol for an alkyl group is an "R." See **problem 23-15.**

- **Alkyl halides** have the generic formulas. Formally, they are derivatives of the hydrohalic acids H—Cl, H—Br and H—I.

- **Alcohols** R—OH and **phenols,** can be regarded as derived from water (H—OH) by replacing one H with an alkyl or aryl side group. In alcohols, the —OH group must be attached to a carbon atom that is saturated, that is, a carbon atom with four single bonds. Alcohols are called **primary, secondary,** and **tertiary** according to the number of other carbon atoms bonded to the carbon atom that bears the —OH group. In phenols, the —OH group is attached to an aromatic group.

- **Ethers** all contain the R—O—R′ functional group where R and R′ are the same or different alkyl groups. They are like water (H—O—H) with both H atoms replaced. Neither of the carbon atoms attached to the ether O may itself be double-bonded to an O atom, because then the molecule would be an ester (see below).

- **Aldehydes** and **ketones** contain a carbonyl group >C=O. The aldehydes have at least one H bonded to the carbonyl. The other group bonded to the carbonyl carbon is —R (an alkyl group) or An aromatic group. The simplest aldehyde is $H_2C=O$ (formaldehyde). The ketones have alkyl or aromatic groups replacing both H atoms in $H_2C=0$.

- **Carboxylic acids** have the formula R—COOH. They can be viewed as deriving from HO—COOH (carbonic acid) by the formal replacement of an —OH group with an —R. They act as weak acids, forming salts upon neutralization with bases.

- **Esters** derive from carboxylic acids when the acid reacts with an alcohol:

$$\mathrm{R - COOH\ (acid) + HO - R'\ (alcohol) \rightarrow R - COOR'\ (ester) + H_2O}$$

- **Amines** can be regarded as derivatives of ammonia NH_3, in which one, two or three H atoms are replaced by alkyl groups. They are sub-classified: RNH_2 is a primary amine, R_2NH is a secondary amine, and R_3N is a tertiary amine.
- **Amides** can also be regarded as derivatives of ammonia NH_3. In amides, either ammonia or a primary or secondary amine has condensed with a carboxylic acid:

$$R - COOH\ (\text{acid}) + NH_2 - R'\ (\text{amine}) \rightarrow R - CONHR'\ (\text{amide}) + H_2O$$

Synthetic Polymers

Polymers are molecules built up by the linking together, in long chains, sheets, or three-dimensional networks, of many identical structural units called **monomer units.** Silicate minerals (Chapter 19) are polymers. Man-made polymers are usually based on organic starting materials. The two major types of polymer growth are **addition polymerization** and **condensation polymerization.**

In addition polymerization, monomer units react to form a polymer chain without net loss of atoms. The polymerization of an alkenes such as ethylene is an addition polymerization. See **problem 23-19.** Addition polymerization often proceeds by free-radical chain reaction (text page 421-2) of molecules that have C=C double bonds. It is started by a suitable **initiator.** In other cases, addition polymerization proceeds by an ionic mechanism.

In condensation polymerization, a small molecule, often water, is split In this type of polymerization, each molecule of monomer has (at least) two different functional groups that react to join the units together in a "head-to-tail" fashion. **Copolymers** form when chemically different monomers are mixed in the polymerization process. If the differing monomer units join the growing polymer chain at random, the result is a **random copolymer.** It is also possible to make **block copolymers,** in which long sequences of each type of monomer; unit are chemically bonded to form the polymer chain, and **graft copolymers,** in which polymer chains of a second sort branch from a polymer chain of the first sort. The polymer chains growing from monomers with more than two reactive sites can be **cross-linked** into sheets and networks, which often have desirable physical properties.

Uses for Polymers

Useful polymers occur in fibers, plastics, and elastomers (rubber).

Fibers. Cellulose, an important natural fiber, is a condensation polymer of glucose. The —OH side-groups along the cellulose chain an be modified to give derivatives like guncotton, nitrocellulose (which is made into celluloid), and cellulose acetate. The first true synthetic polymeric fiber was nylon, a polyamide formed by the condensation of a dicarboxylic acid and a diamine.

Plastics. Plastics are polymeric materials that can be molded or extruded into appropriate shapes and that harden upon cooling or solvent evaporation. Polyethylene, formed by the free-radical addition polymerization of ethylene, can be created as low-density polyethylene (LDPE), high-density polyethylene (HDPE), and linear low-density polyethylene (LLDPE). The difference concerns the number of length of side-chains projected from the polymer molecules. If a methyl group replaces one of the hydrogen atoms in every monomer unit of polyethylene, the result is polypropylene. The relative positions of the methyl groups attached to the carbon backbone may be **isotactic** (all on the same side), **syndiotactic** (alternating in a regular pattern), or **atactic** (distributed at random). Substitution of a chlorine atom for one of the hydrogen atoms in ethylene gives vinyl chloride, which polymerizes to polyvinyl chloride, another very useful plastic. **Plasticizers** are often added to polyvinyl chloride and other plastics to soften them and make them flexible.

Elastomers. An elastomer is a plastic that can be deformed to a large extent and still recover its original form when the deforming stress is removed. Natural rubber is a polymer of **isoprene** (2-methylbutaldiene), in which the geometry at the double bonds along the polymer chain is *cis* in every case. If natural rubber is cross-linked by S—S bonds in a process called **vulcanization,** th product is harder, more resilient, and does not melt. It is this material that is used for automobile tires. Synthetic rubbers include neoprene, in which a chlorine atom replaces the methyl group in natural rubber, and copolymers of butadiene with styrene or butadiene with acrylonitrile.

Natural Polymers

The text discusses two very important classes of natural polymers. These are **proteins** and **nucleic acids.**

Proteins

In proteins, the polymer chain is formed of α-amino acid monomer units. Molecules of α-amino acids contain an amine (—NH_2) and carboxylic acid group (—COOH) attached to the same C atom. Amino acids are amphoteric. See **problem 6-51** All α-amino acids but glycine have a carbon atom that is a **chiral center,** meaning that they can be made in two forms, D and L, that are mirror-image isomers (enantiomers, see text page 609). Twenty different amino acids commonly occur in proteins. Natural proteins contain almost exclusively L-amino acids. The properties of proteins are strongly affected by the nature of the side-groups (see **problem 23-31).** In a protein, the amino group on one amino acid is linked to the acid group on the next in the

peptide linkage, which is a amide group. Every time a link is forged, a molecule of water is split out. Creation of a peptide linkage is non-spontaneous at room conditions. but is driven in organisms by free energy from spontaneous reactions. Protein chains of up to about 50 amino-acid groups are often termed **polypeptides.** No matter how many links there are in such a chain the two ends are distinguishable. One is the amino end, and one is the acid end. There is essentially an infinite number of possible different protein molecules. Not only is the length of the polymer chain variable, but also different amino acids may be linked in different orders.

Although the sequence of side-groups in proteins is exceedingly variable (**problem 23-59**), this is only part of their structural variability. Protein structure also involves the way the chain is folded, coiled, looped back on itself by hydrogen bonds, or cross-linked to other protein chains. **Fibrous proteins** are structural materials that form sheets or fibers. They have regular three-dimensional structures. The polymeric chain may coil into a spiral from which the side-groups of the amino acids extend outward in a helical pattern. Such a right-handed coil is called the **α-helix.** It is maintained by hydrogen bonds. Alternatively, neighboring polypeptide chains may hydrogen-bond together in sheet-like structures. *Globular proteins* have chains that are folded irregularly into a more or less compact globular shape. The folding causes amino acids which are widely separated in the sequence along the chain to lie adjacent to each other in the globule. This kind of folding is vital in the functioning of hemoglobin, the oxygen-carrying molecule in the blood.

The same folding occurs in the structures of *enzymes.* Enzymes are globular proteins that catalyze particular reactions in the cell. The exact folding creates active sites at which steric and electronic factors combine to hold substrate molecules in such as way as to enhance the rates of specific reactions.

Nucleic Acids

Nucleic acids preserve and transmit information about the amino acid sequence in proteins. They are nucleic because they occur in the nuclei of cells. They are acids because they include —$(HO)OPO_2$– groups that are hydrogen-ion donors. Nucleic acids are polymers. The backbone of the polymer chain consists of alternating phosphate and a cyclic sugar. In deoxyribonucleic acid (DNA), the sugar is called deoxyribose. See Figure 23-28, text page 839. Each sugar has as a side-group a nitrogenous base. In DNA, the bases are cytosine (C) guanine (G), adenine (A) and thymine (T). Expressed in somewhat different terms, a nucleic acid molecule is built by linking four different nucleotide units together in a long chain. Recall that a nucleotide consists of a base plus a sugar plus a phosphate.

The order of the bases along the polymeric chain encodes the information which the nucleic acid maintains or transmits. A nucleic acid molecule only 10 nucleotide units long has $4^{10} = 1048580$ possible isomers based on the different sequences of the nucleotides. The concept is the same as the theme of **problem 23-59.** The difference is that the subject now is nucleotides linking together, not amino acids.

In DNA, two polymeric strands intertwine in a double helix. The cystosine side-group on one strand links through hydrogen bonds to a guanine side-group on the other. Each adenine links through hydrogen bonds to a thymine. The base-pairings are quite specific: C...G A...T.

When the DNA double helix is unwound in the present of a supply of nucleotides and a suitable means for their delivery, each single strand serves as a template for the creation of a new complementary strand. The result is two new DNA molecules identical in their base sequences to each other and to their progenitor.

Chemical Processes in Living Cells

A living cell can be pictured as a chemical processing plant. Its operation are subject to all ordinary thermodynamic and kinetic limitations, but living cells employ subtle mechanisms to convert their raw materials into desired products. Certain high free-energy compounds formed spontaneously (deriving their energy from physical processes) on the primitive earth before life began. Two such important prebiotic molecules were adenosine diphosphate **(ADP)** and adenosine triphosphate **(ATP)**. The first organisms evolved ways to use pre-existing supplies of such compounds as sources of free energy. Later, organisms developed the ability to capture light from the sun to make their own high free-energy compounds; this was **photosynthesis,** the primary free-energy source for life as we know it. In anaerobic photosynthesis, light from the sun is absorbed by compounds related to chlorophyll. As the molecules of their compounds return to the ground state, the excitation energy is stored either by the conversion of ADP to ATP or the conversion of the coenzyme $NADP^+$to NADPH, a process that consumes hydrogen present in reducing agents like H_2S or H_2. **Photosystem II** improves this process by capturing higher-energy radiation and by using water as the source of hydrogen. The "waste product" from this type of photosynthesis is oxygen, which accumulated in the atmosphere of the early earth and gradually changed it from reducing to an oxidizing environment. These **light reactions** lead to the accumulation of NADPH and ATP in green plants. They are coupled to **dark reactions** that generate sugars from carbon dioxide and water in a complex multistep process known as the **Calvin cycle.**

Sugars are **carbohydrates,** They have the general formula $C_n(H_2O)_m$. **Monosaccharides,** or simple sugars, can be linked to form long chains, or **polysaccharides.** The monomer unit may be of several types, but are always cyclic. See Figure 23-32, tesxt page 844. Cellulose and starch are both polymers of glucose that differ only the position of the oxygen atoms linking the monomer units. **Metabolism** breaks down complex molecules such as sugars that are formed by photosynthesis and frees the stored chemical energy for the uses of the cells. The **anaerobic** metabolism of glucose, called **fermentation,** does not required oxygen from the air; **aerobic** metabolism does. **Glycolysis,** the anaerobic breakdown of glucose, generates two molecules of ATP per molecule of glucose and gives lactic acid or other products, such as ethanol. Organisms that use aerobic metabolism take oxygen from the air)respire) and, by means of the **citric-acid cycle,** extract enough energy from the oxidation of one molecule of glucose (to CO_2 and H_2O) to fuel the synthesis of 38 molecules of ATP.

Respiration and Metabolism

Anaerobic metabolism (glycolysis) produces only two molecules of ATP per molecule of glucose. Aerobic metabolism is more efficient and produces as many as 38 molecules of ATP per molecule of glucose. The *citric acid cycle* in Figure 23-37 (text page 850) summarizes the central chemistry of aerobic metabolism. The cycle is preceded by the breakdown of the glucose molecule to two pyruvate anions. The citric acid cycle then produces ATP and the reduced forms of *two* enzymes, NADH and $FADH_2$, as it oxidizes the pyruvate the rest of the way to CO_2. FAD stands for flavine adenine dinucleotide. In a final, post-cycle step, the reduced forms of the enzymes are oxidized with oxygen to regenerate the original enzymes (NAD^+ and FAD) and produce additional supplies of ATP.

Detailed Solutions to Odd-Numbered Problems

23-1 a) The standard enthalpy of formation ΔH_f° of gaseous cyclopropane is the enthalpy change in forming $C_3H_6(g)$ in its standard state from $C(s)$ and $H_2(g)$ in their standard states: $3\,C(s) + 3\,H_2(g) \rightarrow C_3H_6(g)$. Imagine this reaction to proceed by the atomization of 3 mol of $C(s)$ and 3 mol of $H_2(g)$ to $C(g)$ and $H(g)$ followed by the formation of 3 mol of C—C bonds and 6 mol of C—H bonds. The sum of the ΔH°'s of these steps equals the desired ΔH°, by Hess's law. Use the bond enthalpies and enthalpies of atomization from Table 8-3 (text page 296):

$$\Delta H^\circ = 3(716.682) + 6(217.96) + 3(-348) + 6(-413) = -64 \text{ kJ}$$

The estimated ΔH_f° of cyclopropane based on average bond enthalpies equals -64 kJ mol^{-1}.

b) The standard enthalpy change for the combustion of cyclopropane is the enthalpy of formation of the products minus the enthalpy of formation of the reactants in the reaction: $C_3H_6(g) + 9/2\,O_2(g) \rightarrow 3\,CO_2(g) + 3\,H_2O(g)$. The enthalpy of reaction is related to the enthalpies of formation:

$$\Delta H^\circ = -1959 \text{ kJ} = 3\underbrace{(-393.51)}_{CO_2(g)} + 3\underbrace{(-241.82)}_{H_2O(g)} - 9/2\underbrace{(0)}_{O_2(g)} - 1\underbrace{(\ x\ )}_{C_3H_6(g)}$$

where x is the exact ΔH_f° of cyclopropane. Solving gives $\Delta H_f^\circ = +53$ kJ mol^{-1}.
c) The calorimetric ΔH_f° of $C_3H_6(g)$ is considerably higher that the ΔH_f° based on bond enthalpies. It requires more energy to make $C_3H_6(g)$ than would be expected simply on the basis of the formation of normal single bonds. Formation of the triangular ring of C atoms forces C—C—C bond angles of 60°, much smaller than the usual tetrahedral angle, 109.5°. The extra energy is the strain energy of cyclopropane. It is about 117 kJ mol^{-1}.

23-3 **a)** The catalytic cracking reaction is:

$$CH_3-(CHH)_8-CH_3 \text{ (each C bearing H above and H below)} \rightarrow CH_3-\overset{H}{\underset{H}{C}}-\overset{H}{\underset{H}{C}}-\overset{H}{\underset{H}{C}}-CH_3 \; +$$

$$CH_3-\overset{H}{\underset{H}{C}}-\overset{H}{\underset{H}{C}}-\overset{H}{C}=CH_2$$

b) Another isomer of the alkene is 2-pentene, in which the double bond is between the second and third carbon atoms in the chain.

$$CH_3-\overset{H}{\underset{H}{C}}-\overset{H}{C}=\overset{H}{C}-CH_3$$

23-5 The structural formulas are:

a) $CH_3-\overset{CH_3}{\underset{H}{C}}-\overset{CH_3}{\underset{H}{C}}-\overset{H}{\underset{H}{C}}-\overset{H}{\underset{H}{C}}-H$

b) $(H)(H_3C)C=C(CH_2-CH_3)(CH_2-CH_3)$

c) cyclopropane ring: one C bearing H, H; one C bearing CH_3, H; one C bearing H, H

d) $H-\overset{H}{\underset{H}{C}}-\overset{CH_3}{\underset{CH_3}{C}}-\overset{H}{\underset{H}{C}}-\overset{H}{\underset{H}{C}}-H$

e) 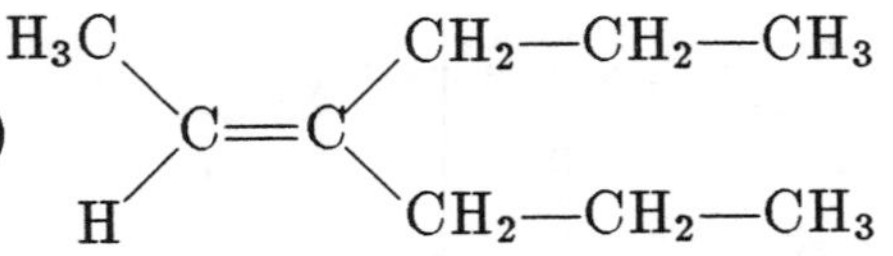
$$\begin{array}{ccc} H_3C & & CH_2{-}CH_2{-}CH_3 \\ & C{=}C & \\ H & & CH_2{-}CH_2{-}CH_3 \end{array}$$

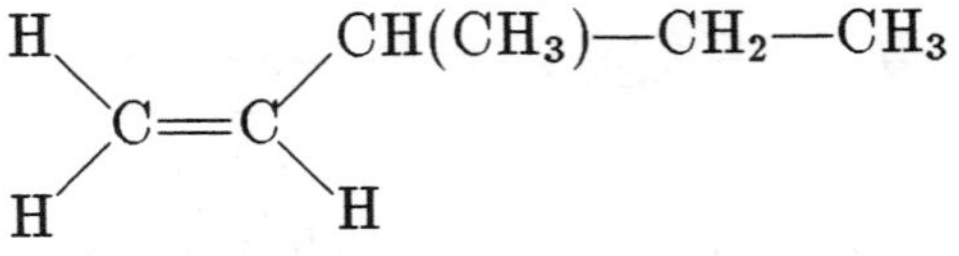

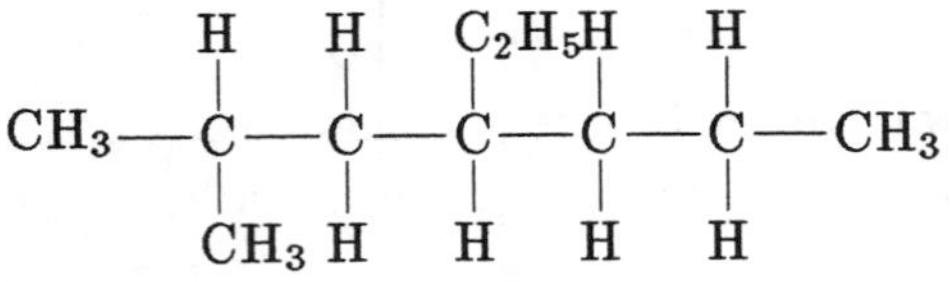

f)
$$\begin{array}{ccc} H & & CH(CH_3){-}CH_2{-}CH_3 \\ & C{=}C & \\ H & & H \end{array}$$

g)
$$\begin{array}{ccccccccccccc} & & H & & H & & C_2H_5 & & H & & H & & \\ & & | & & | & & | & & | & & | & & \\ CH_3 & - & C & - & C & - & C & - & C & - & C & - & CH_3 \\ & & | & & | & & | & & | & & | & & \\ & & CH_3 & & H & & H & & H & & H & & \end{array}$$

h)
$$\begin{array}{ccccc} & & CH_2CH_3 & & \\ & & | & & \\ CH_3{-}C{\equiv}C & - & C & - & CH_2CH_2CH_3 \\ & & | & & \\ & & H & & \end{array}$$

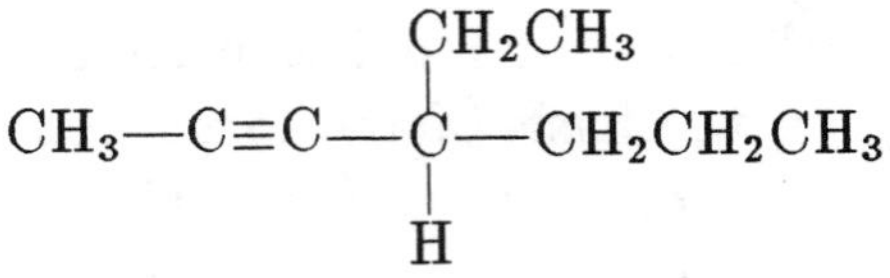

23-7 The *cis* isomers of 3-heptene is at the left and the *trans* is at the right:

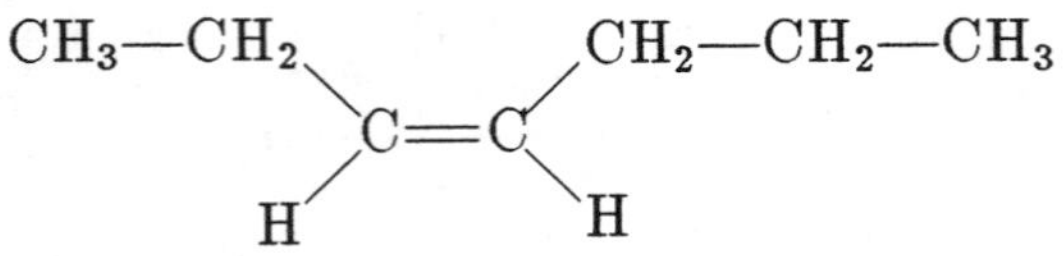

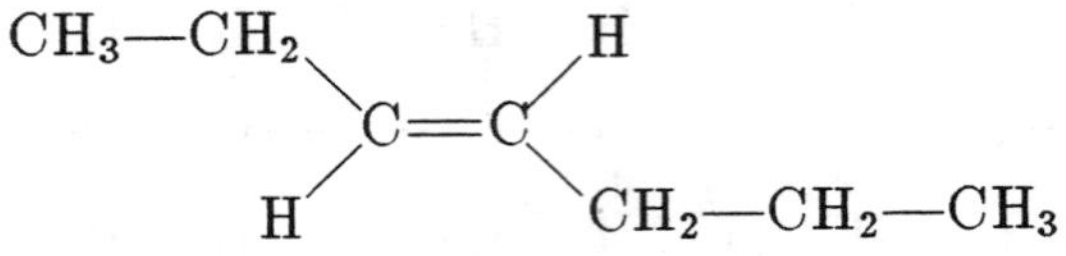

$$\begin{array}{ccc} CH_3{-}CH_2 & & CH_2{-}CH_2{-}CH_3 \\ & C{=}C & \\ H & & H \end{array} \qquad \begin{array}{ccc} CH_3{-}CH_2 & & H \\ & C{=}C & \\ H & & CH_2{-}CH_2{-}CH_3 \end{array}$$

23-9 a) 1,2-hexadiene **b)** 1,3,5-hexatriene (or *trans*-1,3,5-hexatriene)
c) 2-methyl-1-hexene **d)** 3-hexyne

23-11 a) This is an *esterification.* It resembles in form an acid-base neutralization with the alcohol playing the part of the base:

$$CH_3CH_2CH_2CH_2OH + CH_3COOH \rightarrow CH_3COOCH_2CH_2CH_2CH_3 + H_2O$$

The organic product is the ester *n*-butyl acetate.
b) The reaction is a dehydration: $H_3C{-}COO^-\,NH_4^+ \rightarrow H_3C{-}CO{-}NH_2 + H_2O$.
c) The H atom attached to the O atom and one of the H atoms on the neighboring C atom are removed: $H_3CCH_2CH_2OH \rightarrow H_3CCH_2CHO + H_2$.
d) $H_3CCH_2CH_2CH_2CH_2CH_2CH_3 + 11\,O_2 \rightarrow 7\,CO_2 + 8\,H_2O$.

23-13 a) Brominate ethylene to give 1,2-dibromoethane. Then remove HBr:

$$CH_2{=}CH_2 + Br_2 \rightarrow CH_2BrCH_2Br \quad \text{then} \quad CH_2BrCH_2Br \rightarrow CH_2{=}CHBr + HBr$$

b) Treat 1-butene with water (in presence of H_2SO_4):
$CH_3CH_2CH{=}CH_2 + H_2O \rightarrow CH_3CH_2CH(OH)CH_3$.
c) Treat propene with water to given 2-propanol and then dehydrogenate over a catalyst (copper works):

$$CH_3CH{=}CH_2 + H_2O \rightarrow CH_3CH(OH)CH_3 \text{ and } CH_3CH(OH)CH_3 \rightarrow CH_3COCH_3 + H_2$$

23-15 $$\mathrm{R{-}\underset{R}{\overset{R}{C}}{-}OH + HO{-}\underset{H}{\overset{O}{\overset{\|}{C}}}{-}R' \rightarrow HOH + R{-}\underset{R}{\overset{R}{C}}{-}O{-}\overset{O}{\overset{\|}{C}}{-}R'}$$

23-17 It is clear that the molar ratio of the ethylene to the ethylene dichloride is 1 to 1, based on the formulas of the two compounds. Use this ratio to compute the amount of ethylene required to make the 6.26×10^9 kg of ethylene dichloride. Then:

$$6.26 \times 10^9 \text{ kg } C_2H_4Cl_2 \times \left(\frac{1 \text{ mol } C_2H_4Cl_2}{0.09896 \text{ kg } C_2H_4Cl_2}\right) \times \left(\frac{1 \text{ mol } C_2H_4}{1 \text{ mol } C_2H_4Cl_2}\right)$$

$$\times \left(\frac{0.02805 \text{ kg } C_2H_4}{1 \text{ mol } C_2H_4}\right) = 1.77 \times 10^9 \text{ kg } C_2H_4$$

This is 11.2% of the 15.87×10^9 kg total annual production of ethylene. The mass of the chlorine is 4.49×10^9 kg, by a similar computation.

23-19 The addition reaction is $n\, Cl_2C{=}CH_2 \rightarrow \mathrm{+\!(}CCl_2{-}CH_2\mathrm{)\!\!+}_n$.

23-21 The starting monomer must be $H_2C{=}O$, which is formaldehyde.

23-23 **a)** As glycine (NH_2CH_2COOH) polymerizes to the polypeptide, one molecule of water is lost in the formation of each peptide bond.

b) The repeating structure in the polypeptide is:

$$\mathrm{{-}\overset{H}{\underset{}{N}}{-}\underset{H}{\overset{H}{C}}{-}\overset{O}{\overset{\|}{C}}{-}}$$

23-25 The repeating unit in the polyamide has the formula $C_{12}H_{22}N_2O_2$. This formula is the sum of the molecular formulas of adipic acid and hexamethylenediamine minus twice the formula of water, a relationship that derives from the fact that the diacid and diamine polymerize by condensation with loss of one molecule of water for each unit added to the chain. The molar mass of the repeating unit is 226.32 g mol^{-1}. Then the chemical amount of the repeating unit that is needed is:

$$1.00 \times 10^6 \text{ g polymer} \times \left(\frac{1 \text{ mol of units}}{226.32 \text{ g polymer}}\right) = 4419 \text{ mol}$$

This means that the synthesis needs 4419 mol of adipic acid (which has a molar mass of 146.1 g mol^{-1}) and 4419 mol of hexamethylenediamine (molar mass 116.2 g mol^{-1}). These chemical amounts convert to 646 kg of adipic acid and 513 kg of hexamethylenediamine.

23-27 Polyethylene is formed by addition polymerization. This means that the mass of the monomer used to make the polymer equals the mass of the polymer that is formed. No mass is split out in the form of water, for example (in contrast to the polymerization of problem 23-25). Use this fact in a train of unit-conversions:

$$4.37 \times 10^9 \text{ kg polymer} \times \frac{1 \text{ kg monomer}}{1 \text{ kg polymer}} \times \frac{1 \text{ mol monomer}}{0.02805 \text{ kg monomer}} \times \frac{22.4 \text{ L monomer}}{1 \text{ mol monomer}}$$

$$= 3.49 \times 10^{12} \text{ L monomer} = 3.49 \text{ km}^3 \text{ (cubic kilometers!) of gaseous ethylene}$$

23-29 A tripeptide is a chain of three monomer units and has distinguishable ends. Any of the three kinds of building block can go in the first position, any of the three can go in the second position, and any of the three can go in the third position. There are accordingly $3^3 = 27$ possible tripeptides.

23-31 The pentapeptide is:

```
                  H
                  |              (C6H5)
          H3C—C—CH3                |                          CH3
                  |                |                           |
  CH3            CH2              CH2           H             CH2—CH2—CH3
   |              |                |            |              |
H2NC—C—N—C—C—N—C—C—N—C—C—N—C—COOH
   |   ||  |   |   ||  |   |   ||  |   |   ||  |   |
   H   O   H   H   O   H   H   O   H   H   O   H   H
```

The pentapeptide has all non-polar side-groups. It should be more soluble in *n*-octane than in water.

23-33 The empirical formula of phenylalanine is $C_9H_{11}NO_2$. The polypeptide forms with the removal of an H from the amine end of the molecule and an OH from the acidic end. Except for the two monomer units at the two extreme ends, which are negligible, each phenylalanine loses one HOH as the polymer forms. The empirical formula of the polymer is therefore C_9H_9NO. The molar mass of a C_9H_9NO unit is 147.2 g mol^{-1}. If the molar mass of the polypeptide is 17500 g mol^{-1} it contains $17500/147.2 = 119$ monomer units.

23-35 According to the equation for the conversion of ATP→ADP given on text page 844, the ΔG for making 18 mol of ATP's from 18 mol of ADP's is -621 kJ. The conversion of 1 mol ATP to 1 mol ADP therefore has a ΔG equal to -34.5 kJ. This confirms the numerical statement in the problem. In the balanced equation, H^+ (hydrogen ion) is a product. An increased concentration of any product works against the production of products, so the ΔG of the ATP→ADP conversion reaction would be less negative at low pH.

23-37 a) Use the Planck relation, which connects the energy of a photon to its frequency and thence to its wavelength:

$$E = h\nu = \frac{hc}{\lambda} = \frac{(6.626 \times 10^{-34}\ \text{J s})(2.9979 \times 10^{8}\ \text{m s}^{-1})}{4.30 \times 10^{-7}\ \text{m}} = 4.62 \times 10^{-19}\ \text{J}$$

The energy of 1.00 mol of photons is this value multiplied by Avogadro's number because there are Avogadro's number of photons in 1.00 mol of photons. The result is 278 kJ mol^{-1}.

b) The conversion ADP→ATP has ΔG equal to +34.5 kJ. The 278 kJ from a mole of photons is 8.06 times larger than 34.5 kJ. Therefore 8.06 mol of ATP could be produced per mole of the photons. This means that at most 8 individual ATP molecules are produced by a single photon.

23-39 At a pH of 7, the concentration of H_3O^+ is 1.0×10^{-7} M, which is far less than its standard-state value of 1.00 M. The gaseous oxygen and the liquid water are however still in their standard states. The Nernst equation (at 25°C) for the half-reaction then becomes:

$$\mathcal{E} = \mathcal{E}^\circ - \frac{0.0592}{n} \log Q = 1.229\ \text{V} - \frac{0.0592}{4} \log\left(\frac{1}{(1.0 \times 10^{-7})^4}\right) = 0.815\ \text{V}$$

23-41 The compound β-D-galactose has five chiral centers:

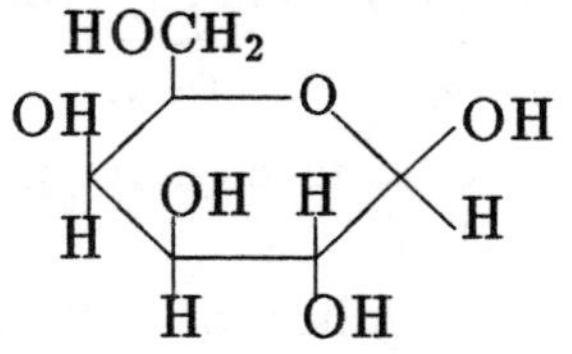

23-43 The free-energy change in the conversion of 2.00 mol of ADP to ATP is:

$$2.00\ \text{mol}(34.5\ \text{kJ mol}^{-1}) = 69.0\ \text{kJ}$$

and the free-energy change in the conversion of 1.00 mol of aqueous glucose to 2.00 mol of aqueous lactic acid is −163 kJ. The fraction of the free energy release in the glucose→lactose conversion that is stored in the ADP→ATP conversion is therefore $69.0/163 = 0.423$.

23-45 The lactic acid (lacH) establishes an equilibrium with its conjugate base (lac^-):

$$\text{lacH}(aq) + H_2O(l) \rightleftharpoons H_3O^+(aq) + \text{lac}^-(aq)$$

At pH 7, the $[H_3O^+]$ is 1.0×10^{-7}. Write the equilibrium-constant expression and insert this value:

$$K_a = 10^{-3.97} = 1.07 \times 10^{-4} = \frac{[H_3O^+][\text{lac}^-]}{[\text{lacH}]} = \frac{(1.0 \times 10^{-7})[\text{lac}^-]}{[\text{lacH}]}$$

It follows that [lac$^-$] = 1072[lacH]. The fraction of the molecules that are presents as lacH is the concentration of that form divided by the sum of the concentrations of the two forms:

$$f_{\text{lacH}} = \frac{[\text{lacH}]}{[\text{lacH}] + [\text{lac}^-]} = \frac{[\text{lacH}]}{[\text{lacH}] + 1072[\text{lacH}]} = 0.00093$$

23-47 a) Take data from Appendix D for the standard enthalpies of formation of the reactants and products of the two reactions, which are combustion reactions. The formula of n-heptane is C_7H_{16} and that of isooctane is C_8H_{18} (see Table 23-1, page 802). The data are:

$$\begin{array}{lccccccl} & C_7H_{16}(l) & +\,11\,O_2(g) & \rightarrow & 7\,CO_2(g) & +8\,H_2O(g) & \\ \Delta H_f^\circ & -187.82 & 0 & & -393.51 & -241.82 & \text{kJ mol}^{-1} \\ & C_8H_{18}(l) & +\,25/2\,O_2(g) & \rightarrow & 8\,CO_2(g) & +9\,H_2O(g) & \\ \Delta H_f^\circ & -224.13 & 0 & & -393.51 & -241.82 & \text{kJ mol}^{-1} \end{array}$$

Combine the ΔH_f°'s to get the ΔH°'s of the two reactions:

$$\Delta H_1^\circ = 8\underbrace{(-241.82)}_{H_2O(g)} + 7\underbrace{(-393.51)}_{CO_2(g)} - 11\underbrace{(0)}_{O_2(g)} - 1\underbrace{(-187.82)}_{\text{heptane}(l)} = -4501.31\ \text{kJ}$$

$$\Delta H_2^\circ = 9\underbrace{(-241.82)}_{H_2O(g)} + 8\underbrace{(-393.51)}_{CO_2(g)} - 11\underbrace{(0)}_{O_2(g)} - 1\underbrace{(-224.13)}_{\text{isooctane}(l)} = -5100.33\ \text{kJ}$$

b) The combustion of 1 mol of isooctane clearly produces much more heat (ΔH_2°) than the combustion of 1 mol of n-heptane (ΔH_1°). The question however concerns the comparison of the ΔH°'s per one gallon of the two fuels:

$$\frac{-4501.31\ \text{kJ}}{1\ \text{mol heptane}} \times \frac{1\ \text{mol heptane}}{100.21\ \text{g heptane}} \times \frac{453.59\ \text{g}}{1\ \text{lb}} \times \frac{5.71\ \text{lb}}{1\ \text{gal}} = \frac{-1.16 \times 10^5\ \text{kJ}}{1\ \text{gal}}$$

$$\frac{-5100.31\ \text{kJ}}{1\ \text{mol octane}} \times \frac{1\ \text{mol octane}}{114.23\ \text{g octane}} \times \frac{453.59\ \text{g}}{1\ \text{lb}} \times \frac{5.77\ \text{lb}}{1\ \text{gal}} = \frac{-1.17 \times 10^5\ \text{kJ}}{1\ \text{gal}}$$

The isooctane produces slightly more heat per gallon.

23-49 Cyclodecene has a 10-membered ring and one C-to-C double bond. The end of every line in the following is understood to be occupied by an H. Every intersection is understood to be occupied by a C:

cis *trans*

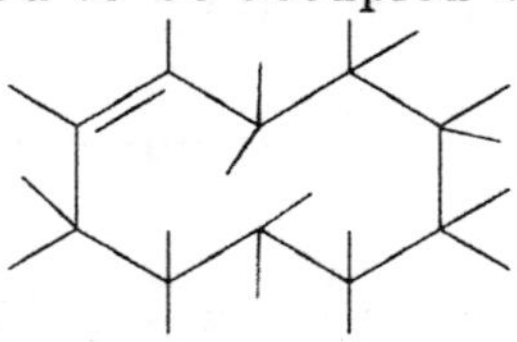

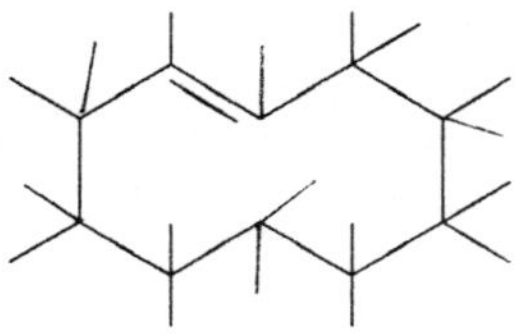

23-51 Dehydrogenation is the removal of hydrogen H_2 from a compound; dehydration is the removal of water H_2O. The compound ethane CH_3CH_3 is dehydrogenated to give ethylene CH_2CH_2. The compound ethanol CH_3CH_2OH is dehydrated to give ethylene.

23-53 The unsaturated hydrocarbon must have a straight-chain skeleton of six carbon atoms because it gives *normal* hexane *n*-hexane when reduced with hydrogen gas. When oxidized it is split in a four carbon acid (butanoic acid) and a two carbon acid (acetic acid). The double bond is therefore in the 2 position: $CH_3—CH_2—CH_2—CH=CH—CH_3$. The compound is 2-hexene. This compound has *cis* and *trans* isomers, but is no way to conclude from the available data which isomer is present. The balanced equations are:

$$C_6H_{12}(l) + H_2(g) \rightarrow C_6H_{14}(l)$$

$$27\,H_3O^+(aq) + 9\,MnO_4^-(aq) + 5\,C_6H_{12}(l)$$
$$\rightarrow 9\,Mn^{2+}(aq) + 5\,C_4H_8O_2(aq) + 5\,C_2H_3O_2(aq) + 43\,H_2O(l)$$

23-55 A catalyst is by definition not consumed in a reaction; it is taken up in one step of a mechanism but regenerated in a subsequent step. In the polymerization of acrylonitrile, the *n*-butyl lithium is consumed and the butyl anion is incorporated into the product. Doing this creates an new anion by which chain-building propagates. The *n*-butyl lithium is irrecoverable and hence is not a catalyst.

23-57 Polyvinyl chloride is $\text{+}CH_2—CHCl\text{+}_n$. The molar mass of its monomer unit is 65.50 g mol^{-1}. Polyvinyl chloride is formed by polymerization of ethylene dichloride CH_2ClCH_2Cl ($\mathcal{M}$ = 98.95 g mol^{-1}) with the loss of one molecule of HCl per monomer unit added to the chain. The theoretical yield from 950 million pounds of monomer is therefore

$$950 \text{ million lb} \times \frac{62.50}{98.95} = 600 \text{ million lb}$$

If the actual yield of polymer is 500 million pounds, then the percent yield is 500/600)× 100% = 83%. If the actual yield gets as high as 550 million pounds, the percent yield is 92%.

23-59 There is a choice of two kinds of amino acid at each of the 22 positions in the polypeptide chain. The two ends of the chain are distinguishable. Therefore, there are 2^{22} or 4.194 million possible isomeric molecules.

23-61 The acid form of alanine has two acid ionization reactions:

$$^+H_3NCH(CH_3)COOH(aq) + H_2O(aq) \rightleftharpoons {}^+H_3NCH(CH_3)COO^-(aq) + H_3O^+(aq)$$
$$^+H_3NCH(CH_3)COOH(aq) + H_2O(aq) \rightleftharpoons H_2NCH(CH_3)COOH(aq) + H_3O^+(aq)$$

The two K's are $K_1 = 10^{-2.3} = 5.0 \times 10^{-3}$ and $K_2 = 10^{-9.7} = 2.0 \times 10^{-10}$. Notice that the although the two K's are both acid-ionization constants two reactions are not consecutive but are instead competitive. The notation K_{a1} and K_{a2} is therefore not used. The formulas of the three forms of alanine are now abbreviated:

$$A \Rightarrow {}^{+}H_3NCH(CH_3)COOH \quad Z \Rightarrow {}^{+}H_3NCH(CH_3)COO^- \quad B \Rightarrow H_2NCH(CH_3)COOH$$

This symbolism shortens the task of writing the acid-dissociation equilibrium expressions:

$$\frac{[H_3O^+][Z]}{[A]} = K_1 \quad \text{and} \quad \frac{[H_3O^+][B]}{[A]} = K_2$$

At pH 7 $[H_3O^+]$ is 1.0×10^{-7} M, and the two expressions become:

$$\frac{[1.0 \times 10^{-7}][Z]}{[A]} = 5.0 \times 10^{-3} \quad \text{and} \quad \frac{[1.0 \times 10^{-7}][B]}{[A]} = 2.0 \times 10^{-10}$$

$$\frac{[Z]}{[A]} = 5.0 \times 10^{4} \quad \text{and} \quad \frac{[B]}{[A]} = 2.0 \times 10^{-3}$$

a) The fraction of the alanine that is in the Z-form is the concentration of that form divided by the sum of the concentrations of all three forms:

$$f_Z = \frac{[Z]}{[A]+[Z]+[B]} = \frac{5.0 \times 10^4[A]}{[A] + (5.0 \times 10^4)[A] + (2.0 \times 10^{-3})[A]} = 0.99998$$

Essentially all of the molecules are in the Z-form at this pH.

b) The fraction of the alanine that is in the B-form is the concentration of that form divided by the sum of the concentrations of all three forms:

$$f_B = \frac{[B]}{[A]+[Z]+[B]} = \frac{2.0 \times 10^{-3}[A]}{[A] + (5.0 \times 10^4)[A] + (2.0 \times 10^{-3})[A]} = 4.0 \times 10^{-8}$$

23-63 The process is: $2\,H_2O + 2\,NADP^+ \rightarrow O_2 + 2\,H^+ + 2\,NADPH$. Note that 4 mol of electrons is transferred when this reaction occurs. The standard potential difference of a cell constructed to run this reaction is the standard reduction potential of the half-reaction at the anode (given in problem 23-39) subtracted from the standard reduction potential of the half-reaction at the cathode (given in problem 23-40):

$$\Delta\mathcal{E}^\circ = \mathcal{E}^\circ(\text{cathode}) - \mathcal{E}^\circ(\text{anode}) = -0.117 - (+1.229) = -1.346 \text{ V}$$

The reaction is not spontaneous. Now use the relation $\ln K = n\mathcal{F}\Delta\mathcal{E}^\circ/RT$:

$$\ln K = \frac{4(96\,485 \text{ C mol}^{-1})(-1.346 \text{ V})}{(8.315 \text{ J K}^{-1}\text{mol}^{-1})(298.15 \text{ K})} = -209.5 \qquad K = e^{-209.5} = 1 \times 10^{-91}$$

23-65 a) Nicotinamide adenine dinucleotide ($NADP^+$) is a coenzyme. It readily takes up electrons that have been photochemically excited. It is thereby reduced to NADPH.
b) Chlorophyll catalyzes the conversion of CO_2 and H_2O in oxygen and glucose using the absorption of light for energy. Chlorophyll absorbs red and green light well to give excited states that readily reduce $NADP^+$. It is itself thereby oxidized, but is re-generated subsequently.
c) The oxaloacetate anion is part of the citric acid cycle. It reacts with acetyl coenzyme A and water to produce citrate anion in an early step in metabolism.
d) Pyruvic acid is part of the citric acid cycle. It is converted to acetyl coenzyme A.
e) Adenosine diphosphate (ADP) is involved in energy storage. It is converted to adenosine triphosphate (ATP) when energy is stored and is regenerated when free energy is needed to drive a nonspontaneous reaction. ed2 **f)** Starch is an energy storage compound for plants. Photosynthesis in plants converts carbon dioxide and water into starch, a polysaccharide.

23-67 Pyruvic acid is $CH_3—CO—COOH$. Oxidation to acetic acid must entail the loss of one carbon atom; the most likely form of this removal is as carbon dioxide. Meanwhile, the product of the reduction of hydrogen peroxide is water:

$$CH_3COCOOH + H_2O_2 \rightarrow CH_3COOH + CO_2 + H_2O$$

Appendices

Answers to Odd-Numbered Problems, App. A

A-1 The trailing zeros in d) and e) must not be omitted when the number is put into scientific notation.
a) 5.82×10^{-5} **b)** 1.402×10^{3}
c) 7.93 **d)** -6.59300×10^{3} **e)** 2.530×10^{-3} **f)** 1.47

A-3 **a)** 0.000537 **b)** 9 390 000 **c)** −0.00247 **d)** 0.006020 **e)** 20 000.

A-5 The number is 746 million kilograms or 746 000 000 kg.

A-7 **a)** Statistical methods for deciding whether to omit a outlier are not developed in the Appendix. Instead the appeal is to use good judgment. The value 135.6 g is grossly out of line with the others.
b) The mean is 111.34 g **c)** The standard deviation is 0.22 g and the 98% confidence limit is

$$\text{confidence limit} = \pm\frac{t\sigma}{\sqrt{N}} = \pm\frac{2.57(0.22\text{ g})}{\sqrt{6}} = \pm 0.23\text{ g}$$

where t comes from Table A-2, text page A-4.

A-9 The measurement of mass in problem A-7 is more precise.

A-11 **a)** five **b)** three **c)** ambiguous (two or three significant figures) **d)** three **e)** four.

A-13 **a)** 14 L **b)** $-0.0034°C$ **c)** 3.4×10^{2} lb **d)** 3.4×10^{2} miles
e) 6.2×10^{-27} J

A-15 Eight, 2 997 215.55

A-17 **a)** −167.25 **b)** 76 **c)** 3.1693×10^{15} **d)** -7.59×10^{-25}

A-19 **a)** −8.40 **b)** 0.147 **c)** 3.24×10^{-12} **d)** 4.5×10^{13}

A-21 The area of the triangle is 337 cm^2. Three significant figures appear in the answer.

Answers to Odd-Numbered Problems, App. B

B-1 **a)** 6.52×10^{-11} kg **b)** 8.8×10^{-11} s **c)** 5.4×10^{12} kg m^2 s^{-3} **d)** 1.7×10^4 kg m^2 s^{-3} A^{-1}

B-3 **a)** 4983°C, but it is very hard to measure such a high temperature to ±1°C. **b)** 37.0°C **c)** 111°C **d)** −40°C.

B-5 **a)** 5256 K **b)** 310.2 K **c)** 384 K **d)** 233 K.

B-7 **a)** 24.6 m s^{-1} **b)** 1.15×10^3 kg m^{-3} **c)** 1.6×10^{-29}A s m **d)** 1.5×10^2 mol m^{-3} **e)** 6.7 kg m^2 s^{-3}.

B-9 One kW-hr is equal to 3.6×10^6 J. Hence, 15.3 kW-hr is 5.51×10^7 J.

B-11 The engine displacement is 6620 cm^3 or 6.62 L.

Answers to Odd-Numbered Problems, App. C

C-1 The slope is 50 miles hr^{-1}.

C-3 **a)** The equation is in the required form with m (slope) equal to 4 and b (y-intercept) equal to −7.

b) The equation is $y = 7/2x - 5/2$. The slope is 7/2, and the y-intercept is −5/2.

c) The equation is $y = -2x + 4/3$. The slope is −2 and the y-intercept is 4/3.

C-5 The graph of y versus x for the equation $y = 2x^3 - 3x^2 + 6x - 5$ is not linear. The value of y rises from −45 at $x = -2$ and +11 at $x = +2$. The graph cuts the x-axis (has $y = 0$) at $x = 1$.

C-7 **a)** $x = -5/7$ **b)** $x = 3/4$ **c)** $x = 2/3$.

C-9 The answers are given to 4 significant figures. **a)** $x = 0.5447, -2.295$ **b)** $x = -0.6340, -2.366$ **c)** $x = +0.6340, +2.366$.

C-11 **a)** Assuming that x is small compared to 2.00 gives $x = 6.5 \times 10^{-7}$. There are also two complex roots.

b) The best method of solution is graphical. There are three roots because this is a third-degree equation: $x = 4.07 \times 10^{-2}, 0.399, -1.011$.

c) The only real root is $x = -1.3732$. It can be arrived at graphically. The other two roots are imaginary. They are of little interest in chemical applications.

C-13 **a)** 4.551 (the three significant figures appear in the mantissa) **b)** To help understand the significant figures, divide the exponent in the number by 2.302585093 to re-express it as a power of 10: $10^{-6.814}$. The "6" plays the role of the characteristic when the antilog is taken and the mantissa has three significant figures. Hence the answer has three significant figures: 1.53×10^{-7}. **c)** 2.6×10^{8} **d)** -48.7264

C-15 The answer is 3.015.

C-17 Few calculators accommodate a number with an exponent exceeding 99 in absolute value. To answer this problem, write

$$\log 3.00 \times 10^{121} = \log(3.00) + \log 10^{121} = 121 + 0.477 = 121.477$$

C-19 Simply change the characteristic from 0 to 7 or from 0 to -3 and add it to the same mantissa: $7 + 0.751 = 7.751$ and $-3 + 0.751 = -2.249$.

C-21 The problem is to find x in the equation $\log \ln x = -x$. One way to proceed is to guess an x, put it into the left side of the equation and see on a calculator if the indicated operations gives back the guess. Adjust the guess and repeat until satisfied. The answer is 1.086.

C-23 **a)** $8x$ **b)** $3\cos 3x - 8\sin 2x$ **c)** 3 **d)** $1/x$.

C-25 **a)** 20 **b)** 78125/7 **c)** 0.0675.